Air Quality

Edited by
Gandikota V. Rao
Sethu Raman
M. P. Singh

2003

Birkhäuser Verlag
Basel · Boston · Berlin

Reprint from Pure and Applied Geophysics
(PAGEOPH), Volume 160 (2003), No. 1-2

Editors:

Prof. G. V. Rao
Dep. of earth and atmospheric Sciences
3507 Laclede Avenue
Saint Louis University
St. Louis, Missouri 63103
USA
e-mail: rao@eas.slu.edu

Prof. Sethu Raman
Professor and Director
State Climate Office of North Carolina
1005 Capability Dr., Suite 213
Campus Box 7236
North Carolina State University
Raleigh, NC 27695-7236
USA
e-mail: sethu_raman@ncsu.edu

Prof M.P. Singh
Director, Ansal Institute of Technology
Sector 55, Gurgaon
Haryana 122003
India
e-mail: mpsingh3@hotmail.com

A CIP catalogue record for this book is available from the Library of Congress,
Washington D.C., USA

Bibliographic information published by Die Deutsche Bibliothek
Die Deutsche Bibliothek lists this publication in the Deutsche Nationalbibliografie; detailed
bibliographic data is available in the internet at http://dnb.ddb.de.

ISBN 3-7643-7005-X Birkhäuser Verlag, Basel – Boston – Berlin

2003 Birkhäuser Verlag, P.O.Box 133, CH-4010 Basel, Switzerland
Member of the BertelsmannSpringer Publishing Group
Printed on acid-free paper produced from chlorine-free pulp. TCF ∞
Printed in Germany

9 8 7 6 5 4 3 2 1   www.birkhauser.ch

Contents

Pure appl. geophys. 160 (2003) 1–2
0033–4553/03/020001–02

▌**Pure and Applied Geophysics**

Preface

Air pollution has become a problem of virtually all countries; it is of serious concern to the developing countries because of rapid industrialization. In the case of the southeast Asian countries, the problem of air pollution has been compounded by the seasonal regional biomass burning. During the El Niño years, air quality in southeast Asia degrades to dangerous levels.

At the University of Brunei Darussalam (UBD), a multi-disciplinary study on *Air Quality* with an emphasis on mathematical modeling was initiated in 1994 by Professor M. P. Singh, an editor of this volume. The faculty from many departments of UBD have collaborated with the various Brunei Government Departments in studying air quality. Noteworthy among such Government Departments were the Meteorological Service Unit of the Ministry of Communication, and the Environmental Health Unit of the Ministry of Health. The University of Brunei conducted a workshop on *Air Quality* in July 1997. Although the air quality in Brunei is generally good, there were episodes of poor air quality stemming from the transboundary migration of smoke and haze caused by the seasonal crop and forest burning. During the period of September 1997 through April 1998, large-scale haze produced by the El Niño weather covered many parts of southeast Asia. Haze-related pollution was severe in Brunei from January to April 1998 as major forest fires spread the haze conditions; resulting in the early closing of schools and the rescheduling of work hours. Simultaneously, there was an increase in respiratory diseases, such as: asthma, bronchitis, emphysema, and pneumonia.

A workshop was organized by Drs. Raman and Singh, the two editors of this volume, in May 1998 which aimed to create a further air quality awareness among user agencies of Brunei and countries of the region. A major recommendation of this workshop was realized when an International Workshop on *Air Quality Management* was organized in November 1999 in Brunei.

The objectives of this 1999 International Conference were to:

a. Highlight and provide a review of the relevant air quality problems of the southeast Asia region.

b. Promote collaborative study and research in environment-related issues of the region, and to bring researchers and policy makers together.

c. Update the existing air quality models by incorporating realistic topography, climatology and other characteristics.

Since the regionally recurring haze affects the health and economic development of millions of people, a half-day symposium on haze was organized as part of the conference. Haze is usually defined as the suspension in the atmosphere of extremely small dry aerosols. These small dry aerosols are invisible but give an opalescent appearance to the sky. In the following articles, some human causes of haze are suggested. The socio-economic problems of the poor who use fire as part of the land management are outlined; effective enforcement of laws is advised and educating the public on safe behavior during haze episodes is suggested to avert large-scale disasters.

The present volume entitled, *Air Quality*, contains papers presented at the November 1999 International Conference on Air Quality Management in Brunei. The volume contains twenty-seven (27) papers organized into three sections: (1) Air pollution (8 papers); (2) Haze and biological effects (9 papers); and (3) Air Quality Modeling (10 papers). Each section contains invited articles from reputed scientists. Many international researchers presented their results in the conference and contributed articles to this volume. Additionally, the Brunei scientists participated enthusiastically in the conference and contributed their research findings for publication. The editors wish to thank the authors for their timely submission of manuscripts with revisions, and the University of Brunei Darussalam for their support. Multidisciplinary research activities are needed to probe the haze phenomenon, which would lead to effective monitoring and management strategies.

This volume, *Air Quality*, contains many original findings pertaining to biomass fires, transboundary pollution and associated haze that impacts on health, transportation, biodiversity, and the economy; thus, it is expected to be a source book for research in this region. Many of the results presented in this volume pertain to southeast Asia and are available under **one roof.** Certain papers could be discussed in graduate level classes dealing with Air pollution, Air Quality, Cloud Physics, and Biophysics. The scientific community will find this book a useful addition to their personal and institutional libraries.

Gandikota V. Rao
Saint Louis University
St. Louis, MO, U.S.A.
E-mail: rao@eas.slu.edu

M. P. Singh
Ansal Institute of Technology
Gurgaon, Delhi, India
E-mail: mpsingh3@hotmail.com

Sethu Raman
North Carolina University
Raleigh, NC, U.S.A.
E-mail: sethu_raman@ncsu.edu

A. Air Pollution

Pure appl. geophys. 160 (2003) 5–16
0033–4553/03/020005–12

▌**Pure and Applied Geophysics**

Operational Advanced Air Pollution Modeling

Tiziano Tirabassi[1]

Abstract — Operational models that use solutions of the advection-diffusion equation based on more realistic assumptions than that of homogeneous wind and eddy diffusivity coefficients are presented. In particular a new parameterization for a model using a solution that accepts wind and eddy diffusivity profiles described by power functions of height is introduced. The performance of the model with the new parameterization was assessed using experimental data sets.

Key words: Analytical solutions, air pollution models, advection-diffusion equation.

1. Introduction

In practice most of the estimates of dispersion from continuous point sources are based on the Gaussian approach. A basic assumption for the application of this approach is that the plume is dispersed by homogeneous turbulence. However, due to the presence of the ground, turbulence is usually not homogeneous in the vertical direction. Moreover, the input parameters of the Gaussian plume model are often related to simple turbulence typing schemes or stability classes. These are utilized where the meteorological state of the atmospheric boundary layer is classified in a simple way based on surface measurements, and where the dispersion from a source is estimated by assuming simple formulae for concentration distribution, in which the dispersion parameters depend simply on downwind distance and the meteorological state of the boundary layer. The problem with such stability classes is that each covers a broad range of stability conditions; they are also very site-specific and biased towards neutral stability when unstable or convective conditions actually exist (WEIL, 1983). In addition, the influence of these factors on the calculated ground-level concentration is considerable (KRETZSCHMAR and MERTENS, 1984).

In the summary by WEIL (1985) it is suggested that the similarity approach replaces the classical Pasquill-Gifford stability classification.

---

[1] Institute ISAC of CNR, via Gobetti 101, I-40129 Bologna, Italy.
E-mail: t.tirabassi@isao.bo.cnr.it

Experimental work and modeling efforts have attempted to parameterize the surface fluxes of momentum, heat and moisture in terms of routinely measured meteorological parameters (e.g., HOLTSLAG and VAN ULDEN, 1983; VAN ULDEN and HOLTSLAG, 1985; TROMBETTI *et al.*, 1986; BELJAARS and HOLTSLAG, 1990). Various organizations world-wide are introducing advanced modeling techniques that make use of the above recent research on the meteorological state of the boundary layer. These advanced modeling techniques (VAN ULDEN, 1978; BERKOWICZ *et al.*, 1986; TIRABASSI, 1989; HANNA and PAINE, 1989; CARRUTHERS *et al.*, 1992; DEGRAZIA *et al.*, 1997) contain algorithms for calculating the main factors that determine air pollution diffusion in terms of fundamental parameters such as the Monin-Obukhov length scale.

Within this framework we have developed several model codes that employ an analytical solution which is not Gaussian, however it allows for variations in the wind and exchange coefficients with height.

2. Analytical Solutions of the Advection-diffusion Equation

Analytical solutions of equations are of fundamental importance in understanding and describing physical phenomena. Analytical solutions (as opposed to numerical ones) explicitly take into account all parameters of a problem, so that their influence can be reliably investigated and it is easy to obtain the asymptotic behavior of the solution, which is usually difficult to generate through numerical calculations.

We provide a short review of the analytical solutions of the two-dimensional advection-diffusion equation:

$$u\frac{\partial C}{\partial x} = \frac{\partial}{\partial z}\left(K_z\frac{\partial C}{\partial z}\right) + S,$$

where u is mean velocity (the wind is assumed along the x axis, while z is the height), C is the mean concentration, S is the source term, and K_z is the vertical eddy exchange coefficient.

Unfortunately, no general solution is known for equations describing the atmospheric transport and dispersion of air pollution. There are some specific solutions, the best-known being the so-called Gaussian solution, which does not, however, realistically describe the concentrations of pollutants in the air; in fact, the models based on it (so-called Gaussian models) use empirical parameters of dispersion in order to force the Gaussian solution to represent the actual concentration field. However, there are models based on non-Gaussian analytical solutions.

ROBERTS (1923) presented a bi-dimensional solution for ground-level sources only, in cases where both the wind speed and vertical diffusion coefficients follow power laws as a function of height. That is:

$$u = u_1(z/z_1)^\alpha,$$

$$K_z = K_1(z/z_1)^\beta,$$

where z_1 is the height where u_1 and K_1 are evaluated.

ROUNDS (1955) obtained a bi-dimensional solution valid for elevated sources, although only for linear profiles of K_z. SMITH (1957a) resolved the bi-dimensional equation of transport and diffusion with u and K_z power functions of height with the exponents of these functions following the conjugate law of Schmidt (i.e., wind exponent $= 1 - K_z$ exponent).

SMITH (1957b) also presented a solution regarding constant u, but K_z following:

$$K_z = K_0 z^a (H - z)^b,$$

where K_0 is a constant and a and b can be:

$a \geq 0$ and $b = 0$,

$a = 0$ and $b > 0$ for $0 \leq z \leq H$,

$a = 1$ and $b > 0$ for $0 \leq z \leq H$,

$a = 1$ and $b = 0$ for $0 \leq z \leq H/2$; $a = 0$ and $b = 1$ for $H/2 \leq z \leq H$,

where H is the height of the atmospheric boundary layer.

SCRIVEN and FISHER (1975) proposed a solution with constant u and K_z as:

$$K_z \equiv z \quad \text{for} \quad 0 \leq z \leq z_s$$
$$K_z = K_z(z_s) \quad \text{for} \quad z_s < z \leq H,$$

where z_s is a predetermined height (generally, the height of the surface layer). This solution allows (as boundary conditions) a net flow of material towards the ground:

$$K_z \frac{\partial C}{\partial z} = V_g C,$$

where V_g is the deposition velocity. The Scriven and Fisher solution has been used in the United Kingdom for long-range transport of pollutant. In FISHER (1975) the deposition of sulphur over the United Kingdom, Sweden and the rest of Europe was compared and it was found that the British contribution to deposition over rural Sweden was about one half of the Swedish contribution.

YEH and HUANG (1975) and BERLYAND (1975) published bi-dimensional solutions for elevated sources with u and K_z following power profiles, but for a unbound atmosphere. That is:

$$K_z \frac{\partial C}{\partial z} = 0 \quad \text{at} \quad z = \infty.$$

DEMUTH (1978) put forward a solution with the same conditions, but for a vertically limited boundary layer. That is:

$$K_z \frac{\partial C}{\partial z} = 0 \quad \text{at} \quad z = H.$$

By applying the Monin-Obukhov similarity theory to diffusion, VAN ULDEN (1978) derived a solution for vertical diffusion from continuous sources near the ground only, with the assumption that u and K_z follow power profiles. His results are similar to Roberts', however he provided a model for non-ground level sources, but applicable to sources within the surface layer.

NIEUWSTADT (1980) presented a solution which was a particular case of SMITH's (1975b) solution noted above. Subsequently, NIEUWSTADT and DE HAAN (1981) extended that solution to the case of a growing boundary layer height. CATALANO (1982), in turn, extended the latter solution to the case of non-zero mean vertical wind profiles. LIN and HILDEMANN (1997) extended the solution of DEMUTH (1978) with boundary conditions suitable for simulating dry deposition to the ground.

Recently, BROWN et al. (1997) derived from the solution of YEH and HUANG (1975), equations for point source releases for the first four moments of the vertical concentration distribution and the magnitude and downwind location of the maximum ground concentration.

In practice, most estimates of dispersion from continuous point sources are based on the Gaussian approach. The input parameters of the Gaussian plume model are frequently related to simple turbulence typing schemes and stability classes. Now advanced modeling techniques have been introduced that make use of recent research on the meteorological state of the boundary layer for calculating the main factors causing air pollution diffusion by ground-level measurements. Within this framework, it is possible to develop operational analytical models that employ some of the above presented solutions. In particular the solutions which allow for variations in the wind and exchange coefficients with height.

3. Analytical Models

We have developed some models (TIRABASSI, 1989; TIRABASSI and RIZZA, 1992, 1993, 1994) that utilize the two-dimensional analytical solutions of the advection-diffusion equation proposed by YEH and HUANG (1975) and BERLYAND (1975) for an unbounded boundary layer and by DEMUTH (1978) for a bounded one, while the cross-wind dispersion is simulated by a Gaussian term. That is:

$$C(x,y,z) = \frac{C_y}{\sqrt{2\pi}\sigma_y} \exp\left[-\frac{y^2}{2\sigma_y^2}\right],$$

where C_y represents the cross-wind integrated concentrations, y is the cross-wind distance and σ_y is the lateral diffusion parameter.

If variations in the wind (u) and exchange coefficient (K_z) with height (z) are:

$$u = u_1 (z/z_1)^\alpha,$$

$$K_z = K_1 (z/z_1)^\beta,$$

where z_1 is the height where u_1 and K_1 are evaluated, the solutions of YEH and HUANG (1975) can be written:

$$C_y = \frac{Q(zh)^p z_1^\beta}{\lambda K_1 x} \exp\left[-\frac{u_1 z_1^r (z^\lambda + h^\lambda)}{\lambda^2 K_1 x}\right] I_{-v}\left(\frac{2u_1 z_1^r (zh)^q}{\lambda^2 K_1 x}\right)$$

while Demuth's solution (DEMUTH, 1978) is written as:

$$C_y = \frac{2Q \cdot q \cdot z_1^\alpha}{H^{\alpha+1} u_1} \left\{ \gamma + (zR/H)^p \sum_{i=1}^{\infty} \left[\frac{J_{\gamma-1}(\rho_{\gamma(i)} R^q) J_{\gamma-1}\left[\rho_{\gamma(i)}(z/H)^q\right]}{J_{\gamma-1}^2(\rho_{\gamma(i)})} \exp\left(-\frac{\rho_{\gamma(i)}^2 q^2 K_1 x}{H^\lambda z_1^r u_1}\right)\right]\right\}$$

where x is the along-wind direction, Q the source emission, h the source effective height, H the mixing height, $\lambda = \alpha - \beta + 2$, $v = (1 - \beta)/\lambda$, $\gamma = (\alpha + 1)/\lambda$, $r = \beta - \alpha$, $R = h/H$, $p = (1 - \beta)/2$, $q = \lambda/2$, J_γ and I_v represent the Bessel function and modified Bessel function of first kind and order γ and v, $\rho_{\gamma(i)}$ the roots of J_γ.

We have set up four different analytical models based on the above solutions: KAPPAG, KAPPAG-LT, CISP and MAOC.

3.1. The KAPPAG Model

The model can handle multiple sources and multiple receptors, simulating time-varying conditions in which each time interval (e.g., 1 hour) is treated as a stationary case. The model output is a statistical summary of the concentrations computed at each receptor, during each time step, and due to each source. Partial and total concentrations are computed for hourly and multi-hour averages. Highest and second- highest values are also evaluated.

The performances of the KAPPAG model have been assessed with success by comparing ground-level concentration estimates of SO_2 relative to ground-level releases (TAGLIAZUCCA et al., 1985) and with SF_6 data relative to elevated releases (TIRABASSI et al., 1986).

3.2. The KAPPAG-LT Model

KAPPAG-LT is the long-term version of KAPPAG which estimates annual ground-level concentrations. Its performance has been assessed with success against SO_2 data from heavily industrialised area (TIRABASSI et al., 1989).

3.3. The CISP Model

CISP (TIRABASSI and RIZZA, 1992) is a screen model that provides a method for estimating maximum ground-level concentrations from a single point source as a function of stability and wind speed. In fact, it is designed for the low-cost, detailed screening of point sources to determine maximum one-hour concentrations and to decide whether use of one of the more sophisticated models is required. CISP model is regarded as a useful tool for a screen analysis, in that it is a relatively simple estimation technique providing conservative estimates of the air quality impact of a specific source or source category.

CISP performance in evaluating maximum ground-level concentrations has been compared with that of the U.S.EPA Regulatory PTPLU2 Gaussian model (TIRABASSI and RIZZA, 1992).

3.4. The MAOC Model

MAOC is a model for elevated point sources in complex-terrain (TIRABASSI and RIZZA, 1993). The simulation of terrain-induced distortion of flow streamlines is accounted for by modifying the effective plume height. That is:

$$h' = Hi + Ah,$$

where Hi is the hill height at the receptor considered and A is an empirical factor. As the plume passes over the mountains and streamlines converge, that effective stack height decreases from h to a new value h'. In stable conditions, the plume may not have enough kinetic energy to climb the mountain. In this case, a critical height is defined so that, if h is less than the critical height, the plume will impinge on the mountain, otherwise it passes over the crest of the hill.

Model performances have been evaluated (TIRABASSI and RIZZA, 1993) using wind tunnel measurements of pollutant concentrations from an elevated source in the presence of rough hills and a neutrally stable flow and have been compared with that of COMPLEX1, the Gaussian model proposed by the U.S.EPA.

4. New Wind and Eddy Diffusivity Profiles Parameterization for KAPPAG Model

We present a new wind and eddy diffusivity profiles parameterization for the KAPPAG model. In the model the wind vertical profiles are parameterized by approximating actual profiles (between ground level and source height) by the least-square's method with the power law

$$u = u_1(z/z_1)^\alpha.$$

If vertical wind data are not available the model uses the above formula, adopting for α the values proposed by IRWIN (1979).

Eddy exchange coefficient profiles are parameterized by fitting the least-square's method with the power law

$$K_z = K_1(z/z_1)^\beta$$

the profiles (between ground level and source height) proposed by PLEIM and CHANG (1992) for neutral and stable conditions ($H/L \geq -10$):

$$K_z = \frac{ku_*z(1 - z/H)^2}{\Phi_h(z/L)}$$

and the K_z profile proposed by DEGRAZIA *et al.* (1997) for unstable conditions ($H/L < -10$):

$$K_z = 0.22w_*H\left(\frac{z}{H}\right)^{1/3}\left(1 - \frac{z}{H}\right)^{1/3}\left[1 - \exp\left(\frac{-4z}{H}\right) - 0.0003\ \exp\left(\frac{8z}{H}\right)\right]$$

where u_* is the friction velocity, L the Monin-Obukhov length, k is the von Karman constant ($k = 0.4$), Φ_h is the universal function of the non-dimensional temperature gradient, and w_* is the convective velocity scale.

5. Model Validation

We evaluated the performance of the KAPPAG model with the new parameterization against measured concentrations, using the Copenhagen and Prairie Grass data sets.

The Copenhagen data set refers to SF_6 tracer data recorded during dispersion experiments carried out in north Copenhagen and is described in GRYNING and LYCK (1984). The tracer was released without buoyancy from a tower at a height of 115 m and was collected at ground-level positions in up to three cross-wind arcs of the tracer sampling units; the sampling units were positioned 2–6 km from the point of release. Tracer releases typically started one hour prior to the commencement of tracer sampling and were suspended at the end of the sampling period, with an average sampling time of one hour. The site was prevalently residential with a roughness length of 0.6 m.

The Prairie Grass data set is composed of dispersion data from a field experiment conducted in open country (the roughness length z_0 was 0.008 m) during the summer of 1956 at O'Neill, Nebraska (U.S.A.). Sulphur dioxide was released from a continuous point source at the height of 0.46 m and the concentration was measured along several arcs. We used the values of the crosswind-integrated concentrations normalized with the tracer release rate as calculated by VAN ULDEN (1978) for arcs 50, 200 and 800 m from the source and the estimates of the surface layer parameter u_* and L reported by VAN ULDEN (1978), as derived from the wind and temperature profiles measured along 16-m masts. The meteorological data used were collected near the ground and, therefore, the validation exercise can be said to simulate the routine application of the model.

Figure 1 shows the cross-wind integrated concentrations normalized with emission at ground level predicted by the model against the observed values of the

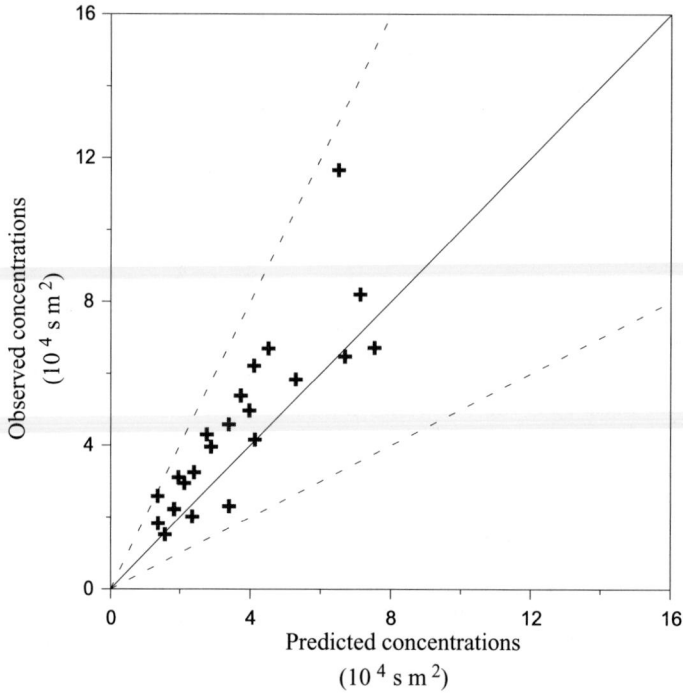

Figure 1
Copenhagen data set. Observed and predicted cross-wind integrated concentrations normalized with emission. Data between dashed lines are in a factor of two.

Copenhagen data set, while Figure 2 shows the concentrations normalized with emission at ground level predicted by the model against the observed values of the Prairie Grass data set.

In Tables 1 and 2 statistical indices of the model performances are reported. The statistical indices were proposed by HANNA et al. (1991) and are widely used in literature (OLESEN, 1995). They are defined as:

$$\text{normalised mean-square error (nmse)} = \frac{\overline{\left(C_o - C_p\right)^2}}{\overline{C_o} \cdot \overline{C_p}}$$

$$\text{correlation coefficient } (r) = \frac{\overline{\left(C_o - \overline{C_o}\right)\left(C_p - \overline{C_p}\right)}}{\sigma_o \cdot \sigma_p}$$

$$\text{fractional bias } (fb) = 2\frac{\overline{C_o} - \overline{C_p}}{\overline{C_o} + \overline{C_p}}$$

factor of two $(fa2)$ = fraction of data for which $0.5 \leq C_p/C_o \leq 2$.

nmse emphasizes the scatter in the entire data set and the normalisation by $\overline{C_o} \cdot \overline{C_p}$ assures that nmse will not be biased towards models that overpredict or underpredict.

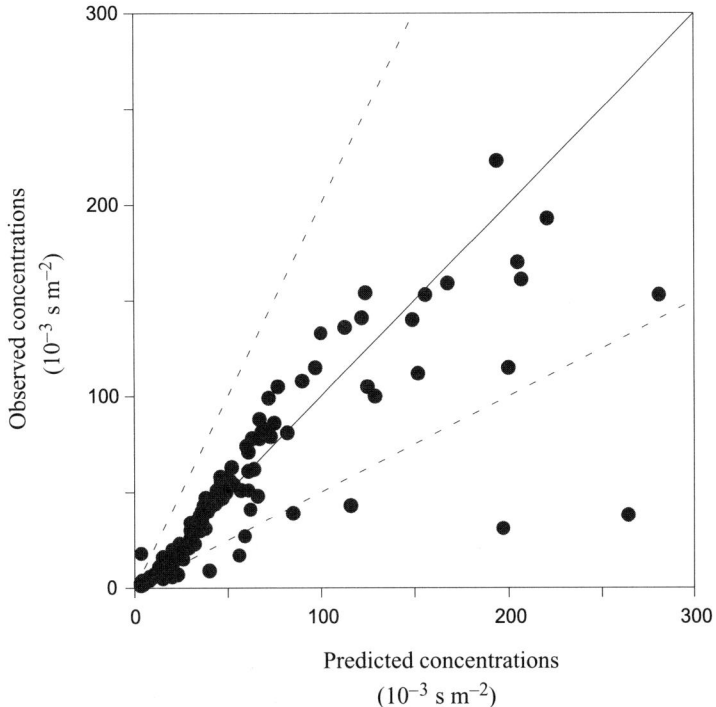

Figure 2

Prairie Grass data set. Observed and predicted concentrations normalized with emission. Data between dashed lines are in a factor of two.

Table 1

Copenhagen data set. Statistical indices for cross-wind integrated concentrations

Model	nmse	r	fa2	fb
Proposed model	0.14	0.86	1	0.22
HPDM	0.16	0.78	1	0.16
IFDM	0.16	0.68	0.96	0.012
INPUFF	0.46	0.36	0.70	0.28
OML	0.52	0.89	0.56	0.57
UK-ADMS	0.34	0.86	0.78	0.41

Table 2

Prairie Grass data set. Statistical indices

Model	nmse	r	fa2	fb
Proposed model	0.46	0.84	0.86	−0.08

The best result is nmse $= 1$. r is the well-known correlation coefficient and fb is the normalized bias with values between -2 and 2 and the best value equals 0. The best value for $fa2$ is 1.

Moreover, to compare the performance of the proposed model with that of other models, Table 1 reports the results obtained on the same data set using other models: HPDM (HANNA and PAINE, 1989), IFDM (COSEMANS et al., 1992), INPUFF (SANDU, 1989), OML (BERKOWICZ et al., 1986), UK-ADMS (CARRUTHERS et al., 1992).

The latter results were obtained as part of a model validation exercise during *The Workshop on Operational Short-range Atmospheric Dispersion Models for Environmental Impact Assessment in Europe* (OLESEN, 1995).

Looking at Table 1 we see that the proposed model has the best nmse, the second best correlation coefficient, the best $fa2$ (with HPDM) and the third best fb.

6. Conclusions

We have presented a short review of non-Gaussian analytical solutions of the advection-diffusion equation. In particular, the solutions accept that wind and eddy diffusivity profiles are power functions of the height.

We have presented four different analytical models based on the above solutions. The models can be applied routinely, using as input the simple ground-level meteorological data acquired by an automatic network. In fact, the fundamental parameters for describing the features of the atmospheric surface and boundary layers are evaluated by ground-level measurements.

We have introduced a new parameterization for wind and exchange coefficients' profiles in the KAPPAG model. The performance of the model with the new parameterization was assessed using the Copenhagen and Prairie Grass data set. In the case of the Copenhagen data set, the proposed model (together with HPDM) proved to have the best performance of the models considered.

REFERENCES

BELJAARS, A. C. M. and HOLTSLAG, A. A. M. (1990), *A Software Library for the Calculation of Surface Fluxes over Land and Sea*, Environ. Soft. *5*, 60–68.
BERKOWICZ, R. R., OLESEN, H. R., and TORP, U., *The Danish Gaussian air pollution model (OML): description, test and sensivity analysis in view of regulatory applications*, Proc. NATO-CCMS 16th Int. Meeting on *Air Poll. Modelling and Its Applications* (eds. De Wispelaere, C., Schiermeier, F. A., and Gillani, N. V.) (Plenum Press, New York, 1986), pp. 453–481.
BERLYAND, M. Y., *Contemporary Problems of Atmospheric Diffusion and Pollution of the Atmosphere*. Translated version by NERC, USEPA (Raleigh, NC, U.S.A., 1975).
BROWN, M. J., ARYA, S. P., and SNYDER, W. (1997), *Plume Descriptors from a non-Gaussian Concentration Model*, Atmos. Environ. *31*, 183–189.
CARRUTHERS, D. J., HOLROYD, R. J., HUNT, J. C. R., WENG, W. S., ROBINS, A. G., APSLEY, D. D., SMITH, F. B., THOMSON, D. J., and HUDSON, B., *UK atmospheric dispersion modelling system. In Air Pollution*

Modeling and its Application IX (eds. van Dop, H. and Kallos, G.) (Plenum Press, New York, 1992), pp. 15–28.

CATALANO, G. D., *An analytical solution to the turbulent diffusion equation with mean vertical wind*. In *Proc. 16th Southeastern Sem. Thermal. Sci.* (Miami, Fl, U.S.A., 19–21 April 1982), pp. 143–151.

COSEMANS, G., KRETZSCHMAR, J., and MAES, G., *The Belgian immission frequency distribution model IFDM*, In *Proc. of the DCAR Workshop on Objectives for Next Generation of Practical Short-range Atmospheric Dispersion Models* (eds. Olesen, H. and Mikkelsen, T) (Riso, Denmark, 6–8 May, 1992), pp. 149–150.

DEGRAZIA, G. A., RIZZA, U., MANGIA, C., and TIRABASSI, T. (1997), *Validation of a New Turbulent Parameterization for Dispersion Models in Convective Conditions*, Bound. Layer Meteor. *85*, 243–254.

DEMUTH, C. (1978), *A Contribution to the Analytical Steady Solution of the Diffusion Equation for Line Sources*, Atmos. Environ. *12*, 1255–1258.

FISHER, B. E. A. (1975), *The Long-range Transport of Sulphur Dioxide*, Atmos. Environ. *9*, 1063–1070.

GRYNING, S. E. and LYCK, E. (1984), *Atmospheric Dispersion from Elevated Sources in an Urban Area: Comparison between Tracer Experiments and Model Calculations*, J. Clim. Appl. Meteor. *23*, 651–660.

GRYNING, S. E., HOLTSLAG, A. A. M., IRWIN, S., and SIVERSTEN, B. (1987), *Applied Dispersion Modelling Based on Meteorological Scaling Parameters*, Atmos. Environ. *21*, 79–89.

HANNA, S. R. and PAINE, R. J. (1989), *Hybrid Plume Dispersion Model (HPDM) Development and Evaluation*, J. Appl. Meteor. *28*, 206–224.

HANNA, S. R., STRIMAITIS, D. G., and CHANG, J. C., *Hazard response modeling uncertainty (a quantitative method)*. In *User's Guide for Software for Evaluation of Hazardous Gas Dispersion Models* (Sigma Reserach Corp., Westford, USA, 1991).

HOLTSLAG, A. A. M. and VAN ULDEN, A. P. (1983), *A Simple Scheme for Daytime Estimation of Surface Fluxes from Routine Weather Data*, J. Clim. Appl. Meteor. *22*, 517–529.

IRWIN, J. S. (1979), *A Theoretical Variation of the Wind Profile Power-law Exponent as a Function of Surface Roughness and Stability*, Atmos. Environ. *13*, 191–194.

KRETZSCHMAR, J. G. and MERTENS, I. (1984), *Influence of the Turbulence Typing Scheme upon the Cumulative Frequency Distribution of the Calculated Relative Concentrations for Different Averaging Times*, Atmos. Environ. *18*, 2377–2393.

LIN, J. S. and HILDEMANN, L. M. (1997), *A Generalised Mathematical Scheme to Analytically Solve the Atmospheric Diffusion Equation with Dry Deposition*, Atmos. Environ. *31*, 59–71.

NIEUWSADT, F. T. M. (1980), *An Analytical Solution of the Time-dependent, One-dimensional Diffusion Equation in the Atmospheric Boundary Layer*, Atmos. Environ. *14*, 1361–1364.

NIEUWSTADT, F. T. M. and DE HAAN, B. J. (1981), *An Analytical Solution of One-dimensional Diffusion Equation in a Non-stationary Boundary Layer with an Application to Inversion Rise Fumigation*, Atmos. Environ. *15*, 845–851.

OLESEN, H. R. (1995), *Datasets and Protocol for Model Validation*, Int. J. Environ. and Pollution *5*, 693–701.

PLEIM, J. E. and CHANG, J. S. (1992), *A Non-Local Closure Model for Vertical Mixing in the Convective Boundary Layer*, Atmos. Environ. *26A*, 965–981.

ROBERTS, O. F. T. (1923), *The Theoretical Scattering of Smoke in a Turbulent Atmosphere*, Proc. Roy. Soc. *104*, 640–648.

ROUNDS, W. (1955), *Solutions of the Two-dimensional Diffusion Equation*, Trans. Am. Geophys. Union *36*, 395–405.

SANDU, I. (1989), *Assessment of the Atmospheric Pollution Degree by Means of the Pollutant Dispersion Models*, Hidrotechnica *34*, 8–16.

SCRIVEN, R. A. and FISHER, B. A. (1975), *The Long-range Transport of Airborne Material and its Removal by Deposition and Washout-II. The Effect of Turbulent Diffusion*, Atmos. Environ. *9*, 59–69.

SMITH, F. B. (1957a), *The Diffusion of Smoke from a Continuous Elevated Poinr Source into a Turbulent Atmosphere*. J. Fluid Mech. *2*, 49–76.

SMITH, F. B., *Convection-diffusion processes below a stable layer* (Meteorological Research Committee, N. 1048 and 10739) (London, 1957b).

TAGLIAZUCCA, M., NANNI, T., and TIRABASSI, T. (1985), *An Analytical Dispersion Model for Sources in the Surface Layer*, Nuovo Cimento *8C*, 771–781.

TIRABASSI, T. (1989), *Analytical Air Pollution Advection and Diffusion Models*, Water, Air and Soil Poll. *47*, 19–24.

TIRABASSI, T. and RIZZA, U., *An air pollution model for complex terrain*. In *Air Pollution* (eds. Zannetti, P., Brebbia, C. A., Garcia, Gardea, J. E., and Ayala Milian, G.) (Computational Mechanics Pub., Southampton and Elsevier, Amsterdam, 1993), pp. 149–156.

TIRABASSI, T. and RIZZA, U. (1992), An Analytical Model for a Screen Evaluation of the Environmental Impact from a Single Point Source, Nuovo Cimento *15C*, 181–190.

TIRABASSI, T., TAGLIAZUCCA, M., and ZANNETTI, P. (1986), '*KAPPA-G, a non-Gaussian Plume Dispersion Model: Description and Evaluation against Tracer Measurements*, JAPCA *36*, 592–596.

TIRABASSI, T., TAGLIAZUCCA, M., and PAGGI, P. (1989), *A Climatological Model of Dispersion in an Inhomogeneous Boundary Layer*, Atmos. Environ. *23*, 857–862.

TROMBETTI, F., TAGLIAZUCCA, M., TAMPIERI, F., and TIRABASSI, T. (1986), *Evaluation of Similarity Scales in the Stratified Surface Layer Using Wind Speed and Temperature Gradient*, Atmos. Environ. *20*, 2465–2471.

VAN ULDEN, A. P. and HOLSTLAG, A. A. M. (1985), *Estimation of Atmospheric Boundary Layer Parameters for Diffusion Applications*, J. Clim. Appl. Meteor. *24*, 1196–1207.

VAN ULDEN, A. P. (1978), *Simple Estimates for Vertical Diffusion from Sources near the Ground*, Atmos. Environ. *12*, 2125–2129.

WEIL, J. C., *Application of advances in planetary boundary layer understanding to diffusion modelling*. In Proc. 6th Symp. *Turbulence Diffusion and Air Pollution*, (AMS, Boston, 22–25 March, 1983) pp. 42–46.

WEIL, J. C. (1985), *Updating Applied Diffusion Models*, J. Clim. Appl. Meteor. *24*, 1111–1130.

YEH, G. T. and HUANG, C. H. (1975), *Three-dimensional Air Pollutant Modeling in the Lower Atmosphere*, Bound. Layer Meteor. *9*, 381–390.

(Received March 1, 2000, accepted July 22, 2000)

 To access this journal online:
http://www.birkhauser.ch

Pure appl. geophys. 160 (2003) 17–19
0033–4553/03/020017–03

❚ **Pure and Applied Geophysics**

Planning for Air Pollution Abatement in the Greater Tehran Area (GTA) in the Islamic Republic of Iran (JICA Project 1995–1997)

O. Yokoyama[1] and K. Takahashi[1]

1. Overview and Present Situation of Air Pollution

The GTA with a current population of over eight millions and an area of approximately 2300 square km, is suffering heavily from life-threatening atmospheric pollution, arising from the rapid urbanization during the last few decades. According to a recent estimate, there are more than 1.4 million vehicles and some 300 thousand industrial factories and offices in Tehran. Although there are few inventories of pollution sources available in Tehran, those available suggest that concentration of CO, NO, NO_2, SO_2, O_3 and SPM in the GTA are well beyond the World Health Organization (WHO) standard. Particularly, the CO concentration often exceeded 80 ppm limit. With this background information, The Government of Japan, in response to the request of the Government of Islamic Republic of Iran (IRI) decided to carry out a study on preparing an integrated master plan for air pollution control in the GTA in the Islamic Republic of Iran and sent a Study Team through the Japan International Cooperation Agency (JICA). While different pollutants are caused by several sources, it is agreed that the main polluter in GTA is the automobile. More than 90% emission is caused by the automobile.

2. A Long-term Average Simulation Model

The model employed by JICA in the GTA was basically the same one used by the Environmental Protection Agency of Japan (JEA, 1993) and was described in a manual for the total amount control of NO_x. The model consists of plume and puff equations and is considered standard in Japan. This model is somewhat different from ISCST3 model used by the United States Environmental Protection Agency

[1] Japan Weather Association, Sunshine City 60, F55, Higashi-Ikebukuro Tokyo Japan.
E-mail: KGG03670@nifty.ne.jp

(EPA-USA). Long-period averaged concentration such as annual is calculated. The plume equation is well known but the continuous puff equation is an analytical solution including calm persistence period. More detail can be seen in JEA (1993). Plume width is estimated by the Pasquill-Gifford chart. More details of the method can be seen in JICA (1997). In the present air pollution simulation, all the main roads and factories in the central area where high pollution sources exist were included.

In order to make various scenarios of air pollution abatement, we used this pollution model. We plotted the observed and calculated annual average concentration of CO in 1996 in Figure 1.

3. Simulation of Future Condition in the Years 2005/2010

The meteorological condition was assumed to continue as it is and remains unchanged from the present. For future inventories of stationary sources the following three scenarios were considered. A 'Do nothing' scenario assumes that the amount of pollutants grows proportionally to the economy. A 'Best' scenario assumes that a pollution reduction plan was followed. Finally a 'Common' scenario is the one that is merely an average of the first two scenarios. All three of them were adopted for the years 2005 and 2010. If we employ the 'Best' scenario, the WHO standard will be satisfied. Some of the definite methods of countermeasure for air pollution include the use of oxygenated gasoline (e.g., MTBE), structural improvement of car manufacturing, introduction of car scraping program, enhancement of public transporting system, and so on.

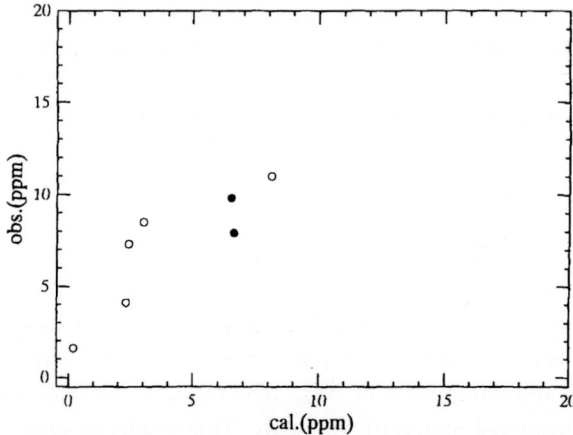

Figure 1

Regression of observed and calculated annual average concentration of CO in 1996 (obs = 0.7 cal + 3.7, $r = 0.69$). (Black and white circle indicates DOE and AQCC stations, respectively. DOE: Department of Environment of Iran. AQCC: Air Quality Control Company).

4. Concluding Remarks

The implementation of air pollution abatement depends on the Iranian Government. Our report shows clearly one direction of remedying the pollution problem. The method used is the one employed in Japan for environmental assessment for air pollution.

Acknowledgement

The present authors would like to express their thanks to the Iranian and Japanese members of the JICA project.

REFERENCES

JAPAN ENVIRONMENTAL AGENCY, ed. (1993), *Environmental Assessment Manual*, Environmental Information Center (in Japanese), 409 pp.

JICA (1997), *The Study on an Integrated Master Plan for Air Pollution Control in the Greater Tehran Area in the Islamic Republic of Iran*, 816 pp.

(Received March 1, 2000, accepted November 12, 2000)

 To access this journal online:
http://www.birkhauser.ch

Pure appl. geophys. 160 (2003) 21–55
0033–4553/03/020021–35

© Birkhäuser Verlag, Basel, 2003

▌Pure and Applied Geophysics

Summertime Characteristics of the Atmospheric Boundary Layer and Relationships to Ozone Levels over the Eastern United States

S. Trivikrama Rao,[*,1,2] Jia-Yeong Ku,[2] Stephen Berman,[3] Kesu Zhang,[1] and Huiting Mao[1]

Abstract — This paper examines the spatial and temporal distributions of the mixing height, ventilation coefficient (defined as the product of mixing height and surface wind speed), and cloud cover over the eastern United States during the summer of 1995, using the high-resolution meteorological data generated by MM5 (Version 1), a mesoscale model widely used in air quality studies. The ability of MM5 to simulate the key temporal and spatial features embedded in the time series of observations of temperature, wind speed, and moisture is assessed using spectral decomposition methods. Also, mixing heights estimated from the MM5 outputs are compared with those derived from observations at a few locations where data with high temporal resolution are available in the Northeast. In addition, the uncertainties associated with the estimation of the evolution of the boundary layer during the morning time are examined. The results indicate that nighttime mixing heights averaged <200 m, rising to 1 km by 10 EST, and to about 2.5 km in the afternoon. Ventilation coefficients followed a similar diurnal pattern, increasing from 500 m^2/s at night to 15,000 m^2/s in the afternoon; the increase due to the growing mixing height and increasing surface wind speeds. Spatial variability of these parameters was relatively small (coefficient of variation = 0.25) at night and in the afternoon when conditions were quasi-stationary, but increased (to 0.5) during morning and evening hours when mixing heights and wind speeds were changing rapidly.

Analyses of surface ozone observations from about 400 sites throughout the eastern United States indicate that days with numerous stations reporting surface ozone concentrations in excess of 80 ppb (i.e., "high ozone" days) generally had less daytime cloud cover, lower surface wind speeds, higher mixing heights, and lower ventilation coefficients than did comparable "low ozone" days. Such meteorological features are consistent with a synoptic anticyclone centered over the mid-south region (Kentucky, Tennessee). Low ozone days were characterized by more disturbed weather conditions (low pressure systems, fronts, greater cloud cover, and precipitation events). Ozone observations at two elevated platforms (∼400 m agl) in Garner, NC, and Chicago, IL, indicated that ozone concentrations aloft were about 40% larger on "high ozone" days than on "low ozone" days. On average, high levels of ozone persist aloft for about 2 to 3 days. Strong vertical mixing in the daytime can bring this pool of upper-level ozone downward to augment surface ozone production. Since ozone can be transported downwind several hundred kilometers from its source region over this time scale, depending on upper-level winds, effective ozone control strategies must take into consideration spatial scales ranging from local to regional, and time scales of the order of several days.

[1] Department of Earth and Atmospheric Sciences, State University of New York, University at Albany, Albany, NY 12222, U.S.A.

[2] New York State Department of Environmental Conservation, Albany, NY 12233, U.S.A.

[3] Earth Sciences Department, State University of New York, College at Oneonta, Oneonta, NY 13820, U.S.A.

* Corresponding author's current address: Dr. S. T. Rao, Director-Atmospheric Modeling Division, U. S. Environmental Protection Agency (MD-E243-02), Research Triangle Park, NC 27711, USA

Key words: PBL, mixing height, ozone, MM5, air quality modeling.

1. Introduction

Meteorological processes within the planetary boundary layer play a critical role in determining the transport, diffusion, and chemical transformation of air pollutants (VUKOVICH, 1995, 1997; CLARK, 1997; ZHANG *et al.*, 1998; ZHANG and RAO, 1999; ANEJA *et al.*, 2000). Three key boundary-layer parameters for studying air pollution events are the mixing height, ventilation coefficient, and cloud cover. The mixing height is the depth, measured upward from the earth's surface, through which air pollutants are vigorously mixed. The ventilation coefficient is defined here as the product of the mixing height and wind speed at 10-m height. Large ventilation coefficients suggest increased dilution, leading to lower pollutant concentration levels. Recent analyses using the Urban Airshed Model (UAM-IV) indicate that ozone concentrations predicted by this model are sensitive to errors and uncertainties in the specification of the mixing-height profile (RAO *et al.*, 1994). Also, the selection of appropriate pollution control strategies appears to depend on the spatial and temporal variability of the mixing height across the model domain (SISTLA *et al.*, 1996). In a recent study of summertime conditions over the northeastern United States, BERMAN *et al.* (1999) showed that average ventilation coefficients during early morning hours were approximately 50% lower on ozone episode days than on non-episode days. In general, it is expected that controls on NO_x (VOC) emissions would be effective under high (low) ventilation conditions (BISWAS and RAO, 2001). Consequently, a reliable depiction — both spatially and temporally — of summer-time mixing height and ventilation coefficients over the eastern United States is useful in understanding the ozone pollution problem in the eastern United States, building confidence in the assessment of the efficacy of emission control strategies, and in forecasting pollution episodes. Under clear skies over land, the mixing height displays a diurnal cycle of growth and decay following the cycle of heating and cooling at the surface. In the hours just after sunrise, the mixing layer grows rapidly, reaching a maximum depth of 1.5 to 2.5 km by the mid-afternoon. As surface convection wanes in late afternoon, the height of the mixed layer is reduced significantly. By sunset, the depth of the mixed layer height is usually <500 m. At night, a stably-stratified layer develops near the ground, extending upward to about 200 m. In general, only the lowest part of this nocturnal boundary layer is fully turbulent, although weak and sporadic turbulence may extend to the top of the surface temperature inversion in urban locations. This is due to increased surface roughness (mechanical turbulence) and heat-island effects (thermal turbulence) which are characteristic of urban areas. Also, nighttime mixing processes occur on longer time scales in which mechanical mixing dominates convective mixing (BEYRICH and WEILL, 1993).

The evolution of the mixed layer is more complicated in the presence of significant cloud cover, precipitation events, frontal passages, etc. Coastal areas, in particular, tend to experience a greater amount of variability in mixing height because of the varying positions of the sea-breeze front. During onshore flow, a convective internal boundary layer (sea-breeze front) develops at the land-sea interface which tends to grow vertically with the square root of distance from the shoreline (STULL, 1988). The interface is marked by sharp changes in the air's vertical temperature, humidity, and wind structure, affecting vertical mixing (ANGEVINE et al., 1996). This feature is a common occurrence along coastlines in the summer.

Surface ozone concentrations are dependent upon the mixing height; early morning hours are characterized by light winds and low mixing heights (i.e., low ventilation coefficients). Under such conditions, motor vehicle emissions from heavy commuter traffic can lead to a significant buildup of ozone precursors (nitrogen oxides [NO_x] and volatile organic compounds [VOC]) near the ground. The growth rate of the mixed layer in the hours following sunrise is critical to the development of high ground-level pollutant concentrations later in the day (UNO et al., 1992). Later in the morning, increased surface heating and wind speed (i.e., larger ventilation coefficients) enhance mixing which facilitates dilution of surface pollutants. If an ozone-rich layer is trapped in the residual layer aloft, however, vertical mixing from below may actually *increase* ground-level ozone concentrations by mixing polluted air downward (i.e., fumigation) as shown by ZHANG and RAO (1999). As the mixed layer continues to grow to the height of about 2.5 km by mid-afternoon, its much greater volume will eventually reduce surface ozone concentrations through greatly increased dilution (ZHANG and RAO, 1999).

Using a network of six intensive rawinsonde stations (6 soundings/day), five radio acoustic sounder system (RASS) sites, and 44 surface stations, BERMAN et al. (1999) studied the behavior of mixing depths and ventilation coefficients in the northeastern United States on 13 ozone episode days and 60 non-episode days in the summer of 1995. Mixing heights were estimated from virtual potential temperature profiles during the daytime and the depth of the nocturnal surface temperature inversion at night. Daytime mixing heights ranged from <500 m offshore to >2000 m inland, with most of the increase occurring within the first 100 km of the coastline. Nighttime mixing depths averaged ~200 m. Morning ventilation coefficients were found to be lower by 50% on ozone episode days than on non-episode days.

The current investigation builds upon the previous study of BERMAN et al. (1999) by using surface and upper-air meteorological data derived from running the MM5 mesoscale model to estimate boundary layer parameters. The area of coverage provided by the MM5 model is considerably larger than that used in the previous study, including nearly the entire eastern United States. Spatial and temporal resolutions are also dramatically increased with a horizontal grid-point spacing of 12 km, vertical profiles of temperature, pressure, humidity, and wind obtained from 12 levels in the boundary layer (from a height of 10 m to 3.1 km), and MM5 output available for every hour of the day and night. In this paper, we provide a comprehensive spatial and temporal depiction of summertime mixing depths and ventilation coefficients over the eastern United States, based on high-resolution meteorological fields obtained from MM5 for the period June 2 to August 30, 1995, and identify boundary-layer characteristics conducive to the accumulation of ozone in the eastern United States.

2. *Estimating Boundary-layer Parameters*

2.1 *MM5 (PSU-NCAR, Version 1) Model*

MM5 is a mesoscale model commonly used to investigate the meteorological processes on a wide range of scales extending from synoptic-scale waves to meso-β scale flows. The model is based on a set of primitive equations developed for a fully compressible atmosphere in a rotating frame of reference. It is designed to use surface measurements and sounding data analyzed on model grids for initialization and boundary conditions. MM5 is also being widely employed to generate the meteorological data needed to drive photochemical models. The model incorporates terrain-following vertical coordinates, realistic topography, radiation-cloud interactions, complete cloud precipitation physics, cumulus parameterizations, a diurnal heating cycle, and surface fluxes of heat, moisture, and momentum. A full description of the MM5 model is given by DUDHIA (1993) and GRELL *et al.* (1994).

In this study, the MM5 model was run on a nested grid domain at 108/36/12 km resolutions using 2-way interactions (see Fig. 1a), with 25 layers in the vertical extending upward from 10 m to about 16 km. The model employs a sigma coordinate system in which the vertical coordinate (σ) is defined as $(p - p_{top})/(p_s - p_{top})$, where p is the pressure at any level, p_s is the surface pressure, and p_{top} is the pressure at the top of the model atmosphere. Therefore, actual heights of the σ surface vary vertically with p and horizontally with the terrain-following surface pressure (p_s). In the present framework the convective boundary layer is located within the lowest 12 levels (approximately 10 m to 3.1 km, agl). The 12-km grid covers a surface area extending roughly over 25° of longitude (70°W to 95°W) and 15° of latitude, (30°N to 45°N). Figure 1b depicts the region of the United States covered by the 12-km grid as well as the locations of approximately 400 surface ozone-monitoring stations included in the U.S. Environmental Protection Agency's Aerometric Information Retrieval System (AIRS) database.

The Blackadar PBL scheme (1979) has been chosen for the present study. Details of the scheme are given by ZHANG and ANTHES (1982). The Kuo/Anthes cumulus parameterization and the Grell scheme (GRELL *et al.*, 1994) have been applied to the 108 km and 36 km grids of the model domain, respectively. There is no convective parameterization scheme for the 12-km domain since it is already in the convective

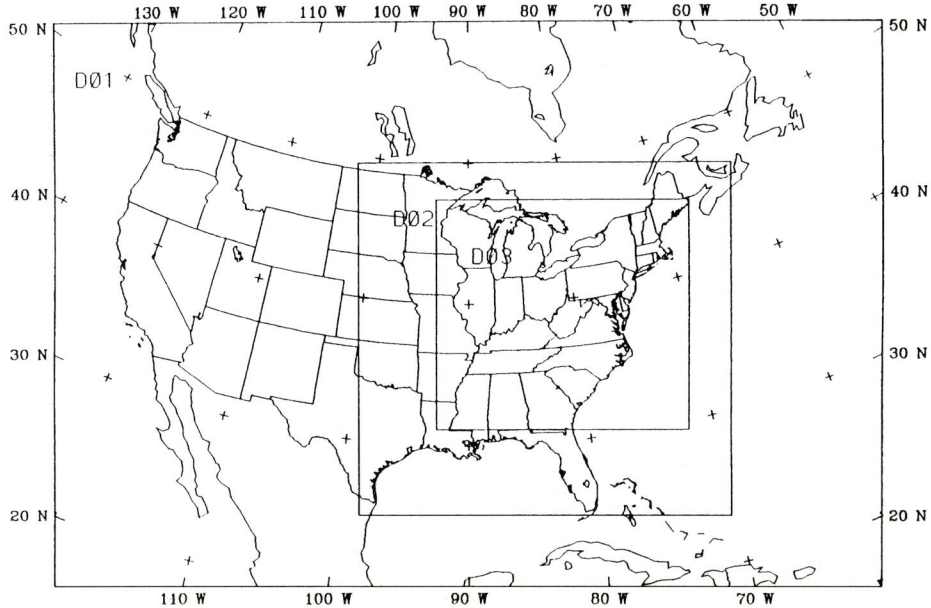

Figure 1a
Map of the entire domain covered by MM5's nested grid system (108/36/12 km).

scale. The grid-resolvable precipitation is also predicted in MM5 simulations by explicitly computing the 5 components of water substance (water vapor, cloud water, rain, snow, and ice) using a mixed-phase scheme. Also, Four-Dimensional Data Assimilation (FDDA) is applied to the model using grid nudging in order to avoid the accumulation of errors in running MM5 for long periods of time. FDDA is a method of incorporating the observations into the time-integration (STAUFFER and SEAMAN, 1990). Thus, the model equations assure the dynamical consistency between variables while observations keep the model close to the true synoptic conditions. The strength of the nudging is determined by the nudging coefficient and the differences between observations and predictions. The nudging coefficient used in this study was set at 2.5×10^{-4} s^{-1} (STAUFFER et al., 1991). In this MM5 seasonal simulation, National Center for Environmental Prediction (NCEP) analyses at the observation times of 0000 and 1200 UTC were used as the first-guess fields. Surface observations available every 3 hours and 12-hourly vertical sounding data were used to further enhance the resolution of the mesoscale analyses. In the vertical direction, ten pressure levels of data from the soundings were added below 500 mb, supplementing the 12 standard pressure levels from the NCEP analysis. We used both 12-hourly 3-D analyses and 3-hourly 2-D analyses as nudging fields during model integration.

In this study, two major developments have been made to modify the original MM5 model for performing the seasonal simulation: (1) update sea-surface

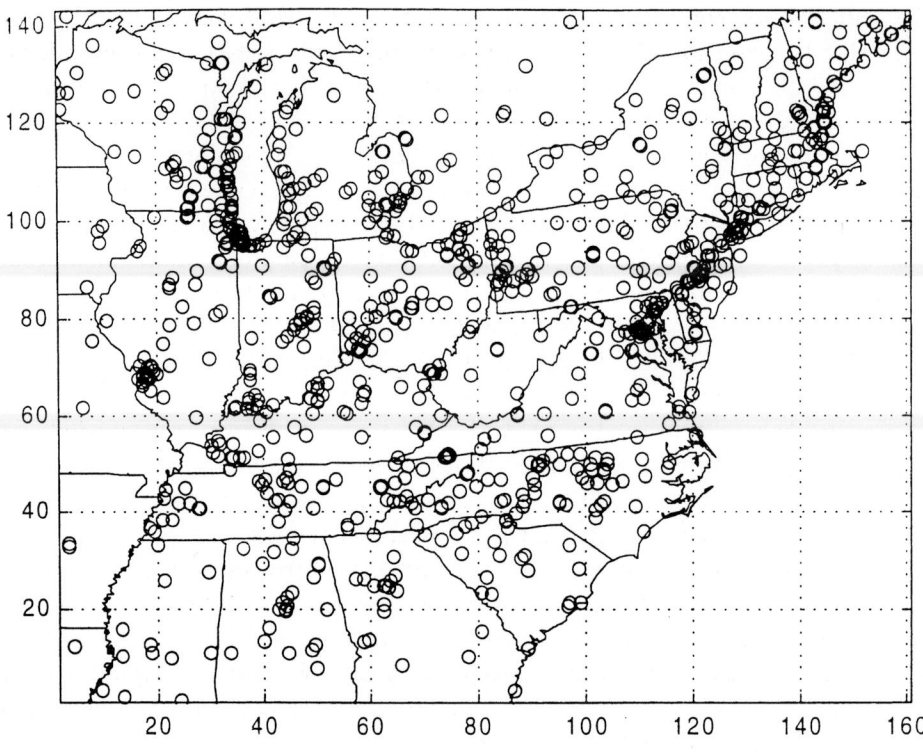

Figure 1b
Map of the eastern United States showing the domain covered by the MM5 12-km grid. Open circles
indicate the locations of surface ozone monitoring stations.

temperature and substrate temperature used in the surface energy balance equation
to describe the seasonal forcing from below, and (2) modify the model input control
system to ingest the new input data at 5–8 day intervals during the extended longer
time integrations.

2.2 Estimating Mixing Heights

VENKATRAM (1978) showed that the turbulence generated by the upward heat
flux in the PBL can be expressed as $\sigma_w = 0.6 \, w^*$ where w^* is the convective scaling
velocity. The convective scaling velocity is given by $w^* = [g q_s z_i / T_s]^{1/3}$ where g is the
gravity, z_i is the mixing height, q_s is the surface heat flux, and T_s is the surface
temperature. The mixing height is given by $z_i = 2/\gamma \int q_s(t) \, dt$, where γ is the potential
temperature lapse rate and t is the time. The mixing height can be calculated from the
heat flux equation by integrating $\partial z_i / \partial t = [w'\theta'_s - w'\theta'_{zi}]/\gamma z_i$ where $w'\theta'$ is the heat flux
(STULL, 1988). Mixing heights for daytime convective conditions can also be
estimated from the HOLZWORTH (1967) profile-intersection technique. This procedure
entails extrapolating the surface temperature adiabatically upward until it intersects

the ambient temperature profile. The top of the mixed layer coincides with the base of an elevated inversion which caps the convective boundary layer. The conventional Holzworth technique does not include the effects of moisture or temperature advection. Advection of moist air from the Atlantic Ocean and Gulf of Mexico is a common feature over the eastern United States during the summer. This tends to raise the air's buoyancy, leading to increased thermal turbulence and a greater mixing height. In order to improve the accuracy of this technique, we chose to estimate the mixing depth from profiles of virtual potential temperature rather than from temperature. Virtual potential temperature was computed from hourly values of temperature, pressure, and relative humidity available from MM5's first 12 levels, ranging in altitude from approximately 10 m to 3.1 km.

In this study, Holzworth's profile-intersection scheme is used to estimate mixing heights for daytime convective conditions. Also, we used MM5 outputs of wind and temperature fields within the boundary layer to calculate the heat flux based on the Monin-Obukhov similarity theory. The hourly values of heat flux were then used to estimate the mixing heights according to the heat-flux method as suggested by STULL (1988). This method may be useful when the observed sounding data are temporally and spatially sparse. To examine the uncertainty associated with the estimation of the mixing height, we compared the mixing heights from the profile-intersection method with those derived from the heat-flux method, focusing on the boundary-layer evolution during the early morning hours, as 06–10 EST is the most critical time from an ozone pollution perspective.

For the stable boundary layer, a number of algorithms have been proposed including h_u (height of the low-level wind maximum), h_i (height of the surface-temperature inversion), h_θ (height of the adiabatic layer), and h_{Ri} (height at which the Richardson number reaches a critical value). A number of studies by YU (1978), ARYA (1981), and MAHRT et al. (1982) have shown that the first three of these (h_u, h_i, and h_θ) considerably overpredict the height of the nocturnal boundary in comparison with h_{Ri} and correlate poorly with the Monin-Obukhov similarity relations u_*/f and $(u_*L/f)^{\frac{1}{2}}$, where u_* is the friction velocity, L is the Monin-Obukhov length, and f is the Coriolis parameter. The weak relationship found with h_u, h_i, and h_θ is believed to be due to their inability to represent the actual depth of the turbulent layer. h_{Ri} appeared to be a more promising indicator, but it is a difficult parameter to determine accurately under calm or light wind conditions which are frequent occurrences at night.

A recent study by LENA and DESIATO (1999) compared ten different algorithms for estimating the nocturnal mixing height with those derived from wind (SODAR-acoustic sounders) and temperature (RASS) profiles. The algorithms consisted of various formulations of boundary-layer parameters: u_*, u_{10}, u_*/f, and $(u_*L/f)^{\frac{1}{2}}$, where u_{10} is the mean wind speed at 10-m height. In general, all the algorithms performed poorly when compared with RASS-based estimates, but reasonably well when compared to SODAR-based estimates. Of the ten algorithms tested, the two that performed best were the CALMET algorithm and the Benkley and Schulman algorithm. CALMET is a preprocessor code designed to prepare meteorological input for simulating air pollution events (SCIRE *et al.*, 1995). The CALMET algorithm uses the expression $h = \min [0.4 \, (u_* L/f)^{\frac{1}{2}}, 2400(u_*)^{3/2}]$ for estimating the stable boundary layer height. This algorithm had the lowest bias of the ten algorithms tested, but tended to underestimate the mixing height in low wind speeds and to overestimate it in high wind speeds. In an earlier study, BERMAN *et al.* (1997) examined the performance of three models and found that CALMET's mixing height estimates were in good overall agreement with profiler measurements for daytime and nighttime hours. BENKLEY and SCHULMAN (1979) proposed the formula $h = 125 \, u_{10}$ for estimating the mixing height at night. This algorithm had the lowest NMSE (normalized mean square error) of the ten algorithms tested and a fairly low bias. Because of its simplicity and overall good performance, we have adopted the Benkley and Schulman algorithm for use in this study. LENA and DESIATO (1999) also suggest that the Benkley and Schulman algorithm adopts a threshold value of 50 m in order to avoid overly small mixing heights under very low wind or calm conditions, a procedure which we have implemented. In summary, for daytime convective conditions, we estimated the mixing height with the adiabatic intersection technique using virtual potential temperature profiles as well as with the heat-flux method using the meteorological fields from the lowest 12 levels of MM5 output. For stable conditions, the Benkley and Schulman algorithm (with a lower limit of 50 m) was used. To determine which stability regime was present, we computed the Monin-Obukhov stability parameter, L, from MM5's wind and temperature profiles using flux-profile formulations provided by BYUN (1990). Estimates of the roughness length were available from MM5 on the basis of the terrain land-use category, determined for each grid point of the model.

2.3 Estimating Ventilation Coefficients

The ventilation coefficient is defined here as the product of the mixing height and a representative boundary-layer wind speed. Surface winds (measured at 10 m agl) are useful for this purpose because of the general interest in the ground-level pollutant concentrations. Dispersion during the early morning hours (06–10 EST) is of particular concern, since this is when the boundary layer is characterized by a combination of low mixing heights, low wind speeds, and high levels of ozone precursor emissions in morning commuter traffic. Ventilation coefficients under these conditions tend to be low, signaling weak dispersion. Later in the day, when the mixing height has grown to 2.5 km and surface wind speeds have increased, the lower atmosphere's dispersive ability is greatly enhanced. These factors are reflected in the much larger ventilation coefficients. It should be noted that even if wind speeds are considerable, factors such as changes in wind directions may form recirculation pattens, and, hence, tend to decrease ventilation.

3. Analysis of Boundary-layer Parameters: June to August, 1995

3.1 Evaluation of the Results of MM5 Model

Since we chose to use the MM5-derived meteorological fields in characterizing the boundary-layer features over the eastern United States in this study, it is prudent to compare the simulated fields with observations and assess the capability of MM5 in reproducing the observed features. Comparison of MM5-predicted fields with observations is not a trivial task since MM5 employs four-dimensional data assimilation (FDDA). We consider observations taken as part of the NARSTO research program (KORC et al., 1996) in model evaluation since they were not used in FDDA. Specifically, we compare MM5 predictions near the grid cells containing the five locations in the Northeast where observations are available from 6 soundings/day on 8 ozone episode days in the summer of 1995. In comparing modeled values with observations, we should keep in mind that there cannot be a perfect agreement between simulated values and observations since the model estimates represent a volume average for the grid cell whereas observations reflect point measurements. Figure 2 depicts the mean difference between simulated and observed values and its standard deviation at specific levels for temperature, mixing ratio, and wind speed as

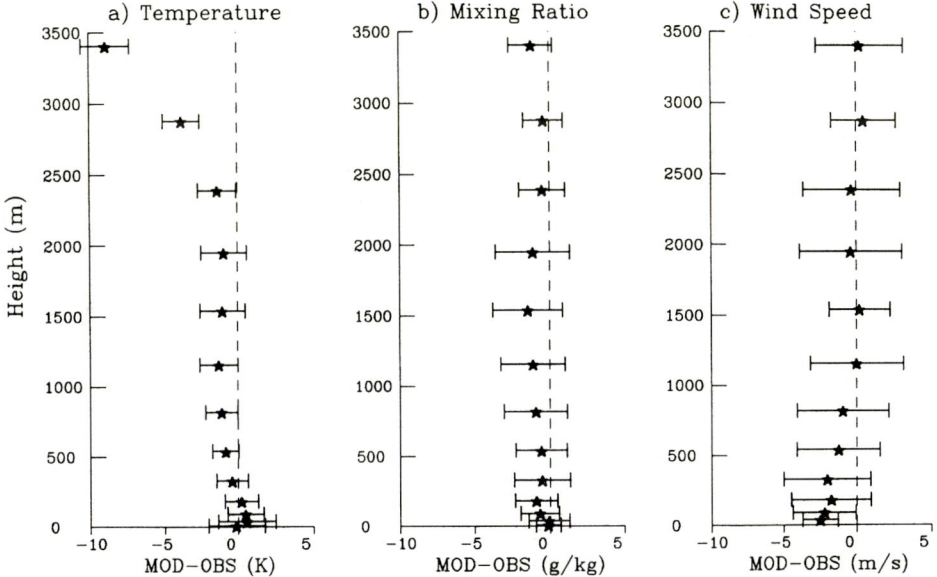

Figure 2
Mean difference and standard deviation of the differences between simulated and observed temperature, mixing ratio, and wind speed as a function of height based on sounding data at five locations in the Northeast (Aberdeen, MD; Atlantic City, NJ; Chatham, MA; Gray, ME; and Sterling, VA) where 6 soundings/day are available for 8 days in the summer of 1995.

a function of height; the agreement between observations and simulated values is good below 2500 m for temperature, mixing ratio, and wind speed. However, temperature was underestimated above the boundary-layer (>2500 m) at the locations where high-resolution soundings were available. A plausible explanation for this might be the reliance of MM5 on 00Z and 12Z soundings for nudging which miss the time of the maximum temperature aloft, particularly along the coastal areas in the eastern United States or the lack of convective parameterization in this version of MM5. Although the mixing heights computed here are not affected by this, such large discrepancies between simulated and observed values warrant a detailed future examination of the reasons for the poor performance of MM5 above 2500 m.

We also evaluate the performance of MM5 using the spectral decomposition technique described by ZURBENKO (1991) on model outputs and observations since it can facilitate a comparison of the key temporal and spatial features imbedded in observations and simulated values. The procedure entails separating the different scales imbedded in the simulated and observed time series of meteorological variables (temperature, wind speed and mixing ratio) for the summer of 1995, namely, intra-day (<10 hours), diurnal (24 hours), synoptic (2–21 days), and longer-term variations. We used the Kolmogorov-Zurbenko (KZ) filter to spectrally decompose the time series of simulated and observed meteorological variables (ZURBENKO, 1991; RAO and ZURBENKO, 1994; ZURBENKO *et al.*, 1996; ESKRIDGE *et al.*, 1997). The time series of meteorological variables can be represented as the sum of different forcings, operating on different space and time scales (RAO *et al.*, 1997; CHAN *et al.*, 1999).

$$X(t) = ID(t) + DU(t) + SY(t) + BL(t)$$

where X is the meteorological variable of interest, ID is the intra-day variation, DU is the diurnal variation, SY is the synoptic-scale variation, and BL is the longer-term variation. These components are separated with the KZ filter with the parameters adopted from HOGREFE *et al.* (2000). Since we use data for 90 days only, the shortest and longest periods that can be resolved are 2 hr and 40 days, respectively. Also, since we do not distinguish between the synoptic and baseline components for wind speed, the synoptic component of wind speed here contains all periods exceeding 2 days.

The correlations between model-predicted and measured temperatures are depicted in Figure 3 using a data interpolation technique as in HOGREFE *et al.* (2000) to facilitate the display of the correlations over the modeling domain. Correlations on the intra-day time scale (<10 hours) are very poor since the spatial scale of the intra-day forcing is on the order of 20 km. Given the 12-km horizontal grid dimension used in the MM5 model, it is not surprising to see that the model does not perform well on the intra-day scale. Certainly, the use of a higher spatial resolution in the MM5 may improve the model's predictions on this time scale if there is a temporally and spatially-dense observational network to support such high-resolution meteorological modeling. From Figure 3, it is evident that the model is

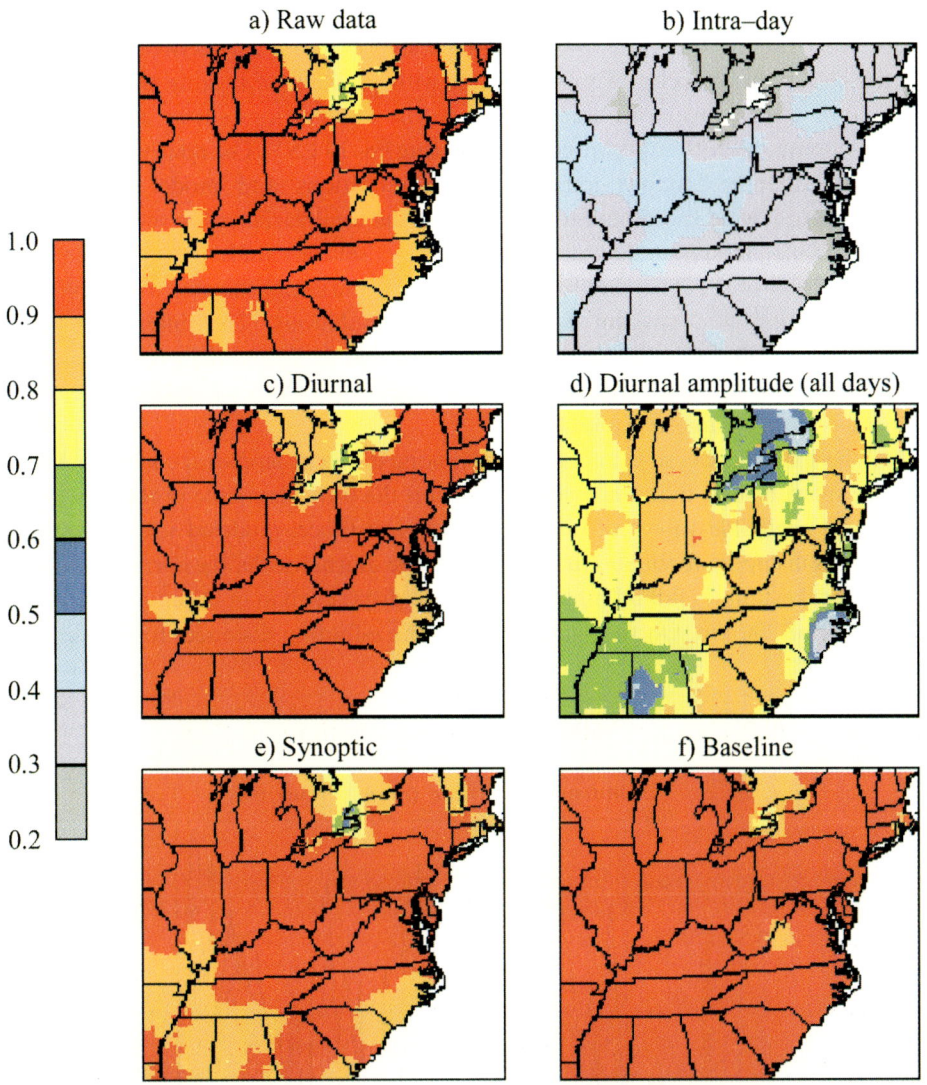

Figure 3
Correlation coefficients between simulated and observed temperatures on different time scales: (a) raw data, (b) for the intra-day time scale, (c) for the diurnal time scale, (d) for the amplitude of the diurnal oscillation, (e) for the synoptic time scale, and (f) for the longer-term (baseline) time scale.

well able to simulate the day-to-day variability of diurnal, synoptic (2–21 days) and longer-term fluctuations in the surface temperature. In regard to the features on the diurnal scale, only the amplitudes and phase shift between the simulated and observed variables are of interest. Although MM5 captures the day-to-day variability in the amplitude of the temperature diurnal forcing, the amplitude is underestimated

(Fig. 4); this is true at all other monitoring stations. Although similar results are evident for the mixing ratio, there is a phase shift of about 2 hr between observed and predicted diurnal cycles of mixing ratio, with predictions lagging the observations (Fig. 5). The mixing ratio diurnal cycle in the observations shows a rapid increase when compared with model predictions during 06–10 EST. The relative contributions of the different forcings imbedded in observations and model predictions of temperature, wind speed, and mixing ratio to the total variance are shown in Table 1. Of particular interest is the fact that the model severely underestimates the relative contribution of the intra-day forcing to the total variance; this is attributable in part to the 12-km grid spacing used in MM5 simulations. The differences between predictions and observations are also attributable to the model's resolution of soil moisture, topography, vegetation, land-water contrast, and uncertainties in the surface-layer parameterization.

The spatial decay in the correlations among different temporal components of the wind speed reveals that while correlations on the intra-day scale in the observations drop to negligible values within about 10-km distance, modeled values depict high correlations up to about 30 km (Fig. 6). On the diurnal scale, the correlations among the model estimates fall off more rapidly with distance than in observations beyond a distance of 200 km. On the synoptic-scale, the *e*-folding distances for the simulated

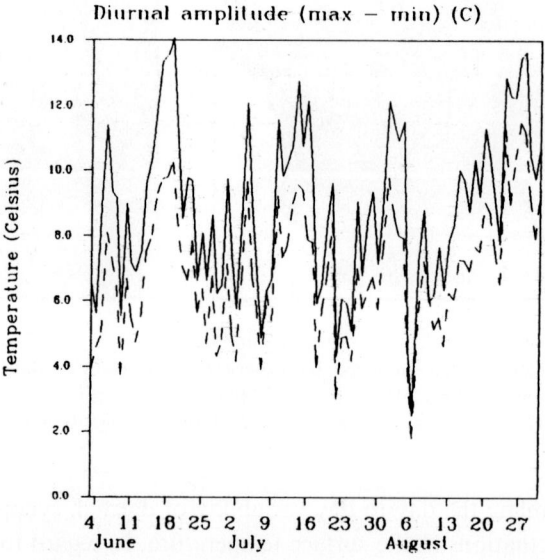

Figure 4

Comparison between the simulated and observed diurnal amplitudes (differences between the maximum and minimum temperatures for the day) in surface temperature at Pittsburgh, PA.

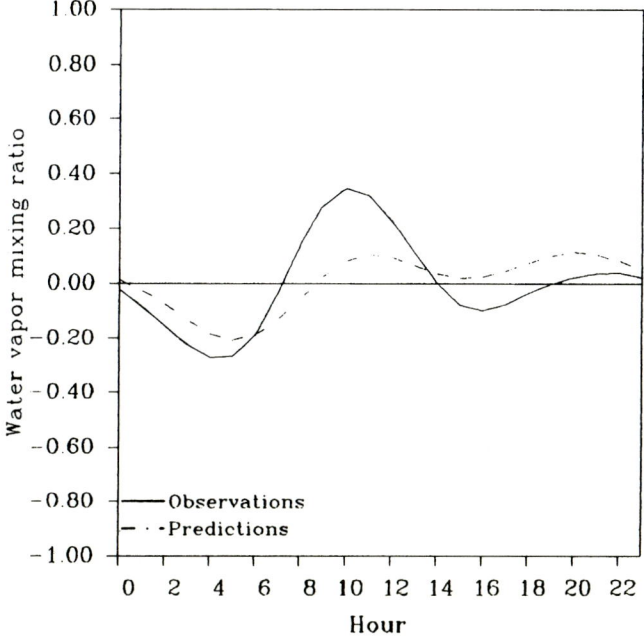

Figure 5

Comparison between the diurnal components of simulated and observed mixing ratios at Pittsburgh, PA.

Table 1

Comparison between the percentage contributions of different temporal components to the total variance in observations and model predictions averaged over 400 sites

Temporal Component	Model Outputs	Observations
Temperature		
Intra-day timescale	1	2
Diurnal timescale	41	59
Synoptic timescale	19	16
Baseline timescale	39	23
Mixing Ratio		
Intra-day timescale	1	3
Diurnal timescale	5	9
Synoptic timescale	35	35
Baseline timescale	60	53
Wind Speed		
Intra-day timescale	4	22
Diurnal timescale	23	40
Synoptic timescale	73	38

values and observations are on the order of 400 km. In general, model estimates are more spatially homogeneous than observations on all time scales, probably a result of the grid resolution employed in the model and inherent smoothing in the simulated

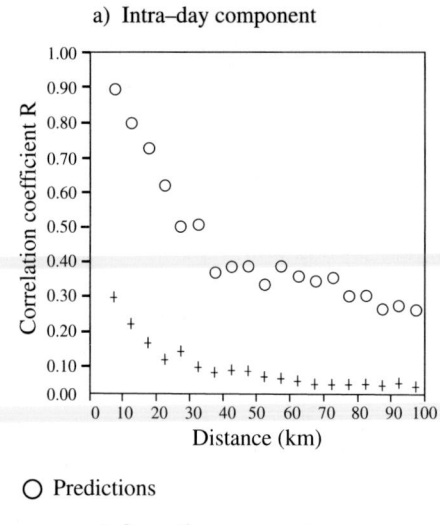

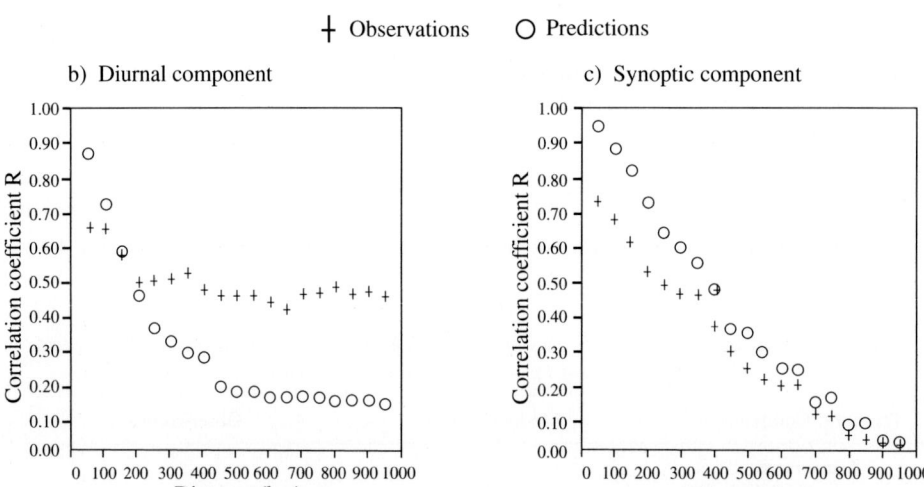

Figure 6

Spatial decay of correlations in observed and simulated wind speed on different time scales.

fields. Thus, these results demonstrate that the meteorological fields simulated by MM5 are not inconsistent with observations and, hence, they can be considered to be a surrogate for reality in estimating mixing heights over the eastern United States. Given the temporal and spatial sparsity of observations, the model outputs can be used to answer questions that cannot be directly addressed through observations.

3.2 *Estimating Mixing Heights*

Hourly mixing heights and ventilation coefficients for 90 summer days (June 2 to August 30, 1995) were estimated at each grid point (12-km horizontal grid dimension) from MM5's three-dimensional meteorological data output using the algorithms discussed in Section 3. The area of data coverage includes most of the eastern United States (Fig. 1b). The mixing heights derived from MM5–predicted fields using two methods, profile intersection and heat flux, are compared to those estimated from soundings taken at 4-hr intervals at five locations during two ozone episodes by the NARSTO experiment (Fig. 7). Although the maximum mixing heights from the three estimates differ somewhat, there is general agreement between the diurnal variations in mixing heights. As noted before, since observations from upper air soundings are not available at a high temporal and spatial resolution, we used the MM5 output to examine the boundary layer evolution over the domain covering the eastern United States. To characterize the uncertainty in estimating the mixing height during the early morning hours, which is the time period of most interest from an ozone pollution perspective, we compared the mixing height estimates of the profile-intersection method with those derived from the heat-flux method at five stations for the entire summer season. The distribution of differences between the two estimates of mixing height during 06–10 EST is presented in Figure 8. In general, mixing heights from the heat-flux method varied considerably when compared with those derived from the profile method. Mixing heights estimated from the profile technique are biased low by about 100 m (mean difference) when compared with those from the heat-flux method; the standard deviation of the differences is about 400 m, which indicates the degree of variability that can be

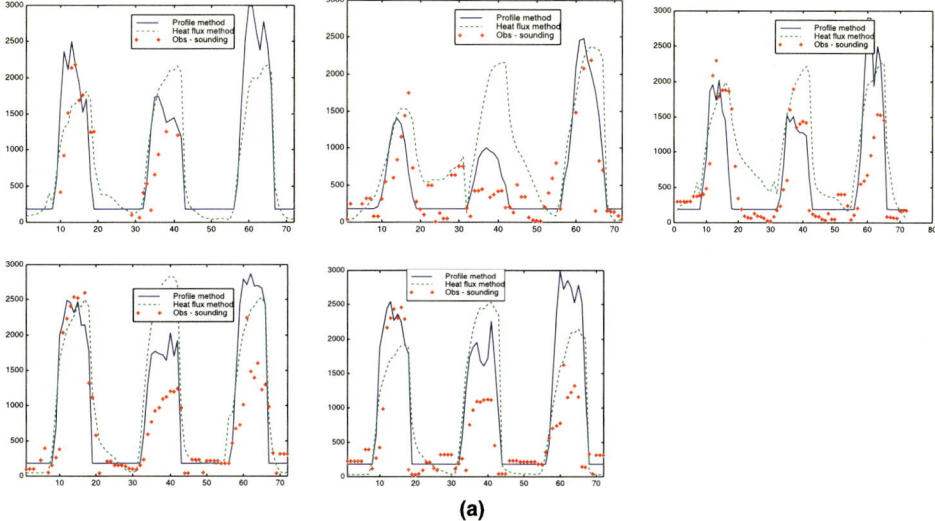

(a)

Figure 7
Comparison of mixing heights estimated from the profile and heat-flux methods using MM5 outputs and using observed sounding data where soundings are available at 4-hr intervals at five stations, (a) July 1995 Episode, (b) August 1995 Episode.

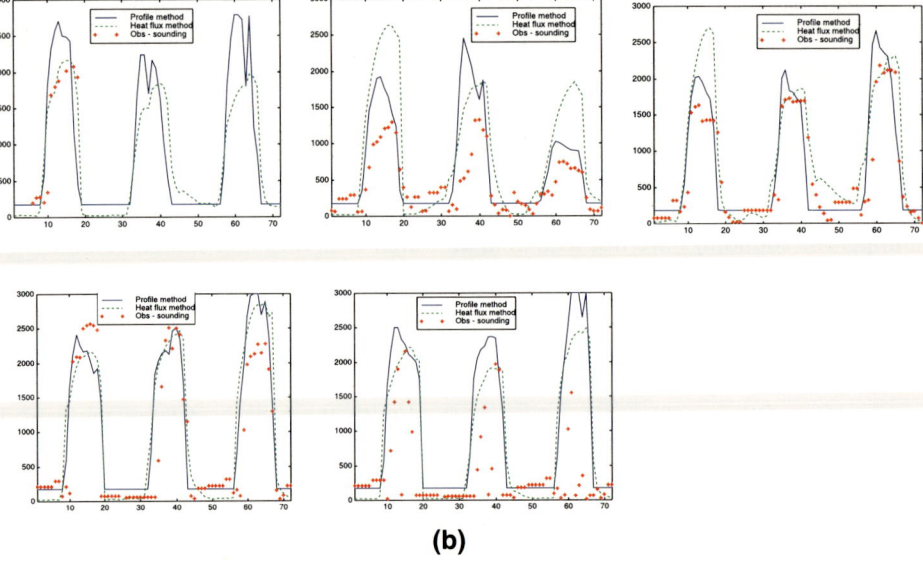

(b)

Figure 7b

expected in the estimates of mixing heights from different methods as the boundary layer evolves during 06–10 EST.

For easier interpretation of the results, the hourly mixing heights estimated from the profile method were averaged over four time periods: 00–04 EST (nighttime), 06–10 EST (morning growth), 12–16 EST (afternoon maximum), 18–22 EST (evening decay). These four time periods were also used for averaging the ventilation coefficients. Since ventilation coefficients are dependent on both mixing height and wind speed, we have also examined the behavior of wind speed as a separate variable.

3.3 Variation of Mixing Height

The variation in mixing heights across the eastern United States averaged for the 90-day summer period is presented in Figure 9. (Note: the same 8 colors are used in all panels of Figure 9, however the colors represent different scales for each panel. This was done to provide the maximum resolution possible, while also accommodating the larger range of mixing heights occurring during midday). Figure 9 shows mixing heights over the Great Lakes to be < 250 m at all times of day, with slightly higher values over Lakes Huron and Erie seen at night (Fig. 9A), perhaps due to their warmer surface waters. Over land, nighttime values (Fig. 9A) average less than 200 m, increasing to 1000 m by 10 EST (Fig. 9B), to maximum values exceeding 2500 m by 16 EST (Fig. 9C), then decreasing to 800 m or less by 22 EST (Fig. 9D). During the convective periods (Figs. 9B and 9C), a narrow band of somewhat higher

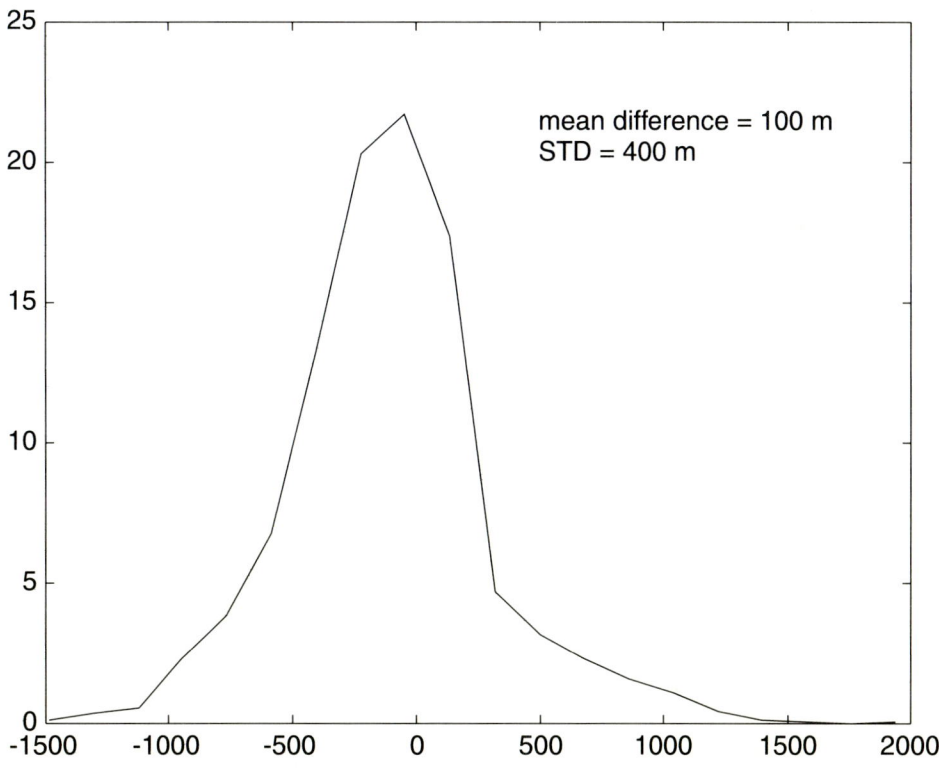

mean difference = 100 m
STD = 400 m

Figure 8
Distribution of differences between the mixing heights estimated from the profile and heat-flux methods
during the 06–10 EST interval using MM5 outputs at the five stations in Figure 2.

mixing heights is seen stretching from New York State southwestward along the Appalachian Mountains to western North Carolina and eastern Tennessee. Over the Atlantic Ocean, a small latitudinal variation in mixing height can be seen ranging from ~400 m at latitude 30°N to about ~200 m at latitude 45°N; this probably reflects the decrease in ocean surface temperatures found extending 1900 km from the warmer Gulf Stream waters in the south to the colder currents off the coast of New England.

Coefficients of variation for mixing height, defined as the ratio of the standard deviation to the mean, were computed for each panel in Figure 9, but are not shown. The coefficient of variation is a statistical parameter that describes the variability about the mean. Larger coefficients indicate greater variability. Coefficients of variation over land were uniformly low at night (<0.25) and in the afternoon (0.25–0.37), indicating relatively little temporal and spatial variability at these times of day. However, the coefficients are twice as large (0.5–0.65) for the morning hours (06–10 EST) and evening hours (18–22 EST). This may be explained by considering the mixing height diurnal cycle. Both nighttime and mid-afternoon periods are relatively

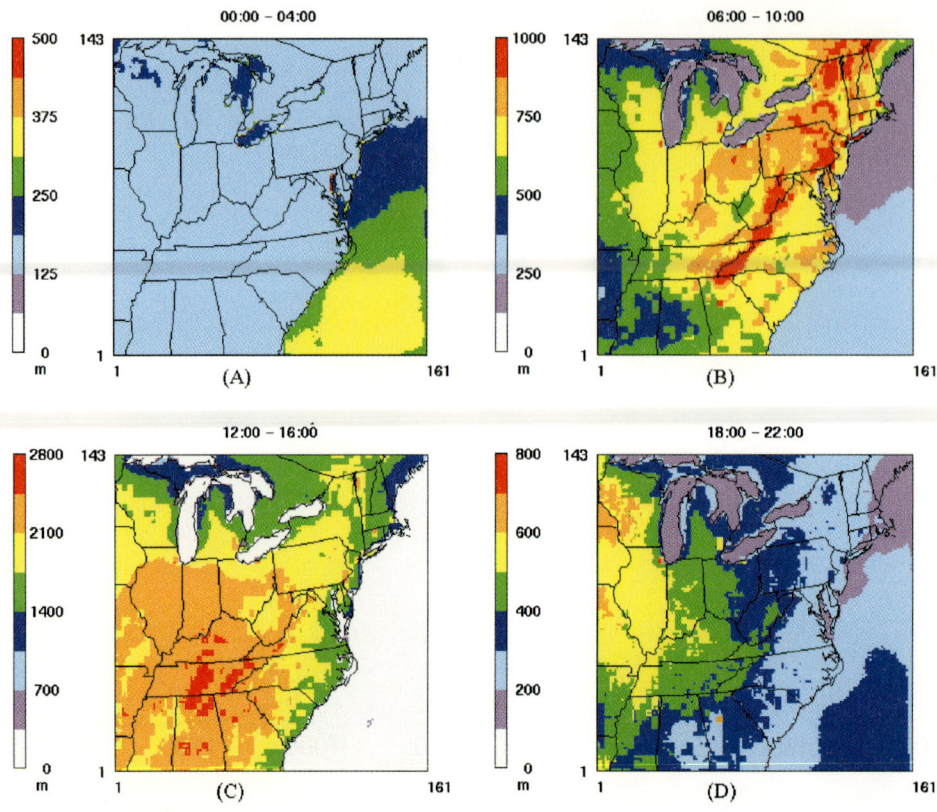

Figure 9
Mixing heights (m) averaged over 90 days (June 2 to August 30, 1995), grouped into four time periods.

quiescent times as this is when the mixing height is either at its minimum (night) or maximum (afternoon), with relatively small variations from hour to hour. In contrast, the morning and evening periods are times of strong transition, with rapid growth occurring between 06 and 10 EST and rapid decay occurring from 18 to 22 EST.

3.4 *Variation of Ventilation Coefficient*

Ventilation coefficients averaged over the 90-day period are depicted in Figure 10 (again, note that the colors in the panels have different values). Not surprisingly, the overall spatial and temporal variation in ventilation coefficients shown in Figure 10 is similar to that found for mixing height in Figure 9. Average ventilation coefficient increases from roughly 500 m^2/s to nearly 5000 m^2/s, a ten-fold increase, between the nighttime hours (Fig. 10A) and the morning hours (Fig. 10B), while for the same periods mixing height increases from roughly 200 m (Fig. 9A) to 1000 m (Fig. 9B), a

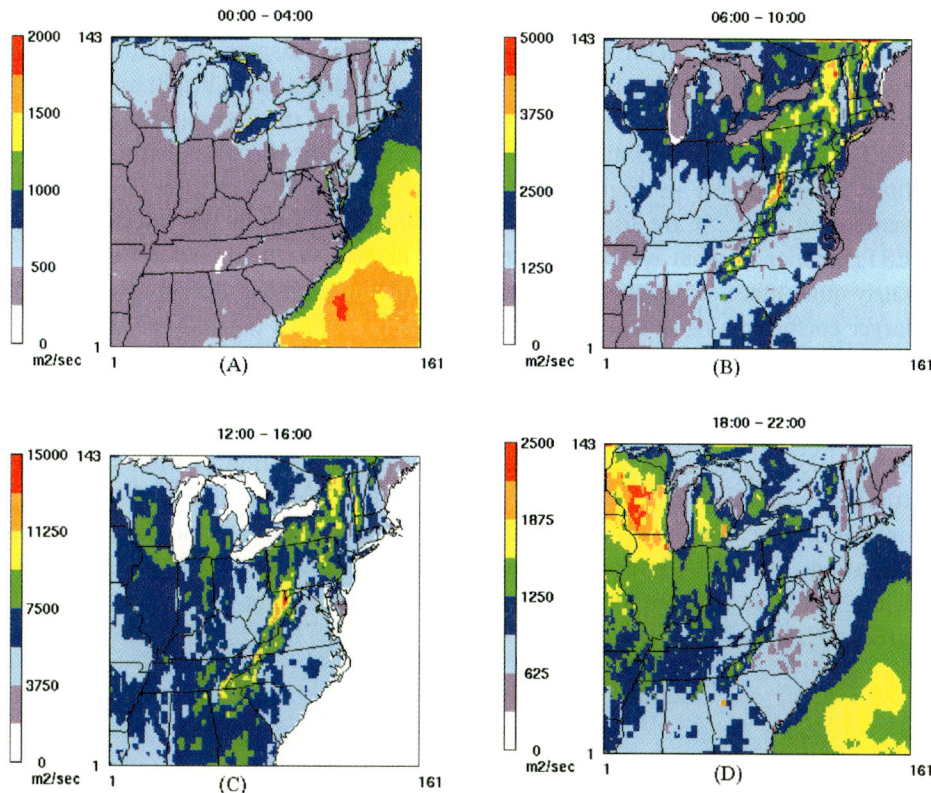

Figure 10
Ventilation coefficients (m²/s) averaged over 90 days (June 2 to August 30, 1995), grouped into four time periods.

factor of 5. Thus, the proportionally much greater increase in the ventilation coefficient is attributed to a near-doubling of average surface wind speed between the two periods. Between the morning and afternoon hours, the rates of increase of mixing height (Figs. 9A and 9B) and ventilation coefficient (Figs. 10A and 10B) are nearly the same, the mixing height increasing by a factor of 2.5 to 2.8 while the ventilation coefficient increases by a factor of about 3.0 to 3.5. This suggests that there is a relatively small increase in surface wind speed between these two periods. Between afternoon and evening, mixing heights fall by a factor of 5 (Figs. 9C and 9D) while for the same periods ventilation coefficients fall by a factor of 10 (Figs. 10C and 10D). This translates to a rough halving of surface wind speeds between the afternoon and evening hours. The increase in the surface wind speed during the growth period of the convective boundary layer is attributed to the downward flux of horizontal momentum from aloft brought about by increased turbulent mixing. Figures 9 and 10 indicate that during periods of strong growth

4.2 Variation of Cloud Cover

Having found (in Section 4.1 above) that mixing heights, their growth rates, and the coefficient of variation tend to be higher on high ozone days than on low ozone days, we next inquire whether this is a result of synoptic conditions favoring higher surface temperatures and stronger convective activity. Figure 13 shows average cloud cover over the eastern United States, obtained from MM5 output, for the same set of low and high ozone days. Figure 13 depicts cloud cover averages for the morning (06–10 EST) and afternoon hours (12–16 EST). Cloud amounts are expressed in tenths of sky cover. On low ozone days morning cloud cover averages 0.30 for most of the eastern United States (Fig. 13A), while it averages to about 0.15 on high ozone days (Fig. 13B). Similar values are seen in the corresponding panels in the afternoon hours (Figs. 13C and 13D). Thus, as expected, high ozone days have about half as much cloud cover as do low ozone days, based on MM5 output data; examination of

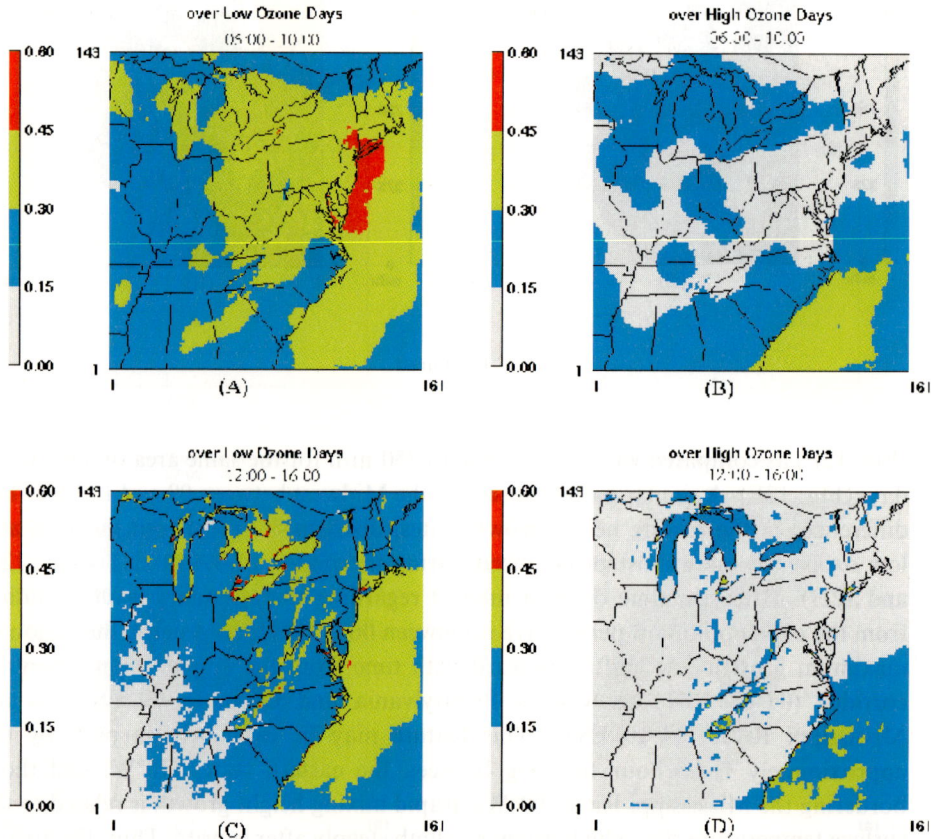

Figure 13
Averaged cloud cover (in tenths) for the morning and afternoon hours on low and high ozone days.

4.2 *Variation of Cloud Cover*

Having found (in Section 4.1 above) that mixing heights, their growth rates, and the coefficient of variation tend to be higher on high ozone days than on low ozone days, we next inquire whether this is a result of synoptic conditions favoring higher surface temperatures and stronger convective activity. Figure 13 shows average cloud cover over the eastern United States, obtained from MM5 output, for the same set of low and high ozone days. Figure 13 depicts cloud cover averages for the morning (06–10 EST) and afternoon hours (12–16 EST). Cloud amounts are expressed in tenths of sky cover. On low ozone days morning cloud cover averages 0.30 for most of the eastern United States (Fig. 13A), while it averages to about 0.15 on high ozone days (Fig. 13B). Similar values are seen in the corresponding panels in the afternoon hours (Figs. 13C and 13D). Thus, as expected, high ozone days have about half as much cloud cover as do low ozone days, based on MM5 output data; examination of

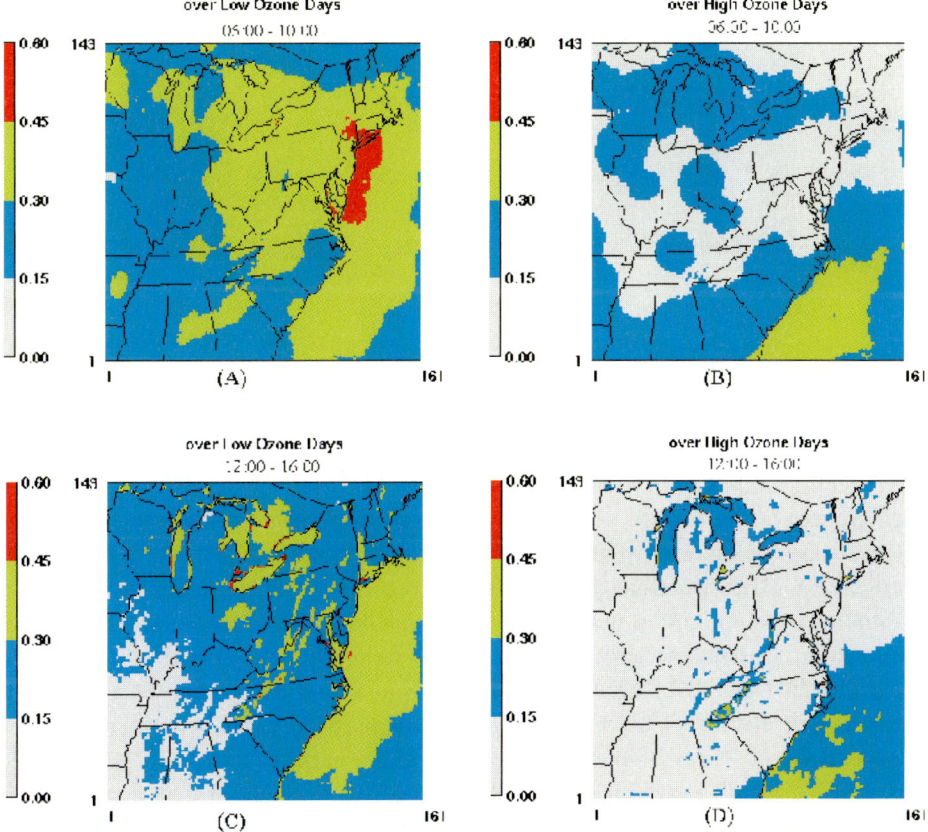

Figure 13
Averaged cloud cover (in tenths) for the morning and afternoon hours on low and high ozone days.

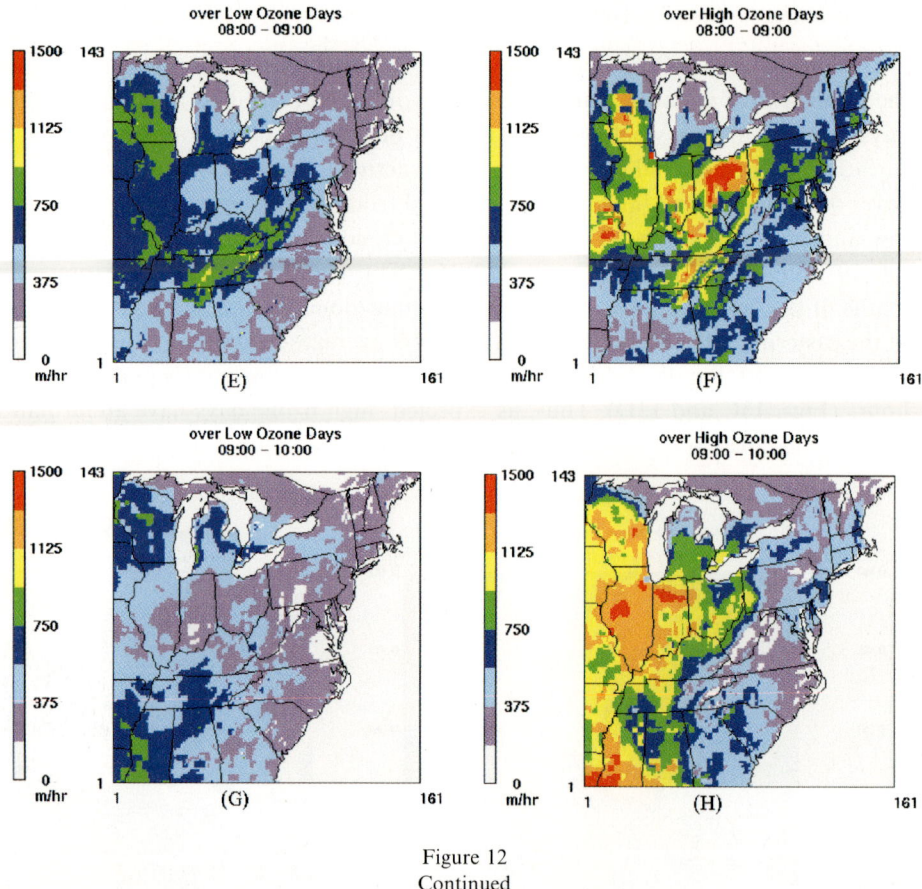

Figure 12
Continued

(Fig. 12F) as compared with rates of 600 to 750 m/h for the same area on low ozone days (Fig. 12E). A similar pattern is seen in the Midwest between 09 and 10 EST. The differences in the mixing height growth rates, distinguishing between the high and low ozone days, are most pronounced during the period of 09–10 EST (Figs. 12G and 12H). Thus, the time of most interest regarding the boundary layer evolution from the ozone pollution perspective is between 06–10 EST. In general, the region of maximum growth rate shifts westward with time from the Appalachian mountain corridor (07–08 EST) to western Pennsylvania and Ohio (08–09 EST) to the Mississippi River (09–10 EST). This feature may be explained, in part, by the approximately 1–1.5 hour time lag between the eastern states (70°W) and those bordering the Mississippi River (90°W). Rapid mixing height growth is related to the surface temperature rise, which begins to climb steeply after sunrise. Thus, the area of maximum morning growth shifts steadily westward, lagging sunrise by approximately 1–2 hours.

days tend to be relatively small (≤150 m), but are somewhat larger (∼300 m) in a narrow band along the Appalachian mountains. By the afternoon (Figs. 11C and 11D), the difference between high and low ozone days is even greater (∼350–500 m), except along the coastline which is likely to feel the effects of the marine layer. As expected, convective activity is generally stronger on high ozone days than on low ozone days.

The growth of the mixing height is particularly critical during the morning hours. Figure 12 shows the rate of growth of mixing height (m/h) hour-by-hour from 06 to 14 EST for low ozone days and high ozone days. The mixing height appears to grow relatively slowly (< 375 m/h) between 06 and 07 EST (Figs. 12A and 12B) with little difference between low and high ozone days. In succeeding hours, however, as the level of convective activity markedly increases, high ozone days show a distinctly higher growth rate than do low ozone days. For example, between 08 and 09 EST growth rates in Ohio range between 1300 and 1500 m/h on high ozone days

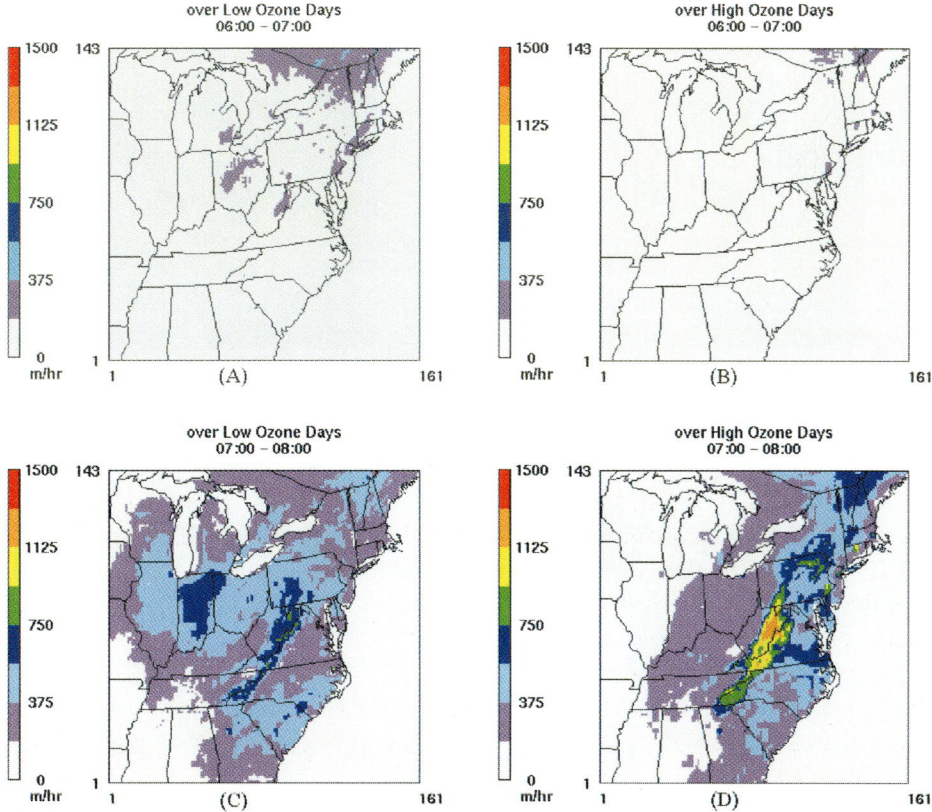

Figure 12
Averaged hourly mixing-height growth rates (m/h) from 06 to 10 EST on low and high ozone days.

first. June 12, 1995 had no stations reporting ozone values < 80 ppb and was ranked last.

4.1 Variation of Mixing Height

Figure 11 compares mixing heights for low (Figs. 11A and 11C) and high ozone days (Figs. 11B and 11D) averaged over two time periods: morning (06–10 EST) and afternoon (12–16 EST). Only the morning and afternoon periods are shown in Figure 11, as these are considered to be the most critical times of day for the formation and accumulation of ground-level ozone. Figure 11 reveals that at most locations mixing heights tend to be higher on high ozone days than on low ozone days. This feature is also evident in mixing heights estimated from the heat-flux method. In addition, the coefficient of variation (standard deviation/mean) for mixing heights is larger for high ozone days than for low ozone days. In the morning hours (Figs. 11A and 11B), mixing height differences between high and low ozone

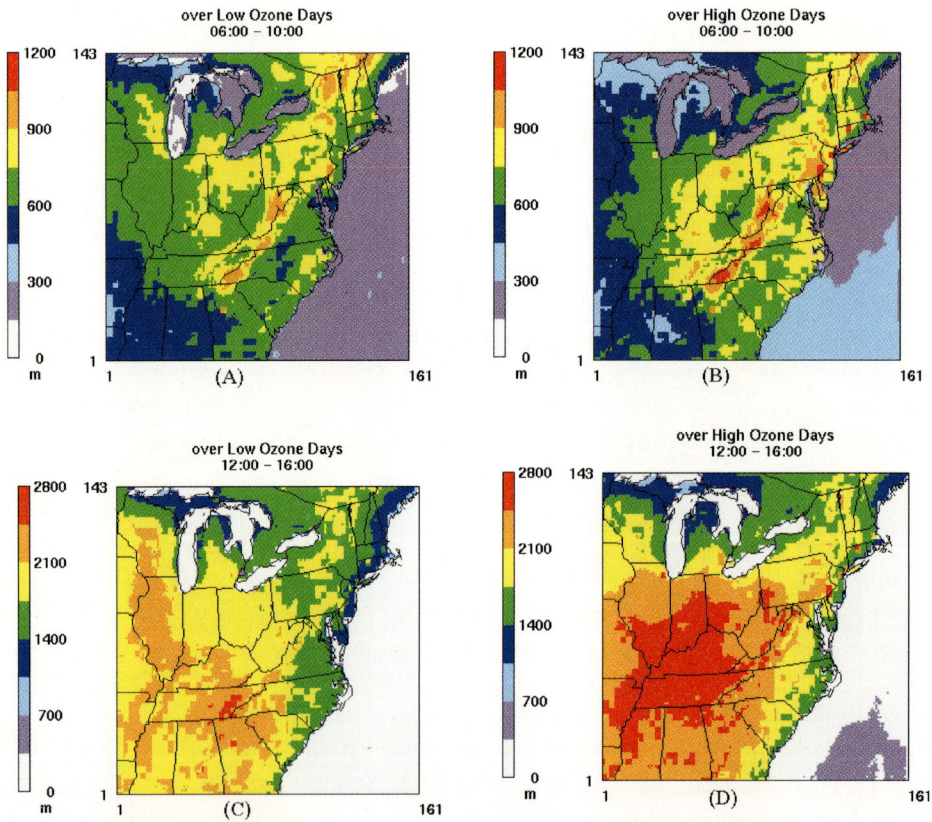

Figure 11
Averaged mixing heights (m) for morning and afternoon hours on low and high ozone days.

(morning) and decay (evening), the surface wind speed varies even faster than the mixing height by a factor of approximately 2. Thus, dispersion/dilution is greatly enhanced during the morning hours by both increasing mixing height and increasing wind speed and, likewise, is greatly diminished during the evening transition hours.

The coefficient of variation for ventilation coefficient was also computed for each time period shown in Figure 10, but is not displayed. Very low coefficients of ventilation (<0.25) are found at night (00–04 EST) and in the afternoon (12–16 EST), similar to what was found for mixing heights. As discussed before, these are fairly quiet times of day with little change in the mixing height or wind speed. Much larger coefficients of ventilation with values between 0.88 and 1.0 are found during the morning (06–10 EST) and evening (18–22 EST) transition periods. The magnitude of the coefficients of variation for the ventilation coefficient are about twice the magnitude of the coefficients of variation for mixing height. The explanation, again, appears to lie in the more rapid variation of wind speed during transition times of day. Between 06 and 10 EST, average surface wind speeds approximately double in magnitude. This produces large wind speed standard deviations over the 4-hour time period. Similarly, between 18 and 22 EST, surface winds drop by half. Again, large standard deviations are the result, which help produce large coefficients of variation.

4. Analysis of Boundary-layer Parameters on High vs. Low Ozone Days

It is of considerable interest to see if boundary-layer parameters vary significantly between days having high concentrations of ozone near the ground and days having low concentrations. It is well known that there is a diurnal variation in emissions loading, and also differences between weekdays and weekends which could introduce a weekly cycle. However, since the day-to-day variation in ozone precursor emissions (due to biogenic and anthropogenic sources) in any given region tends to be small during the summer, one would expect the synoptic forcing to have a major impact on ground-level ozone concentrations. To this end, we analyzed hourly observations of ground-level ozone measured at approximately 400 EPA-approved monitoring stations in the eastern United States for the 90-day period, June 2 to August 30, 1995. Figure 1 shows the locations of these monitoring locations. For each day we tallied the number of sites reporting at least one ground-level ozone value exceeding a value of 80 ppb. The 90 days were then sorted in descending order by the number of stations reporting ozone concentrations exceeding the threshold. The top 25% of the days ($=23$ days) were labeled regionally "high ozone days" and the lowest 25% ($=23$ days) were labeled "low ozone days." July 13, 1995 had the largest number of stations with ozone observations exceeding 80 ppb (390 stations) and was ranked

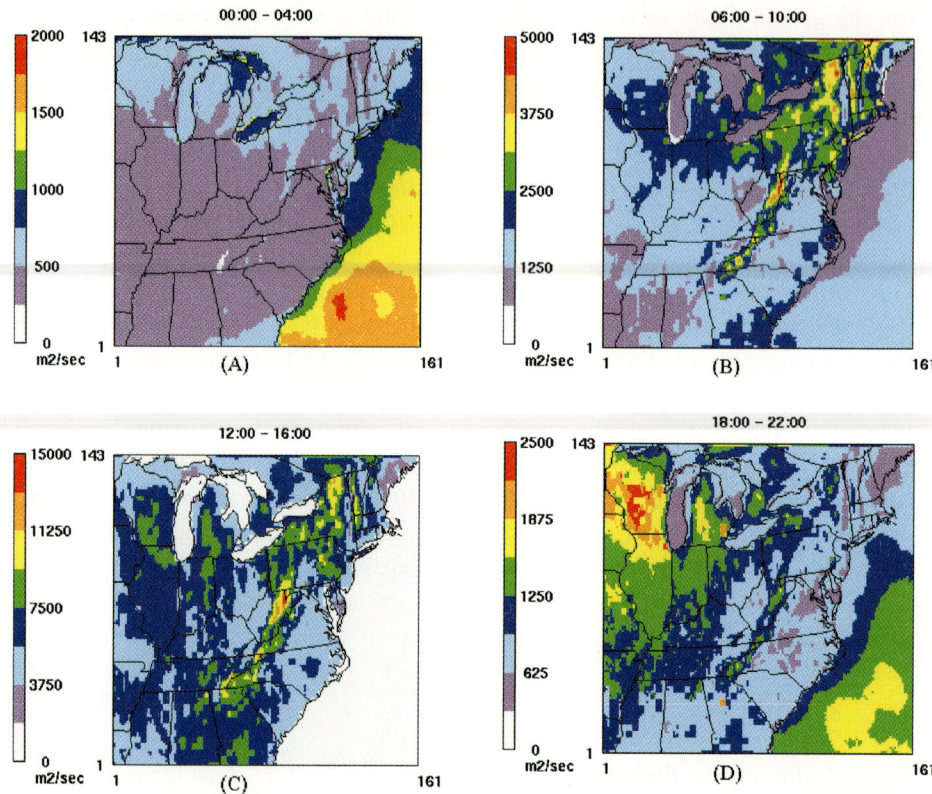

Figure 10
Ventilation coefficients (m^2/s) averaged over 90 days (June 2 to August 30, 1995), grouped into four time periods.

factor of 5. Thus, the proportionally much greater increase in the ventilation coefficient is attributed to a near-doubling of average surface wind speed between the two periods. Between the morning and afternoon hours, the rates of increase of mixing height (Figs. 9A and 9B) and ventilation coefficient (Figs. 10A and 10B) are nearly the same, the mixing height increasing by a factor of 2.5 to 2.8 while the ventilation coefficient increases by a factor of about 3.0 to 3.5. This suggests that there is a relatively small increase in surface wind speed between these two periods. Between afternoon and evening, mixing heights fall by a factor of 5 (Figs. 9C and 9D) while for the same periods ventilation coefficients fall by a factor of 10 (Figs. 10C and 10D). This translates to a rough halving of surface wind speeds between the afternoon and evening hours. The increase in the surface wind speed during the growth period of the convective boundary layer is attributed to the downward flux of horizontal momentum from aloft brought about by increased turbulent mixing. Figures 9 and 10 indicate that during periods of strong growth the daily surface weather maps for the 90-day period confirms that low ozone days were generally cloudier and wetter than high ozone days.

Since ground temperature in MM5 and ozone production in the chemical model are both proportional to short-wave radiation reaching the surface layer, an accurate estimate of cloud cover is essential to both the thermal and chemical processes. MM5 assumes a highly parameterized relationship between cloud cover and relative humidity: the percentage of cloud cover is proportional to (RH–RH_c), where RH is the predicted relative humidity and RH_c is the critical relative humidity ($=60\%$ for high cloud, and 75% for middle and low cloud). WALCEK (1994) compared six different formulations for cloud coverage and suggested a new empirical formulation based on the Air Force Global Weather Central's three-dimensional analysis of cloud cover. According to Walcek's study for April 1981, the simple numerical values selected for RH_c do not seem to be justified and may lead to an underestimation of cloud cover. Therefore, MM5's cloud cover parameterization appears to underestimate cloud cover when $RH < RH_c$, especially in the middle troposphere (850–600 mb) where the largest cloud amounts were observed at relative humidities below 60–80%. Although we did not evaluate MM5's ability to simulate clouds for the entire summer of 1995, a comparison of satellite cloud pictures with model-diagnosed cloud cover for July 14, 1995 indicates that MM5 generated all the major cloud systems over the southeast, northeast, and Great Lakes, but that the total cloud cover percentage was less than 100%. Further examination of cloud cover on other dates showed that MM5 tends to overestimate the areal extent of low clouds which may lead to the underestimation of surface temperature.

4.3 Variation of Ventilation Coefficient

Ventilation coefficients averaged over low and high ozone days for morning (06–10 EST) and afternoon hours (12–16 EST) are presented in Figure 14. Units for ventilation coefficient are m^2/s. Note that the contour interval used for the color scale in Figure 14 differs between the morning and afternoon hours. Again, the reason for this is to allow us to display as much detail as possible, while also accommodating the considerably larger range of values occurring in mid-day. On low ozone days, morning ventilation coefficients (Fig. 14A) average $\sim$2000–3500 m^2/s in the upper Midwest with a narrow zone of higher values ($\sim$3500–5500 m^2/s) found stretching along the spine of the Appalachian Mountains. On high ozone days, morning ventilation coefficients average $\sim$1000–2750 m^2/s in the Midwest and $\sim$2500–4000 m^2/s along the Appalachian corridor (Fig. 14B). Thus, in these areas, ventilation coefficients tend to be lower on high ozone days than on low ozone days. In the Middle Atlantic and New England states, however, mixing heights appear to be somewhat higher on high ozone days, but the pattern is complicated. In the afternoon hours (Figs. 14C and 14D) average ventilation coefficients in the Midwest

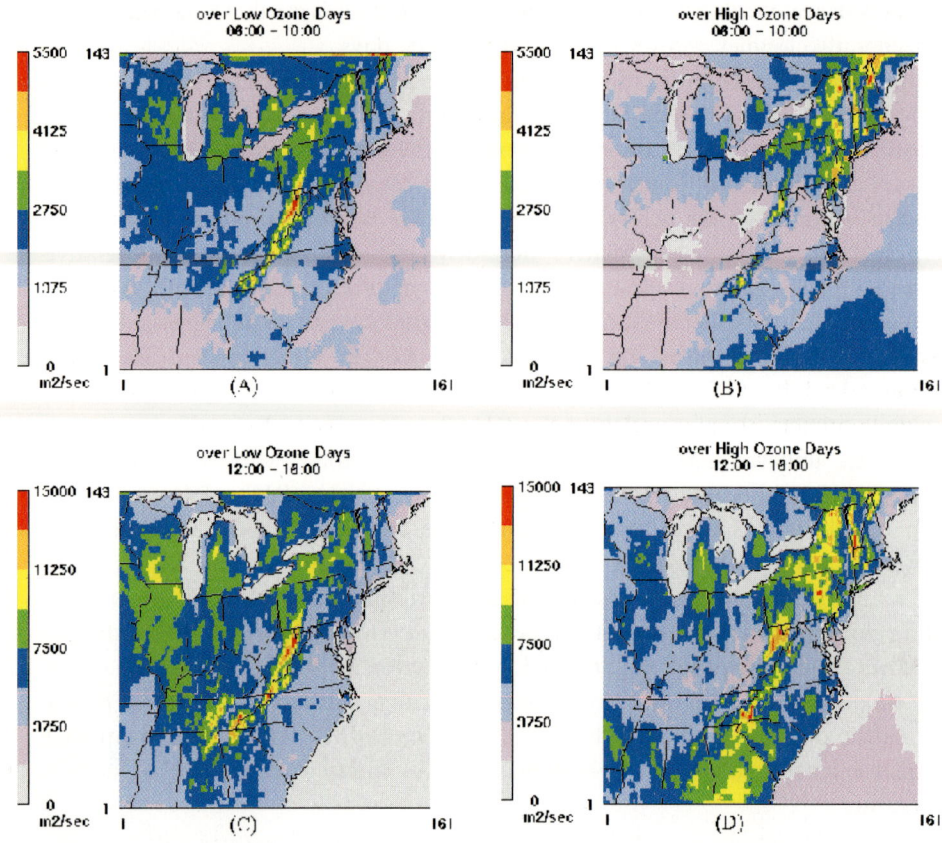

Figure 14
Averaged ventilation coefficient (m^2/s) for morning and afternoon hours on low and high ozone days.

are again somewhat higher on low ozone days ($\sim$4500–9500 m^2/s) than on high ozone days ($\sim$3750–7500 m^2/s). Once again the Middle Atlantic and New England states do not show a simple pattern. In the Southeastern states (North and South Carolina, Georgia, Alabama), however, there is a tendency toward higher ventilation coefficients on high ozone days.

In summary, ventilation coefficients on low and high ozone days display a complicated pattern across the eastern United States. During the morning and afternoon hours in the Midwest, ventilation coefficients appear to be about 10–20% lower on high ozone days than on high ozone days. East of the Appalachians the pattern is more complex, but with a tendency for ventilation coefficients tend to be slightly higher on high ozone days than on low ozone days, particularly in the Northeast. A partial explanation may lie in the fact that storm tracks bringing rapidly-moving weather systems and higher winds tend to converge over the Northeast.

Since we have shown that high ozone days tend to have greater mixing heights (see Fig. 12), surface wind speeds must be lower on high ozone days in order for ventilation coefficients to decrease, or remain relatively unchanged. To check this, we analyzed average surface wind speeds for both high and low ozone days for the 90-day period. During the morning and afternoon hours on high ozone days (Figs. 15B and 15D), surface winds displayed a quasi-circular core of minimum speeds (1–2 m/s) centered over Kentucky. A concentric isotach of 4 m/s, located about 750 km from the center, crosses the southern boundary of the Great Lakes on the north, the Atlantic Coast on the east, and southern portions of Georgia, Alabama, and Mississippi on the south. Synoptic weather maps for this 90-day period show that high ozone events were typically accompanied by large surface anticyclones characterized by high surface temperatures, clear skies, and light winds. These roughly circular isotachs are characteristic of surface anticyclone pressure patterns, associated with high ozone events. On low ozone days (Figs. 15A and 15C), wind

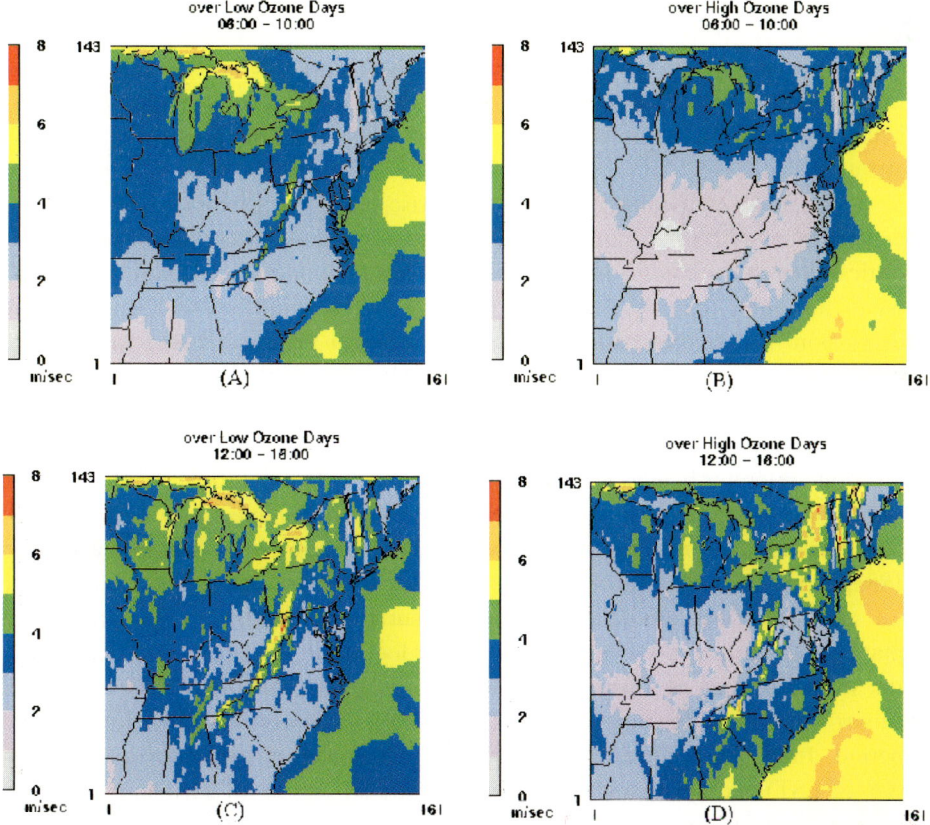

Figure 15

Averaged surface wind speeds (m/s) for morning and afternoon hours on low and high ozone days.

speeds in the Midwest were found to be about 1–2 m/s higher than on high ozone days, but the differences were negligible as one moved toward the Great Lakes and Atlantic Ocean. Low ozone days also tended to have more disturbed weather conditions (e.g., low pressure centers, stronger isobaric gradients, frontal boundaries, greater cloud cover, precipitation events, etc.) than did high ozone days. These findings are consistent with those of MUKAMMAL *et al.* (1982), CARDALINO and CHAMEIDES (1991), NRC (1991), WOLFF and LOIY (1980), VUKOVICH (1995), and other researchers.

5. Relationship of Boundary-layer Parameters to the Vertical Ozone Distribution

The analysis of boundary-layer parameters for high and low ozone episode days, described in Section 4 above, shows that the meteorological factors favoring high ozone days in summer are consistent with the model of a surface anticyclone centered over the mid-south region of the eastern United States (e.g., Kentucky and Tennessee). Skies tend to be clear and surface winds light in the core region. Strong daytime insolation warms the ground, increasing convective turbulence, and raising mixing heights. Ventilation coefficients are slightly reduced in the anticyclone's center because of the lighter winds, but tend to increase on the periphery where winds are stronger.

In the hours immediately following sunrise, ozone precursors (NO_X and VOC's) emitted from morning commuter traffic tend to accumulate near the ground. This is a critical time of the day for pollutant levels, since the mixing height is low and surface winds are light; the combination produces small ventilation coefficients and weak dispersion. As the sun rises higher, photochemical processes begin to convert the precursors into ozone, raising ambient ozone levels near the ground. With increased vertical mixing, we would expect surface ozone concentrations to decrease as the ozone becomes mixed with cleaner air from aloft if there is no ozone source aloft. If an ozone reservoir is present in the nighttime residual layer (about 0.5–1.0 km above the ground) vertical mixing from below may actually *increase* ground-level ozone concentrations by entraining this upper-level ozone downward (CLARKE and CHING, 1983; SILLMAN *et al.*, 1990; RYAN *et al.*, 1998; ZHANG and RAO, 1999). To study this, we analyzed a set of hourly ozone measurements made on two instrumented platforms, one atop a 411-m television tower located in Garner, North Carolina, and the second located at a height of 366 m, about two-thirds of the way up the Sears Tower office building, in Chicago, IL. Measurements were available for the 90-day period, June 2 to August 30, 1995. Ozone values were averaged over the morning (06–10 EST) and afternoon (12–16 EST) hours for the set of 23 high ozone and 23 low ozone days used in this study. Table 2 shows the averages obtained for both locations. Although the North Carolina site is rural and the Illinois site is urban, both locations show about 40% higher ozone concentrations aloft on high ozone

Table 2

Means of upper-level ozone observations (ppb)

Sears Tower, Chicago, IL	Low Ozone Days	High Ozone Days	Percentage Change
Morning Hours: 06–10 EST	25	42	68%
Afternoon Hours: 12–16 EST	47	60	28%
Average	**36**	**51**	**42%**
North Carolina Tower			
Morning Hours: 06–10 EST	41	57	39%
Afternoon Hours: 12–16 EST	45	63	40%
Average	**43**	**60**	**40%**

days than on low ozone days. With only two sites available for analysis, however, these results should be regarded as preliminary. Nevertheless, they suggest that a pool of ozone tends to be present aloft during the night which may contribute to the build-up of surface ozone levels through vertical mixing and fumigation as the boundary layer starts to grow in the following morning. The presence of an ozone-rich layer aloft during high surface ozone episodes is confirmed by aircraft observations as discussed by ZHANG *et al.* (1998).

In a related study, ZHANG and RAO (1999) examined hourly ozone observations from two levels (1.5 m and 411 m) on the same television tower at Garner, North Carolina, for the period July 7 to 18, 1995. During this period, hourly ozone concentrations at 1.5 m showed a strong diurnal pattern of variation which was not evident at the upper level. Surface ozone concentrations decreased sharply to very low values after sunset owing to NO titration and deposition precesses. The ozone gradient between 411 m and ground-level was largest at night. After sunrise, as turbulent mixing became re-established, the vertical gradient began to decrease and by noon the ozone gradient had disappeared completely, or reversed, so that ozone levels at the ground then matched or exceeded ozone levels aloft. ZHANG and RAO (1999) suggest that vertical mixing in the daytime convective layer is an important process in bringing ozone aloft down to the ground in the morning hours.

Combining the results obtained by ZHANG and RAO (1999), VUKOVICH (1995), and others with those found in the present study using the MM5 output, the following scenario of surface ozone behavior can be suggested for high ozone days. At night, ground-level ozone concentrations are small, but may be considerably greater in the nocturnal residual layer several hundred meters aloft. In the hours following sunrise (06–10 EST), surface concentrations increase due to two factors: local ozone production through conversion of ozone precursors by photochemical processes and downward entrainment of ozone from the upper residual layer. As the convective regime is established, vigorous vertical mixing begins to reduce the vertical ozone gradient. By noon, ozone concentrations at the ground tend to match or exceed ozone concentrations aloft. However, surface ozone concentrations continue to rise into the afternoon hours. This is due to the complex interplay between the processes of ozone production, vertical mixing, and horizontal advection. Photochemical production peaks around noon when the sun is highest, vertical mixing is most effective in midmorning when growth of the mixing height is strongest, and horizontal advection of ozone from upwind sources is at its maximum in the afternoon when surface winds are strongest (ZHANG and RAO, 1999). After sunset, surface ozone levels begin to fall rapidly as NO titration and deposition processes become dominant. In contrast, low ozone days are characterized by cooler, cloudier conditions, and stronger surface winds. The greater wind speeds tend to increase dispersion of ground-level ozone precursors during the morning rush hours. Increased cloud cover reduces the intensity of solar radiation reaching the ground, leading to reduced photochemical production of ozone. Less insolation also results in lower surface temperatures and a slower rate of growth of the mixing height, making vertical mixing less effective. As a result, ground-level ozone concentrations remain low. From Table 2, we see that low ozone days also tend to have lower ozone concentrations aloft so that even if mixed downward, they would not be able to significantly raise ground-level ozone concentrations.

Since ozone aloft appears to play such a significant role in contributing to high surface ozone concentrations, further research is needed in determining the source of this ozone reservoir aloft, its chemical transformation over time, its horizontal transport, and its residence time. Ozone aloft appears to originate in surface photochemical processes and is then mixed upward in the rapidly growing convective boundary layer. When the surface temperature inversion reforms around sunset, the pool of upper-level ozone is cut off from the surface and remains trapped aloft in the so-called "residual layer," about 0.5 to 1.0 km above the surface. An estimate of the persistence of high-level ozone concentrations is provided in Figure 16. The frequency of high ozone occurrences aloft is plotted against the duration of each event, ranging from 1 to 6 consecutive days. The observations come from the 411-m tower at Garner, NC, and the 366-m observation on the Sears tower in Chicago, IL, referred to earlier. The data in Figure 16 are for the period (June 2 to August 30, 1995) when the upper-level ozone exceeded 70 ppb and include 34 days for Garner and 36 for Chicago. Because of the relatively small number of days available and the normally lower ozone concentrations found aloft, we used a cutoff value of 70 ppb for this comparison (instead of the 80 ppb used for surface ozone) to ensure enough data points for a meaningful analysis. The figure shows that, in general, the frequency of long-duration events (4 to 6 days) is lower than for short-duration events (1 to 3 days). The low frequency ($<10\%$) shown in Figure 16 for two consecutive days does not follow the overall trend, and may be due to the small sample size used. Thus, it appears that ozone can persist for about 2–3 days aloft before NO titration and deposition processes remove it. Within this time frame, horizontal transport by upper-level winds can advect the ozone hundreds of kilometers downwind from its source region. Using two years' of surface ozone observations from stations in New Jersey and Connecticut, RAO (1988) showed that

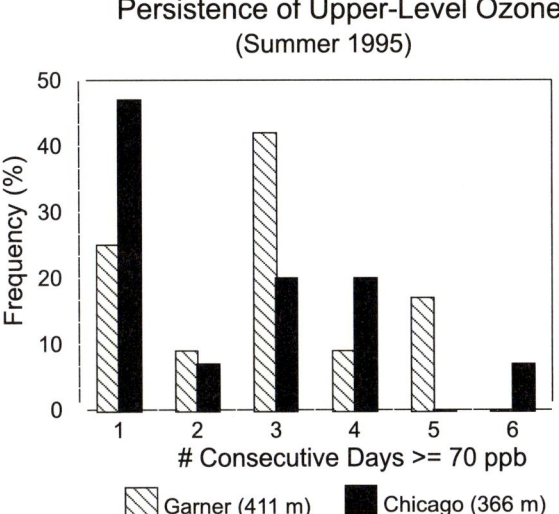

Figure 16
Percentage of high ozone events (> 70 ppb) aloft plotted against the duration of each event (consecutive days) at Garner, NC and Chicago, IL.

once the surface ozone level exceeds 80 ppb, the probability of that exceedance lasting for 2–3 days is very high. The persistence of high ozone aloft was also reported by LOGAN (1989). Time scales of 1–2.5 days and space scales of 600 km for surface ozone concentrations in the eastern United States were found in a later study by RAO *et al.* (1997). In summary, studies of ambient ozone at the surface and aloft suggest that an effective ozone control strategy needs to address both local and regional spatial scales and time scales of the order of several days.

6. Summary

Meteorological fields from the MM5 mesoscale model were used to prepare a comprehensive analysis of summertime (June 2 to August 30, 1995) mixing heights and ventilation coefficients for the eastern United States. Daytime mixing heights were estimated using the profile intersection method based on HOLZWORTH's (1967) and heat-flux techniques. Under stable conditions, the depth of the nocturnal boundary layer was estimated from an algorithm based on surface wind speed suggested by BENKLEY and SCHULMAN (1979).

For the entire 90-day summer period, nighttime mixing heights averaged less than 200 m, but increased to ~1 km by 10 EST, reaching 2.5 km or greater by mid-afternoon, then falling back to ~0.8 km by 22 EST. During convective times of day, mixing heights were found to be about 10% higher along the Appalachian Mountain chain stretching south-south-westward from New York State to eastern Tennessee. This feature can be attributed to greater buoyancy associated with orographic lifting. Over the Great Lakes the effects of surface heating are notably absent and mixing heights there remain below 250 m at all times of day. Coefficients of variation indicate relatively low variability (0.25–0.30) at night and in mid-day when there is little change in mixing heights from hour to hour, but are twice as high (0.50–0.65) in mid-morning and evening when growth and decay rates are strong.

Ventilation coefficients increase by a factor of 10 between the nighttime (500 m^2/s) and morning hours (5000 m^2/s) and by another factor of 3 (to 15,000 m^2/s) in the afternoon. These increases are attributed to both increasing mixing heights and increasing wind speeds occurring between night and day. Ventilation coefficients likewise fall in the evening hours as mixing heights collapse and winds diminish. As with mixing height, coefficients of variation tend to be small at night and in the afternoon, when conditions change little from hour to hour, but are larger during times of growth and collapse.

We used surface ozone observations from 400 sites in the eastern United States covering the same 90-day period (June 2 to August 30, 1995) to study the differences between "high ozone" days and "low ozone" days. Each summer day was ranked according to how many stations reported hourly surface ozone concentrations > 80 ppb. The top 25% ($= 23$ days) were labeled "high ozone" days; the bottom 25% were labeled "low ozone" days. In general, mixing heights were higher on high ozone days than on low ozone days with the largest differences occurring in the afternoon hours. Also, since the uncertainty in estimating the mixing height on high ozone days is larger than that on low ozone days, there is a real problem in specifying the boundary-layer evolution properly in simulating ozone concentrations with photo-chemical models for episodic days.

Regionally, differences in the mixing height between high and low ozone days were largest over the Midwestern states with much smaller differences found along the Great Lakes and on the Atlantic seaboard. Sea-breeze circulations bringing cooler marine air inland may partly be responsible for the smaller differences seen there. Cloud cover during the morning and afternoon hours averaged 0.15 over most of the eastern United States on high ozone days, but was twice as large (0.30) on low ozone days. On high ozone days, isotachs showed a core of minimum speed (1–2 m/s) centered over Kentucky and Tennessee, increasing to 4 m/s located $\sim$750 km out from the center. On low ozone days, wind speeds were 1–2 m/s higher overall, and showed little spatial variation. Ventilation coefficients tended to be about 50% lower in the Midwest on high ozone days than on low ozone days. Other regions showed more variability. The Northeast region displayed a complicated spatial pattern, probably a result of complex topography and vigorous synoptic-scale circulations that frequent the region (RAO *et al.*, 1997; BRANKOV *et al.*, 1998). Slightly higher ventilation coefficients on high ozone days were also seen along the southeast coast, perhaps due to higher winds associated with sea-breeze circulations.

The characteristic low cloud cover, light wind speeds, and large daytime mixing heights seen on high ozone days is consistent with a large surface anticyclone centered over the mid-south near Kentucky and Tennessee. An examination of surface weather maps for the summer of 1995 shows anticyclonic pressure systems commonly occurring over the eastern United States on high ozone days. These anticyclones tend to be centered over the south-central states. More disturbed synoptic weather patterns (e.g., low pressure, fronts, greater cloud cover, precipitation areas) occur more frequently on low ozone days.

Ozone measurements on elevated platforms ~400 m agl at two locations (Garner, North Carolina, and Chicago, IL) reveal that concentrations of ozone aloft were about 40% greater on high ozone days than on low ozone days. In a previous study, Zhang and Rao (1999) showed that the vertical ozone gradient on the North Carolina tower (between 411 m and ground-level) reached a maximum value at night and then gradually disappeared during the morning hour, or even reversed direction. These results suggest that high surface ozone concentrations may be attributed to both local photochemical processes involving vehicle emissions and also to entrainment of ozone-rich air downward by the well-mixed convective boundary layer. An analysis of ozone aloft using the North Carolina and Chicago tower data for the summer of 1995 shows a mean residence time of 2–3 days during high-ozone events. Since ozone aloft can persist for several days before being removed by NO titration and deposition processes, its source region may lie several hundreds of kilometers upwind, depending on wind speeds aloft. Hence, emission control strategies need to look at both local and regional spatial scales and time scales of the order of several days for ozone.

Acknowledgements

This research was supported by the U.S. Environmental Protection Agency under Grant No. R8263731476 and the New York State Energy Research and Development Authority under Contract Nos. 4914ERTERES99 and 6085ERTERS00.

References

Aneja, V. P., Arya, S. P., Murray Jr. G. C., and Manuszak, T. L. (2000), *Climatology of Diurnal Trends and Vertical Distribution of Ozone in the Atmospheric Boundary Layer in Urban North Carolina*, J. Air and Waste Manag. Assoc. *50*, 54–64.

Angevine, W. M., Trainer, M., McKeen, S. A., and Berkowitz, C. M. (1996), *Mesoscale Meteorology of the New England Coast, Gulf of Maine, and Nova Scotia: Overview*, J. Geophy. Res. *101*, 28,893–28,901.

Arya, S. P. S. (1981), *Parameterizing the Height of the Stable Atmospheric Boundary Layer*, J. Appl. Meteor. *20*, 1192–1202.

Benkley, C. W. and Schulman, L. L. (1979), *Estimating Hourly Mixing Depths from Historical Meteorological Data*, J. Appl. Meteor. *18*, 772–780.

BERMAN, S., KU, J. Y., ZHANG, J., and RAO, S. T. (1997), *Uncertainties in Estimating the Mixing Depth — Comparing Three Mixing-depth Models with Profiler Measurements*, Atmos. Environ. *31*, 3023–3039.

BERMAN, S., KU, J. Y., and RAO, S. T. (1999), *Spatial and Temporal Variation of the Mixing Depth over the Northeastern United States During the Summer of 1995*, J. Appl. Meteor. *38*, 1661–1673.

BEYRICH, F. and WEILL, A. (1993), *Some Aspects of Determining the Stable Boundary Layer Depth from Soear Data*, Boundary-Layer Meteor. *63*, 97–116.

BISWAS, J. and RAO, S. T. (2000), *Uncertainties in Episodic Ozone Modeling Stemming from Uncertainties in the Meteorological Fields*, J. Appl. Meteor. *40*, 117–136.

BLACKADAR, A. K., *High resolution models of the planetary boundary layer*. In *Advances in Environmental Science and Engineering* (J. Pfafflin and E. Ziegler, eds.) vol. 1 (Gordon and Breach Science Publishers 1979), pp. 50–85.

BRANKOV, E., RAO, S. T., and PORTER, P. S. (1998), *A Trajectory-clustering-correlation Methodology for Examining Long-range Transport of Air Pollutants*, Atmos. Environ. *32*, 1525–1534.

BYUN, D. W. (1990), *On the Analytical Solutions of Flux-profile Relationships for the Atmospheric Layer*, J. Appl. Meteor. *29*, 652–657.

CARDALINO and CHAMEIDES (1990), *Natural Hydrocarbons, Urbanization, and Urban Ozone*, J. Geophys. Rev. *95*, 13,971–13,979.

CHAN, D., RAO, S. T., ZURBENKO, I. G., and PORTER, P. S., *Linking Changes in Ozone to Changes in Emissions and Meteorology*. In Air Pollution 99 VII (eds. C. A. Brebbia, M. Jacobsen, and H. Power) (WIT Press 1999) pp. 663–675.

CLARK, R. D., *Vertical profiles of meteorological variables and ozone concentrations in the nocturnal boundary layer at Gettysburg, PA*, Proc. 12th Symp. *Boundary Layers and Turbulence* (Vancouver, BC 1997) pp. 417–418.

CLARKE, J. F. and CHING, J. K. S. (1983), *Aircraft Observations of Regional Transport of Ozone in the Northeastern United States*, Atmos. Environ. *17*, 1703–1712.

DUDHIA, J. (1993), *A Nonhydrostatic Version of the Penn State-NCAR Mesoscale Model: Validation Tests and Simulation of an Atlantic Cyclone and Cold Front*, Monthly Wea. Rev. *121*, 1493–1513.

ESKRIDGE, R. E., KU, J. Y., RAO, S. T., PORTER, S. P., and ZURBENKO, I. G. (1997), *Separating Different Scales of Motion in Time Series of Meteorological Variables*, Bull. Am. Met. Soc. *78*, (7), 1473–1483.

GRELL, G. A., DUDHIA, J., and STAUFFER, D. R. (1994), *A Description of the Fifth-generation Penn State/ NCAR Mesoscale Model (MM5)*, NCAR Technical Note, NCAR/TN-389 + STR. 138 pp.

HOGREFE, C., RAO, S. T., ZURBENKO, I. G., and PORTER, P. S. (2000), *Interpreting the Information in Ozone Observations and Model Predictions Relevant to Regulatory Policies in the Eastern United States*, Bull. Am. Meteor. Soc. *81*, 2083–2106.

HOLZWORTH, G. C. (1967), *Mixing Depths, Wind Speeds and Air Pollution Potential for Selected Locations in the United States*, J. Appl. Meteor. *6*, 1039–1044.

KORC, M. E., ROBERTS, P. T., and BLUMENTHAL, D. L. (1996), *NARSTO-Northeast Data Management Plan*, Version 3.1, Report prepared for Electronic Power Research Institute, Palo Alto, CA, by Sonoma Technology, Inc., Petaluma, CA, STI-95141-1537, Research Project EPRI WO9108-01, 70 pp.

LENA, F. and DESIATO, F. (1999), *Intercomparison of Nocturnal Mixing Height Estimate Methods for Urban Air Pollution Modelling*, Atmos. Environ. *33*, 2385–2393.

LOGAN, J. A. (1989), *Ozone in Rural Areas of the United States*, J. Geophys. Rev. *94*, 8511–8532.

MAHRT, L., ANDRE, J. C., and HEALD, R. C. (1982), *On the Depth of the Nocturnal Boundary Layer*, J. Appl. Meteor. *21*, 90–92.

MUKAMMAL, E. I., NEUMANN, H. H., and GILLESPIE, T. J. (1982), *Meteorological Conditions Associated with Ozone in Southwestern Ontario, Canada*, Atmos. Environ. *16*, 2095–2106.

NRC (1991), *Rethinking the Ozone Problem in Urban and Regional Air Pollution*, National Research Council, National Academy Press, Washington, DC.

RAO, S. T. (1988), *Prepared Discussion : Ozone Air Quality Models*, JAPCA, *38*, 1129.

RAO, S. T., SISTLA, G., KU, J. Y., ZHOU, N., and HAO, W. (1994), *Sensitivity of the Urban Airshed Model to mixing height profile*. Proc. Eighth AMS/AWMA Joint Conference on the *Applications of Air Pollution Meteorology*, Nashville, TN, January 1994.

RAO, S. T. and ZURBENKO, I. G. (1994), *Detecting and Tracking Changes in Ozone Air Quality*, J. Air and Waste Mgt. Assoc. *42*, 1204–1211.

RAO, S. T., ZURBENKO, I. G., NEAGU, R., PORTER, P. S., KU, J. Y., and HENRY, R. F. (1997), *Space and Time Scales in Ambient Ozone Data*, Bull. Am. Meteor. Soc. *78*, (10), 2153–2166.

RYAN, W. F., DODDRIDGE, B. G., DICKERSON, R. R., MORALES, R. M., and HALLOCK, K. A. (1998), *Pollutant Transport During a Regional Ozone Episode in the Mid-Atlantic States*, J. Air Waste Mgt. Assoc. *48*, 786–797.

SCIRE, J. S., INSLEY, E. M., YAMARTINO, R. J., and FERNAU, M. E. (1995), *A User's Guide for the CALMET Meteorological Model*, EARTH TECH.

SILLMAN S., LOGAN, J. A., and WOFSEY, S. C. (1990), *A Regional-Scale Model for Ozone in the United States with a Subgrid Representation of Urban and Power Plant Plumes*, J. Geophys. Res. *95*, 5371–5748.

SISTLA, G., ZHOU, N., HAO, W., KU, J. Y., RAO, S. T., BORNSTEIN, R., FREEDMAN, F., and THUNIS, P. (1996), *Effects of Uncertainties in Meteorological Inputs on Urban Airshed Model Predictions and Ozone Control Strategies*, Atmos. Environ. *30*, 2011–2025.

STAUFFER, D. R. and SEAMAN, N. L. (1990), *Use of Four-dimensional Data Assimilation in a Limited-area Mesoscale Model. Part I: Experiments with Synoptic-scale Data*, Mon. Wea. Rev. *118*, 1250–1277.

STAUFFER, D. R., SEAMAN, N. L., and BINKOWSKI, F. S. (1991), *Use of Four-dimensional Data Assimilation in a Limited-area Mesoscale Model. Part II: Effects of Data Assimilation within the Planetary Boundary Layer*, Mon. Wea. Rev. *119*, 734–754.

STULL, R. B., *An Introduction to Boundary Layer Meteorology*, (Kluwer Academic Publishers, Dordrecht, The Netherlands 1988) 666 pp.

UNO, I., WAKAMATSU, S., UEDA, H., and NAKAMURA, A. (1992), *Observed Structure of the Nocturnal Boundary Layer and its Evolution into a Convective Mixed Layer*, Atmos. Environ. *26B*, 45–47.

VENKATRAM, A. (1978), *Estimating the Convective Velocity Scale for Diffusion Applications*, Boundary-Layer Meteor. *15*, 447–452.

VUKOVICH, F. M. (1995), *Regional-scale Boundary Layer Ozone Variations in the Eastern United States and their Association with Meteorological Conditions*, Atmos. Environ. *29*, 2259–2273.

VUKOVICH, F. M. (1997), *Time Scales of Surface Ozone Concentrations in the Regional, Non-urban Environments*, Atmos. Environ. *31*, 1513–1530.

WALCEK, C. J. (1994), *Cloud Cover and Its Relationship to Relative Humidity During a Springtime Midlatitude Cyclone*, Mon. Wea. Rev. *122*, 1021–1035.

WOLFF, G. T. and LOIY, P. J. (1980), *Development of an Ozone River Associated with Synopic-scale Episodes in the Eastern United States*, Envir. Sci. Technol. *14*, 1257–1260.

YU, T. (1978), *Determining Height of the Nocturnal Boundary Layer*. J. Appl. Meteor. *17*, 28–33.

ZHANG, D. L. and ANTHES, R. A. (1982), *A High-resolution Model of the Planetary Boundary Layer, Sensitivity Tests and Comparisons with SESAME-79 Data*, J. Appl. Meteor. *21*, 1594–1609.

ZHANG, J., RAO, S. T., and DAGGUPATY, S. M. (1998), *Meteorological Processes and Ozone Exceedances in the Northeastern United States during the 12–16 July 1995 Ozone Episode in the Northeastern United States*, J. Appl. Meteor. *37*, 776–789.

ZHANG, J. and RAO, S. T., (1999), *The Role of Vertical Mixing in the Temporal Evolution of the Ground-level Ozone Concentrations*, J. Appl. Meteor. *38*, 1674–1691.

ZURBENKO, I. G. (1991), *Spectral Analysis of Nonstationary Time Series*, Int. Stat. Rev. *59*, 163–173.

ZURBENKO, I. G., PORTER, P. S., RAO, S. T., KU, J. Y., GUI, R., and ESKRIDGE, R. E. (1996), *Detecting Discontinuities in Time Series of Upper Air Data: Development and Demonstration of an Adaptive Filter*, J. Climate *9*, 3548–3560.

(Received March 1, 2000, accepted July 25, 2000)

To access this journal online:
http://www.birkhauser.ch

Pure appl. geophys. 160 (2003) 57–74
0033–4553/03/020057–18

Planning for Air Quality Concerns of the Future

BRUCE B. HICKS[1]

Abstract — In recognition of growing needs for forecasts of air quality and atmospheric deposition to accompany classical weather forecasts, a new generation of atmospheric prediction models is slowly evolving. These share the common feature that atmospheric chemistry will be directly incorporated into advanced forecast schemes. It is argued that in most practical applications, the over-riding need is not for accurate prediction of some quantifiable air quality component, but rather a forecast of the probability of harmful consequences to exposure. In this event, it is not concentrations that need to be forecast, but the probability that concentrations will exceed some predetermined level at which consequences could be harmful. This argument extends from emergency response applications to ecosystem decline.

Key words: Air quality, forecasting, prediction, deposition, eutrophication.

1. Introduction

Demands for improved quality of life are imposing increasing pressures on both human health and the environments in which people live. Expanding industrialization increases the likelihood of accidents that can release large quantities of hazardous chemicals into the air, dispersing downwind to areas where people are affected whenever concentrations exceed dangerous levels. Forecasting the concentration field following such an event is exceedingly difficult, because a central requirement is to correctly calculate the local wind direction and over short time intervals the wind direction varies considerably in ways that are largely stochastic. In complex terrain and especially in urban areas, the presence of surface obstacles complicates matters further by injecting an additional level of randomness not yet handled by deterministic models. It is argued that models should not be expected to forecast concentrations at a specific place and time, but rather the probability that a particular location will experience dangerous levels within prescribed time periods.

Similar considerations arise in the context of air pollution. Concerns about declining air quality are growing, not only from the perspective of human health but also as they relate to effects of deposition of pollutants from the atmosphere to ecosystems. The human health impacts of the problem are well acknowledged. The

[1] Air Resources Laboratory, National Oceanic and Atmospheric Administration, 1315 East West Highway, Silver Spring, MD 20910, U.S.A. E-mail: bruce.hicks@noaa.gov

ecosystem consequences of atmospheric deposition are frequently not considered at the same level of importance, yet these can have strong effects on the productivity of agricultural areas and coastal waters, for example, and hence might well constitute a limiting factor in the availability of food. All such considerations are necessarily influenced by variability in the prevailing climate. The problems that arise are not only due to the long-term effects of a continuing dose of harmful pollutants, but also the sudden short-term exposure to dangerous events or concentrations. Thus the problem has both chronic and acute aspects. At this time, air pollution control strategies usually target major sources with the goal of reducing downwind exposure levels to below some critical level. It is becoming increasingly apparent, however, that the role of meteorology is central in the occurrence of specific episodes. These might arise, for example, when the winds blow a parcel of pollution across a sensitive area, causing a local unacceptable "exceedance" even though the long-term average exposure is acceptable. Once again, there appears to be need for a new approach to the problem. What is needed is not only an accurate depiction of how conditions will change with time over long periods, but also a prediction over considerably shorter time scales of the probability that any given ecosystem will be subject to stresses that will cause its decline or death.

There is slow but continuing improvement in the ability to forecast weather, due not only to improved understanding of the important processes but also to the recent rapid increase in computing power. Far better descriptions of processes can now be accommodated in the codes designed to run operationally. The consequences of these advances take many forms. For example, periods of extreme temperatures or of a high probability of rain are now forecast with some considerable confidence. The prediction of where tropical storms come ashore has been refined, and the amount of warning before severe storms (e.g., tornados) has been reduced. In parallel with these improvements, there has been increasing interest in the development of capabilities to forecast air quality.

Air pollution has been a worldwide concern for centuries. In many places, the problem has been addressed with some considerable success by implementing severe restrictions on the emissions of the chemicals that lead to air pollution—primarily particulate matter and the gaseous oxides of sulfur and nitrogen. In areas where societal pressures on the atmospheric environment are greatest, there is need to predict future air quality for several reasons. First, the aging of the population is accompanied by a growing sensitivity to air pollution, especially particulate matter and worrisome trace gases like ozone. Sensitive elements of the population could take protective measures if warnings were available of the likelihood of aggravating air pollution periods. Second, accurate forecasting would potentially permit authorities to take protective steps, such as limiting the use of private transportation.

Four areas of air quality forecasting concern are discussed here—emergency response, air quality forecasting, assessment and scenario development for regulatory purposes, and ecological protection. In each case it is argued that the familiar focus on trying to derive accurate predictions on a point-by-point basis might prove far less productive than a concerted attempt to predict the likelihood that concentrations (and/or atmospheric deposition) will exceed some specified value. That is, that the effort would better be directed at forecasting the probability of exceedances rather than predicting actual concentrations.

The matter is of current relevance in North America, where coordinated efforts to improve air quality forecasting capabilities are underway. In the context of the present series of papers concerning environmental problems in Southeast Asia, the matter is proposed as an item of potentially accute interest because local scientists have the vital advantage of selecting their own paths to follow, avoiding unproductive paths followed elsewhere and so accelerating their achievements of the common goal.

2. Predicting Dispersion—Emergency Response

For any specific place and for a specified averaging period, it is possible to define an average wind velocity. However, the instantaneous velocity field is highly variable, such that there is considerable uncertainty about the direction any particular parcel of air will take. It is highly unlikely that any single small parcel will move in the direction of the average wind, since some level of turbulence is mostly present. As sequential parcels of air originating at a single location move in response to the turbulence that is an inherent part of the velocity field, they describe an envelope that is described in classical dispersion theory by a plume with a Gaussian distribution of crosswind mean concentrations. In practice, the common assumption of a straight line average trajectory (or plume centerline) holds only if there are no local confounding influences, such as the presence of buildings or other surface inhomogeneities. Figure 1 shows an example of how the average wind field varies in the vicinity of an inland industrial plant in the United States, as determined by a network of wind-measuring towers. Even in this inland case, far from the effects of coastal circulations, emissions from the plant will follow trajectories that are quite complex. It is obvious that an assumption that emitted pollutants will continue to move in a direction indicated by any single set of meteorological sensors will be substantially misleading, even if the source of such wind information is very close to the location of the emissions. Yet it is standard practice to maintain a single meteorological tower near potential source locations, and to predict downwind dispersion on the basis of its measurements alone.

The conceptual inability of a straight line plume or puff transport model to describe reality can be handled either statistically or deterministically. Either the effects of topographic (or meteorological) complexity can be included in the statistical description of the plume, or detailed understanding of the flow field can be used to specify the way in which the plume will meander. The statistical nature of the

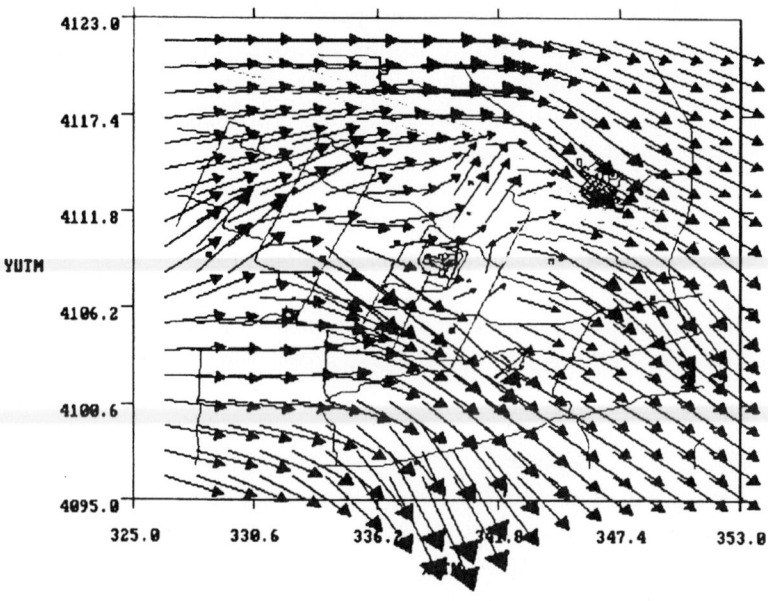

Figure 1

An example of the hourly-average wind field near an industrial plant near Paducah, Kentucky, as derived from a tower network around the plant. Emergency response plans for this plant call for reliance on a single tower observation of the wind, with and assumed straight-line trajectory.

plume is described by a crosswind standard deviation, σ_y, which will be substantially different in the two cases. This is the source of considerable confusion in plume modeling. Either the crosswind standard deviation contains the effects of terrain complexity on the mean flow fields, or it does not. In the latter case, these effects are handled deterministically by the modeling technique that is employed. In this case of a meandering plume, the concept is again one of an envelope that contains the individual puffs of material dispersing at specific instants. Any single puff will grow at a rate described by an expression for the crosswind width, $\sigma_y(1)$; this puff could be located anywhere within the bounds of the ensemble plume described by the meandering plume model, with crosswind width $\sigma_y(2) > \sigma_y(1)$, or within the much wider boundaries of the statistical straight line plume when the effects of topography are not considered explicitly, with $\sigma_y(3) > \sigma_y(2)$.

In any event, the movement of any single parcel of air will be complicated by random turbulence even in the most simple of situations, and further complicated by terrain inhomogeneities in most real-world circumstances. Regardless of the uncertainty introduced by turbulence, for horizontal, spatially homogeneous terrain in stationary meteorological conditions, plume and puff dispersion model predictions can be quite accurate. In urban areas, however, the accuracy of the predictions is unfavorable. Figure 2 illustrates this for a historic series of experimental tracer

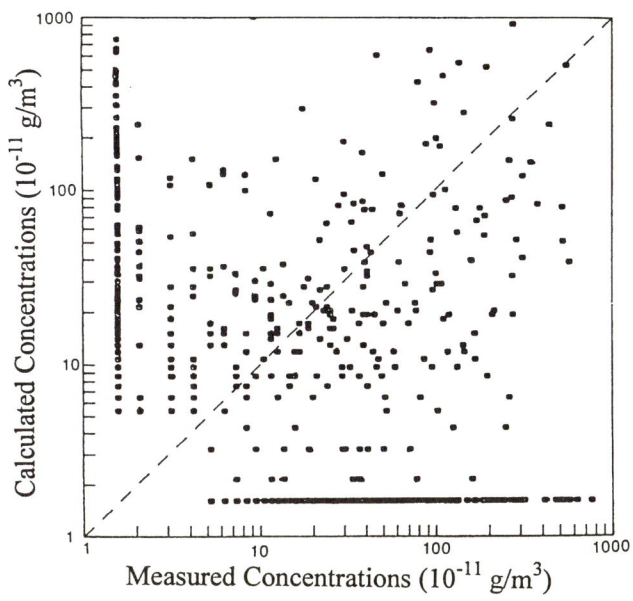

Figure 2

Results obtained in a year-long series of tracer experiments conducted in Washington, DC, in 1987. Note that many occasions occur for which either observed concentrations indicated a plume "hit" whereas the model predicted a "miss," or *vice versa* (after DRAXLER, 1987).

releases in Washington, DC (the METREX experiment of 1983/4; see DRAXLER, 1987). There is slight agreement between actual concentrations and predictions made by models even when these models are driven by actual prevailing meteorological information gathered simultaneously with the releases. Today, models would probably do insignificantly better, because of the need in such circumstances to take local details into account, and to allow for the consequences of random variations (turbulence) that cannot be predicted explicitly. However, it should be noted that even in this early case the range of observations is much the same as the range of predictions, even though the point-to-point comparison is far less than encouraging.

In practice, the conditions in which emergency response forecasts of dispersion are most likely to be required are urban, where people live and work. Buildings add a level of complexity not included in classical plume and puff dispersion models. There has been a long history of studies of building wake effects, starting with workers such as McCORMACK (1963). The downwind dispersion of pollutants that enter the atmosphere in a city or urban environment will likely be dominated by the way in which buildings interact with the prevailing wind field. A slight deviation in wind direction can cause a major difference in the wake downwind of specific structures. There may be funneling of the wind that will carry pollutants emitted locally into spaces where people may be exposed considerably more than predicted by simple models. Circulation cells may exist, in which concentrations of pollutants could continue to build up if emissions are within them. However in much the same way there may be areas essentially protected from the impact of pollutants from local sources. In many practical instances, it is strongly desired to identify such favored areas in advance, so that these may be used as congregation points in the event of some feared accidental release. In such instances, the need is for a climatological description of where the highest and lowest concentrations are likely to be experienced, resulting from emissions at a specified place and time, and relative to the distribution of buildings in a specific area. The answers, once again, are necessarily probabilistic.

The above considerations lead to the conclusion that accurate prediction of the concentration at any particular point downwind of a near-surface release of substances into the air for any specific time is beyond the capabilities of contemporary deterministic models, and is likely to remain so. In recognition of this, many models aim to describe the statistical distribution of the field of concentrations arising from many replications of releases in similar circumstances. In practical applications, it is usually extreme conditions that are of most interest to managers and emergency response teams. The most important information that is required of concentration prediction models is the likelihood that concentrations at specific locations and times will exceed dangerous levels. When couched in these terms, the results of Figure 2 are not as negative as visual inspection suggests. In Figure 3, the matter is addressed probabilistically. The comparison is now between predicted and observed probabilities that exposures will be below specified values. For this analysis, occasions for which the model predicted a gravely incorrect wind

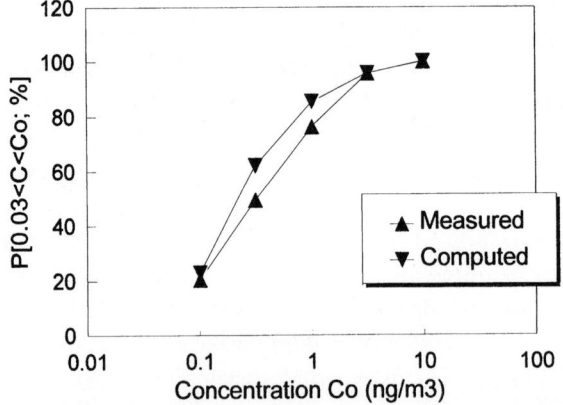

Figure 3
Distributions of the concentrations shown in Figure 2. For this examination of how well the models mirror the distribution of actual concentrations, the analysis is limited to occasions for which the plume direction was predicted accurately, by omitting all data for which either the measured or the predicted concentration was below 0.03 ng/m^3.

direction are excluded, so that Figure 3 is a test solely of how well the dispersion routines work. The two curves are reassuringly close. This can be interpreted to be evidence, once again, that correctly gauging the plume direction is the key step in any plume dispersion prediction exercise.

The suggestion that the matter would be better addressed probabilistically is far from new. In fact, the classical approach has been to look at the issue in this way. The well-known Gaussian plume methodology is based on the recognition that one cannot hope to predict with certainty the path and diffusion of dispersing materials in any specific instance, but instead one can describe the overall behavior of many such releases occurring in similar conditions. The Gaussian plume model describes the ensemble of many releases. Clearly, comparing predictions made by a Gaussian model to any particular set of observations from a single field study is likely to be quite unrewarding. Figure 4 illustrates this. The diagram shows a comparison between actual observations made in repeated transects across a plume of tracer material and a Gaussian-plume prediction for the same condition, after alignment with the observed

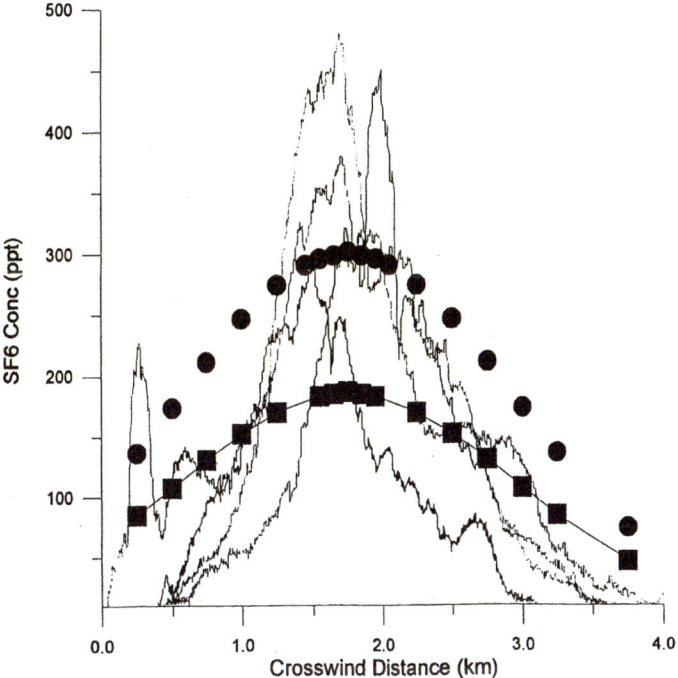

Figure 4
A comparison between predictions of a Gaussian plume model, with wind direction adjusted to agree with observations, and concentrations of tracer released during a field test conducted in Florida in 1998. Two Gaussian products are illustrated, both corresponding to slightly stable atmospheric conditions, but with different assumptions regarding the height of the release and the prevailing wind speed. A number of sequential transects are shown, each obtained using a real-time monitor carried in a mobile laboratory.
(Data provided by R. P. Hosker and colleagues, Oak Ridge, Tennessee.)

wind direction. (Note that without this alignment, the predicted plume would have missed the observations by a considerable distance. The experiment was conducted in a coastal region where wind direction is difficult to predict.) It is obvious that the average of a number of replications of such transects may indeed approach the prediction, nonetheless that reliance on the prediction for any such specific occurrence may be considerably misleading.

The need for accurate predictions to assist in the management of emergencies will require a new way of focusing on the problem. Models capable of predicting concentrations downwind of releases from any location are unlikely to be developed, although clearly all possible attempts should be made to most close by approach this ideal as is possible. Adoption of a probabilistic approach seems likely to be most profitable. It seems better for the science to be asking how best to predict the probability of danger at a specific place and time instead trying to predict, with quantifiable accuracy, the concentrations that will be encountered.

One simple consequence of this is that the way in which the acceptability of models is assessed might profitably be rethought. In the past, all methodologies for evaluating model performance have been based on point-to-point comparison (in space and time) of observations against predictions (Fox, 1981). For applications of the future, a more appropriate test would be based on the ensemble of all tests, and the accuracy of predictions of the probability of exceeding specified concentrations.

3. Forecasting Urban Air Quality

Many cities have tested methodologies for forecasting tomorrow's air pollution. There are two basic approaches being developed: one addressing the desire to forecast actual concentrations, and the second focusing on the need to provide warnings of periods when concentrations will exceed preset critical levels, i.e., on exceedances. In most instances, the techniques involved have been based upon a statistical examination of the meteorological factors influencing pollution levels, and then forecasting those properties revealed by multiple correlation. The regression relationships (or neural networks) that are used necessarily represent the interactions among averages. Use of these relationships to predict extremes is therefore demanding, yet it is prediction of these extreme circumstances that constitutes the main practical goal. It is not surprising, therefore, that comparisons of predictions and observations typically demonstrate a failure to reproduce the full dynamic range of the observations (e.g., KOHLEHMAINEN et al., 1999). The underprediction phenomenon notwithstanding, most contemporary air quality forecasting activities adopt a statistical approach in which forecasts of key meteorological variables are used as input to regression expressions (see COPE et al., 1998). Such approaches are being tested in the United Kingdom, Sweden, Australia, and several states in the USA.

States, for example, the acid rain debate of the 1970s and 1980s resulted in a major reduction in the emissions of sulfur and nitrogen oxides from fossil-fuel power plants. It is clear that the chemistry of the lower atmosphere has since changed, with nitrogen oxides now being more important than sulfur oxides in many situations. Figure 7 delineates the trend with time of the total sulfur-to-nitrogen molar ratio in

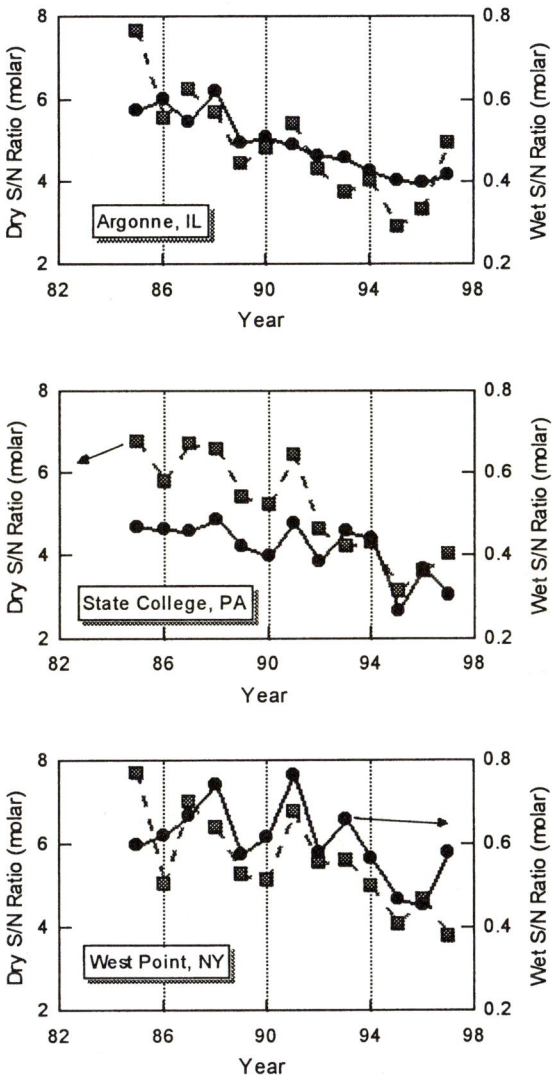

Figure 7

The change with time of the molar ratio of sulfur species to nitrogen species (oxidized plus reduced) in air and in precipitation at selected stations of the U.S. Atmospheric Integrated Research Monitoring Network (AIRMoN) see (HICKS et al., 2000). http://www.arl.noaa.gov/research/themes/aq.html#3).

precipitation and in near-surface air at selected sites in the eastern United States. Here, a trend with time is clear, with nitrogen increasing in importance. The role of nitrogen is becoming dominant in concerns such as acid deposition and visibility. This is important in many ways, as will become apparent below. It must be anticipated that similar changes in the nature of air chemistry and atmospheric deposition will occur elsewhere in the world. Consequently an overriding consideration is that whatever prediction capabilities are used must take the range of pollutants into account and not focus solely on those that are currently of dominant concern.

For developing regulations and guiding societal development, predictions are needed of the consequences of different scenarios. There have been many instances in which poor regulatory choices have been made because of the limitations of available models. One such example is the tall-stack approach favored in the United States during the 1960s and 1970s, when the operating philosophy was that "the solution to pollution is dilution." In practice, this approach turned a local problem into a regional one. During the same period, filtration systems were added to smokestacks, to remove particulate matter from the effluent gas stream and hence to eliminate the objectionable visible aspects of smokestack plumes. These particles are mostly basic. Their removal left the emissions dominated by acidifying trace gases, without the inherent neutralizing capabilities of the predominantly alkaline particles. Long-range transport of acids and downwind acid deposition resulted. Long after the period of concern about acid rain in North America it is now evident that the acid rain problem was exacerbated by pollution control decisions based on models that only told part of the story. To protect against future problems of this kind, the models used to guide the development of regulatory and/or growth strategies must be capable of addressing both local and distant repercussions of different emission control scenarios, and must also extend to consideration of ecosystem and human health consequences.

To protect against occurrences of "regulatory surprises," the early simple single-phenomenon predictive models have made way for far more complex and comprehensive simulations. The regional atmospheric dispersion and deposition models constructed during the acid rain decade of the 1980s serve as perhaps the best examples of the kind of advanced model now available. Even for these complex codes, the predictions carried a probabilistic overtone. The main product was estimates of the deposition of acidic and acid-forming chemicals far downwind of source regions. Model outputs were used for this purpose for two main reasons. First, running the models is far less expensive than mounting the widespread measurement campaigns necessary to provide the spatial information yielded by the models. Second, observations only reveal information looking at the present and backwards in time; what regulators need is a look forward in time – prediction, not analysis of the past. Recognition that the models are inherently incapable of precise prediction was manifested in a call for quantification of the levels of uncertainty associated with each model product.

In European work, levels of atmospheric deposition estimated by predictive models are using within a "critical loads" regulatory framework. In this approach, critical loads are associated with all ecosystems, across the entire domain of interest. A critical load is the value of atmospheric deposition beyond which damaging ecological effects are expected. The concept parallels that of the air quality exceedances mentioned above. U.S. scientists have not always sympathized with the critical load approach, because critical loads cannot be measured and because modeling based on them could fail to protect highly sensitive ecosystems that are much smaller than grid cells. Regardless of the North American concern about the utility of the critical loads concept, it has been used with great profit by regulators in Europe, working on a multinational basis. Once again, the concept lends itself to the application of probabilistic thinking. In fact, the issue of subgrid sensitivities is now addressed in European critical loads models on a probabilistic basis.

5. Forecasting for Ecological Protection

Along with the increasing importance of nitrogen species in the North American pollution regime, there is an underlying need for more complete understanding of the interaction between air pollution and the various factors leading to ecosystem decline. An example has arisen along the mid-Atlantic coast of the United States, where the environment is being threatened by society and its pollution. The region at risk stretches from the foothills bordering the coastal plain to the edge of the continental shelf. This region is the focal area for a major part of the population, economy, industry, and food production of the United States. Emissions from most aspects of society migrate towards the coastal water bodies on which the population relies, and these water bodies (and the watersheds that serve them) are deteriorating as a consequence. The sources of the damaging pollutants may be local or quite distant. The pathway by which they enter the water bodies (or their surrounding catchment areas) may involve transport through ground water, rivers, and the atmosphere, as well as direct discharge. There is no universal rule.

It has recently been shown that the contribution of atmospheric deposition can be a considerable and at times a dominant factor affecting the decline of coastal and marine ecosystems. Internationally, the role of the atmosphere as a contributor to the world's oceanic biosphere has been well recognized (DUCE et al., 1991), and within North America there is a long history of aquatic biological effects that are related to atmospheric deposition (SCHINDLER, 1988). Ten years ago, it was claimed that about one third of the nitrogen nutrients entering the Chesapeake Bay were of atmospheric origin (FISHER and OPPENHEIMER, 1991). Subsequent work has supported that approximation, and has extended the geographic area of its applicability from the far northeastern coast to South Carolina. However, it is the mid-Atlantic coast where most of the atmospheric nutrient loading seems to be focused. At this time there are urgent multi-agency efforts under way to improve the quantification of the role of atmospheric deposition in estuarine nitrogen loadings. The nitrogen compounds of interest derive from burning fossil fuel (in automobiles, as well as in power plants and industry), and from agricultural practices.

Coastal areas are especially susceptible to pollution, since major population centers are often located along coasts, and frequently on rivers draining large watersheds. These watersheds collect pollutants from a wide variety of sources, among which atmospheric deposition features prominently.

Table 1 is a summation of recent information from open literature sources assembled by VALIGURA et al. (1996), addressing a number of East Coast U.S. estuaries and bays. Among these, the Chesapeake Bay system is the most studied example. Figure 8 shows the general location of the bay, the extent of its watershed, and the spatial spread of the airshed from which emissions into the air affect it (with highest probability) through atmospheric transport and deposition. The geographic extent of the airshed is defined using a computer model (DENNIS, 1997) simulating the processes by which nitrogen compounds are carried through the air from varied sources (power plants, industry, automobiles, agriculture, etc.) and deposit to the Chesapeake Bay watershed. In defining the airshed, the focus is on nitrogen compounds, for which it is not only deposition directly to the water surface itself that

Table 1

Recent estimates of the contribution of N deposition from the air as a contributor to East Coast U.S. ecosystem enrichment (after VALIGURA et al., 1996)

	Narragansett Bay	Long Island Bright	Delaware Bay	Chesapeake Bay	Albemarle-Pamlico Sound
Watershed:					
Area (km^2)	4708	43481	36905	165886	59197
N Deposition (Tg)	4.2	45	53	17	39
Loading transferred to water body	0.3	7	5	29	6.7
Tidal waters					
Area (km^2)	328	4820	1846	11400	7754
N Deposition (Tg)	0.3	5	3	16	3.3
Total N load from the atmosphere	0.6	12	8	45	10
Total load from all sources	5	60	54	170	23
Percentage due to atmospheric deposition	12	20	15	27	44

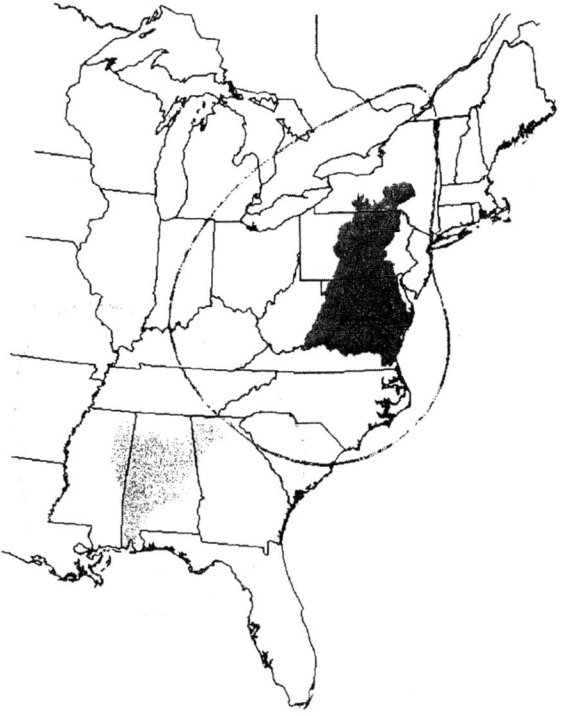

Figure 8
The watershed and airshed of the Chesapeake Bay (on the Atlantic Coast), as defined by DENNIS (1997).

is of importance, but also the deposition to the surrounding watershed (some of which is transmitted through ground water and streamflow to the Bay). This transmission of pollutants through the terrestrial biosphere is a key area of uncertainty in assessing the relative impact of atmospheric deposition on aquatic ecosystems.

Table 1 summarizes current understanding pertaining to the chronic, long-term nutrient enrichment problems of east coast U.S. aquatic ecosystems. However, living organisms respond individually to whatever combination of stressors is affecting them at any specific time. Furthermore, a highly stressed ecosystem will be more vulnerable to sudden insults than will an unstressed ecosystem. Atmospheric precipitation constitutes an irregular stressor of double significance—not only do extreme events cause flooding, but all precipitation carries with it a dose of nitrogen nutrients, both reduced and oxidized. The rainfall certainly causes runoff to occur, carrying pollution from the surrounding countryside to all of the water bodies in the region. But also the rainfall carries biologically active nutrients that could serve as triggers of rapid growth of organisms into shallow waters where such growth is indeed found. The matter requires further exploration. For the moment, it is clear that the utility of forecasts extends far beyond the conventional considerations of severe weather, and that once again the utility of the forecasts is most likely in warning regulators and environmental managers of the likelihood of an outbreak of potentially harmful algal blooms or other micro-organism effect.

In this area of concern, it is relevant to note that all ecosystems are adjusted to natural levels of atmospheric deposition that are not zero, even in then absence of humans. Concerns arise whenever deposition levels rise so that ecosystems will be stressed beyond their relevant (and spatially confined) "comfort zones." In this application as well, it is not forecasting of average values that is most needed, but accurate prediction of the likelihood that safe levels will be exceeded at times when these exceedances could be damaging. Once again, a probabilistic approach seems demanded by the science involved.

6. Discussion and Conclusions

Southeast Asia is characterized by a complex mix of land and water, by a challenging and geologically active landscape, by a steadily increasing population, and by a strong reliance on the seas for food. There is need for forecasts in each of the areas of concern discussed above. However, the geographic complexity of the region imposes severe difficulties in developing accurate forecast tools, at any scale. It is therefore proposed that the need for a probabilistic approach is more important for Southeast Asia than for other regions.

For planners to protect sensitive people (and ecosystems) from episodes of severe air pollution, there needs to be a warning system in place that parallels the conventional weather forecasting process. Many attempts are being made to institutionalize such systems. Some workers rely on the use of multiple regression relationships derived from past data to generate predictions of what the near-term future air quality will be. Other workers are developing deterministic methods to achieve the same goals, arguing that the regression approaches are fundamentally limited because of their inherent inability to describe extremes. However, these deterministic methods are strongly limited by their need for source term information that is often lacking. Data assimilation is required, with forecasting of air quality and related properties on a probabilistic basis. Finally, no matter what strategy is adopted for forecasting future air quality, it is essential that air be considered as a part of the total ecosystem in which people reside, and not as a medium that can be regulated in isolation from the other media.

To help protect the population and the environment, methods are needed to predict the probability that people (for example) will be exposed to alarmingly high concentrations of pollutants such as ozone, forest fire smoke, and inhalable particulate material from other sources. Availability of such information will permit response strategies to be developed and refined. At this stage, in the absence of forecasts, all that is possible is the constant roll-back of all emissions.

The need for emergency response capabilities is widespread, and is increasing as industrial facilities age and urban areas expand. In many situations, potential origins of leaks and accidents can be identified, and response strategies developed accordingly. In urban areas, surface obstructions impose complexity that defeats current deterministic predictive models. Hence, a probabilistic approach is optimal.

In the case of environmental protection, contemporary regulatory emphasis is on long-term, chronic exposure to pollution, whereas in practice living organisms also respond to the impact of stressors on short time scales. To develop remedial strategies, there is an inherent need to take episodic insults into account, and to structure protective (or remedial) actions on the forecasts that are available.

In none of these circumstances can precise predictions be made. Rather, it is proposed that the needs of the Southeast Asian region can best be addressed by adopting a probabilistic approach. In this, the intent would not be to predict the exposure level or the doses to be delivered to some receptor at a given time and place, but rather to predict the probability that exposure and/or dose will exceed some prescribed level.

In any event, it must be remembered that the lack of accurate source term information plagues all air quality and emergency response prediction schemes. Adoption of a probabilistic approach to air quality forecasting will not reduce the need for improved source term information of all kinds.

REFERENCES

COPE, M. E., MANNINS, P., HESS, G., MILLS, G., PURI, K., DEWUNDEGE, P., TILLY, K., and JOHNSON, M. (1998), *Development and Application of a Numerical Air Quality Forecasting System*, Proc. 14th International Conference on Clear Air and Environment, Melbourne, 18–22 October, 1998, Clean Air Society of Australia and New Zealand, pp. 353–358.

DENNIS, R. L. *Using the Regional Acid Deposition Model to determine the nitrogen deposition airshed of the Chesapeake Bay watershed*. In *Atmospheric Deposition of Contaminants to the Great Lakes and Coastal Waters* (ed. J. E. Baker) (SETAC Press, Pensacola, Florida 1997), pp. 393–413.

DRAXLER, R. R. (1987), *Accuracy of Various Diffusion and Stability Schemes over Washington, D.C.*, Atmos. Environ. *21*, 491–499.

DRAXLER, R. R. (2000), *Meteorological Factors of Ozone Predictability at Houston, Texas*, J. Air and Waste Managem. Assoc. *50*, 259–271.

DUCE, R. A., MERRILL, P. S., Atlas, E. L., BUAT-MENARD, P., HICKS, B. B., MILLER, J. M., PROSPERO, J. M., ARIMOTO, R., CHURCH, T., ELLIS, M., GALLOWAY, J. N., HANSON, L., JICKELLS, T. D., KNAP, A. H., REINHARDT, K. H., SCHNEIDER, B., SOUDINE, A., TÖKOS, J. J., TSUNOGAI, S., WOLLAST, R., and ZHOU M. (1991), *The Atmospheric Input of Trace Species to the World Ocean*, Global Biogeoche. Cycles *5*, 193–259.

FISHER, D. C., and OPPENHEIMER, M. (1991), *Atmospheric Nitrogen Deposition and the Chesapeake Bay Estuary*, Ambio *23*, 102–108.

FOX, D. G. (1981), *Judging Air Quality Model Performance*, Bull. Am. Meteorol. Soc. *62*, 599–609.

HURLEY, P. J. (1999), *The Air Pollution Model (TAPM) Version 1: Technical Description and Examples*. CSIRO Atmos. Res. Tech. Paper *43*, 41 pp.

KOHLEHMAINEN, M., MARTIKAINEN, H., HILTUNEN, T., and RUUSKANEN, J. (1999), *Forecasting Air Quality Parameters Using Hybrid Neural Network Modelling*. Available at http://www.uku.fi/laitokset/ ympkem/airquality/hybrid/poster/abs.html.

PIELKE, R. A. and ULIASZ, M. (1998) *Use of Meteorological Models as Input to Regional and Mesoscale Air Quality Models—Limitations and Strengths,* Atmos. Environ. *32,* 1455–1466.
SCHINDLER, D. W. (1988), *The Effects of Acid Rain on Fresh-water Ecosystems,* Science *239,* 149–157.
VALIGURA, R. A., LUKE, W. T., ARTZ, R. S., and HICKS, B. B. (1996), *Atmospheric Nutrient Input to Coastal Areas—Reducing the Uncertainties.* NOAA Coastal Ocean Program Decision Analysis Series Number 9, 24 pp. plus appendices.

(Received April 30, 2000, accepted August 31, 2000)

 To access this journal online:
http://www.birkhauser.ch

Pure appl. geophys. 160 (2003) 75–80
0033–4553/03/020075–06

© Birkhäuser Verlag, Basel, 2003

▌Pure and Applied Geophysics

Indoor Radon Radioactivity at the University of Brunei Darussalam

TAN KHA SHENG[1] and HU SHZE JER[1]

Abstract — Indoor radon radioactivity in the rooms on the ground floor and first floor of the Physics Department, Faculty of Science, Universiti Brunei Darussalam was measured using a system that consists of an air filter pump, ZnS detector, photomultiplier tube and counter. Ground floor rooms' radon radioactivity was found to be about three times higher than that of the first floor. The maximum ground floor indoor radioactivity is only 0.39 Bqm^{-3}, a value relatively low and safe compared to the mean outdoor radon concentration of 1.41 Bqm^{-3} measured (HU and TAN, 2000). The main source of radon emanation originates from the ground soil rather than the building materials.

Key words: Radioactivity, radon, indoor radioactivity.

Introduction

Radon and its short-lived decay products form a major issue in the exposure of man to natural sources of radioactivity. The radon isotope, ^{222}Rn, has a short half-life of 3.82 days. It belongs to the primordial decay series of ^{238}U. It is a noble gas that exhales from the topsoil and building materials into the air. ^{222}Rn is produced by the decay of the parent nuclei ^{226}Ra. ^{226}Ra may be found in soil particles and has a half-life of 1622 years. Depending on the type of mineral in the soil, it may be distributed either uniformly throughout the soil particles or primarily at the particle surfaces (MOHAWKS and JEFFRIES, 1994). The decay series of ^{222}Rn ends with the stable lead isotopes ^{206}Pb. However, as far as airborne ^{222}Rn progeny is concerned, the isotope ^{210}Pb, with a half-life of 22.3 years is usually considered the metastable end product. All intermediate decay products are heavy metals, which tend to attach to aerosols or to plate out onto surface areas. The airborne radon progeny is thus, in general, not in secular equilibrium with its parent.

Naturally occurring radon and its short-lived progeny form a significant environmental problem as they represent a dominant source for exposure of the human population to ionizing radiation (UNSCEAR, 1993). Inhalation is the most

---
[1] Physics Department, University Brunei Darussalam, Brunei Darussalam BE1410.
E-mail: tanks@ubd.edu.bn and sjhu@ubd.edu.bn

critical pathway for exposure to radon progeny. The radon isotopes decay and give rise to short-lived decay products that easily attach to aerosols and dust particles. These may be inhaled and result in a radiation dose to the lungs through alpha decay. The radon dose due to the inhalation of ^{222}Rn and progeny is mainly caused by energetic alpha particles emitted by the short-lived decay products, ^{218}Po and ^{214}Po. The total amount of alpha-energy that will be delivered to the lung area by the inhaled and deposited decay products of ^{222}Rn is of primary concern in assessing the radon dose. The trachea-bronchial area of the lungs is the most sensitive area.

Building materials originate within the earth's crust and thus contain radionuclides from the uranium and thorium series that give rise to external radiation. Building materials like concrete, cement, sand and brick contain radionuclides that may enhance exposure levels indoors. These sources of enhanced natural radiation may result in a higher dose to the population in general since most people spend 80% of their time indoors. Radon infiltrates buildings from the soil underneath through exhalation from building materials and through ventilation with outdoor air. Radon has the greatest effect on one- or two-storey buildings and it also originates from the soil and ground water besides the building materials. Due to relatively low ventilation rates in buildings, a build-up of radon occurs.

In Brunei Darussalam, an increasing number of buildings are air-conditioned such that windows are seldom opened for ventilation purposes. There is a possibility that radon build-up may occur. It is with this objective that the present investigation was performed.

Experimental

An investigation to determine the indoor radon radioactivity in the rooms of the Physics Department of the Faculty of Science building was carried out. These rooms are located either on the ground floor or on first floor of the Faculty of Science building. The radon measurements were carried out at different hours throughout the day.

Radon and its progenies were measured by the filter method. The set-up for ^{222}Rn measurement consists of an air filter pump, ZnS detector that detects the alpha particles emitted, photomultiplier tube (OKEN SP-20) and counter (OKEN RC-1O1A). All the equipment was put on a trolley for ease of movement. The system was calibrated with a known sample to give readings in the Bqm^{-3} activity units. The calibration was performed at the Department of Nuclear Sciences, University of Nagoya. Japan. The calibration graph obtained is as shown in Figure 1. In the calibration curve, the activity is plotted against the counts per liter per minute. The count is the 40 minutes timer counts with the background counts subtracted.

The air was sampled at a height of about 1 m above the floor by pumping it through a piece of filter paper of pore size 1.2 µm for 15 minutes as determined by

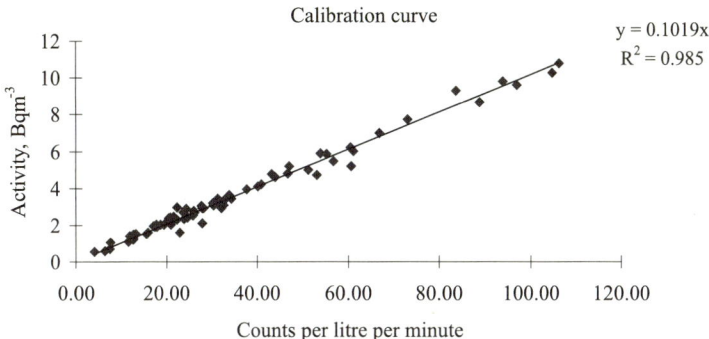

Figure 1
Calibration graph of activity versus counts per liter per minute.

the calibration procedure. The pump rate was set at 30 liter per minute. The filter was then inserted into a holder facing the ZnS detector and the counter set to count emission of α-particles emitted by the decay of ^{222}Rn. The activity was read off from the calibration curve obtained.

Results and Discussion

The readings obtained for the rooms on the first floor of the building are generally very low and they vary from 0.06 Bqm^{-3} to 0.13 Bqm^{-3}. The readings obtained for the ground floor laboratory rooms measured over a period of two weeks, both in the morning and late afternoon were found to range from 0.13 Bqm^{-3} to 0.39 Bqm^{-3}. The mean value of these readings from the laboratory rooms on the ground floor is 0.35 Bqm^{-3}, a value that is relatively low and comparable to that obtained in the US (UNSCEAR, 1982), where the ^{222}Rn concentration in the air at 1–3 m above the ground ranges over two orders of magnitude (0.37–37 Bqm^{-3}). This mean value is also low compared to the outdoor radon concentrations obtained for the same building over a four month period that ranges from 0.40 Bqm^{-3} to 1.41 Bqm^{-3} (HU and TAN, 2000). The ground floor rooms' readings being higher (about three times higher) than the readings for first floor rooms are also logical since the main source of radon gas obviously has to be from the earth (ground).

Measurements in one of the laboratory rooms on the ground floor at different hours over a period of about three weeks showed that there is periodic hour-to-hour variation of the radon radioactivity (Fig. 2). The maximum radioactivity concentration, which is about 0.39 Bqm^{-3}, is observed in the morning hours. These readings decrease as the hour increases. Lower radioactivity values are found in the afternoon and they are about 0.1 Bqm^{-3}. Overall, readings in the late afternoon are about 1/3 the value of the morning readings. This variation of activity at a given locality at

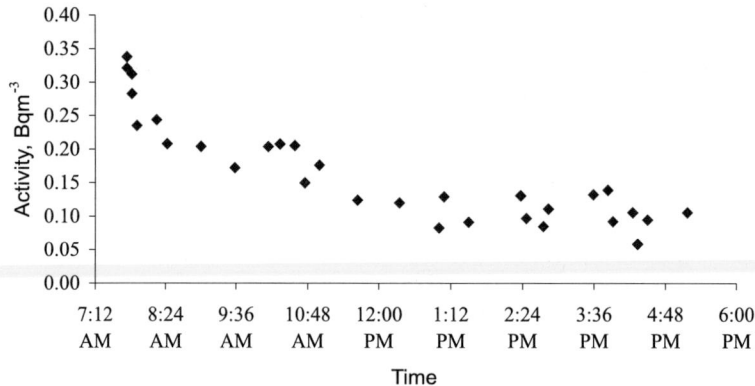

Figure 2
Activity versus time.

different hours is primarily due to the different emanation rate of radon gas from the soil. To explain this, we need to understand the meaning of the radon emanating factor which is defined as the ratio of radon emanating from the soil particles to the air and water-filled pores to the total amount of radon produced.

When radium decays, the radon atom will recoil because of the emission of α-particle. The radon atom will move a distance of 20–70 nm in common minerals due to this recoil. Thus, the emanation of the radon atoms from soil particles can cause the radon to recoil (Fig. 3) (B) within the particle itself, (A) from one particle to another or (C) into the interstitial pores of the soil (BLAAUBOER and SMETSERS, 1996). Those recoils that occur through process (A) and (B) will be trapped in the soil particles while those that occur through process (C) will escape into the air. If the interstitial pores are partially filled with water, the pore water may absorb the radon atoms recoiling from a soil particle and the recoil distance will be reduced. The low solubility of the radon in water enables most of the radon to reach the air filled pores by diffusion. The water content in the soil thus plays an important part in determining the emanation factor.

After radioactive decay of radium, the radon isotopes will be trapped in the soil particles, unless some mechanism is available to allow them to escape (emanate) into the interstitial air. In deep soil (2 m downward), the emanated radon gas will still be in secular equilibrium with the parent radionuclide radium. In the upper soil layer the interstitial radon concentration will decrease because of exhalation, a net transport of soil gas to the atmosphere. The rate of exhalation of radon from soil to the atmosphere is dependent on the emanation factor and the moisture content of the soil. When the radon exhalates from the soil and reaches the atmosphere, it will diffuse. A high water content of the soil pores increases the emanation factor. The pore water may absorb the radon atoms recoiling from a soil particle. A low water content (dry soil) will reduce this absorbing effect and thus increase the number of

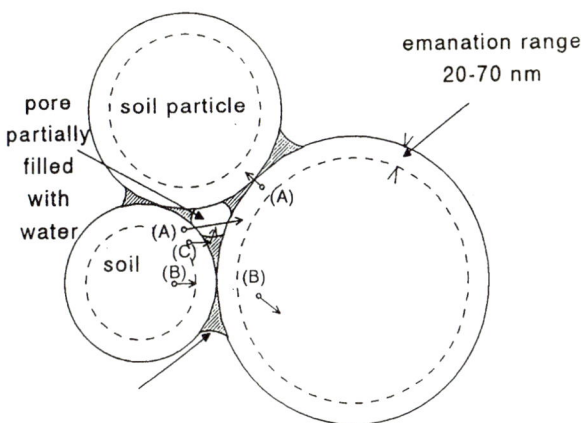

Figure 3
Schematic diagram of emanation of radon atoms from soil particles.

radionuclides recoiling into other soil particles, thereby reducing the emanation factor. The temperature drop at night causes condensation in the soil/concrete structure of the floor and the wall. The interstitial and pore water will absorb the radon atoms recoiling from a soil particle. Since radon solubility in water is low, the emanation rate will increase and thereby gives rise to a higher radioactivity level in the morning. In the later half of the day as the temperature warms, the water content in the soil and concrete structures will be reduced. This will then cause the radon emanation rate to decrease, giving rise to a lower radioactivity level.

No such changes were observed for the radon measurements obtained throughout the day for the rooms on the first floor. The variation in radon radioactivity levels from morning until late afternoon for the ground floor rooms implies that the main source of radon emanation in the rooms of the Physics Department originates from the ground (soil) rather than the building materials of the concrete and wall structures. The maximum value obtained is 0.39 Bqm^{-3}. The outdoor radon concentrations measured outside the same building (HU and TAN, 2000) range from 0.40 Bqm^{-3} to 1.41 Bqm^{-3}. The outdoor radon average concentration according to PUT et al. (1986) is 3 Bqm^{-3} and the indoor radon range of concentration varies from 8–140 Bqm^{-3} according to VAAS et al. (1993). Since the US Action Level of the radon concentration is 150 Bqm^{-3} while that of the UK is 200 Bqm^{-3}, therefore, the radon radioactivity level in the Physics Department of the Faculty of Sciences building, University Brunei Darussalam can be considered relatively low and safe.

Acknowledgement

The authors would like to thank Dr. S. Minato, NIRIN, Japan for lending his Radon monitoring equipment for this work.

REFERENCES

BLAAUBOER, R. O. and SMETSERS, R. C. G. M., *Variations in Outdoor Radiation Levels in the Netherlands* (Elinkwijk BV, Utrecht 1996) 263 pp.

HU, S. J. and TAN, K. S. (2000), *Radon and its Progeny in Outdoor Air*, Asian J. on Sci. and Technol. Developm. *17*(2), 65–70.

MOHAWKS, L. and JEFFRIES, C. (1994), *Distribution of Radium in Mineral Sand Grains and its Potential Effect on Radon Emanation*, Radon Protection Dosimetery *56*(1–4), 199–200.

UNSCEAR (1993), *Sources, Effects and Risks of Ionizing Radiation (Annex A)*, UNSCEAR Report 1993 to United Nations, NY.

UNSCEAR (1982), *Ionizing Radiation: Sources and Biological Effects*, UN Publication, NY.

PUT, L. W., DE MEIJER, R. J., and VELDHUIZEN, A. (1986), *Radon Concentrations in Netherlands*, Ministry of VROM, Leidschendam, Report No. *14*, 36 pp.

VASS, L. H., KAL, H. B., JONG, P. DE and SLOOFF, W. (1993), *Integrated criteria document radon*. National Institute of Public Health and Environm. Protection, Bilthoven, NL RIVM Report No. 710401021, 25 pp.

(Received April 30, 2000, accepted August 15, 2000)

 To access this journal online:
http://www.birkhauser.ch

Pure appl. geophys. 160 (2003) 81–105
0033–4553/03/020081–25

© Birkhäuser Verlag, Basel, 2003

Pure and Applied Geophysics

The Airshed for Ozone and Fine Particulate Pollution in the Eastern United States

KEVIN L. CIVEROLO,[1] HUITING MAO,[2] and S. TRIVIKRAMA RAO[2]*

Abstract — To examine the spatial scales associated with atmospheric pollutants such as ozone (O_3) and fine particulate matter ($PM_{2.5}$), we employ the following five techniques: (1) Analysis of the persistence of high O_3 concentrations aloft; (2) spatial and lag correlations between the short-term components (i.e., weather-induced variations) in the time series of O_3 and $PM_{2.5}$ throughout the eastern United States; (3) analysis of mixed-layer forward trajectories compiled at different locations on a climatological basis to identify the potential region covered in 1-day of atmospheric transport; (4) analysis of three-dimensional Lagrangian trajectories of tracer particles for three high-O_3 episode events in the summer of 1995; and (5) analysis of the spatial extent over which emissions have an impact through photochemical model simulations. Regardless of the method chosen, the results demonstrate that pollutants such as O_3 and $PM_{2.5}$ have the potential to affect regions having spatial scales of several hundred kilometers. This finding has implications to regulatory policies for addressing the pollution problem, and for optimally designing monitoring networks for such pollutants.

Key words: Ozone, fine particulate matter, trajectory analysis, space and time scales, photochemical modeling.

1. Introduction

Ozone (O_3) and fine particulate matter ($PM_{2.5}$) are ubiquitous pollutants, and analyses of such air quality data are complicated by the fact that the time series exhibit anthropogenic and natural forcings on different temporal and spatial scales (RAO *et al.*, 1997). Using the trajectory clustering and time-lagged intersite correlation analysis method, BRANKOV *et al.* (1998) have demonstrated that O_3 in the eastern United States is a regional-scale problem. Furthermore, the potential shift from the 1-hour to an 8-hour ozone National Ambient Air Quality Standard

[1] New York State Department of Environmental Conservation, Division of Air Resources, Bureau of Air Research, Albany, New York 12233-3259, U.S.A.

[2] Department of Earth and Atmospheric Sciences, University at Albany, State University of New York, Albany, New York 12222, U.S.A.; now at EOS, Morse Hall, University of New Hampshire, Durham, NH 03824 U.S.A.

* Corresponding author: Currently at U.S. Environmental Protection Agency, Atmospheric Sciences Modeling Division, Research Triangle Park, NC 27711, U.S.A. E-mail: rao.st@epa.gov

(NAAQS) (USEPA, 1996a) is expected to greatly expand the spatial extent of the O_3 problem in the eastern United States (CHAMEIDES *et al.*, 1997; SAYLOR *et al.*, 1998). In addition to the 8-hour O_3 standard, the USEPA has proposed new daily and annual standards for $PM_{2.5}$ (USEPA, 1996b). Hence, as these new pollutant standards are considered, it becomes evident that we need to examine the spatial extent, or the airshed, for such pollutants, in assessing population exposure and in designing effective emission control strategies.

The concept of the spatial extent of an airshed is less well-defined than that for a given watershed. While watersheds have definite physical dimensions, the atmosphere is not bounded in such a way. Whereas transport through a watershed is limited to rivers and other bodies of water, and the surrounding land surfaces, transport through the atmosphere can occur over much longer distances. The airshed associated with nitrogen deposition into the Chesapeake Bay has been defined as the geographic region that contributes about 70–80% of the nitrate to the Bay watershed (APPLETON, 1995; DENNIS, 1997); this region is about five times as large as the area of the watershed itself, defining the region of influence of nitrate to be greater than 700 km.

While it may be difficult to precisely determine the spatial scales associated with O_3, $PM_{2.5}$, acidic deposition, and toxic pollutants, the fact remains that they exist. In the case of O_3, the meteorological features associated with high O_3 events in the eastern United States have been examined by VUKOVICH (1994), RYAN *et al.* (1998), and ZHANG *et al.* (1998). A common synoptic-scale feature associated with O_3 episodes over the eastern United States is the presence of a high pressure system aloft (500 mb) which is usually accompanied by subsidence, clear skies, strong shortwave radiation, high temperatures, and stagnant air masses near the ridge line of the sea-level high pressure region. In addition, westerly and southwesterly low-level jets during these episodic events facilitate the transport of pollutants over long distances (ZHANG *et al.*, 2001). These synoptic conditions augment local photochemical production and contribute to elevated levels of pollutant concentrations which blanket much of the northeastern United States for several days (ZHANG *et al.*, 1998).

In this paper we discuss the temporal and spatial scales and attempt to define the airshed associated with O_3 and other pollutants, using five different methods. First, we examine O_3 observations to characterize the persistence of high concentrations aloft. Next, we use linear regression between the synoptic-scale or weather-induced variations (characteristic time scale of 2 to 21 days) embedded in the time series of $PM_{2.5}$ and O_3 at different locations to determine the distances over which a high degree of correlation is maintained. We then use mixed-layer forward trajectory analysis on a climatological basis to define the airshed for 1-day pollutant transport, the potential spatial extent over which pollutant concentrations are affected by a given source region. We next attempt to corroborate these findings by quantifying atmospheric transport using three-dimensional particle trajectories with a Lagrangian particle dispersion model for three O_3 episodic periods that occurred in the summer of 1995. Finally, we compare a base case photochemical model simulation using the UAM-V model (SAI, 1995) with hypothetical emission reduction scenarios to examine the spatial extent over which ozone benefits can be realized when emissions in a source region are reduced.

2. Databases

The first two analyses—characterization of high pollutant concentrations aloft, and spatial correlation analysis—involve historical measurements of O_3 and $PM_{2.5}$. The databases used here are described below.

To characterize high O_3 concentrations aloft, we considered O_3 measurements atop the World Trade Center in New York City (430 m MSL) and the Big Meadows site in Shenandoah National Park, VA (1097 m MSL). We compiled hourly concentration data during the April through October photochemical season from 1983 through 1998, calculated daily maximum 8-hour concentrations, and counted the number of times that the daily maximum 8-hour concentration exceeded 70 ppbv (i.e., O_3 persistence). High-resolution historical data are sorely lacking to perform a similar analysis for $PM_{2.5}$; however, we obtained daytime (0700–1700 LT) $PM_{2.5}$ data from the Appalachian Mountain Club (AMC) from 1988–1997 for the June through September period (B. Hill, 1999, personal communication).

For the regression analysis we first extracted hourly O_3 concentrations from the EPA Aerometric Information Retrieval System (AIRS) database for the 1992–96 summer seasons, defined operationally as June 1 to August 31, for each O_3 monitor in the eastern United States. To be considered in the analysis, a monitoring location had to include summertime data during at least two of the five years; nearly 500 monitoring locations were included in this analysis (see Fig. 1(a)). To moderate the effects of intra-day variations in hourly O_3 time series, we calculated the daily maximum 8-hour O_3 concentrations. Given daily pollutant concentration data, we can obtain information on time scales greater than two days only (Nyquist frequency).

We performed a similar analysis with the daily average $PM_{2.5}$ mass data from fifteen locations throughout the eastern United States (Fig. 1(b)). Seven sites belong to the National Park Service Interagency Monitoring of PROtected Visual Environments (IMPROVE) network. These sites are still operational, but we have only considered data through August, 1998. Twenty-four hour samples are reported each Wednesday and Saturday. Seven sites, located mainly in New England and northern New York, comprised the NorthEast States for Coordinated Air Use Management (NESCAUM) network, in operation from 1988 to 1993. Data were

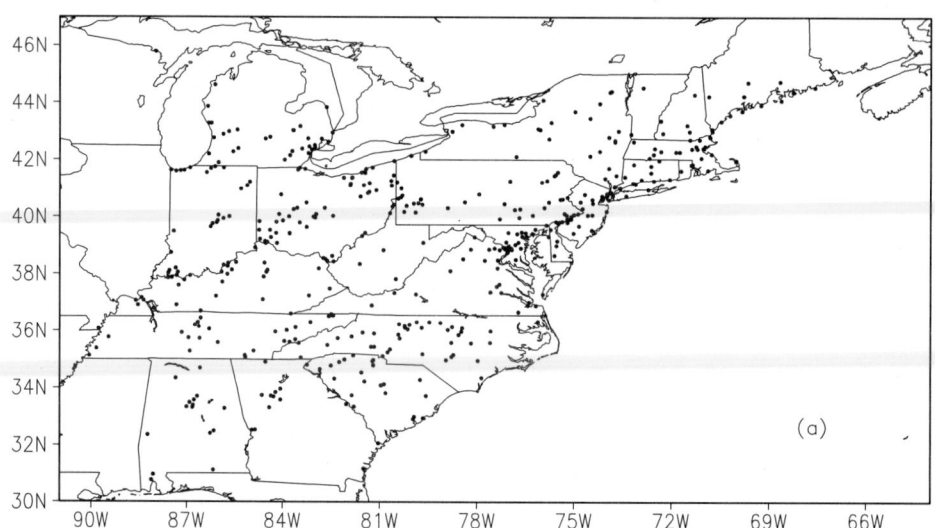

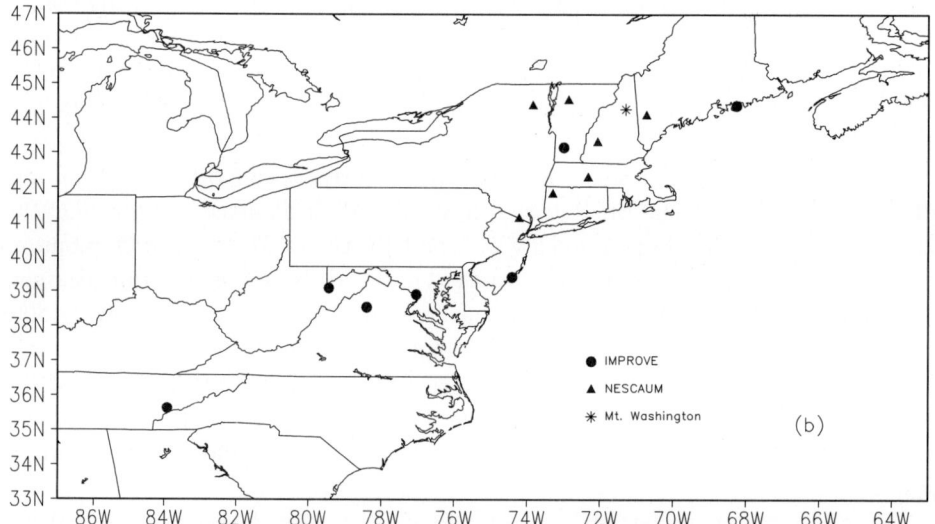

Figure 1
(a) Monitoring network of daily maximum 8-hour O_3 data. (b) Monitoring network of the daily $PM_{2.5}$ mass data.

reported every Wednesday, Saturday, and every sixth day that was not a Wednesday or Saturday. Finally, we also used the AMC data from Mt. Washington from 1988 through 1997.

3. Results

a. The Problem

Figure 2 shows the 1-hour design values (defined as the fourth highest daily maximum concentration over three consecutive years) and 8-hour design values (defined as the fourth highest daily maximum concentration average over three consecutive years) for O_3 at various locations throughout the eastern United States for the 1997–1999 period. Locations identified as yellow, orange, and red indicate where the design values exceed the current 1-hour NAAQS for O_3 and the proposed 8-hour NAAQS. Figure 2 indicates that while noncompliance with the 1-hour standard appears to be confined to a few urban areas (e.g., Atlanta, GA; the Richmond, VA to Boston, MA urban corridor) and the immediate downwind regions, the longer-term O_3 standard will cause much of the eastern United States to become noncompliant. From Fig. 2, whereas 13% of the monitors have design values which are noncompliant with the 1-hour standard, 71% of the monitors are noncompliant with the 8-hour standard. Also evident from Fig. 2 is that while there is a sufficient number of monitors in urban and suburban locations, the coverage in rural areas is rather sparse.

Because of the adverse health effects of particulate matter, the USEPA has also proposed a revision to the particulate matter standard (USEPA, 1996b), including an annual $PM_{2.5}$ mass standard of 15 μg m^{-3} averaged over three consecutive years. Figure 3 shows the 3-year average $PM_{2.5}$ during 1991–1993 at 15 locations. Although these data are sparse and not the most currently available, Figure 3 suggests that rural locations throughout the southeast and mid-Atlantic regions experience average $PM_{2.5}$ concentrations in the 11–13 μg m^{-3} range, comparable to concentrations at more urban locations. However, it should be noted that the reference method proposed by EPA for $PM_{2.5}$ mass measurements is different from the methods used in the $PM_{2.5}$ mass database analyzed here.

The above results illustrate that high pollutant concentrations are a regional issue, not simply a local or an urban-scale problem. In the next five sections we will present the results of five different sets of analyses to quantify the spatial and temporal extents of the O_3 and $PM_{2.5}$ pollution problem.

b. Persistence of High Pollutant Concentrations Aloft

We first focus on O_3 measurements aloft rather than near ground level, since nighttime pollutant concentrations measured at high elevation sites are more typical of free tropospheric values and are less affected by pollutant removal processes at the surface. Figure 4 displays the histogram of the number of consecutive days over which high O_3 concentrations (daily maximum 8-hour O_3 > 70 ppbv) persisted at the World Trade Center and Shenandoah National Park sites. Such high concentrations aloft have been shown to contribute substantially to ground-level O_3

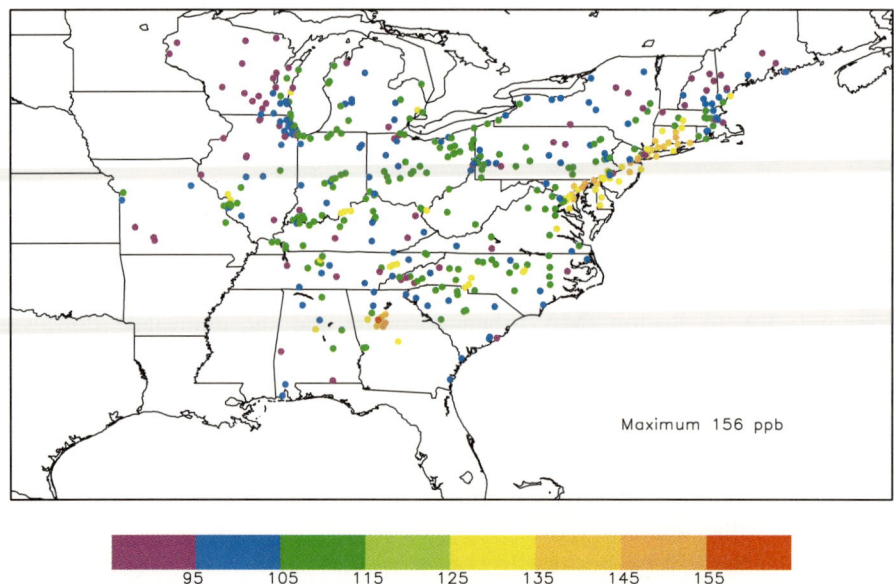

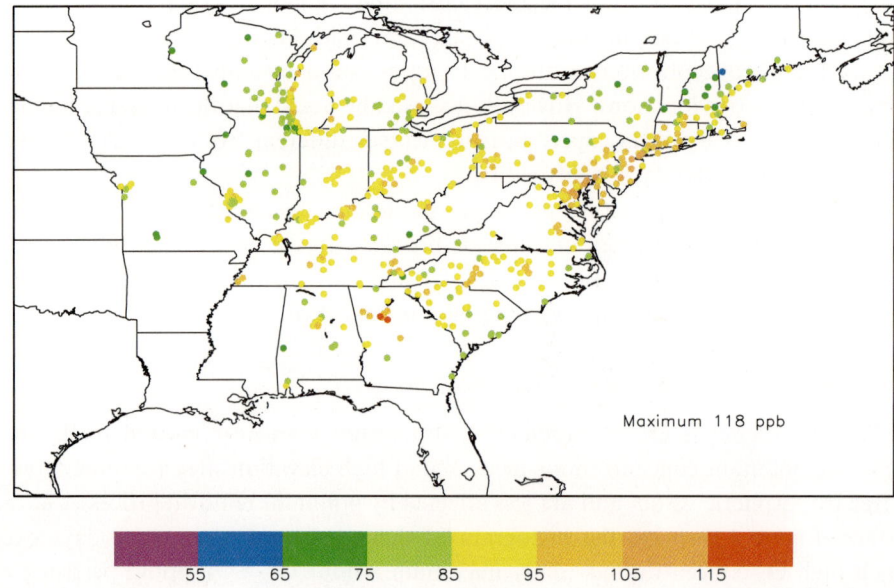

Figure 2
Ozone 1-hour and 8-hour design values (ppb) during the 1997–1999 period for locations throughout the eastern United States.

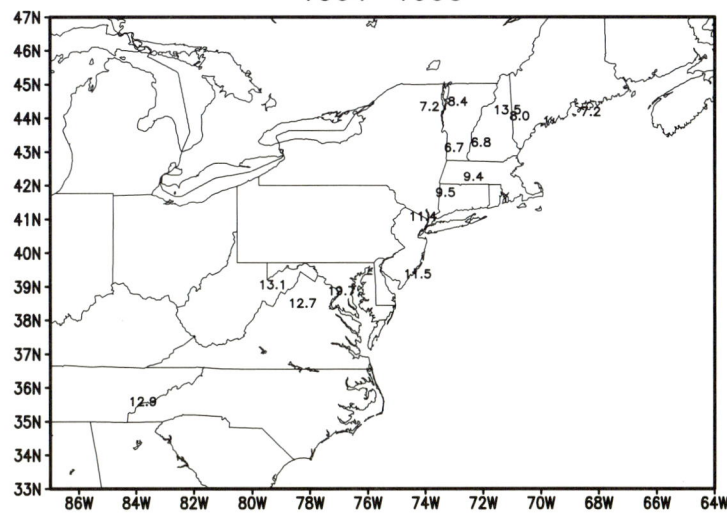

Figure 3
Average daily PM$_{2.5}$ mass (μg m^{-3}) from 1991–1993 at 15 locations in the eastern United States.

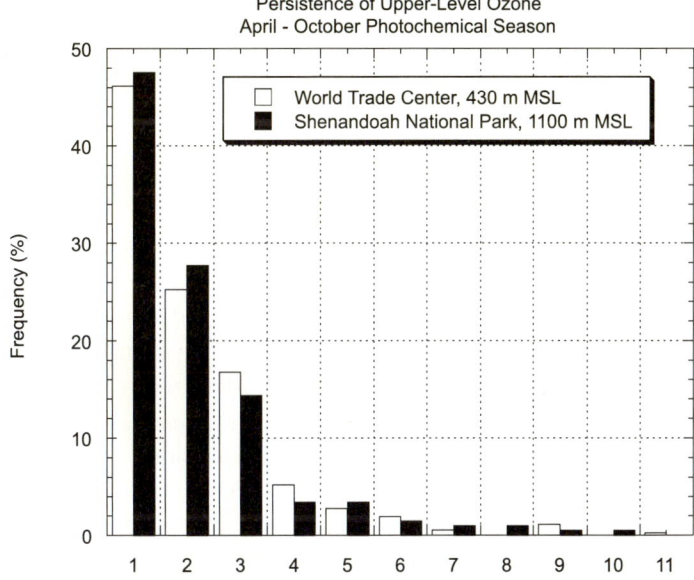

Figure 4
Histogram of the persistence of elevated (> 70 ppbv) daily maximum 8-hour O$_3$ concentrations aloft at the World Trade Center in New York City, NY and Shenandoah National Park, VA during the April to October ozone season from 1983–1998.

through vertical mixing processes, as the surface-based inversion starts to break-up in the morning (ZHANG and RAO, 1999; BERKOWITZ *et al.*, 2000). At both sites high O_3 levels are most likely to persist for 1–3 days, consistent with the results of BERMAN *et al.* (1999) who examined observations from the Sears Tower in Chicago, IL (366 m AGL) and a television tower in North Carolina (411 m AGL). Although the mean duration of O_3 persistence was about two days at the World Trade Center and Shenandoah National Park sites, high O_3 can persist for 10 or more days at both sites. Aloft wind speeds of 10–20 m s^{-1} imply that inert pollutants which are removed from the atmosphere slowly could be transported over distances on the order of 1000 km during the course of 1–2 days. Since O_3 and $PM_{2.5}$ are reactive pollutants, however, their characteristic transport distances may not be as extended.

From 1988–1997, the mean and the standard deviation daytime $PM_{2.5}$ mass at Pinkham Notch (1910 m MSL) at Mt. Washington, NH were 15.8 and 12.6 μg m^{-3}, respectively. Defining concentrations more than one standard deviation above the mean to be "elevated," there were five multi-day periods from 1988–1997; four were 2 days long, one was 4 days long. However, this number could be higher, since during this period there were an additional 15 single "elevated" days with no data available either on the previous or following day. This highlights the need for long-term, high-resolution monitoring (i.e., daily sampling) of $PM_{2.5}$.

There is no *a priori* reason to expect that $PM_{2.5}$ will be associated with shorter time and space scales than O_3. Thus, both O_3 and $PM_{2.5}$ can persist aloft for 2–3 days. One way to refine the estimate of the scales of these pollutants is to examine the degree of correlation between observations over hundreds of kilometers. Such a methodology is presented in the following section.

c. Spatial Analysis of Ozone Observations

To successfully interpret meteorological or ambient pollutant concentration time series data one must extract only those scales appropriate for a given analysis (ESKRIDGE *et al.*, 1997; RAO *et al.*, 1997). Pollutant time series data consist of forcings operating on different time and space scales; they include intra-day, diurnal, synoptic, seasonal, and long-term (trend) components (CHAN *et al.*, 1999; HOGREFE *et al.*, 2000; PORTER *et al.*, 2001). The intra-day component, with a time scale of <10 hours, consists of hour-to-hour changes in emissions, vertical mixing, and chemical reactions, while the diurnal component reflects the day and night differences in the atmospheric photochemical and physical processes. The synoptic-scale component includes the effects of weather systems on pollutant concentrations. The longest scales reflect the meteorological and environmental changes that occur over the course of a year (seasonal), or long-term climate/policy changes (trends).

ESKRIDGE *et al.* (1997) demonstrated the need to extract the short-term components from surface temperature time series data to isolate the effects due to weather-related forcings. By correlating in space the synoptic-scale components of surface temperature data from stations across the eastern United States, the authors showed that such correlations decay with distance and the spatial scale—defined as the e-folding distance in the correlation coefficient from a central location (Washington, DC)—for temperature is about 1000 km, which is consistent with the length scale of the synoptic forcing on the order of 1000 km (HOLTON, 1979). Using the daily maximum 1-hour O_3 concentrations, RAO et al. (1997) showed that along the direction of the prevailing wind, the characteristic length scale for O_3 is on the order of 600 km. We performed a similar analysis, using both daily average $PM_{2.5}$ mass and daily maximum 8-hour O_3 concentrations from various locations throughout the eastern United States. This methodology provides a measure of the coherence in pollutant levels among different monitors embedded within the same synoptic weather pattern.

Since O_3 concentrations are strongly influenced by seasonality, we extracted the synoptic-scale component, which is characterized as having time scales ranging from about 2–21 days, from the O_3 time series data using the Kolmogorov-Zurbenko ($KZ_{m,n}$) filter technique (RAO et al., 1995, 1997; ZURBENKO et al., 1996). This is a low-pass filter technique consisting of n iterations of a moving average with window length m. The parameters m and n are chosen to optimize the separation of different frequency components. Figure 5 shows the transfer function for the $KZ_{13,3}$ filter used in this analysis to extract the different frequencies. For a given frequency, when the

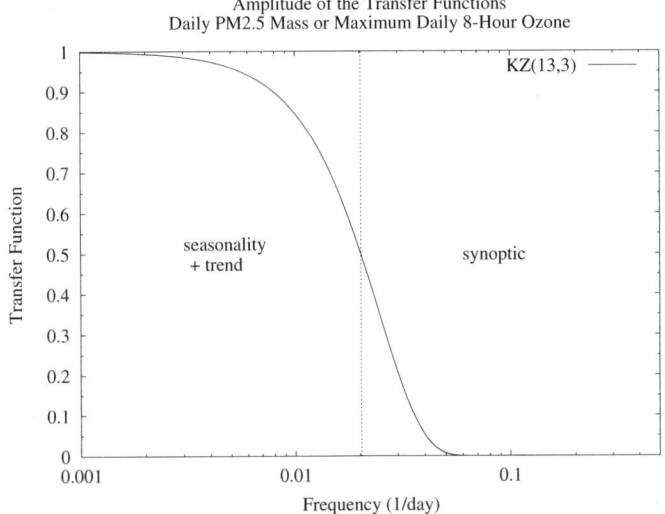

Figure 5

Transfer function for the $KZ_{13,3}$ filter (a window length of 13 days with three iterations leads to a 50% cutoff frequency of about 50 days), used to analyze the synoptic forcings in the daily $PM_{2.5}$ mass and daily maximum 8-hour O_3 data. The dotted line indicates the 50% cutoff frequency of the filter.

value of the transfer function equals one, the frequency components are completely passed through; if the transfer function is zero, the frequencies are filtered out.

The components of the synoptic-scale forcings for O_3 are, then, given by:

$$O_3(\text{synoptic-scale}) = O_3(\text{original}) - KZ_{13,3} \,, \tag{1}$$

where $O_3(\text{original})$ is the original 8-hour daily maximum time series and $KZ_{13,3}$ includes only the seasonal and trend components. The 50% cutoff frequency for this filter is about 50 days. An ideal filter would be a square function tuned to the appropriate cutoff frequency; nonetheless, it is clear from Figure 5 that the filter cannot entirely remove the effects of intra-seasonal scale forcings. However, it is also clear that forcings on time scales exceeding one year are effectively removed, and that forcings on time scales of less than about 20 days are passed through.

Once the five-year long O_3 time series data were spectrally decomposed, we performed a linear regression between the synoptic forcing in the O_3 data from two locations—Pittsburgh and Philadelphia, PA—and every other site included in this analysis. Only the isolines which correspond to an e-folding decrease in the correlation coefficient are shown in Figure 6 for the two sites. Ignoring the contours over Ontario and the ocean, as well as the isolated features over western New York and the Virginias since they are artifacts of the contouring scheme, the correlation coefficients generally fall off along an axis that is oriented from southwest to

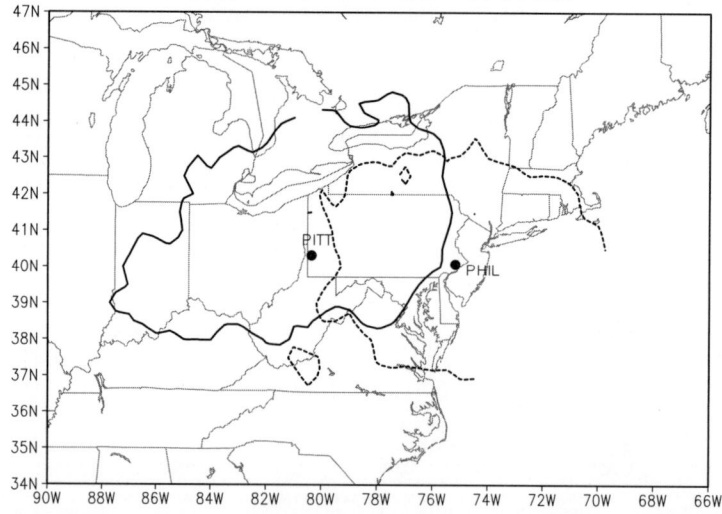

Figure 6

Spatial correlation of the summertime synoptic-scale (2–21 days) components of the O_3 time series between Pittsburgh (solid line) and Philadelphia (broken line) and various locations throughout the eastern United States. Only the isolines corresponding to the e-folding distance in correlation coefficient are shown.

northeast along the urban corridor, the predominant summertime flow in the mid-Atlantic region. There is substantial overlap between the two ozone spatial footprints which cover parts of southern Canada and 14 eastern states. This implies that source attribution for receptor areas such as central New York state can be a complicated problem.

In order to assess the characteristic 1- to 2-day transport distances associated with the synoptic-scale O_3, we performed a time-lagged correlation analysis using the data from Pittsburgh and every other monitor. We lagged each time series with respect to that at Pittsburgh by 0 and 1 days, then determined the number of lags needed to maximize the correlation. Figure 7 shows the locations that maximized the correlation when Pittsburgh lagged by 1 day (triangles) or Pittsburgh led by 1 day (squares). For clarity, a lag of zero, indicative of same day transport, is not shown. Also, only correlation coefficients that are statistically significant at the 95% confidence level are shown in Figure 7. These results suggest that O_3 levels in a region from Virginia to Maine can potentially be affected by emissions in the Pittsburgh area within 1 day; similarly, air masses from as far away as Michigan or the Carolinas have the potential to affect pollutant levels at Pittsburgh. From Figures 6 and 7, it is evident that the airshed for 1- to 2-day transport can encompass the entire north eastern United States.

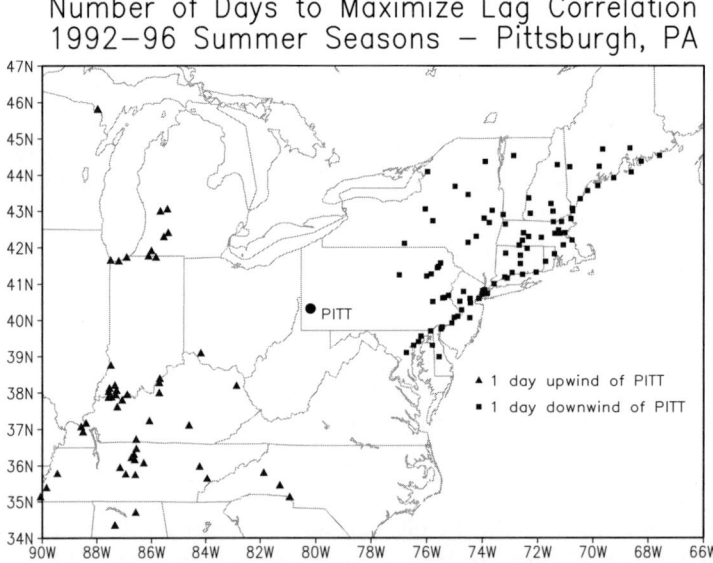

Figure 7

Number of days needed to maximize the summertime synoptic-scale O_3 correlations between Pittsburgh (large dot) and various locations throughout the eastern United States. Only the sites which Pittsburgh lags by 1 day (triangles) or leads by 1 day (squares) are shown, and only the statistically significant (95%) correlation coefficients were considered.

The above results demonstrate that the synoptic-scale components of the O_3 time series at different locations retain a high degree of correlation over hundreds of kilometers. In other words, observations of O_3 from nearby monitors cannot be thought of as statistically independent observations. Thus, individual observations in the immediate vicinity of a city such as Pittsburgh or Philadelphia, while informative from an exposure assessment point of view, will be difficult to interpret from an emissions management perspective.

One limitation associated with the analysis of daily $PM_{2.5}$ mass data is that, until more monitoring locations come on-line and more data become available, $PM_{2.5}$ data are currently both temporally and spatially sparse. On average, these data sets consist roughly of twice per week measurements. Considerable information on the synoptic time scales of about 2–5 days is missing in the $PM_{2.5}$ database. Therefore, we refer to the weather-induced variations in the $PM_{2.5}$ data time series as intra-seasonal, rather than synoptic-scale variations.

In order to extract the intra-seasonal scale forcings from these time series, we applied the same $KZ_{13,3}$ filter, and the intra-seasonal forcings embedded in these pollutant data are given by:

$$PM_{2.5}(\text{intra-seasonal scales}) = PM_{2.5}(\text{original}) - KZ_{13,3} \ . \tag{2}$$

Since we only included fifteen sets of $PM_{2.5}$ time series in this analysis, we performed a pairwise correlation between all locations, rather than a spatial correlation from one central point. Figure 8(a) shows the correlation coefficients for O_3 as a function of distance from Philadelphia, while Figure 8(b) displays the pairwise correlation coefficients for $PM_{2.5}$ as a function of distance from each location. Best fit lines are also shown for both O_3 and $PM_{2.5}$. Only O_3 data that fell along the southwest-to-northeast orientation (prevailing wind direction in the summer) are plotted, while all pairwise correlations for $PM_{2.5}$ are shown, since the $PM_{2.5}$ monitoring locations are generally oriented along this prevailing direction.

From Figures 8(a) and 8(b), it is evident that the spatial extent (e-folding distance for the correlation coefficient) for the synoptic forcing in O_3 and $PM_{2.5}$ data is on the order of 500–600 km. With a limited sampling schedule and so few monitoring locations, the synoptic information in the $PM_{2.5}$ data on time scales of about 2–3 days has been lost. Hence, the spatial scale for $PM_{2.5}$ should be viewed as a lower estimate.

The analysis presented in this section is statistically-based (i.e., correlations), and as such does not explicitly take the physical and chemical processes (i.e., cause-and-effect) that affect pollutant concentration levels into account. Although we can associate correlation with pollutant transport, this is purely a statistical analysis, and, therefore, does not provide a mechanism for the pollutant transport. Therefore, we adopt a mixed-layer modeling approach which represents a simple physical basis for the pollutant transport.

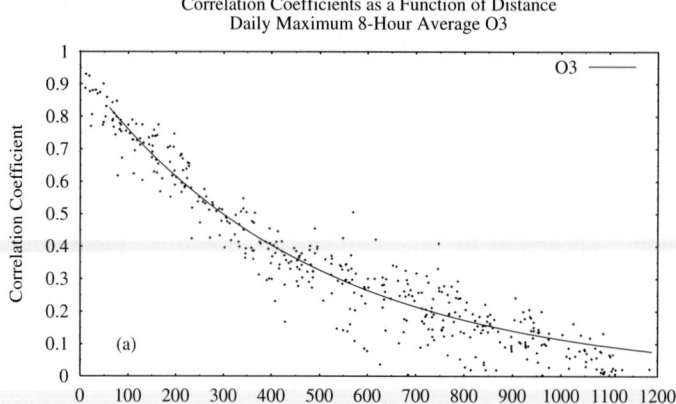

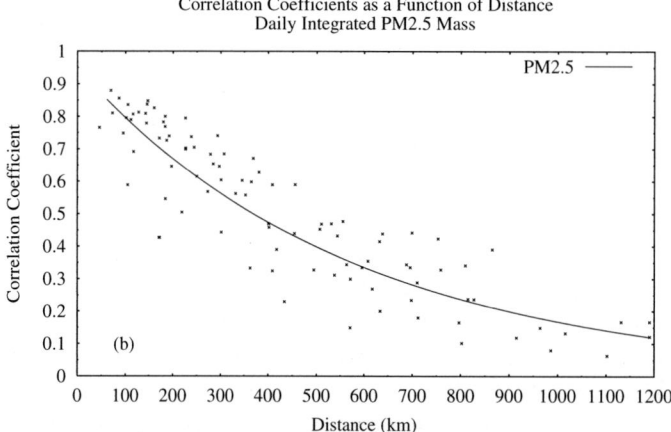

Figure 8

(a) Correlation coefficients between summertime synoptic forcings in O_3 between Philadelphia and all other sites along prevailing flow direction, as a function of distance from Philadelphia. Both the data points and a best-fit line are shown. (b) Pairwise correlation coefficients of intra-seasonal $PM_{2.5}$ mass as a function of distance from each location. Both the data points and a best-fit line are shown.

d. Mixed-layer Trajectory Analysis

For each summertime day during the 1992 to 1996 period, forward trajectories from Philadelphia and Pittsburgh were computed with the HYbrid Single Particle Lagrangian Integrated Trajectory (HY-SPLIT, Version 3.0) model (DRAXLER, 1992), using the National Weather Service's Nested Grid Model (NGM) data, archived every two hours with a horizontal grid dimension of approximately 180 km. All 1-day mixed-layer trajectories were started at 1200 EST at a height of 500 m above the surface. After all trajectories were generated, a 2.5 longitude × 2.0

latitude grid was superimposed upon the eastern US region, and the number of times that a trajectory passed through each grid cell was counted. The distribution of trajectories at each grid location was then normalized to the grid cell surrounding Pittsburgh and Philadelphia (the respective cells containing the maximum number of trajectory "hits").

These trajectories were calculated with coarsely-gridded and temporally-sparse meteorological data, and as such, the absolute error associated with these trajectories increases roughly linearly with time and/or distance from the source (STUNDER, 1996). The errors associated with individual trajectories can typically be on the order of 100–200 km, or on the order of 30% of the trajectory path length, even for 1–2 day trajectories (STOHL, 1998). By considering five years of summertime trajectories, the approximate summertime airshed for 1-day transport in the mixed layer can be identified. In other words, we define the approximate region that can be influenced by 1-day transport from the two locations by focusing on the entire set of trajectories on a climatological basis because of the inherent uncertainty associated with individual trajectories.

Figure 9 displays the spatial extent over which the trajectories originating from Pittsburgh and Philadelphia were at least 10% as likely to fall; such a region will encompass most of the trajectories without showing the outliers. For example, the region about Philadelphia includes parts of 12 states extending from northern

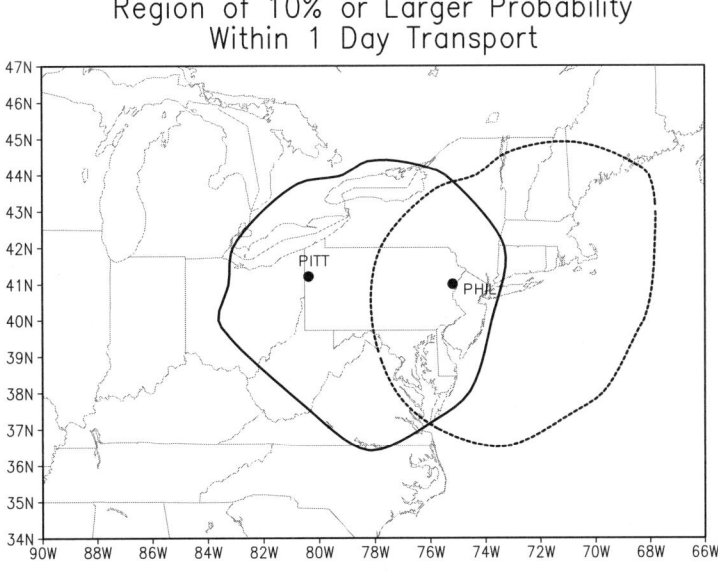

Figure 9

Regions within which 1-day HYSPLIT forward trajectories during the summer seasons (June 1 through August 31) of 1992–96, originating from Philadelphia (broken line) and Pittsburgh (solid line), are at least 10% as likely to fall with respect to their maxima.

Virginia to southern Maine. As with the correlation analysis in the previous section, there is substantial overlap between the two airsheds, and a general southwest-to-northeast orientation of the trajectory distributions, reflecting the prevailing flow during the summer. Because of the substantial overlap of the airsheds of pollutant transport, source attribution would be a difficult task for the mid-Atlantic region through southeastern New York state.

The distance from Philadelphia to the edge of the contour along the major axis is approximately 600 km. This finding suggests that emission sources from a city such as Philadelphia have the potential to affect pollution levels across a considerably larger area within 1-day of transport and that the pollutant transport problem is a regional, national, and international issue. From a regulatory point of view, a minimum of 11 surrounding states should have a vested interest in any emissions control program envisioned for the Philadelphia area. Similarly, emissions from Pittsburgh can affect pollution levels in southern Canada and at least seven surrounding US states, making it especially difficult to attribute pollutant levels in receptor areas of central Pennsylvania, New York, and Ontario to specific emission sources.

After compiling the five summers worth of trajectories, we clustered them according to the methodology used by BRANKOV *et al.* (1998) to further refine the intersite correlation methodology in the previous section. In short, average trajectory flow directions are computed by first assigning all trajectories to representative "seed" clusters, merging the closest average trajectories, then iterating until all trajectories have been assigned to their most representative cluster. The purpose was to correlate the synoptic-scale components of O_3 only on days when the synoptic flow pattern was generally west to east, a condition conducive to precursor transport and O_3 formation. While this procedure is more rigorous than simply correlating the synoptic-scale components of O_3 during all transport regimes, the results were remarkably similar to those from the previous section, and, therefore, are not shown here.

The above mixed-layer trajectory modeling approach simply describes the large-scale air movement only. As with the correlation analysis in the previous section, this methodology does not take into account complex, high-resolution physical and chemical processes affecting the pollutant concentrations. To do so, one needs to use three-dimensional modeling of transport and dispersion to better characterize the relevant atmospheric processes.

e. Three-dimensional Mesoscale Dispersion Modeling

One limitation associated with the mixed-layer trajectory approach outlined above is that the model relies on coarse-resolution input meteorological fields, namely, the NGM data for the historical period of 1992–96, archived with two-hour time resolution and ~ 180 km horizontal grid spacing. Since terrain features and atmospheric flows can vary over much smaller scales, linkage to a fine-grid mesoscale meteorological model would be advantageous.

The Mesoscale Dispersion Modeling System (MDMS) is a Lagrangian model (ULIASZ, 1994) which utilizes three-dimensional meteorological fields from the Regional Atmospheric Modeling System (RAMS) (PIELKE *et al.*, 1992). The MDMS releases tracer particles at a fixed rate and tracks them in three-dimensional space over time. For this analysis, we rely on the season-long meteorological modeling performed by KALLOS *et al.* (1997) using RAMS for the eastern United States with 12 km horizontal grid dimensions to drive the dispersion model, focusing on three O_3 episodic periods in 1995—June 18–20, July 13–15, and July 31–August 2. In this analysis, we attempt to define the potential region that a city such as Philadelphia can affect within 1–2 days of transport through forward trajectory modeling, as well as define the region which can potentially impact air quality in Philadelphia through backward trajectory modeling.

We initialized each forward trajectory at 1200 EST on June 18, July 13, and July 31 for the three episodes, and allowed the model to proceed for 48 hours. The particle release rate was set to 2 s^{-1}. Figure 10(a) displays the particle distributions in the

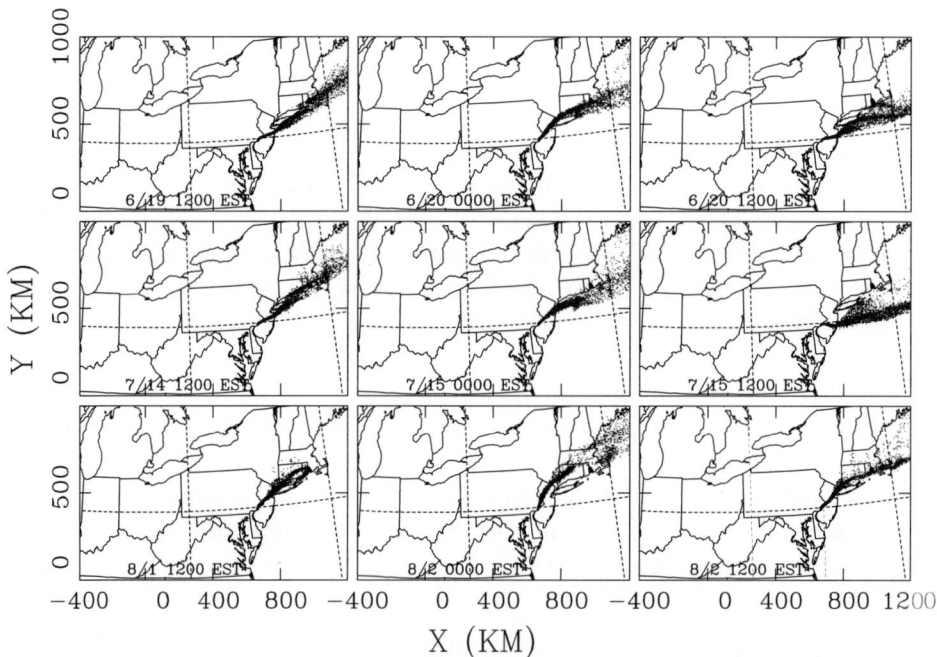

Figure 10a

(a) Distribution of tracer particles released from Philadelphia in the near-surface (0–500 m). The columns show the particle plumes 24, 36, and 48 hours after release, from left to right. The rows represent the three episodes (6/18–20, 7/13–15, and 7/31–8/2), from top to bottom. Trajectories were started at 1200 EST on the first day and allowed to proceed for 48 hours. (b) Same as (a), except for the 500–1000 m layer.

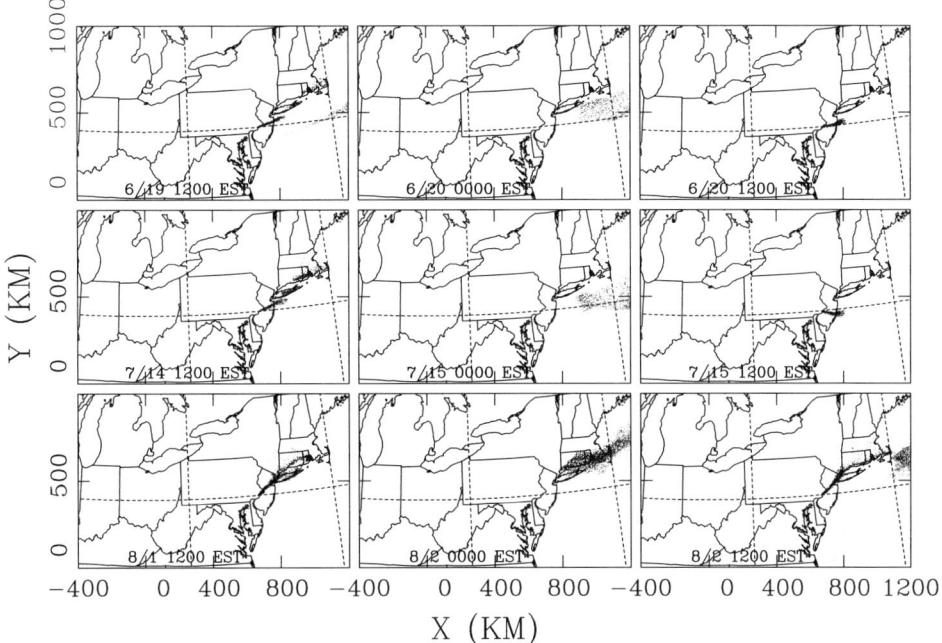

Figure 10b

lower part of the boundary layer (0–500 m)—shown from left to right—after 24, 36 and 48 hours. The three periods are shown with the June episode along the top row and the July/August episode along the bottom row. Figure 10(b) presents the same information for the residual (500–1000 m) layer. In the lower layer, the distribution of particles does not change appreciably throughout the day, and follows the prevailing flow direction. Within 12–24 hours, tracer particles from Philadelphia have already reached Massachusetts, and can be observed about 600 km downwind, off the coast of southern Maine.

The situation is different for the particle distributions in the upper part of the boundary layer. During the nighttime and early morning hours this residual layer has become detached from the surface layer, and particles in the vicinity of the source region are confined to the surface layer. There are effectively no particles in the residual layer over the immediate Philadelphia area; any particles in the residual layer have been transported far downwind of Philadelphia. During the afternoon hours, the two layers become well-mixed and coupled, so that surface emissions can reach the upper parts of the planetary boundary layer.

Previous studies by KLEINMAN et al. (1994), ZHANG and RAO (1999), BERKOWITZ et al. (2000), and RAO et al. (2003) demonstrated that surface concentrations of pollutants such as O_3 are highly dependent on pollutant levels aloft. ZHANG and RAO (1999) used various observational and modeling techniques, including a similar dispersion modeling approach to investigate the role of vertical mixing on surface O_3 concentrations, and found that during the morning evolution of the boundary layer, O_3-rich air trapped in the residual layer mixes downward. RAO *et al.* (2003) found that on days with high surface O_3 concentrations, the O_3 levels aloft were about 40% higher than corresponding days with low surface O_3, and that O_3 aloft has an approximate atmospheric residence time of 2–3 days. It appears that the dispersion model used here is able to simulate the boundary layer processes which affect surface layer pollutant concentrations described by these authors.

The predicted tracer particle concentration fields depend upon the locations of the emission sources, local meteorological fields, and terrain/surface features. To examine how these concentration fields can vary in space, we released particles from seven additional locations—Pittsburgh; Richmond, VA; Washington, DC; Columbus and Cincinnati, OH; Detroit, MI; Toronto, Ont., Canada. The release rate at each location was kept at 2 s^{-1}.

As expected, within 24 hours, the Detroit/Toronto, the Ohio/Pittsburgh, and the Richmond/Washington/Philadelphia plumes tend to merge together. Figure 11 shows the simulated particle concentrations, averaged over the height of the boundary layer (from 0–1000 m) over 24-hour intervals, for the three episodes. From left to right, the panels display the locations, the average during hours 0–24, and the average during hours 24–48. Figure 11 illustrates the difficulty in attributing the particle concentrations in New England to any of the eight point sources individually by the second day. Each tracer particle plume extends over distances of > 500 km by the second day of transport, and, thus, has the potential to contribute to observed pollutant levels over such distances. Note that there is a substantial overlap between the individual particle airsheds.

The above analysis is source-oriented, however the MDMS can also be used for receptor-oriented analysis. As further evidence that urban areas such as Philadelphia are not isolated, we performed corresponding backward trajectories from Philadelphia, keeping the initial time and particle release rate the same as with the forward trajectories. By dispersing particles backward in time from Philadelphia, one can estimate the influence function (ULIASZ, 1994), which depicts the areal extent of the emission sources which have the potential to impact a city such as Philadelphia.

Figure 12 displays the influence function averaged from 0–1000 m over 24-hour intervals, from left to right, during the three episodes—0–24 hours earlier and 24–48 hours earlier. Trajectories were started on June 20, July 14, and August 2, respectively, at 1200 EST. Within two days of transport, sources 600 km south of Philadelphia have the potential to contribute to pollution levels observed over Philadelphia. This region includes the urban centers of Richmond, Washington, Baltimore, MD, and Wilmington, DE. Hence, while local emissions will certainly contribute to O_3 and $PM_{2.5}$ levels observed in a city such as Philadelphia, one cannot simply attribute pollutant levels downwind of Philadelphia to emissions from

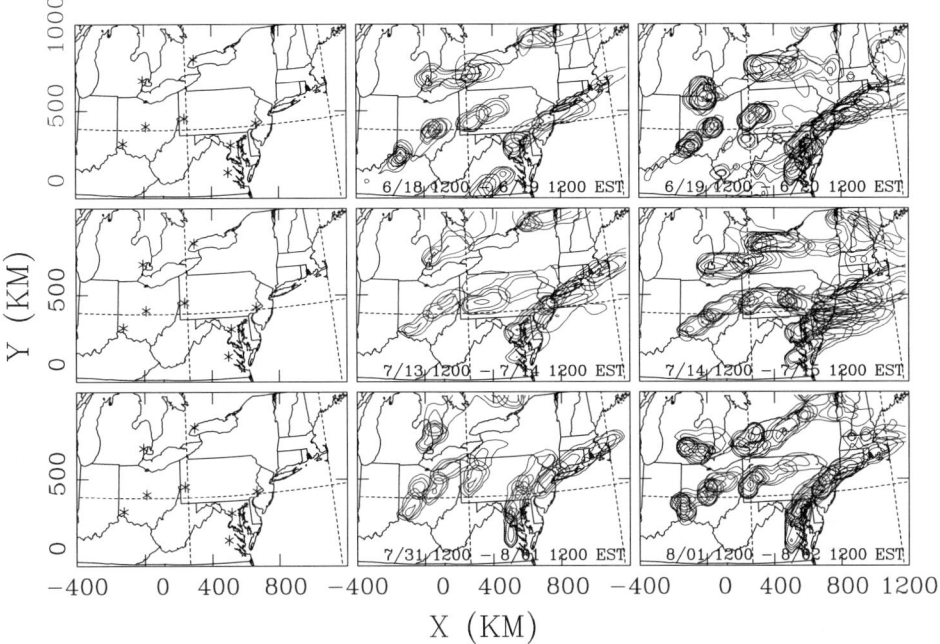

Figure 11

Simulated tracer particle concentrations of forward trajectories, averaged over 24 hours from 0–1000 m, for the three episodes. Particles were released on 6/18, 7/13, and 7/31 at 1200 EST from eight sites: Philadelphia and Pittsburgh, PA; Columbus and Cincinnati, OH; Detroit, MI; Toronto, Ont.; Richmond, VA; and Washington, DC. The panels from left to right show the locations, the average particle plumes 0–24 hours after release, and the average particle plumes 24–48 hours after release.

Philadelphia alone. Again, emissions management policies implemented some 600 km upwind will have the potential to affect air quality in Philadelphia itself.

f. Impact of Emission Reductions through a Three-dimensional Photochemical Model Simulation

The analysis presented above is most pertinent to passive, non-reactive pollutants. For reactive pollutants, such as O_3 and $PM_{2.5}$, the problem is considerably more complex because of the photochemistry. To quantitatively describe the potential airshed for a variety of pollutants, one needs to perform three-dimensional modeling, reflecting atmospheric physical and chemical processes.

In addition to linking the seasonal RAMS simulation to the MDMS, we used the RAMS meteorological fields to drive seasonal photochemical simulations with the Urban Airshed Model (UAM-V) (SAI 1995), a three-dimensional photochemical model that uses the Carbon Bond IV chemical reaction mechanism. Figure 13 displays the 36-km model domain, and details of the UAM-V simulation and model performance can be found in SISTLA *et al.* (2001) and HOGREFE *et al.* (2000).

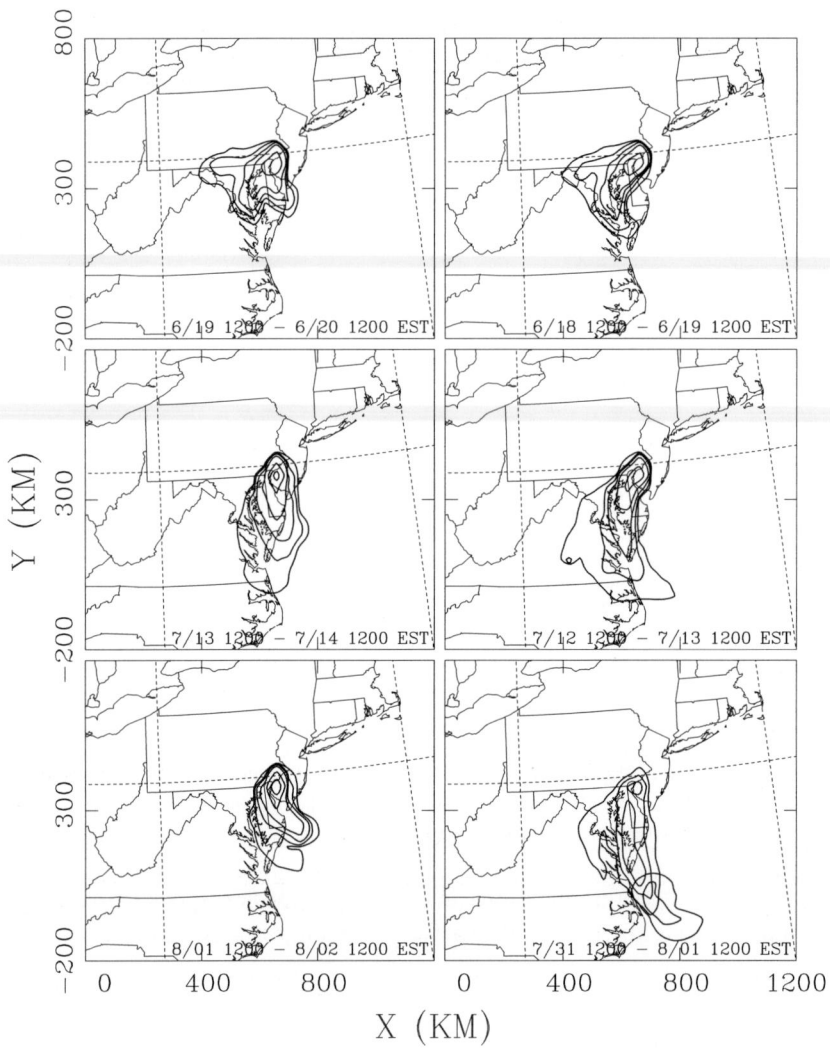

Figure 12

Simulated tracer particle concentrations of backward trajectories (influence functions), averaged over
24 hours from 0–1000 m, for the three episodes. Particles were released on 6/20, 7/14, and 8/2 at 1200 EST
from Philadelphia. The panels show the average influence function 0–24 hours after release, and 24–48
hours after release.

Employing the same 1995 anthropogenic and biogenic emissions inventory as
HOGREFE *et al.* (2000), a base case photochemical simulation was carried out,
covering much of the eastern United States for the entire summer of 1995. To
assess the effects of hypothetical reductions in anthropogenic emissions, we then
removed all anthropogenic emissions from two 26 grid cell × 22 grid cell regions
(shaded areas in Fig. 13) that covered the greater Philadelphia and Pittsburgh

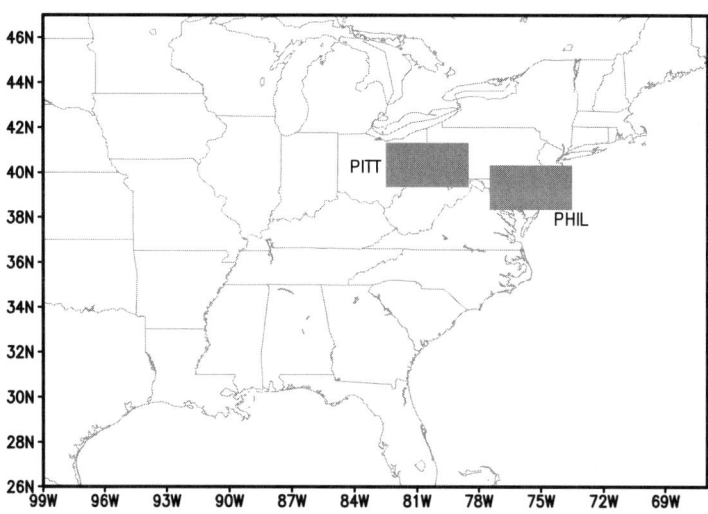

Figure 13

The UAM-V 36-km modeling domain. The shaded boxes indicate where anthropogenic emissions were removed from the Pittsburgh and Philadelphia areas, before proceeding with the two sensitivity simulations.

areas. Emissions were removed individually from these two regions. RAO *et al.* (1998) used a similar approach to compare the spatial extent of the ozone footprint derived from observations, with that of the hypothetical emission reductions from photochemical modeling analysis. The purpose of these sensitivity simulations was to determine the spatial extent to which these "across-the-board" or uniform anthropogenic emission reductions in the greater Philadelphia and Pittsburgh areas could potentially affect O_3 concentrations throughout the region, while keeping the meteorology constant.

It has been shown that in both observational and model-predicted pollutant time series data, only those components whose temporal scales are greater than or equal to one day are relevant from a policy-making perspective (HOGREFE *et al.*, 2000). Over the entire 89-day modeling period and at each grid cell, we computed the mean and standard deviation of the difference between the base case daily maximum 8-hour O_3 and each sensitivity case daily maximum 8-hour O_3 ($\Delta O_3 = O_3(base) - O_3(sensitivity)$), and the upper 95% confidence interval for ΔO_3 was computed. We then normalized by the largest differences—24 ppb in the Pittsburgh region, 27 ppb in the Philadelphia region. Figure 14 shows only the contours denoting the 95% probability region within which O_3 improvements of at least 10% of the maximum decreases, corresponding to improvements of about 2–3 ppb, would be realized.

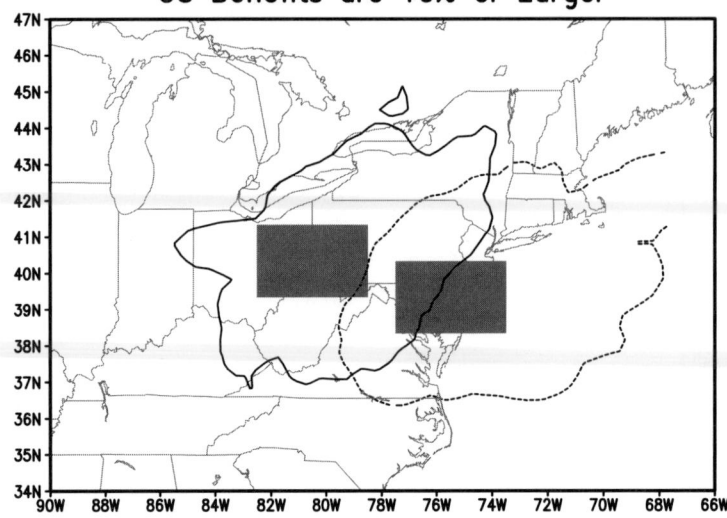

Figure 14

The 95% probability region within which the O_3 improvement (base case minus sensitivity case) is at least 10% of the maximum improvement, over the 1995 summer season. The maximum O_3 improvement was 24 ppb in the Pittsburgh region (solid line), and 27 ppb in the Philadelphia region (broken line).

As expected, the largest O_3 reductions are evident in the near-field where anthropogenic emissions were eliminated, and the O_3 reductions fall off along a general southwest-to-northeast orientation. Even though these anthropogenic emission reductions are small across the entire domain (roughly 10% reductions in NO_x and VOC emissions), their effects can be felt as far away as the central and Cape Cod regions of Massachusetts. Throughout much of southeastern New York and southern New England, average O_3 improvements are in the range of 10–20% of the maximum changes in the near-field. This implies that the airshed for O_3 precursor emissions released from cities such as Philadelphia and Pittsburgh is on the order of 600 km.

The northeastern urban corridor, from Richmond to Boston, MA (with Philadelphia as the approximate center) contains many pollution sources which cannot be isolated from each other, as evident from the previous analysis involving three-dimensional dispersion modeling. The emissions reduction scenario presented in this photochemical modeling analysis is, obviously, hypothetical. However, these photochemical simulations demonstrate that the effects of emissions from one location can be seen hundreds of kilometers downwind. HOGREFE *et al.* (2000) demonstrated that the emissions control strategy proposed by the USEPA as part of its NO_x state implementation plan (SIP) call (USEPA, 1999) provides the largest reductions in the synoptic-scale and longer-term forcings embedded in the O_3 time series.

4. Discussion

Using various approaches, we have demonstrated that the spatial scales associated with ozone and other pollutants are on the order of several hundreds of kilometers in the northeastern United States, as evident from the overlapping airsheds. While each methodology has its own limitations, the different analyses—a combination of observational and modeling approaches—consistently show that O_3 and $PM_{2.5}$ pollution is a regional, multi-state, and even international issue, not simply an urban or local problem. That is, the areal extent of the airshed is about 600 km in this region. Such scales imply that over distances of less than about 100 km, it is difficult to unequivocally attribute sources or distinguish pollution levels observed at upwind, city center, and downwind locations. For example, emissions from Baltimore, MD and Washington, DC have the potential to affect Philadelphia, PA which, in turn, can impact pollution levels downwind throughout New York state and New England.

From an epidemiological perspective, a spatially-dense observational database at the local level or neighborhood scale would yield important information, since emissions, traffic and population patterns, and local meteorological or surface features are city-specific. However, observations of pollutant concentrations from such a dense network are neither temporally nor spatially independent of each other. In terms of O_3, the existing routine monitoring network focuses primarily on urban and suburban areas, with scant coverage in rural areas throughout the eastern United States. For $PM_{2.5}$, high resolution (daily) observations and more monitors in all areas are needed to address the regionality of these issues.

Since emissions from one locality have the potential to affect a vastly larger region, hundreds of kilometers downwind, individual consolidated statistical Metropolitan areas (CSMAs) or states cannot solve these pollution problems by themselves. Given the overlapping regions of influence or airsheds, only through regional air management partnerships (RAMPs) would we be able to devise effective means of reducing pollution in the eastern United States. Until such a regional perspective is adopted and RAMPs are implemented, it will be difficult to evaluate the relative efficacies of individual, low-level emission reduction strategies that vary from one location to another in improving air quality.

Acknowledgements

This work was supported in part by the New York State Energy Research and Development Authority under agreement numbers 4914-ERTER-ES99 and 6085-ERTER-ER-00, and the United States Environmental Protection Agency under grant number R826731476. The views expressed in this paper do not necessarily reflect those of the supporting agencies. The authors wish to thank R. Henry,

E. Brankov, C. Hogrefe and W. Hao for their technical assistance, as well as B. Hill of the Appalachian Mountain Club for providing the data from Mt. Washington, NH.

REFERENCES

APPLETON, E. L. (1995), *A Cross-media Approach to Saving the Chesapeake Bay*, Environ. Sci. and Technol. *29*, 550A–555A.

BERKOWITZ, C. M., FAST, J. D., and EASTER, R. C. (2000), *Boundary Layer Vertical Exchange Processes and the Mass Budget of Ozone: Observations and Model Results*, J. Geophys. Res. *105*, 14,789–14,805.

BERMAN, S., KU, J.-Y., and RAO, S. T. (1999), *Spatial and Temporal Variation in the Mixing Depth over the Northeastern United States during the Summer of 1995*, J. Appl. Meteorol. *38*, 1661–1673.

BRANKOV, E., RAO, S. T., and PORTER, P. S. (1998), *A Trajectory-clustering-correlation Methodology for Examining the Long-range Transport of Air Pollutants*, Atmos. Environ. *32*, 1525–1534.

CHAMEIDES, W. L., SAYLOR, R. D., and COWLING, E. B. (1997), *Ozone Pollution in the Rural U.S. and the New NAAQS*, Science *276*, 916.

CHAN, D., RAO, S. T., ZURBENKO, I. G., and PORTER, P. S., *Linking changes in ozone to changes in emissions and meteorology*. In *Air Pollution VI* (C. A. Brebbia, M. Jacobson, and H. Power, eds.) (WIT Press, 1999) pp. 663–675.

DENNIS, R. L. (1997), *Using the regional acid deposition model to determine the nitrogen deposition airshed of the Chesapeake Bay Watershed*. In *Atmospheric Deposition of Contaminants to the Great Lakes and Coastal Waters* (J. E. Baker, ed.), Proceedings from the Society of Environmental Toxicology and Chemistry (SETAC) 15th Annual Meeting, November, 1994, Denver, CO, pp. 293–304.

DRAXLER, R. R. (1992), *Hybrid single-particle Lagrangian integrated trajectories (HY-SPLIT): Version 3.0—User's guide and model description. NOAA Technical Memorandum* ERL-ARL-195, Air Resources Laboratory, Silver Spring, MD.

ESKRIDGE, R. E., KU, J. Y., RAO, S. T., PORTER, P. S., and ZURBENKO, I. G. (1997), *Separating Different Scales of Motion in Time Series of Meteorological Variables*, Bull. Am. Met. Soc. *78*, 1473–1483.

HOGREFE, C., RAO, S. T., ZURBENKO, I. G., and PORTER, P. S. (2000), *Interpreting the Information in Ozone Observations and Model Predictions Relevant to Regulatory Policies in the Eastern United States*, Bull. Am. Met. Soc., to appear.

HOLTON, J. R., *An Introduction to Dynamic Meteorology (2nd ed.)* (Academic Press, San Diego, CA, 1979) 391 pp.

KALLOS, G., LAGOUVARDOS, K., and KOTRONI, V. (1997), *Atmospheric Modeling Simulations over the Eastern United States with the RAMS3b Model for the Summer of 1995*. Final report to Electric Power Research Institute (EPRI), Palo Alto, CA.

KLEINMAN, L., LEE, Y.-N., SPRINGSTON, S. R., NUNNERMACKER, L., ZHOU, X., BROWN, R., HALLOCK, K., KLOTZ, P., LEAHY, D., LEE, J. H., and NEWMAN, L. (1994), *Ozone Formation at a Rural Site in the Southeastern United States*, J. Geophys. Res. *99*, 3469–3482.

PIELKE, R. A., COTTON, W. R., WALKO, R. L., TREMBACK, C. J., NICHOLLS, M. E., MORAN, M. D., WESLEY, D. A., LEE, T. J., and COPELAND, J. H. (1992), *A Comprehensive Meteorological Modeling System—RAMS*, Meteor. Atmos. Phys. *49*, 69–91.

PORTER, P. S., RAO, S. T., ZURBENKO, I. G., DUNKER, A. M., and WOLFF, G. T. (2001), *Ozone Air Quality over North America: Part II – An Analysis of Trend Detection and Attribution Techniques*, J. Air and Waste Manage. Assoc. *51*, 283–306.

RAO, S. T., KU, J. Y., BERMAN, S., ZHANG, K., and MAO, H. (2003), *Summertime Characteristics of the Atmospheric Boundary-layer Evolution and Relationships to Ozone Levels over the Eastern United States*, Pure and Appl. Geophys., this issue.

RAO, S. T., ZALEWSKY, E. E., and ZURBENKO, I. G. (1995), *Determining Temporal and Spatial Variations in Ozone Air Quality*, J. Air and Waste Manage. Assoc. *45*, 57–61.

Rao, S. T., Zalewsky, E. E., Zurbenko, I. G., Porter, P. S., Sistla, G., Hao, W., Zhou, N., Ku, J.-Y., Kallos, G., and Hansen, D. A., *Integrating observations and modeling efforts in ozone management efforts*. In *Air Pollution Modeling and Its Application XII* (S.-E. Gryning and N. Chaunerliac, eds.) (Plenum Press, New York, 1998) pp. 115–124.

Rao, S. T., Zurbenko, I. G., Neagu, R., Porter, P. S., Ku, J. Y., and Henry, R. F. (1997), *Space and Time Scales in Ambient Ozone Data*, Bull. Am. Met. Soc. *78*, 2153–2166.

Ryan, W. F., Doddridge, B. G., Dickerson, R. R., Morales, R. M., Hallock, K. A., Roberts, P. T., Blumenthal, D. L., Anderson, J. A., and Civerolo, K. L. (1998), *Pollutant Transport During a Regional O₃ Episode in the Mid-Atlantic States*, J. Air. and Waste Manage. Assoc. *48*, 786–797.

Saylor, R. D., Chameides, W. L., and Cowling, E. B. (1998), *Implications of the New Ozone National Ambient Air Quality Standards for Compliance in Rural Areas*, J. Geophys. Res. *103*, 31,137–31,141.

Sistla, G., Hao, W., Ku, J.-Y., Kallos, A., Zhang, K., Mao, H., and Rao, S. T. (2001), *An Operational Evaluation of Two Regional-scale Ozone Air Quality Modeling Systems over the Eastern United States*, Bull. Am. Met. Soc. *82*, 945–964.

Stohl, A. (1998), *Computation, Accuracy, and Applications of Trajectories—A Review and Bibliography*, Atmos. Environ. *32*, 947–966.

Stunder, B. J. (1996), *An Assessment of the Quality of Forecast Trajectories*, J. Appl. Meteorol. *35*, 1319–1331.

Systems Applications International (SAI) (1995), *Users Guide to the Variable Grid Urban Airshed Model (UAM-V)*, San Rafael, CA.

Uliasz, M. (1994), *Lagrangian particle dispersion modeling in mesoscale applications*. In *Environmental Modeling II* (P. Zannetti, ed.), Computational Mechanics Publications, pp. 71–102.

United States Environmental Protection Agency (USEPA) (1996a), *National Ambient Air Quality Standards for Ozone: Proposed Decision*, Federal Register *61*, 65,717–65,750.

United States Environmental Protection Agency (USEPA) (1996b), *National Ambient Air Quality Standards for Particulate Matter: Proposed Decision*, Federal Register *61*, 65,638–65,713.

United States Environmental Protection Agency (USEPA) (1999), Draft report on the use of models and other analyses in attainment demonstrations for the 8-hour ozone NAAQS. EPA-44/R-99-004, May 1999, Research Triangle Park, NC.

Vukovich, F. M. (1994), *Boundary Layer Ozone Variations in the Eastern United States and their Association with Meteorological Variation: Long-term Variation*, J. Geophys. Res. *99*, 16,838–16,850.

Zhang, K., Mao, H., Civerolo, K., Berman, S., Ku, J. Y., Rao, S. T., Doddridge, B., Philbrick, C. R., and Clark, R. (2001), *Numerical Investigation of Boundary-Layer Evolution and Nocturnal Low-Level Jets: Local VERSUS Non-Local Schemes*, Environ. Fluid Mech. *1*, 171–208.

Zhang, J. and Rao, S. T. (1999), *The Role of Vertical Mixing in the Temporal Evolution of Ground-level Ozone Concentrations*, J. Appl. Meteorol. *38*, 1674–1691.

Zhang, J., Rao, S. T., and Daggupaty, S. M. (1998), *Meteorological Processes and Ozone Exceedances in the Northeastern United States during the 12–16 July 1995 Episode*, J. Appl. Meteorol. *37*, 776–789.

Zurbenko, I., Porter, P. S., Rao, S. T., Ku, J. Y., Gui, R., and Eskridge, R. E. (1996), *Detecting Discontinuities in Time Series of Upper Air Data: Development and Demonstration of an Adaptive Filter Technique*, J. Climate *9*, 3548–3560.

(Received March 1, 2000, accepted July 25, 2000)

To access this journal online:
http://www.birkhauser.ch

Pure appl. geophys. 160 (2003) 107–116
0033–4553/03/020107–10

© Birkhäuser Verlag, Basel, 2003

▌Pure and Applied Geophysics

Atmospheric Particulate Concentration Measured in an Urban Area Bandung

Puji Lestari[1] and Savitri[1]

Abstract — Atmospheric particulate concentration for total suspended particles (TSP) and for PM_{10} (particulate matter under 10 micron) was measured in Jalan Braga and ITB campus, Bandung. Six samples were collected over one- or two-day time periods using High Volume Sampler (HVS) for TSP and Low Volume Sampler (LVS) or Anderson Cascade Impactor for PM_{10}. Samples were further analyzed to determine concentrations of metals, sulfate and nitrate. Concentration of NO_x (NO and NO_2) was also measured hourly and simultaneously during the sampling period. The results from this study show that the atmospheric particulate concentration in Jalan Braga for TSP ranged from 304.04 to 363.17, and for PM_{10} concentration ranged from 277.02 to 336.44 $\mu g/m^3$. The lead concentrations were 1.42–2.37 $\mu g/m^3$ in the TSP and 0.81–1.57 $\mu g/m^3$ in the PM_{10}. The nitrate concentrations were 5.89–6.51 $\mu g/m^3$ and 2.27–3.45 $\mu g/m^3$ for the TSP and PM_{10}, respectively. The hourly NO_x concentration varied between 0.14–0.35 ppm. The total elements (metals, sulfate and nitrate) found in the samples contribute from 20 to 25% of the total particulate concentration.

Key words: Suspended particulate (TSP), PM_{10}, nitrate, sulfate, lead, atmospheric particulate concentration.

Introduction

Atmospheric particulate concentration is one of the major pollutants concerned in the ambient air due to the effect of the material to the environment and human health. Particulate with size particle less than 10 μm diameter could penetrate the human respiration system and cause lung damage. Particles larger than 10 μm diameter generally do not penetrate deep into the lungs. However, these size particles cause the major dry deposition to the earth surface (Lin *et al.*, 1993; Noll and Fang, 1989; Noll *et al.*, 1997) which cause the problems to the environment.

The air pollution problem in developing countries such as Indonesia is becoming a major issue in major cities such as Bandung, Jakarta and Surabaya, due to the urban activities. The major sources of the air pollution problem in an urban area are mainly from the transportation and industrial sectors. From previous study in Jakarta by (PCI Report 1997), the transportation sector contributed 83% of the fine

---

[1] Department of Environmental Engineering, Bandung Institute of Technology (ITB), JL.
Ganesha No. 10 Bandung – Indonesia. E-mail: pujilest@indo.net.id

particles concentration in Jakarta. There have been many measurements concerning the total suspended particulate matter in Indonesia. However, the study that measured the PM_{10} (< 10 μm diameter) simultaneously with the TSP measurement and other pollutants caused by transportation (Pb, CO and NO_x) have not been done yet.

The objective of this study was to characterize the atmospheric particulate concentration for total suspended particulate and PM_{10}. Samples of airborne particles from two sampling locations were collected with high volume sampler, low volume sampler, and Anderson cascade impactor. NO_x and CO concentration was also measured simultaneously with the airborne concentration.

Results from this study demonstrate that the concentration of atmospheric particulate in Jalan Braga exceeds the ambient air quality standard (260 μg/m^3) and the PM_{10} contribute from 75 to 95% of the total suspended particulate matter in both sampling sites. The results of this study are compared to the measurement of Particulate Matter concentration for the cities of Jakarta and Surabaya.

Experimental Method

Six of the TSP and PM_{10} samples were collected simultaneously in the dry season (June–July, 1999) during the period of no rain or trace of rain in Jalan Braga and the ITB campus. Atmospheric particulate concentrations were measured with the High Volume Sampler (HVS), Low Volume Sampler (LVS), and the Andersen 1 Actual Cubic Foot per Minute (ACFM) nonviable Ambient Particle Sizing Sampler (AAPSS) with a preseparator. Meteorological conditions such as wind speed, wind direction and temperature were also measured during the sampling period.

Total Suspended Particulate (TSP) concentrations were measured using the HVS sampler. HVS drew a large volume of air through the fiberglass filter (exposed area 17.5×22.5 cm) with flow rate 1.13 to 1.7 m^3/min. This sampling device collects particles from 0.1 to 100 μm diameter on the filter.

For PM_{10} (particulate < 10 μm diameter) were measured using an LVS sampler. This equipment used a horizontal elutriator to separate the particle larger than 10 μm based on the flow rate. The air flow rate was adjusted to a 10 μm-cut point diameter. The multi stages Andersen 1 ACFM cascade impactor with preseparator were also employed to measure particle less than 10 μm diameter and fraction size distribution. This sampling device also separates particle into the following size range, > 10 μm, 5.8–9 μm, 2.1–3.3 μm, 0.65–2.1 μm, 0.43–0.65 μm, and 0.1–0.43 μm. Greased mylar films were used as media to avoid particle bounce.

Particles collected on the filter for LVS, HVS, and Cascade Impactor were analyzed using a gravimetric method by weighing the filters before and after sampling to determine ambient particulate concentration.

Chemical Analysis

After final weighing, samples were further analyzed using atomic absorption and spectrophotometer to determine the chemical compositions. For metals analysis, samples were digested in 2 mL of ultrapure HNO_3 and heated at 150–200 °C for about 4 hours and made up to 50 mL with distilled-deionized water. Filtrates were analyzed using atomic absorption to determine the metals compositions. For nitrate and sulfate analysis, samples were analyzed using the spectrophotometer method. Background contamination was monitored by using operational blanks (unexposed filter) which were processed simultaneously with field samples. Based on the sampled air volume, ambient concentrations were calculated.

Sampling Site

Jalan Braga site is located in the commercial area of downtown Bandung. Jalan Braga is one of the busiest streets (traffic), consisting predominantly of shopping and office buildings in the center of the city. The side of the street is used as a parking area. ITB campus sampling site is located on the roof of a four-story building located in a mixed institutional and residential area on the North side of Bandung.

Results and Discussion

Table 1 provides the concentration of particulate matter, lead and nitrate collected from the HVS and LVS sampler in the Jalan Braga sampling location. The concentration of total ambient particulate ranges from 304.04 to 385.05 $\mu g/m^3$. These concentrations exceeded the limit of ambient air quality standard (260 $\mu g/m^3$). The PM_{10} concentration measured from the LVS range from 277.02 to 336.44 $\mu g/m^3$.

Table 1

Ambient concentrations of particulate, Pb, and NO_3 measured on Jl. Braga, Bandung ($\mu g/m^3$)

Sample	TSP			PM_{10}		
#	Particulate	Pb	NO_3^-	Particulate	Pb	NO_3^-
1	363.17	2.37	5.99	314.47	1.16	3.45
2	337.37	1.88	6.26	285.28	1.06	3.08
3	385.05	1.58	6.03	331.36	1.01	3.27
4	340.06	2.22	5.89	310.79	1.57	3.24
5	342.71	1.86	6.01	295.45	1.01	3.45
6	304.04	1.42	6.51	268.75	0.81	2.27
Average	345.40 ± 27.20	1.89 ± 0.36	6.12 ± 0.23	301.02 ± 22.44	1.10 ± 0.26	3.13 ± 0.44

These concentrations were extremely high compared to the U.S. EPA standard (150 $\mu g/m^3$). However, the Indonesian Government has not regulated the standard for PM_{10} yet.

Table 2 presents the ambient concentration measured from the ITB-campus site. The ambient particulate concentrations were low and ranged from 79.4–102.2 and 70.9–80.2 $\mu g/m^3$ for both TSP and PM_{10}, respectively. The contribution of the particulate matter less than 10 μm diameter in the ambient particle concentration was very significant. Figures 1 and 2 show the comparison of the ambient particulate concentration from TSP and PM_{10} from the Jalan Braga and ITB campus sites. Particulate matter less than 10 microns contributes from 75 to 95% of the total suspended particle in both sampling locations. The lead concentrations existed in

Table 2

Ambient concentrations of particulate and lead measured on Jl. Ganesha, Bandung ($\mu g/m^3$)

Sample	TSP		PM_{10}	
	Particulate	Pb	Particulate	Pb
1	95.2	0.97	71.6	0.71
2	79.4	1.49	70.9	1
3	86.2	1.2	77.2	0.72
4	102.2	1.34	80.2	0.99
5	92.4	1.3	74.7	0.96
6	86.5	1.6	78.5	1.11
Average	90.32 ± 8.00	1.32 ± 0.22	75.52 ± 3.77	0.92 ± 0.16

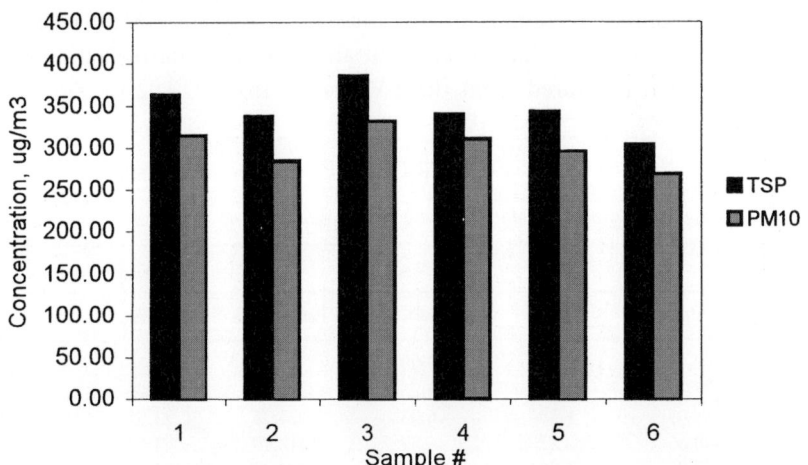

Figure 1
Comparison of particulate concentrations in the TSP and PM_{10} in Jl. Braga, Bandung.

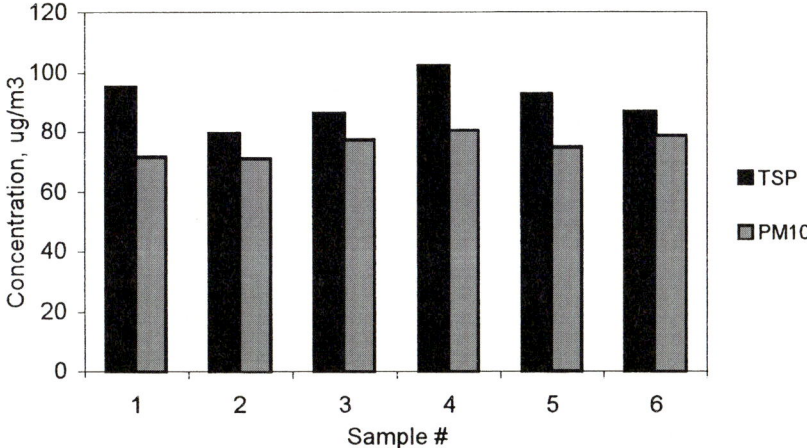

Figure 2
Comparison of particulate concentration in the TSP and PM_{10} in Jl. Ganesha, Bandung.

both the PM_{10} and TSP in significant amounts, however this concentration does not exceed the Indonesian ambient air quality standard which is 60 $\mu g/m^3$. Figure 3 presents the comparison of the ambient lead concentration from the HVS and LVS sampler in Jalan Braga, and Figure 4 shows lead concentration from the HVS and Cascade impactor at the ITB-campus sampling location. It indicates that most of the time more than 60% of the lead concentration existed in the PM_{10} particle. The main source of lead is from the transportation sector since the leaded gasoline is still used in Indonesia.

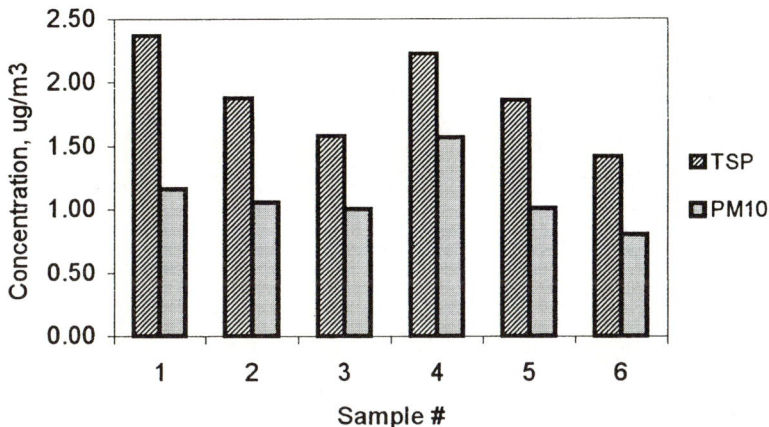

Figure 3
Comparison of Pb concentrations in the TSP and PM_{10} in Jl. Braga, Bandung.

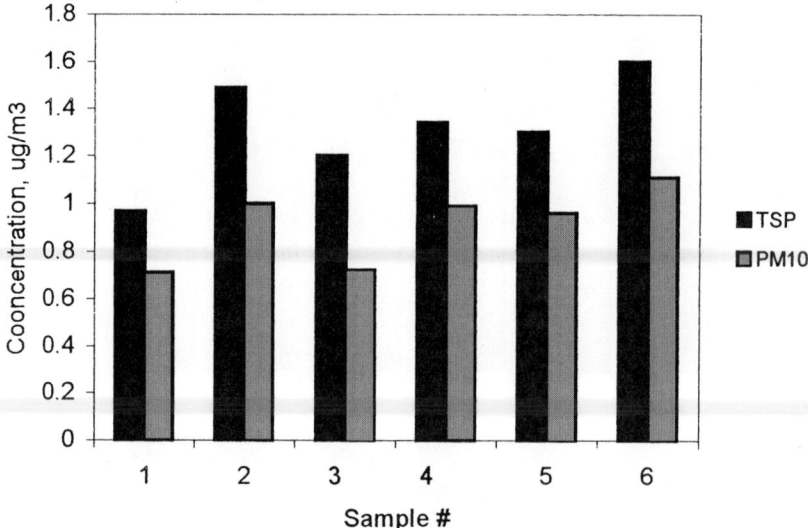

Figure 4
Comparison of Pb concentrations in the TSP and PM_{10} in Jl. Ganesha, Bandung.

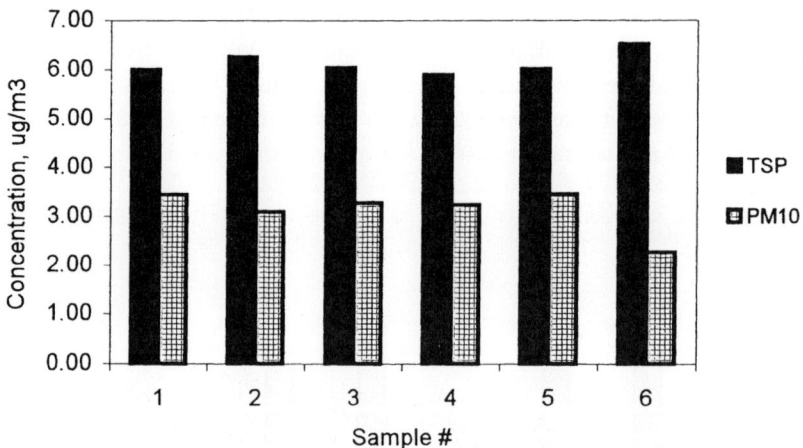

Figure 5
Comparison of the nitrate (NO_3^-) concentrations in the TSP and PM_{10} in Jl. Braga, Bandung.

Figure 5 shows the nitrate concentration in the TSP and PM_{10} samples. The nitrate was also observed in both the PM_{10} (< 10 μm diameter) and TSP with the average value 3.13 and 6.11 μg/m^3, respectively. About 50% of the nitrate concentration exists in the particles less than 10 μm diameter and the other 50% exists in the particle larger than 10 micron diameter. It seems that nitrate was dominant in the coarse fraction (coarse particle is particle larger than 2.5 micron diameter). The finding agrees with a previous study in Chicago that at temperature above 25 °C more than 50% of nitrate exists in coarse fraction (LESTARI, 1996).

The NO_x concentrations were measured hourly only in the Jalan Braga location. The ambient NO_x concentrations exceeded the ambient air quality standard. Figure 6 shows the hourly NO_x concentration for six samples. The concentrations were measured starting at 8.00 am to 7.00 pm. The data (samples # 1–5) show that concentrations of NO_x continue to increase during the day on weekdays. However, concentrations measured on Sunday (sample # 6) when the traffic was low indicate otherwise. As for the NO_x concentration, the NO concentrations show a similar pattern, and continue to increase with time during the day on weekdays, and show

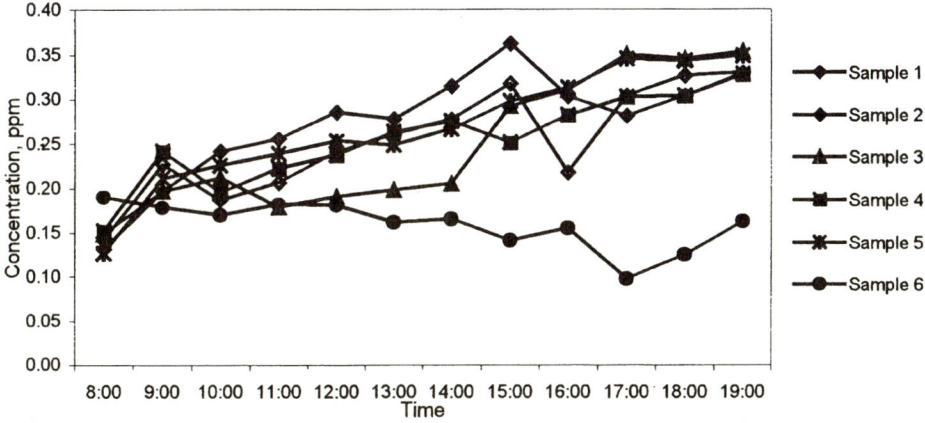

Figure 6
Concentration of NO_x on Jl. Braga, Bandung.

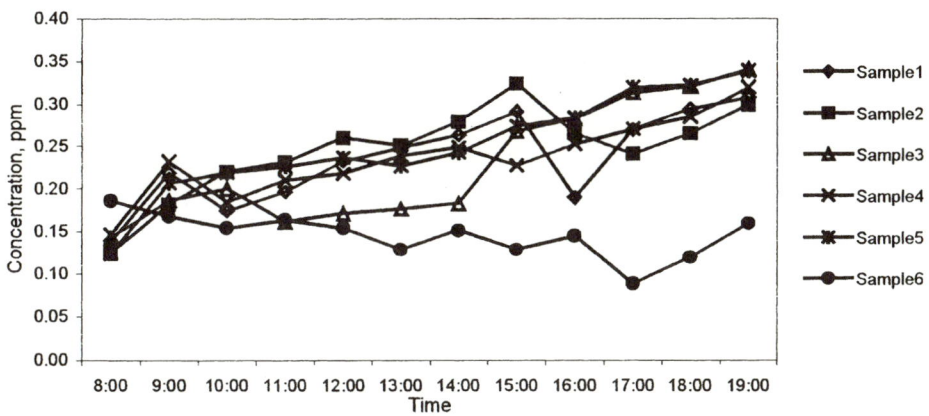

Figure 7
Concentration of NO on Jl. Braga, Bandung.

otherwise on Sunday. This phenomenon was shown in Figure 7. The concentration of NO_2 is presented in Figure 8. It shows that the concentration of NO_2 (samples # 1–5) started to increase in the morning through afternoon, and started decreasing after 5.00 pm. From literature, NO_2 in the atmosphere mostly is formed from the oxidation of NO (SEINFELD, 1986). The data conclude that the formation of NO_2 in the atmosphere is far better during daylight than at nighttime. Therefore, during the evening the concentration of NO_2 starts decreasing.

Table 3 presents the concentration of CO measured simultaneously in Jalan Braga. The concentration of CO in Jalan Braga most of the time was relatively high.

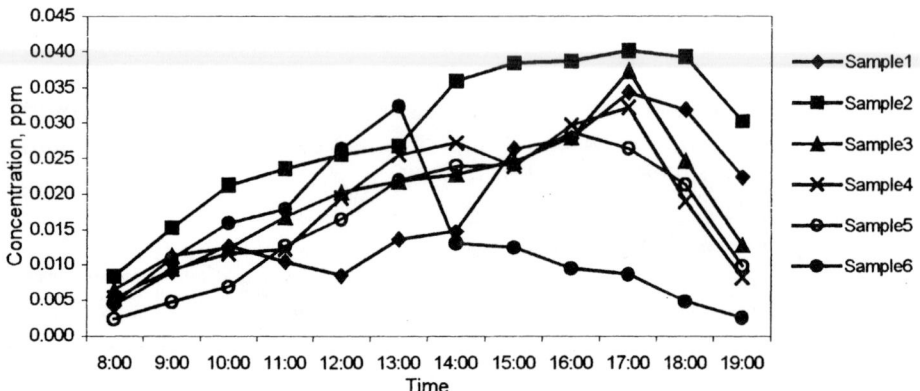

Figure 8
Concentration of NO_2 on Jl. Braga, Bandung.

Table 3

Concentration of CO on Jl. Braga, Bandung (ppm)

Time	Sample #					
	1	2	3	4	5	6
7:00	7.838	10.563	12.184	8.59	14.86	8.791
8:00	12.127	15.915	16.23	15.973	17.071	4.394
9:00	14.06	14.52	10.94	12.46	14.903	3.913
10:00	12.784	10.19	8.67	10.78	13.122	5.284
11:00	10.642	14.18	12.332	9.024	10.97	8.087
12:00	13.6	12.632	9.535	14.108	12.334	15.31
13:00	14.527	14.75	15.384	8.64	14.82	14.684
14:00	16.531	10.32	11.812	14.21	13.67	11.459
15:00	9.208	6.44	8.06	10.8	8.71	6.503
16:00	9.934	9.25	13.683	12.976	9.562	12.49
17:00	16.049	18.299	17.66	18.592	19.034	8.63
18:00	12.746	15.301	11.374	10.083	13.691	8.833
19:00	8.932	13.62	8.914	6.923	10.396	7.751
Average	12.23	12.8	12.06	11.78	13.3	8.933

However, this concentration did not exceed the Indonesian standard regulation which is 20 ppm/over 8 hours time. The main source of the CO is from the transportation sector.

The results from this study are compared to the measurement of ambient concentration from other urban areas, Jakarta and Surabaya. Table 4 presents the comparison of the ambient concentration for particulate matter, NO_x, CO and lead from Jakarta, Surabaya and Bandung. The concentration of particulate matter varied in different cities depending on the sampling locations. However in certain locations from all of the urban cities, the concentrations mostly far exceed the ambient air quality standard. The NO_x concentrations from Jakarta and Bandung exceed the ambient air quality standard, and for Surabaya in some of the sampling locations the concentrations are also higher than the standard. For all of the urban areas, the CO concentration did not exceed the Indonesian government standard. Lead concentrations are mostly low for Jakarta, Surabaya and Bandung. The chemical compositions of metals, sulfate and nitrate in the particulate TSP are shown in Table 5.

Table 4

Comparison of the Air Quality in Urban Areas

Urban city	TSP ($\mu g/m^3$)	NO_x (ppm)	CO (ppm)	Pb ($\mu g/m^3$)	Note
Jakarta, 1991[1]	76.3–993.8	0.019–0.224	0.80–4.82	–	Ambient air quality
Surabaya, 1998[2]	241–734	0.005–0.0701	0–4.38	0–16.8	standard in Indonesia:
Bandung, 1999	79.4–385	0.0963–0.3616	3.91–18.3	0.71–2.366	TSP = 260 $\mu g/m^3$ (24 hr)
					NO_x = 0.05 ppm (24 hr)
					CO = 20 ppm (8 hr)
					Pb = 60 $\mu g/m^3$ (24 hr)

Sources: [1] LP – ITB and BAPEDAL Jakarta, Indonesia
[2] BAPEDALDA East Java, Indonesia

Table 5

Concentration of metals, sulfate and nitrate in TSP samples on Jl. Braga, Bandung ($\mu g/m^3$)

Sample	TSP	Pb	NO_3^-	Ca	Cu	Zn	Fe	K	Na	SO_4^{2-}	Total element
1	363.17	2.37	5.99	13.77	3.58	0.55	9.97	12.06	39.75	5.23	93.27
2	337.37	1.88	6.26	18.77	3.64	0.47	9.48	11.23	37.80	7.05	96.56
3	385.05	1.58	6.03	13.42	2.35	0.45	10.97	8.45	32.15	4.74	80.15
4	340.06	2.22	5.89	10.19	3.11	0.66	8.54	7.24	34.49	4.66	76.99
5	342.71	1.86	6.01	17.08	2.10	0.46	9.15	9.93	36.22	5.34	88.13
6	304.04	1.42	6.51	14.68	3.18	0.35	6.98	10.23	36.54	5.56	85.45
Average	345.40	1.89	6.12	14.65	2.99	0.49	9.18	9.86	36.16	5.43	86.76

Conclusions

1. The ambient concentrations of particulate matter and NO_x in Jalan Braga exceed the ambient air quality standard.
2. PM_{10}, contribute 75–95% of the total ambient particulate concentration for both locations (Jalan Braga and ITB campus).
3. The concentration of CO is below the Indonesian government standard regulation.
4. More than 60% of the lead concentration exists mostly in the PM_{10} particles.
5. Nitrate concentrations exist in both PM_{10} and TSP.

REFERENCES

LESTARI, PUJI (1996), *Atmospheric Particulate Sulfate and Nitrate: Distribution, Formation, and Deposition*, Ph.D. Thesis, Illinois Institute of Technology, Chicago, U.S.A.

LIN, J., FANG, G., HOLSEN, T. M., and NOLL, K. E. (1993), *A Comparison of Dry Deposition Modeled from Size Distribution Data and Measure with a Smooth Surface for Total Mass, Lead, and Calcium in Chicago*, Atmosph. Environm. *27A*, 1131–1138.

NOLL, K. E. and FANG, K. Y. P. (1989), *Development of a Dry Deposition Model for Atmospheric Coarse Particles*, Atmosph. Environm. *23*, 585–594.

NOLL, K. E., LESTARI, PUJI, and HOLSEN, T. M. (1997), *Dry Deposition of Sulfate and Nitrate in Chicago*, Presentation at the Air and Waste Management Association's 90th Annual Meeting and Exhibition, Toronto, Ontario Canada.

POLLUTION CONTROL IMPLEMENTATION (PCI) (1997), *Study of Fine Atmospheric Particle and Gas in the Jakarta Region*, Project Report no. 38.

SEINFELD, J. H. (1986), *Atmospheric Chemistry and Physic of Air Pollution*, A Wiley Interscience Publication, John Wiley & Sons, New York.

(Received March 1, 2000, accepted September 6, 2000)

 To access this journal online:
http://www.birkhauser.ch

Pure appl. geophys. 160 (2003) 117–141
0033–4553/03/020117–25

© Birkhäuser Verlag, Basel, 2003

❙ Pure and Applied Geophysics

Assessing Seasonal Transport and Deposition of Agricultural Emissions in Eastern North Carolina, U.S.A.

Jamie R. Rhome,[1,2] Dev Dutta S. Niyogi,[1]
and Sethu Raman[1]

Abstract — There is an increasing interest regarding the fate of nitrogenous compounds emitted from agricultural activities in the southeastern United States. Varying climate, topography and proximity to the Atlantic Ocean particularly complicates the problem. An increased understanding of the interaction of synoptic scale flow with mesoscale circulations would constitute a significant improvement in the assessment of regional scale transport and deposition potential. This knowledge is necessary to facilitate current and future modeling attempts in the region as well as for planning future monitoring sites to develop a cohesive regional policy for the abatement strategies. The eastern portion of North Carolina is used as a case example due to its high, localized emission of nitrogen compounds from agricultural waste. Three periods: July 2–7, 1998, October 5–11, 1998, and December 12–19, 1998, corresponding to three different seasons were studied. Surface wind and thermodynamic patterns were analyzed using surface observing stations and archived-model analysis results centered over eastern North Carolina. Diurnal and seasonal patterns were identified for dispersion and concentration values obtained using an air pollution transport and dispersion model. This mesoscale information was used to draw qualitative conclusions regarding the possible trends and deviations in the dynamic trajectories as well as the resulting near-surface concentrations and deposition potential in eastern North Carolina. Results show that highly variable seasonal and diurnal atmospheric circulations characterize the study domain. These variations can significantly impact the transport and fate of pollutants released in this region. Generally, summer provides the highest potential for localized deposition, while fall can provide opportunity for long-range transport. The results also suggest that mean climatological or seasonally averaged flow patterns may not be sufficient for analyzing the fate of the agricultural releases in this region. At the very least, mean and variance based analysis is required to capture the climatology of the dispersion and deposition patterns. These patterns in eastern North Carolina appear to be sensitive to the strength and location of air mass boundaries along the coastal plain, indicating diverse scale interactions affecting the variability and uncertainty in the regional pollutant transport.

Key words: Air pollution, atmospheric deposition, North Carolina, trajectory analysis, nitrogen compounds.

---

[1] State Climate Office of North Carolina, and Department of Marine, Earth, and Atmospheric Sciences, North Carolina State University, Raleigh, NC, U.S.A. E-mail: dev_niyogi@ncsu.edu

[2] Present Affiliation: Tropical Analysis and Forecast Branch, National Hurricane Center

Dr. Devdutta S. Niyogi, State Climate Office of North Carolina and Department of Marine, Earth, and Atmospheric Sciences, North Carolina State University, Raleigh, NC 27695 - 7236 U.S.A.

Introduction and Background

Processes associated with many environmental problems such as air quality, acid rain, hydrology, and water quality are strongly influenced by different aspects of the circulation of the atmosphere (YARNAL, 1993). Field monitoring studies have long suggested the potential importance of regional and long-range atmospheric transport to the distribution of agricultural chemicals in the environment (KURTZ, 1990; MAJEWSKI and CHAPEL, 1995). However, application research is severely hampered by uncertainties in the mesoscale and diurnal patterns of near-surface winds and thermodynamic structure. Therefore, local scale transport and deposition of surface chemical or agricultural waste emissions are not well understood.

North Carolina exhibits a large variation in its temporal and spatial climate due to complex topography and proximity to the Atlantic Ocean. Recent research has also shown that a significant teleconnection exists between climate in North Carolina and global patterns such as El Niño-Southern Oscillation (ROSWINTIARTI *et al.*, 1998; RHOME *et al.*, 2000). In addition, the presence of sea-breeze circulation along southeastern North Carolina further complicates the regional wind patterns (GILLIAM *et al.*, 1999). This circulation consists of a shallow low-level onshore flow and an opposing return flow aloft. Often associated with this phenomenon is a mesoscale front with a narrow convergence zone and upward motion along the leading edge. Surface convergence zones and the associated vertical motions have been shown to have a significant impact on the parcel trajectory and wet deposition (COOTER *et al.*, 1997). These interactions among atmospheric motions at different spatial scales in eastern North Carolina and the implications on meteorology and pollutant transport and deposition are discussed in this paper. To date little emphasis has been placed on the impacts of mesoscale variability on climatic or mean wind fields and the transport and deposition patterns in coastal North Carolina.

One pollutant that is gaining considerable attention in North Carolina is the large-scale emissions of nitrogen compounds such as ammonia from agricultural swine-based farm activity (ANEJA *et al.*, 1997). In a study reported by WALKER *et al.* (2000) there has been a substantial increase in the swine population and related nitrogenous emissions and deposition in southeastern North Carolina in recent years. Figure 1 shows the observed annual precipitation — weighted ammonium concentration (mg/L) obtained from the National Atmospheric Deposition Program/National Trends Network (NADP/NTN) site in eastern North Carolina (Site NC35). Overall, there is a significant increase in the amount of airborne nitrogen collected from rainfall in the form of ammonium (formed as a result of ammonia linking with sulfates, nitrates, and chlorides). ANEJA *et al.* (1997) suggest, these emissions have a diurnal maximum during the afternoon as well as a seasonal maximum during the summer primarily due to maximum lagoon water temperatures. For instance, in 1998, the NC35 site in eastern North Carolina, showed a seasonal mean ammonium deposition of 0.54 kg.ha^{-1} in fall as against 2.06 kg.ha^{-1} in summer. It is hence

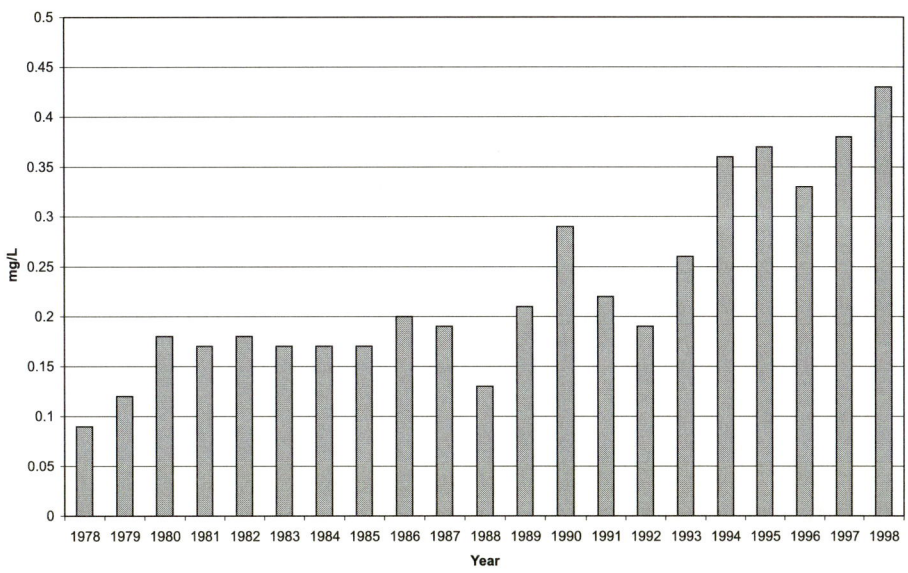

Figure 1
Annual ammonium concentration at the NADP/NTN site NC35 in Clinton, North Carolina. The increasing ammonium and related reduced nitrogen concentrations and deposition are considered to be significantly related to the increase in the agricultural emissions from animal waste lagoons in eastern North Carolina.

important to study the seasonal variations in the thermodynamic and wind patterns to assess the potential for transport and deposition in eastern North Carolina. Identification of critical meteorological regimes that can create deviations in transport and deposition patterns will constitute a significant improvement in the understanding of atmospheric loading of nitrogen to sensitive areas along the mid-Atlantic region.

Previous studies have investigated allied features for regions elsewhere in the United States. For instance, soil emissions and re-volatilization of chemical residue for agriculturally based chemicals have been modeled as responding to wind speed, humidity, temperature, solar radiation, cloud cover, precipitation and soil temperature (COOTER et al., 1997). Chemical transport has been studied as a function of turbulent motions as well as horizontal and vertical advection by the regional scale wind field (CHANG, 1990). Wet and dry chemical depositions are known to be influenced by rain volume, temperature, wind speed, local and regional turbulence, radiation, and humidity (COOTER et al., 1997). Hence, the effects of seasonal and different scale interactions on pollutant concentrations and deposition fields are of great relevance and interest for agricultural environmental analysis.

In the following section, we will describe a methodology for assessing the regional transport of agricultural emissions in eastern North Carolina. In the analysis, priority is given to readily available meteorological data sets. Based on the varying synoptic and mesoscale processes over eastern North Carolina, we will provide a qualitative assessment of the transport and dispersion potential for reduced nitrogen compounds. For this analysis, three different case scenarios are presented in section 3 corresponding to three different seasons (Winter, Fall, and Summer). In particular, the evolutions of the synoptic and mesoscale aspects of the seasonal case studies are presented in this section. In section 4, the spatial and temporal patterns for concentration and deposition in relation to varying (seasonal/diurnal) flow patterns for the three seasonal cases are discussed. Section 5 summarizes the results concerning the dependence of transport and deposition on synoptic and mesoscale wind fields, and how differing scales can interact in eastern North Carolina.

2. Methodology

Mesoscale wind and thermodynamic patterns for the summer, fall, and winter seasons over eastern North Carolina are investigated. Summer, fall, and winter periods were chosen for seasonal analysis as they represent varying complexity of the interactions of mesoscale and synoptic scale forcing. Spring patterns are somewhat similar to those during the fall, and so spring was not included in the seasonal analysis. These seasons were also chosen based on analysis of trends in precipitation–weighted concentration of ammonium obtained from National Atmospheric Deposition Program/National Trends Network site NC35 for 1998. The seasonal average values were 0.48 mg.l^{-1} and 0.41 mg.l^{-1} for spring and fall, as against 0.26 mg.l^{-1} and 0.76 mg.l^{-1} for winter and summer, respectively (cf., WALKER *et al.*, 2000). Thus, winter, fall, and summer were considered for analysis and one period was chosen to represent each of these seasons. The study periods were July 2–7, October 5–11, and December 12–19, 1998. These were chosen by factors such as data availability, presence of dominant synoptic and mesoscale features, as well as presence of both significant anomalies and dominant seasonal patterns that typically frequent the region.

The first part of this analysis was the interpretation of seasonal features in the wind and thermodynamic field over eastern North Carolina. Surface meteorological observations from over twenty weather stations across eastern North Carolina for the three periods were analyzed (NIYOGI *et al.*, 1997). The National Oceanic and Atmospheric Administration's (NOAA) Real-time Environmental Applications and DisplaY (READY) system was then used to analyze surface wind and moisture patterns over the study region. We chose 3-hourly archived analysis from the National Center for Environmental Prediction (NCEP) Eta Data Assimilation System (EDAS) for this study. EDAS is an intermittent assimilation system consisting of successive model forecasts and Optimum Interpolation (OI) analyses for a pre-forecast period (12-h for the early Eta) on a 38 level, 48 km grid.

The next step was the use of a tracer model to analyze transport and dispersion patterns. The HYSPLIT (HYbrid Single-Particle Lagrangian Integrated Trajectory, Draxler and Hess 1997) model was used for computing air parcel trajectories for dispersion and deposition simulations. The dispersion of a pollutant is calculated by assuming either puff or particle dispersion. In the puff model, puffs expand until they exceed the size of the meteorological grid cell (either horizontally or vertically) and then split into new puffs, each with a share of the pollutant mass. In the particle model, a fixed number of initial particles are advected through the model domain by the mean and turbulent wind field components. The model's default configuration assumes a puff distribution in the horizontal and particle dispersion in the vertical direction. In this way, the greater accuracy of the vertical dispersion parameterization of the particle model is combined with the advantage of an ever-expanding number of puffs to represent the pollutant distribution. The Hysplit system also includes a detailed deposition module which explicitly accounts for dry and wet deposition along with resuspension following WESLEY (1989) and WALMSLEY and WESLEY (1996), similar to the approach adopted in RADM (Regional Acid Deposition Model, CHANG, 1990). Additional information pertaining to the Hysplit modeling system can be found in DRAXLER and HESS (1997).

To determine a potential source region for eastern North Carolina, the 1997 estimated inorganic nitrogen deposition from nitrate and ammonia data available from the NADP/NTN was analyzed. Figure 2 shows the annual mean for 1997, with a local maximum in deposition over southeastern North Carolina. Based on this, a source region was considered centered over 35.1 N and 77.9 W, in our analysis. A constant unit emission (mass units per hour) was used to facilitate a qualitative analysis of transport/dispersion based on wind and thermodynamic patterns. The units in this case are considered unimportant, as output air concentration units will be the same as input units. Further, since a non-reactive tracer is considered in the transport and dispersion model, the actual concentration will be a direct function of the source strength. A 10-m surface layer was used for the first source column in the Hysplit system, and the analysis was performed over 0.3-degree (about 33-km) horizontal grid resolution. The deposition option was active in all the transport and concentration runs in this study.

Assessment is then made regarding the seasonal variation in mesoscale climate and its impacts on transport/dispersion, and potential deposition of surface emitted pollutant species in eastern North Carolina. An interesting point for applications related to extension and policy planners is the access available to the analysis and plotting system over the internet. The model can be run interactively on the internet or installed on a local computer through the READY system (http://www.arl.noaa.-gov/READY). Thus the applied methodology has the additional benefit of facilitating future studies by research/academic personnel as well as local governments interested in policy and regulation planning.

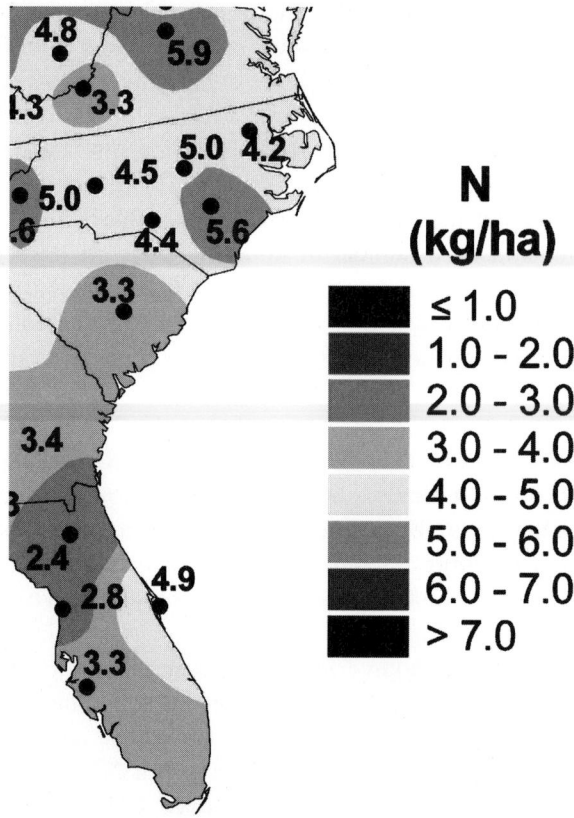

Figure 2
Observed inorganic nitrogen deposition for nitrate and ammonium for 1997 over North Carolina. Note the
relatively high deposition values in southeast North Carolina.

3. Case Discussion

The meteorological setting for the three cases corresponding to the three seasons
(winter, fall, and summer) are discussed in this section.

a) December 12–19

Winter in eastern North Carolina is characterized by land-sea gradients in
temperature due primarily to the effects of the Gulf Stream on the coastal region.
This area is a favored region for the formation of frontal boundaries that interact
with synoptic scale patterns (RAMAN *et al.*, 1998). These boundaries can significantly
affect the transport and deposition especially when superimposed with larger scale
flow regimes. Winter flow patterns are unusual in comparison to the other seasons of
interest and therefore concentrations and dispersion values may vary significantly.

Climatologically it is expected that the highest concentration values would be east of the source region since the mean winds are westerly over eastern North Carolina. However, eastern NC can be dominated by prolonged strong low-level northeasterly flows in the winter months due to a common pattern known as cold air damming (CAD). This pattern may typically occur 3–4 times per month during the winter season with a local peak during March (BELL and BOSART, 1988). Event duration may vary from hours to days. However, a local source of emission over southeastern NC would allow significant deviations in transport and deposition patterns in even the weakest and shortest CAD events. In addition, the strong static stability associated with this pattern and enhanced precipitation along coastal fronts can present near optimal conditions for enhanced local deposition fluxes.

The period December 12–19, 1998 was characterized by a CAD event. A cold dome of high pressure became entrenched over the mid-Atlantic on 12 December. Damming began during the day on the 12th as surface winds became increasingly northeast. Figure 3a shows the surface wind vectors overlaid with relative humidity (percentage) for this period. A convergence zone, typically associated with this pattern, is seen near the southeast Carolinas. A surface low formed over the northwestern Gulf of Mexico in response to an upper level trough over the

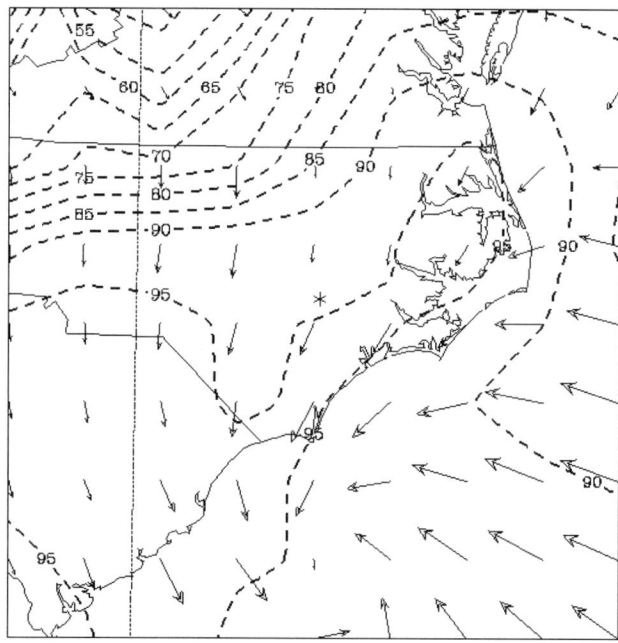

Figure 3a
00 UTC December 13, 1998 surface wind vectors (arrows) and relative humidity (dashed lines) for the study domain. A persistent northeasterly flow exists for this case. A convergence zone was also present which influenced local wind fields and transport.

central United States. The low strengthened as it interacted with strong baroclinicity along the mid-Atlantic coast. As shown in Figure 3b, rapid cyclogenesis occurred along the frontal boundary resulting in northwesterly flow over the domain and drainage of the cold dam near the end of the period. A strong anticyclone then moved over the region as the cyclone moved northeast and away from the domain. Resulting winds were light and variable. Thus the period from December 12–19, 1998 was characterized by the evolution of a common winter weather pattern in eastern North Carolina. The particular case demonstrated here was quite long and intense. A weaker and shorter case will be examined for the Fall in the following section.

b) October 5–11, 1998

Fall is characterized by an increasing southward invasion of Canadian air. Weak disturbances tend to move rapidly through the Carolinas. As a result, this season represents a somewhat dry period over eastern NC excepting landfalling tropical disturbances. Seasonally averaged winds are generally from the west, however flow patterns are highly variable depending on the frequency and strength of cold fronts passing through the region. Fall represents a period where high values of seasonally averaged wet deposition flux are less likely, but significant daily values are

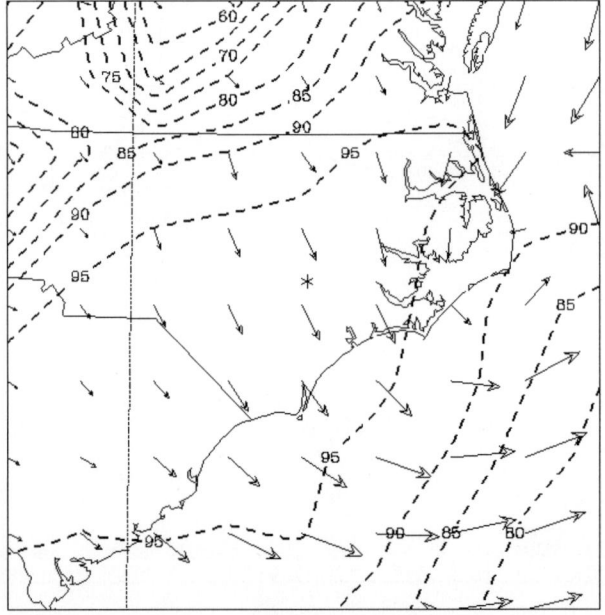

Figure 3b

Same as Figure 3a, except that the frontal boundary coincides with cyclogenesis that results in a northwesterly flow over the domain and drainage of the cold dam on December 13, 1998.

intermittently possible depending on local as well as diurnal scale interactions with larger scale forcing.

A strong anticyclone over New England dominated the first part of this period. The slow movement of this system allowed northeasterly flow to persist over eastern NC for several days. A strong disturbance and an associated cold front located in the central U.S. approached the region during the middle part of the case. Consequently, winds became increasingly southwesterly. Figure 4a shows the wind vectors and relative humidity variation over the domain for 8 October, 1998. Strong southerly flow dominates eastern North Carolina that changes to a more southwesterly flow ahead of the approaching cold front. Such low-level wind variations can cause significant changes in the pollutant transport and the resulting concentration patterns over the region. The cold front moved through the region and stalled off the coast. In addition, a series of weak surface lows formed and moved along the cold front. Figure 4b shows the frontal boundary slowly moving off the coast resulting north to northwest winds over the source region. This period, was thus characterized by highly variable flow patterns over eastern North Carolina as would be expected during Fall season. However, these daily variations are due primarily to rapidly evolving synoptic scale motions.

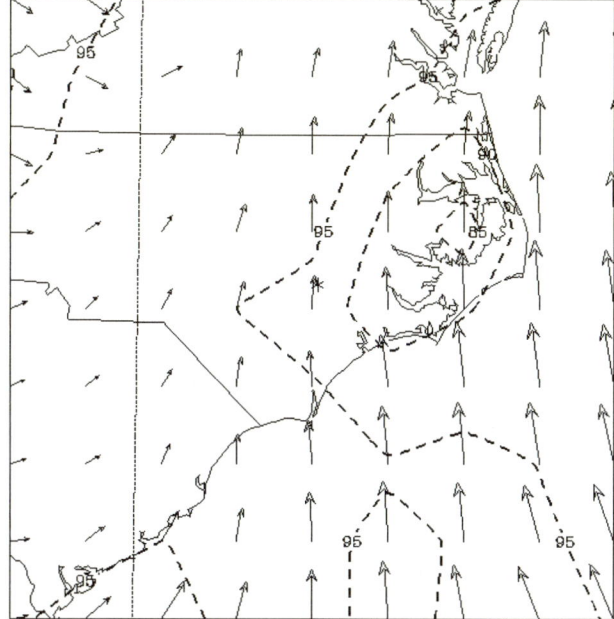

Figure 4a
Wind vectors and relative humidity variation (dashed lines) over the domain for 8 October, 1998. A strong southerly flow dominates the eastern North Carolina that changes to a southwesterly flow ahead of an approaching cold front.

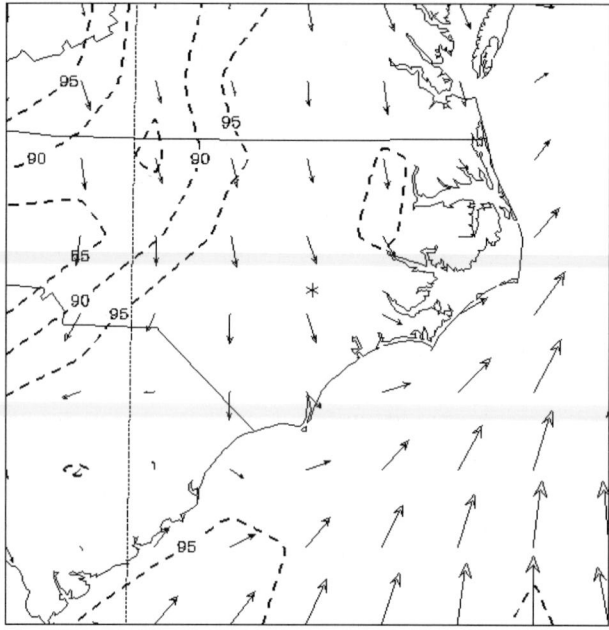

Figure 4b
Slow passage of the frontal boundary (shown in Fig. 4a) offshore resulting in northwest transport on
October 8, 1998.

c) July 2–7, 1998

July is characterized by weak winds owing to decreased latitudinal temperature gradient. As a result, weather systems affecting the region are weak and slow moving. In addition, air masses originating from Canada usually make little southward progress and have limited effect on eastern North Carolina. The region becomes increasingly under the influence of a nearly permanent high-pressure system commonly known as the Bermuda High. South to southwesterly flow associated with this weather system can persist for several days. This flow regime favors the inland propagation of the diurnal sea breeze thus setting the stage for complex mesoscale circulations over eastern North Carolina. Blocking patterns are also more frequently observed during the summer months. Although summer is characterized by weak large-scale forcing, it represents the wettest time of the year (typically about 16 inches precipitation out of 48 inches annually, over the source region) due to the frequent occurrence of convection. In addition, mesoscale processes associated with land-sea interaction and physiological differences in soil characteristics can enhance convection over southeastern NC. Therefore, significant diurnal variation in transport and deposition of pollutants could exist. In addition, this season represents the best opportunity for potentially large daily deposition values due to a seasonal maximum in locally emitted nitrogen compounds and locally heavy rainfall associated with convection (cf., ANEJA *et al.*, 1997).

No significant synoptic scale weather events affected the region during the case period. A weak surface disturbance moved through the region and so significant daily variation in wind patterns existed. Figure 5 shows a case where northwesterly surface wind flow dominated the area, which would transport material over the ocean. Note also the large humidity gradient that parallels the coast in this flow regime. For the next day, as shown in Figure 6, there were southeasterly winds over the domain, which shift the moisture gradient inland. The pattern would favor transport over central portions of North Carolina with possible larger wet deposition values along the moisture gradient due to enhanced precipitation. The period July 2–7, 1998 was thus typical of summer conditions, characterized by highly variable flow conditions principally due to weak synoptic forcing leading to increasing importance of diurnal mesoscale circulations. Therefore, we will focus more on mesoscale processes and diurnal variations in the analysis for this period.

In summary, the three seasonal patterns examined here show distinct features and variability over the study domain (eastern North Carolina). In the following section, we will discuss how these patterns affect the distribution of model-simulated transport and deposition. Additionally, we also discuss the interaction between

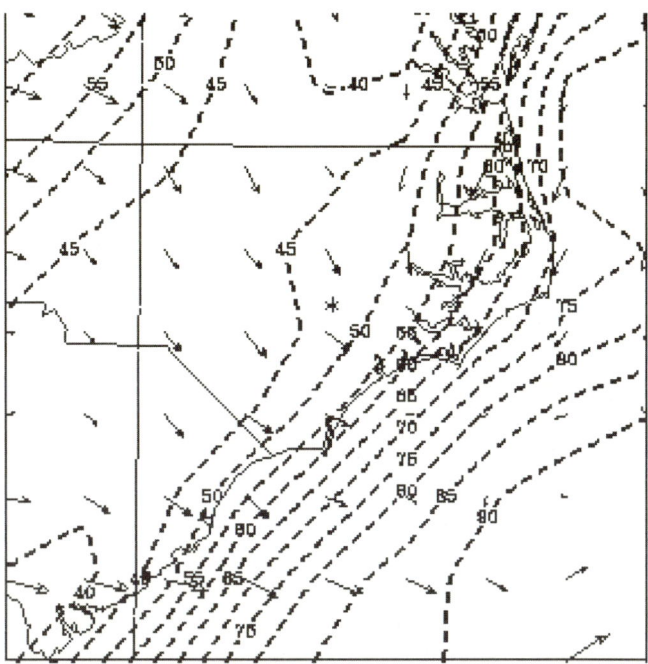

Figure 5
Strong offshore winds associated with a relative humidity gradient (dashed lines) parallel to the coast for 2–7 July 1998 case.

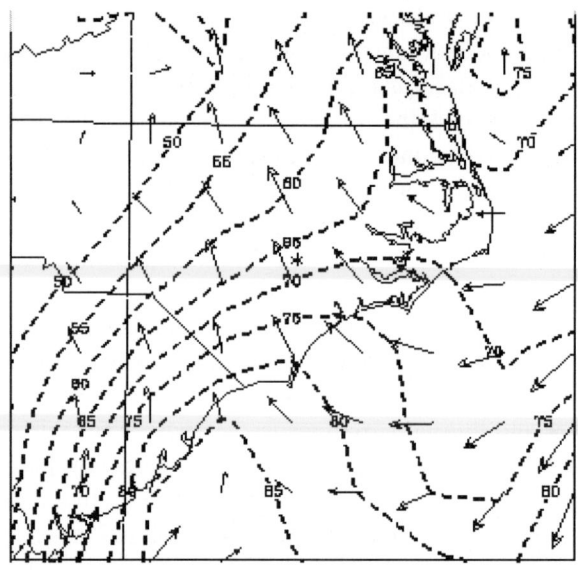

Figure 6
For the same case as in Figure 5, except that the surface winds show inland intrusion. The day-to-day
shifting of the winds for the summertime scenario are significant source of uncertainty and variability for
eastern North Carolina.

different scales of motions and the implications on transport/dispersion and
deposition over eastern North Carolina.

4. Transport and Dispersion

The transport and dispersion patterns corresponding to the three cases discussed
above are presented in this section.

a) December 12–19, 1998

This period encompasses the initiation, peak, and dissipation of a cold air
damming pattern. In addition, cyclogenesis along the coastal front during this period
allows analysis of the impacts of transport and deposition during rapid drainage of
the dam as the winds back to the northwest. This period thus represents different
potential patterns that could influence transport and deposition during the winter
season. We will discuss implications on transport and deposition during the
initiation, mature, and dissipation stages of the cold air-damming event.

Transport during the initiation was highly variable due to rapidly veering surface
winds. However, some transport to the southwest was noted, as winds are
predominately northeasterly (not shown). Cold air damming initiation is not always

associated with precipitation and so wet deposition values are expected to be variable depending upon the individual cases. However, due to prevailing northeasterly flow often associated with this pattern, higher concentration values are expected to the southwest of the source. Therefore, the potential for larger wet deposition values is also to the southwest of the source region. In absence of wet deposition, the increasing static stability and a thermal inversion associated with this pattern could increase the potential for dry deposition velocities with unstressed surface conditions (PLEIM et al., 1999). Trajectory analysis during this period shows that the parcel had limited vertical movement. Model results initialized on 00 UTC 12 December show transport to southwest increasing with time, as winds become more northeasterly. Figure 7 shows that concentration distribution patterns become parallel to the coast (streak-like). That is, the surface convergence associated with the developing surface front increases low-level concentrations along the front. However, the highest concentrations remain stagnated in southeastern North Carolina during the initial stage as shown in Figure 7.

As the cold dam reaches maturity, the pollutants are oriented to the southwest. Figure 8 shows a sharp eastward gradient in the concentration patterns while the westward gradient is weak. This pattern is likely due to unstable air east of the coastal front causing vertical diffusion, and stable air (thermal inversion) west of the boundary inhibiting vertical diffusion. Surface emissions such as nitrogenous gases

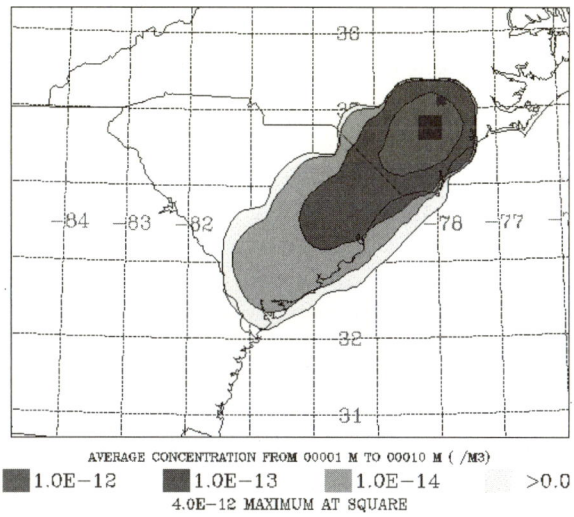

AVERAGE CONCENTRATION FROM 00001 M TO 00010 M (/M3)

■ 1.0E−12   ■ 1.0E−13   ■ 1.0E−14   >0.0

4.0E−12 MAXIMUM AT SQUARE

Figure 7

Concentration distribution for a ground-level source (closed square) in southeastern North Carolina from 24-hour model simulation for a wintertime scenario with a persistent northeastly flow beginning 00UTC December 12, 1998. Maximum concentration is shown at larger square just southwest of source region. Contour values are unit-less indicating mass units per cubic meter.

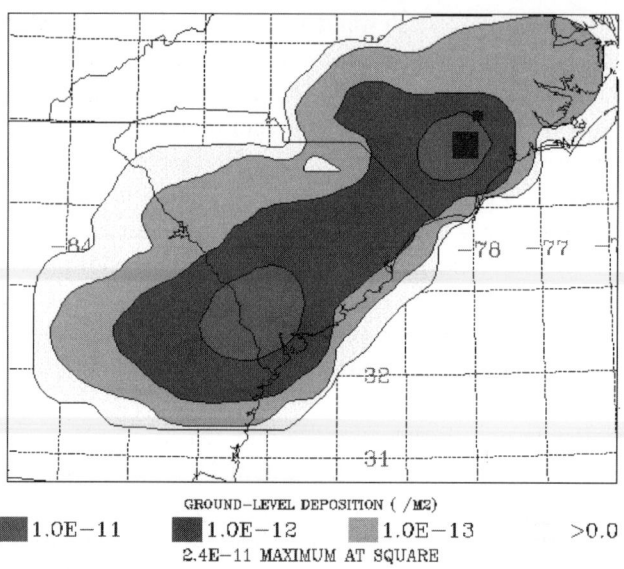

GROUND–LEVEL DEPOSITION (/M2)

■ 1.0E−11   ■ 1.0E−12   ▨ 1.0E−13   >0.0
2.4E−11 MAXIMUM AT SQUARE

Figure 8

Same as in Figure 7 except the contours are for surface deposition. Note the local maxima in extreme southeastern North Carolina and along the South Carolina/Georgia border. Values are unit-less indicating mass units per cubic meter.

from the agricultural swine-lagoons in the eastern North Carolina can therefore be trapped in the low-level stable environment associated with the cold air dam. Such a winter case thus represents possible optimal conditions for long-range transport and deposition of chemical emitted in eastern North Carolina, to eastern South Carolina and Georgia as shown in Figure 8. Transport and deposition this far southwest of the source region is unlikely in other situations. Since upper-level winds are rarely oriented from the northeast during the winter, transport to the southwest must be governed by near-surface winds. Therefore surface winds need to be oriented from the northeast to facilitate pollutant transport to South Carolina and Georgia. The winter season appears to present the best opportunity for this type of near-surface wind pattern with a long duration.

As the pattern reached maturity, a weak surface low formed along the coastal front and moved rapidly out to sea. Pollutant concentration distribution patterns remained nearly uniform during this period. The only exception was the breakdown of the strong eastward gradient in concentration indicating that the low pressure associated with the front allowed some leakage through the front. However, the surface low was too weak and moved rapidly to initiate drainage of the dam and so the persistent flow regime and pollutant concentration distribution patterns remained similar for several additional days (not shown). However this pattern was eventually dissipated by a stronger cyclone along the coastal front which allowed the statically stable air to drain rapidly eastwards. In other such cold air damming cases, drainage may take several days as no significant synoptic forcing facilitates such dissipation. Thus pollutant concentrations can remain relatively high over eastern North Carolina even during winter. Additionally, such a meteorological setting can also be associated with light lingering precipitation, which can enhance regional wet deposition values. In this particular case, a low formed near the northern Florida coast and rapidly intensified as it encountered strong baroclinicity along the Carolina coast. However, initial northward movement of the surface low was slow and therefore initially enhanced the northeasterly flow over the area. The low deepened off the NC coast and drained the dam as it moved northeast. Resulting pollutant concentration patterns associated with this scenario are shown in Figure 9. Thus, during dissipation of the cold air dam, pollutant transport was primarily restricted to eastern and northeastern North Carolina. Pollutant transport towards the end of the event was also quite interesting as shown in Figure 10. As the low continued to move off the coast, the region began to be dominated by a strong migratory anticyclone. Higher concentration values remained almost entirely over eastern North Carolina due to weak steering currents as well as strong subsidence associated with the anticyclone. Precipitation is usually absent during such a pattern, thus reducing the wet deposition potential. However, since the period prior to this event is often

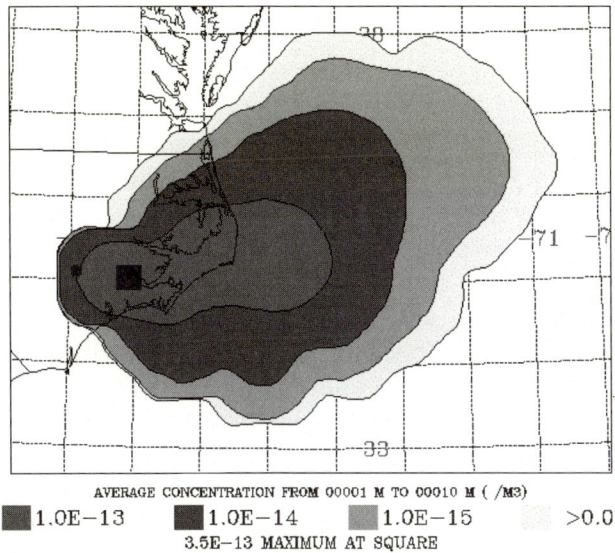

AVERAGE CONCENTRATION FROM 00001 M TO 00010 M (/M3)

■ 1.0E−13   ■ 1.0E−14   ■ 1.0E−15   >0.0

3.5E−13 MAXIMUM AT SQUARE

Figure 9

Concentration distribution for a ground-level source (small box) in southeastern North Carolina from 24-hour model simulation for a wintertime scenario beginning 00UTC December 17, 1998. Concentration pattern indicates rapid draining of the surface concentrations with the dissipation of the cold air damming pattern.

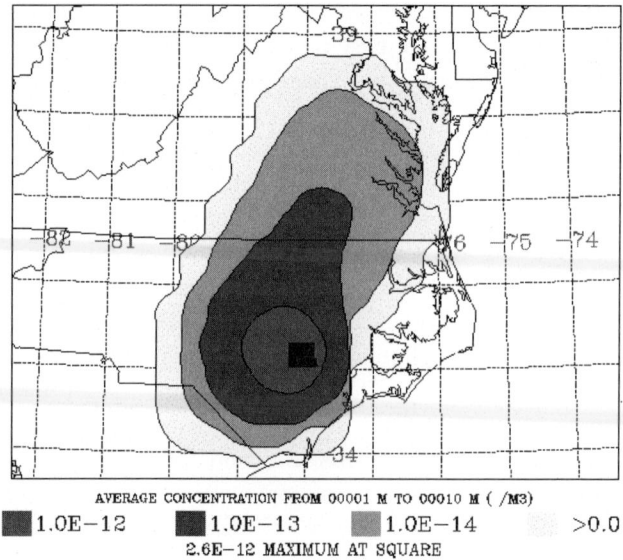

AVERAGE CONCENTRATION FROM 00001 M TO 00010 M (/M3)

■ 1.0E−12   ■ 1.0E−13   ■ 1.0E−14   >0.0

2.6E−12 MAXIMUM AT SQUARE

Figure 10

Same as in Figure 9 except 24-hour model simulation beginning 00UTC December 19, 1998 with the region being dominated by a anticyclone. Resulting winds are weak with highest concentration values remaining near the source region. Such a scenario can exhibit high-surface concentration and deposition possibilities for northern and central North Carolina.

marked by intermittent precipitation, the surface resistance for dry deposition is often the lowest (cf., PLEIM *et al.*, 1999) and can lead to significantly high dry deposition values (cf., FINKELSTEIN *et al.*, 2000; NIYOGI *et al.*, 2000).

In summary, the persistent northeasterly flow associated with cold air damming over the mid-Atlantic appears to present optimal conditions for transport of surface emitted pollutants from sources in southeastern NC to portions of South Carolina and Georgia. This transport also appears to be sensitive to the duration of the event as strong static stability curtails vertical mixing and stronger upper-level winds are not generally oriented from the northeast. Therefore, it appears regions to the south and west of the source region (in southeastern North Carolina) can be vulnerable to higher concentration and deposition values during the winter season when cold air damming events are more frequent and strongest. Conversely, cyclogenesis along the southeastern United States coast appears to present the best opportunity for drainage of pollutants from eastern North Carolina, sweeping them off the coast. Alternatively, the presence of strong anticyclones over the southeastern United States will allow concentration to build in near the source regions.

A weaker and shorter case of the northeasterly flow regime described above is presented in next section for the Fall case.

b) October 5–11, 1998

For this case, as discussed earlier, an anticyclone situated over New England governs transport during the first part of the period. This establishes a prolonged northeasterly flow. This pattern is similar to the initialization stage of the cold air damming event mentioned above. However, a well-defined frontal boundary and associated surface convergence was not as evident as for the Winter case. The northeasterly flow does facilitate some transport south and west but, unlike the stronger cold air-damming event, long-range transport was not evident (not shown). The high pressure system slowly drifts offshore, and the winds veer to a more easterly southeasterly component. Figure 11 shows an inland surge of the plume that results as the high pressure system migrates offshore. For this scenario, high concentration buildup can be seen over northern South Carolina and the foothills of North Carolina as shown in Figure 11. However, this pattern of transport does not last long as the winds begin to veer to a southwesterly direction in response to an approaching cold front. The resulting concentration patterns show transport to north of the source region (not shown). The corresponding deposition pattern is shown in Figure 12. It will be generally dominated by wet deposition along and ahead of front because of precipitation. Note that significant deposition can be seen as far north as

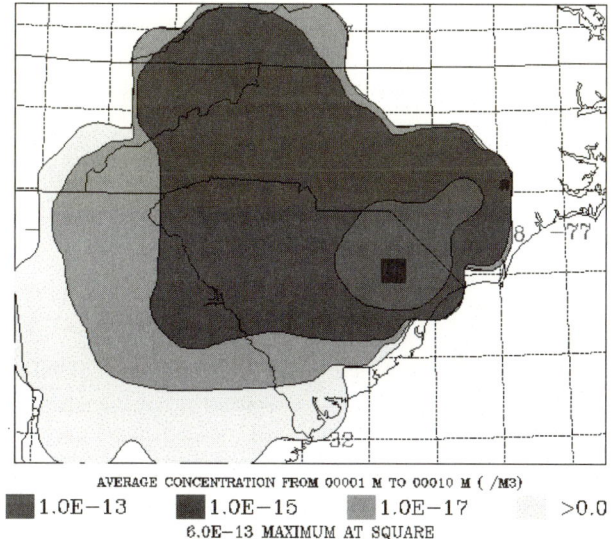

AVERAGE CONCENTRATION FROM 00001 M TO 00010 M (/M3)

$\blacksquare$ 1.0E−13 $\blacksquare$ 1.0E−15 $\blacksquare$ 1.0E−17   >0.0

6.0E−13 MAXIMUM AT SQUARE

Figure 11

Concentration distribution for a ground-level source (small box) in southeastern North Carolina from 24-hour model simulation for a wintertime scenario beginning 00UTC October 6, 1998 indicating the transport resulting from a migratory high off the northeastern United States. Results delineate a scenario for which maximum concentration and deposition can be over extreme northern South Carolina and the foothills of North Carolina despite the source being in southeastern coastal NC.

the Chesapeake Bay area. With the approaching cold front, winds at the surface increase from the southwest thus enhancing the transport potential to the northeast of the source. In addition, enhanced static instability and vertical motions ahead of the front allow greater opportunity for the parcels to mix vertically and possibly escape the boundary layer. Such a scenario can lead to a greater potential for long-range transport. This is seen in the analyses of the deposition patterns shown in Figure 13. Significant ground-level deposition values are seen as far north as New England. As expected, the corresponding concentration values are much smaller as compared to the near-source values (not shown). Thus, the vertical mixing and stronger flow aloft represent near optimal conditions for long-range transport to the northeast. As the front passes through eastern North Carolina and stalls off the coast, winds over central and eastern North Carolina are primarily from the north and northeast while winds off the coast are much stronger and have a significant southerly component. The pollutant plume for this regime is now to the southeast of the source region representing a rapid change in transport (not shown). The enhanced surface convergence increases both surface pollutant concentrations as well as wet deposition potential along the front. Behind the front, winds are from the northwest and so transport is to the southeast. Concentration and deposition patterns contours become linearly oriented along the frontal boundary (not shown).

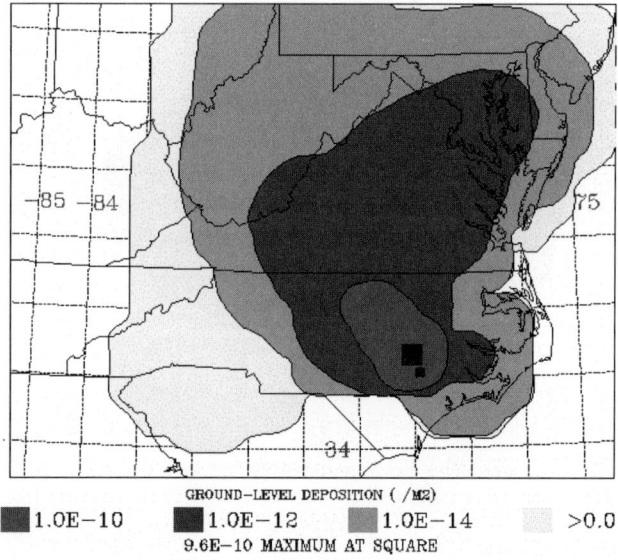

Figure 12

Ground-level deposition from 24-hour model simulation beginning 00UTC October 7, 1998. The highest deposition values have spread northward in association with veering winds ahead of an approaching cold front. Results suggests that higher concentration values and possible significant wet deposition values can affect almost entire North Carolina and points north for such a scenario.

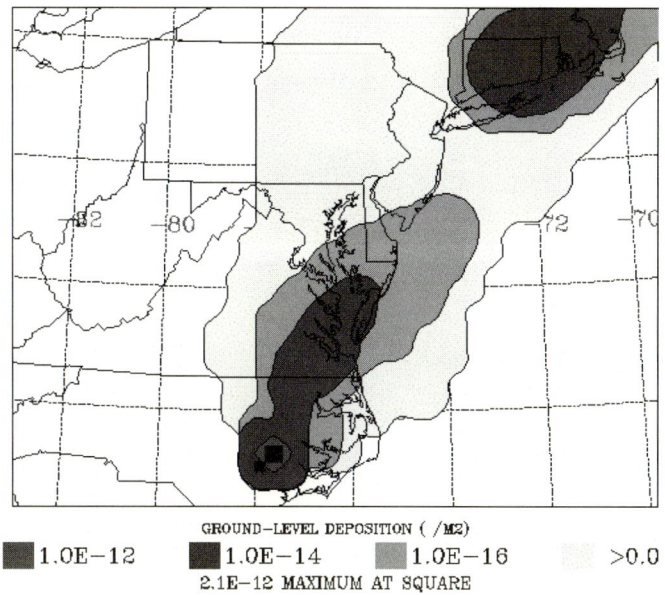

GROUND–LEVEL DEPOSITION (/M2)

■ 1.0E−12   ■ 1.0E−14   ■ 1.0E−16   >0.0

2.1E−12 MAXIMUM AT SQUARE

Figure 13

Same as in Figure 12 except from 24-hour model simulation beginning 00UTC. Scenario with a cold front to transport material even up to New England with the passage of cold front off NC coast.

This is primarily due to surface convergence and precipitation along the boundary. The front slowly moves offshore resulting in offshore transport.

Transport and deposition during the Fall season appears to be sensitive to the frequency and strength of cold frontal passage. In addition, precipitation ahead of and along the frontal boundary allows greater wet deposition potential. It also appears that the Fall season presents some potential for long-range transport and deposition of materials released in eastern North Carolina to points northeast of the source.

c) July 2–7, 1998

For the summer case, due to weak steering currents, significant long-range transport is unlikely and the pollutants are primarily confined to near source areas. During the first day of the period, transport is to the east with the strongest deposition values occurring along the coast as shown in Figure 14. In this regime significant wet deposition can occur in the event of convective precipitation. For day two, transport is more to the southeast as a weak low affects the surface wind pattern. Correspondingly, as shown in Figure 15, southeastern North Carolina and northeastern South Carolina show highest deposition potential. This pattern continues through day three. By day four of the period, as an effect of the Bermuda High, significant concentration values move due north of the source region. Such a

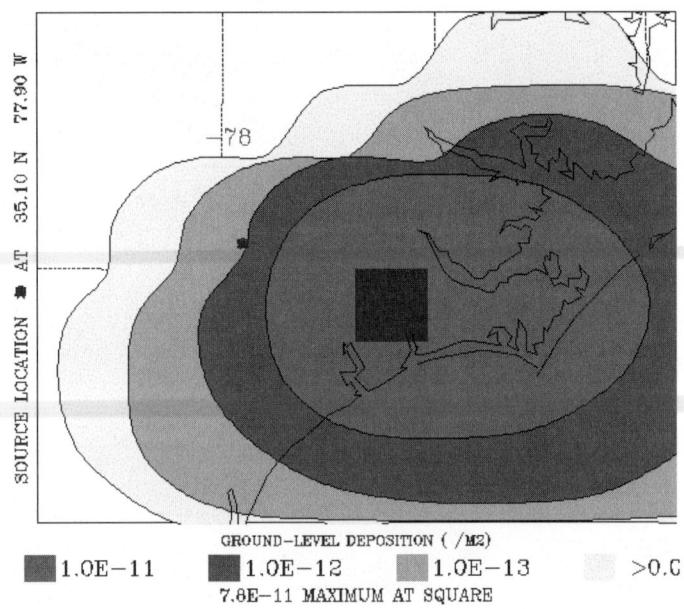

Figure 14
Ground-level deposition from 24-hour model simulation beginning 00UTC July 2, 1998. The source region is indicated by a small star with highest values indicated by a square. Transport under this regime is toward the southwest with high potential for significant wet deposition over extreme eastern North Carolina.

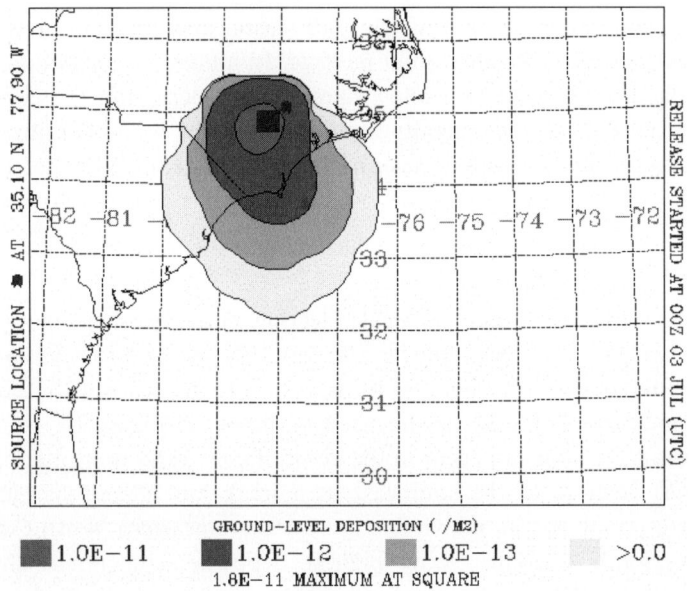

Figure 15
Same as in Figure 14 except from 24-hour model simulation beginning 00UTC July 3, 1998 for a summertime scenario providing peak concentrations in the southeast North Carolina region.

flow regime may persist for several days with uniform wind field leading to significant potential for concentration and deposition buildups over the region. However, the largest deposition values remain closer to the source region (Figure 16) as compared to the Fall case which can show similar southwesterly flow patterns (see Figure 13). This is due to several reasons. First, upper-level winds are typically weaker during the summer. Also, on a larger scale, strong subsidence associated with the Bermuda High may act to decrease the lifetime of the parcel above ground. This was evident in the trajectory analysis (not shown) as the parcels rapidly descended to the surface. The released material may be deposited and then re-emitted (see, for example, DRAXLER and HESS, 1997). This local deposition decreases the potential for long-range transport. As the pollutant moves further away from the source, concentrations are also significantly lowered due to dilution and mixing within the atmospheric boundary layer. Moreover, the summer boundary layer is typically deeper as compared to Fall. Another feature important for the summer condition is that the transport patterns often exhibit day-to-day variability in the concentration and deposition patterns. As seen in Figures 17a and 17b, there are distinct variations in the concentration values for 5 and 6 July. Figure 17a shows the pollutant deposition east and southeast of the source off the coast, while for the next day, the pollutant is

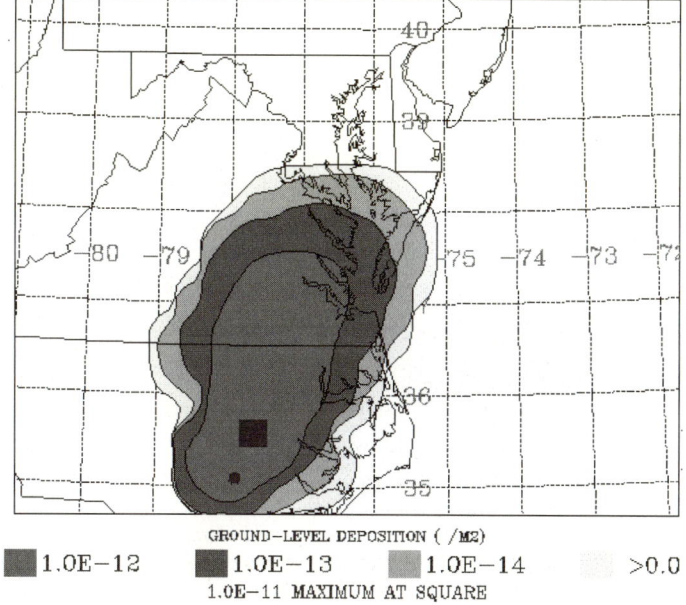

GROUND–LEVEL DEPOSITION (/M2)

■ 1.0E−12   ■ 1.0E−13   ■ 1.0E−14   >0.0

1.0E−11 MAXIMUM AT SQUARE

Figure 16

Same as in Figure 14 except from 24-hour model simulation beginning 00UTC July 4, 1998 representing transport and deposition dominated by the presence of the Bermuda High. The plume moves north of the source region possibly affecting southeastern Virginia.

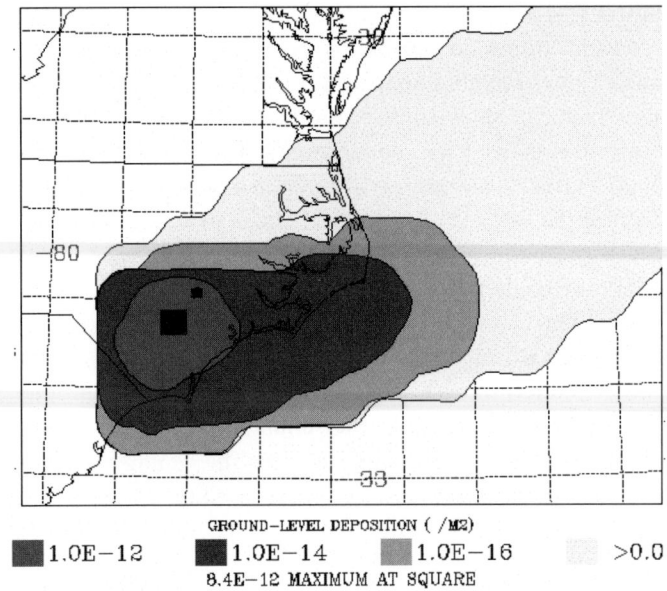

Figure 17a
Same as in Figure 14 except from 24-hour model simulation beginning 00UTC July 5, 1998. The plume is advected over portions of southeastern NC and southwest SC.

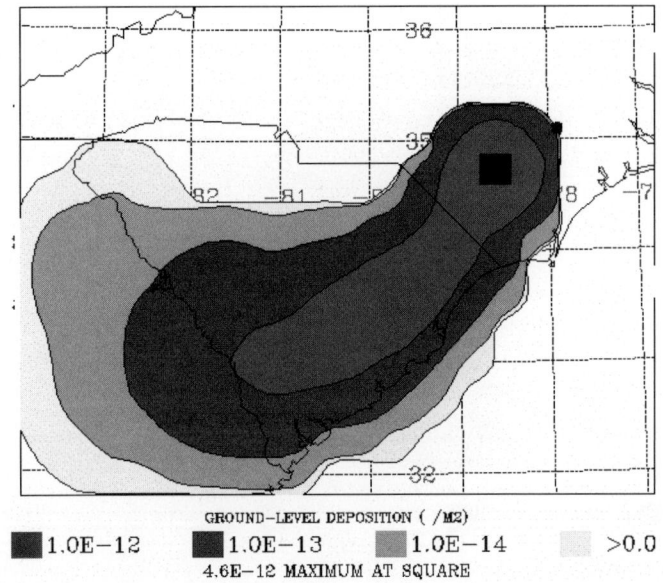

Figure 17b
Concentration distribution for a ground-level source (small box) in southeastern North Carolina from 24-hour model simulation beginning 00UTC July 6, 1998. Note the significant difference in patterns between Figure 17a despite only 24-hour difference in model initialization indicating that diurnal effects of the local forcings in the summertime scenario are dominant.

transported and deposited inland as shown in Figure 17b. Such a variation can be due to the interaction between large-scale winds and local land–sea breeze effects (and the return flow aloft) expected during the summer. This summer time land–sea breeze interaction along with the large-scale wind changes needs to be studied further. Thus the transport and dispersion associated with the summer scenario demonstrated significant day-to-day variability which can result in larger uncertainty for pollutant concentrations and deposition values over eastern North Carolina. These patterns, also appear to be sensitive to the track of weak surface disturbance that move through the region and to the location of the Bermuda High. Even though the wind field associated with the Bermuda High shows good temporal and spatial continuity, long-range transport is not likely due to weak upper-level winds and subsidence. Large concentration and deposition values are therefore typically restricted to near-source areas during the summer. Localized heavy rain associated with slow moving thunderstorms thus present the potential for significant localized nitrogen loading in eastern North Carolina.

5. Conclusions

This study assesses the potential for regional transport and deposition in eastern North Carolina for pollutant species such as reduced nitrogen typically emitted from agricultural sources. Using model analyses, mesoscale wind and thermodynamic patterns associated with three seasons are analyzed. Concentration and deposition patterns are then compared with respect to the wind and thermodynamic patterns during varying scenarios for the three periods.

Analysis shows there is a significant seasonal effect in the transport pattern over eastern North Carolina. There were distinct predominant or "climatological" wind patterns from different directions. However, in addition to the seasonal base flow, the study also identifies and highlights the role of local scale features such as coastal fronts and sea-breeze circulations interacting with large-scale events. Thus, a climatological analysis should consider the mean and the variance associated with the dispersion and deposition patterns, at the very least. Third, the analysis delineates significant day-to-day variability in the transport and deposition pattern especially during the summer case. All three case studies show these features to some extent in the outcome with summer (July case) demonstrating the largest variability. Additionally, it was shown that though seasons can be, in general, described by mean flow patterns, significant variations and dominant anomalous weather patterns inherently exist in this region, mandating a case by case analysis of the concentration and deposition patterns. With these interactions, there is also significant source of uncertainty and variability in the local scale concentration and deposition patterns. These uncertainties are compounded by the diurnal land-sea breeze cycle especially during the summer scenario when the background flow is weak.

In general, summer appears to present the best opportunity for significant local pollutant deposition. Weak winds, convection (often slow moving), and a seasonal peak in emissions may however, cause significant daily variability. Overall, Fall presented the best conditions for long-range transport and deposition. For all the three seasonal cases considered, in addition to the basal climatological flow pattern, day-to-day variations affect transport and deposition patterns in eastern North Carolina. Hence, analysis of seasonal means may not accurately depict this pattern, and more detailed weekly or daily analysis may be necessary to capture the regional pollutant loading.

One of the long-term goals of any pollution assessment study is to be able to develop a comprehensive budget of the atmospheric loading. In eastern North Carolina, for instance, such a budget would account for sources and sinks of nitrogen, and the associated uncertainty. To address this issue, the present study undertook the meteorological viewpoint pertinent to transport and regional deposition. Using routinely available observations and model results, an assessment is attempted. Although some general conclusions can be made regarding the vulnerability and uncertainty of the loading (source/sink) estimates, a budget was not possible. We believe interpretation of the local versus large-scale loading of nitrogen compounds over eastern North Carolina is still unresolved and needs to be addressed to help the task of assessing the regional budget.

Acknowledgements

The study benefits from the NC Agricultural Research Services agricultural meteorological network (AgNet) maintained by the State Climate Office of North Carolina (http://www.nc-climate.ncsu.edu) at North Carolina State University. Nitrogen deposition estimates were obtained from the National Atmospheric Deposition Program (NRSP-3)/National Trends Network. (1998). NADP Program Office, Illinois State Water Survey (http://nadp.sws.uiuc.edu/). The Hysplit trajectory and concentration analysis was performed using the READY resources (http://www.arl.noaa.gov/READY). The authors wish to acknowledge helpful discussions with Dr. Ellen Cooter, NOAA/ARL, Professors Thomas Hopkins at MEAS, NCSU, and Wayne Robarge, Soil Science at NCSU on transport and atmospheric deposition. Special appreciation is extended to Professors G. V. Rao at St. Louis University for constructive suggestions, which helped improve the presentation of this paper.

References

ANEJA, V. P., LEIGH, Y., WALKER, J., and CHAUHAN, J. (1997), *Atmospheric Ammonia/Nitrogen Compounds Emissions and Characterizations*, Workshop on Atmospheric Compounds Emissions, Transport, Transformation, Depositions, and Assessment, March 10–12.

BELL, B. D. and BOSART, L. F. (1988), *Appalachian Cold-Air Damming*, Monthly Weather Review *116*, 137–162.

CHANG, J. S., *The regional acid deposition model and engineering model*, NAPAP SOS/T Rpt 4, in *Nat. Acid Precip. Assess. Prog., Acidic Dep.: State of Sci. And Tech.*, vol 1, pp. 44-1–1-F42, (USGPO, Washington, D.C. 1990).

COOTER, E. J., RHOME, J. R., and HILL, J.B. (1997), *Spring and Summer 1995 Regional Climate Conditions and the Assessment of Atrazine Exposure in and around Lake Michigan*, 10th Conf. Appl. Clim., Reno, NV, Amer. Meteor. Soc. Boston, Oct. 20–23.

DRAXLER, R. R. and HESS, G. D. (1997), *Description of the Hysplit_4 Modeling System*, NOAA Tech Memo ERL ARL-224, Dec., 24 pp.

FINKELSTEIN, P., ELLESTAD, T., and NEAL, J. (2000), *Ozone and Sulfur Dioxide Dry Deposition to Forests: Observations and Model Evaluation*, J. Geophys. Res. *105*, 15,365–15,379.

GILLIAM, R., RAMAN, S., and NIYOGI, D. (1999), *Seabreeze Frontogenesis in North Carolina: Coastline Shape, Synoptic Flows, and Land Use Pattern*, 3rd Conf. Coastal Atmos. Ocean. Predic. Proc., 3–5 November 1999, New Orleans, Amer. Meteor. Soc., Boston, Mass.

KURTZ, D.A. (ed.), *Long Range Transport of Pesticides* (Lewis Publishers, Inc., Chelsea, 1990), 462 pp.

MAJEWSKI, M. S. and CHAPEL, P. D., *Pesticides in the Atmosphere: Distribution, Trends, and Governing Factors* (Ann Arbor Press, Inc., Chelsea, 1995) 214 pp.

NIYOGI, D., RAMAN, S., and FUNK, K. (1997), *North Carolina Coastal Climatology and the Potential for Pollution*, 10th Conf. Appl. Clim., Reno, NV, Amer. Meteor. Soc. Boston, Oct. 20–23.

PLEIM, J., FALKESTEIN, P., CLARKE, J., and ELLESTAD, T. (1999), *A Technique to Estimating Dry Deposition Velocities Based on Similarity with Latent Heat Fluxes*, Atmos. Environ. *33*, 2257–2268.

RAMAN, S., REDDY, N., and NIYOGI, D. (1998), *Mesoscale Analysis of a Carolina Coastal Front*, Bound.-Layer Meteorol. *86*, 125–145.

RHOME, J., NIYOGI, D., and RAMAN, S. (2000), *Mesoclimatic Analysis of ENSO and Severe Weather in North Carolina*, Geophys. Res. Lett. *27*, 2269–2272.

ROSWINTIARTI, O., NIYOGI, D. S., and RAMAN, S. (1998), *Teleconnections Between Tropical Pacific Sea-surface Temperature Anomalies and North Carolina Precipitation Anomalies during El Nino Events*, Geophys. Res. Lett. *25*, 4201–4204.

WALKER, J., ANEJA, V., and DICKEY, D. (2000), *Atmospheric Transport and Wet Deposition of Ammonium in North Carolina*, Atmos. Environ. *34*, 3407–3418.

WALMSLEY, P. and WESELY, M. (1996), *Modification of Coded Parameterizations of Surface Resistances to Gaseous Dry Deposition*, Atmos. Environ. *30*, 1181–1196.

WESLEY, M. (1989), *Parameterization of Surface Resistance to Gaseous Dry Deposition in Regional Scale Numerical Models*, Atmos. Environ. *23*, 1293–1304.

YARNAL, B., *Synoptic Climatology in Environmental Analysis* (Belhaven Press 1993) 195 pp.

(Received August 16, 2000, accepted April 18, 2001)

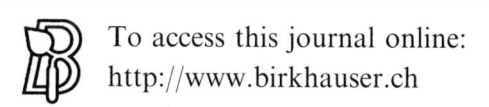

To access this journal online:
http://www.birkhauser.ch

B. Haze and Biological Effects

Pure appl. geophys. 160 (2003) 145–156
0033–4553/03/020145–12

© Birkhäuser Verlag, Basel, 2003

❙Pure and Applied Geophysics

Environmental Emission of Mercury During Gold Mining by Amalgamation Process and its Impact on Soils of Gympie, Australia

HARKIRAT S. DHINDSA,[1,3] ANDREW R. BATTLE,[1]
and SVENNING PRYTZ[2]

Abstract—The aims of this study were to estimate the total amount of mercury released to the environment during 60 years of gold mining (1867–1926) at Gympie, Queensland, Australia and to measure the mercury levels in soil samples surrounding the mining activity. We estimated that 1902 tonnes of mercury was released to the environment and about 1236 tonnes of which was released to the air. The mean mercury in the soil samples in the vicinity of the Scottish battery varied from 1.07 to 99.26 $\mu g\ g^{-1}$ as compared to 0.075 μg g^{-1} as background mercury concentrations. The maximum mercury concentration measured in sediments of the Langton Gully was 6.12 $\mu g\ g^{-1}$. These results show that large amount of mercury was used in this area during gold mining. Since mining is active in the area and Langton Gully flows into Mary River, we therefore, recommend that mercury concentration in air and fish should be monitored.

Key words: Mercury, soil, sediment, cold vapour atomic absorption spectrometry, environment.

Introduction

Mercury contamination is considered one of the worst hazards among anthropogenic impacts upon the global environment. Mercury is one of the few metal pollutants that causes human health problems; even death due to inhalation of vapour and ingestion of contaminated food has been reported. It has been well documented that bacterial methylation process can convert inorganic mercury to methyl mercury which accumulates in fish (MASON and MOREL, 1993). The consumption of mercury contaminated fish was associated with Minamata disease in Japan and Felt hat disease in England was associated with inhalation of mercury vapour that caused many deaths (MITRA, 1986). It is estimated that worldwide more than 1400 human have died and over 20,000 have suffered from mercury poisoning over the last 40 years, giving an illness with mortality rate range of 7–11% (D'ITRI, 1992).

---

[1] Chemistry Department, University of New England, Armidale, NSW 2351, Australia.
[2] Maroochy Research Center, Queensland Department of Primary Industries, Queensland, Australia.
[3] Present address: DOSME, SHBIE, Universiti Brunei Darussalam, Brunei Darussalam.
Corresponding author: Harkirat S. Dhindsa, E-mail: hdhindsa@ubd.edu.bn

Mercury emission to environment is either from natural sources (estimated to be 2500 tonnes per year (tyr^{-1}), NRIAGU and PACYNA, 1988) or from anthropogenic sources (estimated to be 4000 tyr^{-1}, PORCELA *et al.*, 1996 cited in LACERDA, 1997). The use of mercury in gold and silver mining has been one of the major sources of anthropogenic emission of mercury to the environment before cyanidation process replaced amalgamation process. A recent study shows that in Brazil, 77.9 tyr^{-1} of mercury was released to the environment through amalgamation process used in gold mining in the Amazon region, which was 67% of the total mercury emission to the atmosphere in the country (LACERDA and MARINS, 1997).

The amount of mercury released to the environment during gold mining from mines that are presently not in operation is difficult to calculate accurately. However, data from mining areas where amalgamation is still in use have been used to calculate the mercury Emission Factor (EF) for the amalgamation process. The Emission Factor is the amount of mercury emitted to produce 1.0 kg of gold. Although EF values are influenced by the quality of gold ore and the climate of the mining area, they are useful for providing reasonable estimates of the total mercury emitted to the environment during gold mining in the past if the total amount of gold recovered is known. MALLAS and BENECDITO (1986) interviewed miners in Para State in the northeastern Amazon region and reported that EF values could range from 2.0 to 4.0 kg Hg kg^{-1} of gold. LACERDA and SOLOMONS (1991) reported EF to be 1.7 based on calculations by actually determining the Hg balance throughout the entire gold production process. WISE (1966) estimated that the mass of mercury consumed in the amalgamation process is of the same order as the mass of gold recovered (EF = 1.0) for 19th century mines in Victoria, Australia and MELLOR (1952) assumed EF higher than 2.0 for this technique. LACERDA (1997) used EF value of 1.5 to compare the rate of input of mercury in tyr^{-1} to the environment from different countries. There is overall agreement among researchers that 65–87% of the total emission of mercury is released to the atmosphere (LACERDA, 1997; MITRA, 1986).

In Australia, the amalgamation process for the extraction of gold was used from late 19th until middle of the 20th century (SMYTH, 1869; BYCROFT *et al.*, 1982). During this period, large amounts of mercury were released to the environment that resulted in contamination of many sites in Australia. The tailings from an abandoned gold mine on the Thomson River Victoria, Australia, contained 40–90 µg g^{-1} of mercury (MELBOURNE AND METROPOLITAN BOARD OF WORKS, 1975). Moreover tailings contained 88 µg g^{-1} of mercury at Wood Point and 120 µg g^{-1} of Hg at Blackwood, Victoria, Australia (BYCROFT *et al.*, 1982). There are studies from all over the world about soil contamination from mercury used in gold mining (GLOVER *et al.*, 1975; BYCROFT *et al.*, 1982; EPA Victoria, 1982; PFEIFFER and LACERDA, 1988; LACERDA, 1997). No published data were available for the Gympie gold mining area, Queensland.

According to Gympie Museum and Gympie and District Historical Society information sheets, alluvial gold was first discovered in October 1867 at Gympie,

Queensland, Australia by James Nash. Gympie is about 160 km from Brisbane. It was estimated that all the alluvial gold was exhausted by the end of 1868. Amalgamation was used to recover gold until all mining in Gympie ceased in 1923. A total of 1207.4 tonnes of gold was recovered during 60 years from 1867 to 1926, of which 2.6 tonnes was alluvial gold.

The aims of this study were (i) to estimate the total amount of mercury released to the environment during 60 years of gold mining (1867–1926) at Gympie, Queensland, Australia and (ii) to determine the mercury levels in Gympie soils and sediments from Langton Gully 75 years after the amalgamation process in the area ceased.

Experimental

Total mercury in samples was measured using the method described by Adeloju et al. (1994) The details of the procedure are given below.

Sampling

Samples from 25 sites around the Scottish Gympie Battery were collected. Four sites (1–4) at some distance from the Battery were expected not to be directly contaminated; one site (5) from the shaft down to the above sites; two sites (6 and 7) from tailing heaps between the shaft and the battery; eight sites (8–15) around the Scottish Gympie Battery; two sites (16–17) from a dam close to the battery; six sites (18–23) along the Langton Gully; site one (24) from Old Victoria Battery and one site (25) from Deep Creek fossicking area. The study area is illustrated in Figure 1.

To collect a sample 4–8 holes, depending upon the area, were dug to 20 cm depth and approximately 4 kg of soil was thoroughly mixed and sampled by dividing samples into half each time until the required sample of about 100 g was obtained. Samples were transported to the laboratory in air-tight clean plastic containers at low temperature. Samples were air dried and ground to a fine powder to ensure sample homogeneity. Moisture in soil samples was determined by drying the samples at 80°C for 5 hours and water content in samples varied from 4 to 25%.

Reagents and Standards

All reagents used were of analytical grade and water was deionised. Mercury stock solution was prepared by dissolving 1.3535 g of $HgCl_2$ in 20% (v/v) HCl. Organic stock solution was prepared by dissolving an appropriate amount of methyl mercury chloride in ethanol (12 mL of 96% v/v) and diluting it to 100 mL with water. Stannous chloride solution (30% w/v) in 20% HCl was prepared daily and stabilised by adding a piece of tin. Working standards were prepared daily in 2% HCl. Hawkesbury River sediment (AGAL-10) reference material high in mercury was purchased from the Australian Government Analytical Laboratories (AGAL).

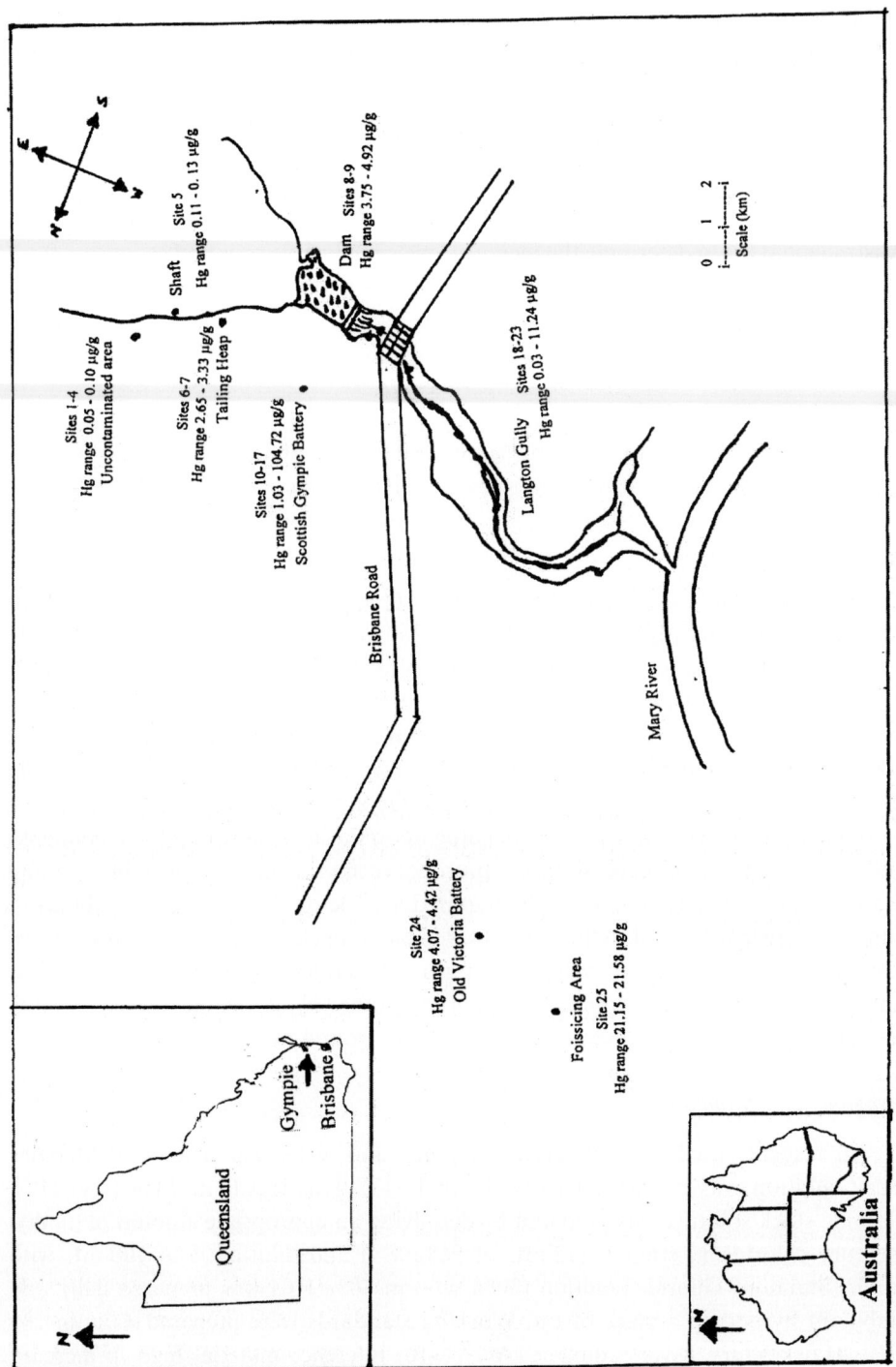

Figure 1

Study and sampling areas.

Sample Preparation

Soil sample 0.2 g of soil sample (after correction for moisture) was digested using 2.5 mL of digestion mixture (1 part of concentrated H_2SO_4 and 2.25 parts of concentrated HNO_3) in a 100 mL Erlenmeyer flask at 90°C in a water bath for one hour. The digest was diluted with water to a known volume by transferring it to a 100 mL volumetric flask. 1.0 mL of the diluted digest was transferred to a 100 mL Erlenmeyer flask containing 2.5 mL of the digestion mixture and 16.5 mL of deionised water in order to make the total volume of 20.00 mL.

CV-AAS Measurement

A Teflon coated small magnet was added to the above Erlenmeyer flask containing 20 mL of the sample solution and it was placed on a magnetic stirrer. The flask was then connected to a water reservoir and the mercury cell using a dreschel head. In all measurements, a total of 20.00 mL volume of the standard/ digest and 1.0 mL of 30% (w/v) stannous chloride were stirred using magnetic stirrer for three minutes to concentrate mercury vapour in headspace. The mercury vapour concentrated in headspace was displaced at a rate of 10 mLs^{-1} from the flask into the cell using water displacement method. After recording the peak height signal, the cell was purged with compressed air to clean out any residual mercury in the cell.

Instrumentation and Glassware

A GBC Model 902 atomic absorption spectrometer (GBC, Australia) operated in a double-beam mode was used for mercury determination for all samples under the following conditions: hollow cathode lamp current 3 mA, slit width 0.5 nm, wavelength 253.7, signal integration time 10 s, with an open ended quartz cell at room temperature.

All glassware and plastic containers were soaked in nitric acid (2 M) for at least 24 hours and rinsed three times with distilled water and two times with deionised water before use, then dried in an oven at 100°C to remove all traces of water.

Results and Discussion

Amount of Mercury Emitted to the Environment

Table 1 shows that about 1902 tons of mercury was emitted to the environment at a rate of about 30 tyr^{-1} for sixty years from 1876–1926. The EF proposed by LACERDA (1997) was preferred over EF of 1.0 proposed by WISE (1966) for Victoria, Australia because climatically Gympie (Lattitude 22°) Queensland is

Table 1

Amount of mercury emitted to environment

Gold Mined (t)	Emission Factor (EF)	Estimated Hg loss to environment (t)	Rate of Hg loss to environment (tyr^{-1})	Estimated Hg emitted to air (t)
1267.9	1.0 (WISE, 1982)	1267.9	21.2	824–1078
1267.9	1.5 (LACERDA, 1997)	1901.8	31.7	1236–1617

hotter by $7 \pm 2°C$ than Victoria (Lattitude 37°). Vaporisation of mercury to air increases exponentially with temperature and the vaporisation rate doubles for every 10°C increase of temperature. EF of 1.5 used in this study was lower than 1.7 calculated by LACERDA and SOLOMONS (1991) by actually determining the Hg balance throughout the entire gold production process. The rate 31.8 tyr^{-1} was 2–3 times the rates calculated for Bendigio (11 tyr^{-1}) and 4—6 times the rate for Victoria (5 tyr^{-1}), both in Australia (LACERDA, 1997). Moreover over 67% (1236.3 tons) of it was released into air inhaled by miners and populations in the surrounding areas. However, one should keep in mind that these estimates have been based on the amount of gold recovered from the area, which was obtained from official records. There is a strong possibility that all the gold metal mined from this area might have not been officially recorded. Based on this limitation, the total amount of mercury released into the environment is expected to be more than the estimated amount in this study. Moreover, the value of EF used in this study is an estimate based on other studies. The actual value was not known. The EF value is influenced by many factors such as the climatic conditions, management of the mining operations, quality of ore, etc. The EF value estimates might have influenced (increased or decreased) the total amount of mercury released into the environment.

Analysis Technique Variables: Detection Limit, Mercury in Reagents, and Recoveries

The absolute detection limit of 1 ng was obtained for the method. The mean background mercury concentration in reagents used for a measurement was 1.5 ± 0.5 ng. The reliability and accuracy of the method was further evaluated by doing recovery studies on the reference material and a soil sample. The recovery of inorganic mercury from the reference material was $101.2 \pm 4.3\%$ (n = 3) and of organic mercury was $96.9 \pm 5.6\%$ (n = 3). The amount of mercury recovered from the reference material 12.68 ± 0.75 µg g^{-1} was close to the certified value of 11.79 ± 1.1 µg g^{-1}. From the soil sample, the recovery for inorganic mercury was $93.9 \pm 2.3\%$ whereas for organic mercury was $81.2 \pm 0.2\%$. The recovery for organic mercury, although low, was within the acceptable range. Moreover, 98% of mercury in soils is in inorganic form (BARNETT *et al.*, 1995 and DAVIS *et al.*, 1997).

Background Mercury Levels in Soil Samples

Table 2 shows background mercury levels at four sites (1–4) which were at a considerable distance from the mining activity, therefore were considered to be less prone to mercury contamination. Mean mercury level of 0.075 µg g^{-1} of soil was considered as background mercury in soil samples. The background mercury concentrations in the samples were consistent with the results reported by DAVIS *et al.* (1997). According to them mercury levels in nonmercuriferous crustal soils and sediments in background areas, not directly impacted by anthropogenic discharges or volcanic emissions, range from 0.05 to 0.20 µg g^{-1}. However, it is believed that mercury levels at sites 1–4 might also have been slightly elevated due to atmospheric precipitation of passively dispersed mercury in air from surrounding mining activity that involved amalgamation process. Nevertheless, the mean value of 0.075 µg g^{-1} of soil provided an excellent standard for comparing mercury contamination in the area.

Mercury in Soil Samples

The mean mercury concentration at site 5, Oriental Console Shaft, 0.12 µg g^{-1} of soil was slightly higher than the background, but was still low and within the acceptable background level. A slight increase in mercury concentration at this site over the background concentration (0.075 µg g^{-1}) might be the result of transport activity involved in moving tailings from the shaft to the battery. The samples collected from sites 6 and 7 were sand samples from tailing heap after cyaniding. Mercury levels in samples from these sites were much higher than the background levels. The mercury concentrations in these tailings were lower than reported by WISE

Table 2

Background mercury in Gympie soils

Site	1	2	3	4
No. of analyses	3	3	3	3
Mercury (µg g^{-1})	0.10	0.07	0.08	0.05
Mean Deviation (µg g^{-1})	0.01	0.01	0.00	0.00

Table 3

Mercury levels between the background and the battery sites

Site	5	6	7
No. of analyses	4	4	4
Mercury (µg g^{-1})	0.12	2.80	3.16
Mean Deviation (µg g^{-1})	0.01	0.13	0.18

(1966) in Victoria because some mercury from tailings is expected to be removed during the cyanidation process.

Mercury levels in the vicinity of Scottish Gympie Battery are reported in Table 4. The mean mercury levels in this area ranged from 1.07 to 99.26 µg g^{-1} of soil. Extremely high levels were found around the battery indicating that large quantities of mercury were used in the direct vicinity of the battery. Site 8 was at some distance form the battery, therefore, it contains lower concentration of mercury than sites (9–11) adjacent to the battery. Moreover, at sites 8, 9 and 12, it appeared to have some land filling done that might have happened afterwards. Despite land filling, mercury levels were much higher than the background levels. Samples collected from sites 13 and 14, close to retort house were also high in mercury 12.82 µg g^{-1} and 4.17 µg g^{-1}, respectively. There was no vegetation around this area which may be the result of mercury and/or cyanide contamination.

Mercury in Sediments

Mercury levels in the vicinity of water dam are reported in Table 5. Samples from site 15 consisted of sediments from a huge water tank above the battery, used for water supply for crushing ore at the battery. Higher levels of mercury in sediments from the water tank than in sediments from the dam indicate the accumulation of mercury in the water tank due to recycling of the water used for crushing the ore. Sites 16 and 17 were directly from the dam and the samples collected were muddy. A mean mercury level 4.3 µg g^{-1} was found in samples collected from the dam. This concentration of mercury is about 57 times the background concentrations. Mean mercury concentrations in soil around the water reservoir were lower than those measured around the Scottish Gympie Battery (Table 5).

Table 4

Mercury levels around Scottish Gympie battery

Site	8	9	10	11	12	13	14
No. of analyses	4	4	4	4	4	4	4
Mercury (µg g^{-1})	8.50	19.14	99.26	87.04	1.07	12.82	4.17
Mean Deviation (µg g^{-1})	0.34	1.17	4.62	3.55	0.04	0.77	0.25

Table 5

Mercury in sediments of water reservoir used during mining

Site	15	16	17
No. of analyses	3	4	4
Mercury (µg g^{-1})	10.79	4.72	3.93
Mean Deviation (µg g^{-1})	0.41	0.21	0.23

Table 6

Mercury in sediments of Langton Gully

Site	18	19	20	21	22	23
No. of analyses	4	4	3	4	3	3
Mercury ($\mu g\ g^{-1}$)	4.04	3.47	2.68	6.12	0.13	0.04
Mean Deviation ($\mu g\ g^{-1}$)	0.26	0.05	0.09	0.11	0.01	0.01

Table 6 shows the mercury levels in the sediments collected from the Langton Gully. Langton Gully samples were collected from sites downstream from the dam behind the bridge on the Brisbane Road. Mean mercury levels in sediments from sites 18 to 21 ranged from 2.68 $\mu g\ g^{-1}$ to 6.12 $\mu g\ g^{-1}$. The mercury concentrations in the sediments suggest that the gully is contaminated. These concentrations are considerably higher than those reported in sediments of the Lerdererg River next to the Blackwood gold mine in Victoria, Australia (BYCROFT *et al.*, 1982). Sites 20 and 21 contained less mercury because the samples were clay samples and sampling at site 21 close to site 20 was done at a deeper level. Mean mercury concentrations in Langton gully sediments and soil samples from the water reservoir were comparable but lower than the samples from Scottish Gympie Battery (Table 4).

Mercury at Another Location

There were 600 heads of batteries in the Gympie region that used mercury coated copper plates to trap gold. Mercury concentrations in soils around these batteries are also expected to be high. To ensure the use of mercury at other sites we included samples from sites 24 and 25. Site 24 covered Deep Creek bank mud accumulated on ground bar below the Old Victoria Battery. The presence of 4.22 $\mu g\ g^{-1}$ of Hg in the mud indicated the movement of mercury from the Old Victoria Battery. Site 25 was a fossicking area immediately below the Old Victoria battery. The mean mercury concentration at site 25 was greater than 21 $\mu g\ g^{-1}$. The mercury levels at this site may further increase because the area is still open to the general public for gold fossicking, and some amateur fossickers have been reported to use mercury.

Implications

In general, soils in the vicinity of Scottish Gympie Battery were found to be contaminated with mercury even after 75 years since the use of mercury in the area ceased. The natural decontamination process for mercury is very slow (MITRA, 1986), since most of the mercury in the soil is elemental and in an inorganic form, it

Table 7

Mercury from Deep Creek fossicking area and from Old Victoria Battery

Site	24	25
No. of analyses	4	4
Mercury (μg g^{-1})	4.22	21.37
Mean Deviation (μg g^{-1})	0.17	0.24

vaporises and disperses in air. Therefore, the mercury concentration in air might increase as compared to the background mercury concentration in ambient air, which is 1.5 to 1.8 ng m^{-3} (BEAUCHAMP and TORDON, 1998). The concentrations of mercury vapour in air over contaminated soils and precious metal mines have been reported to reach 1500 ng m^{-3} at ground level and dust particles are also known to contain high concentrations of mercury (MITRA, 1986). Mercury is a neurotoxin. It is known that inhaled mercury accumulates in brain. The rate of accumulation of mercury in the brain is much higher than the disposal rate from the brain by the natural body mechanism (MITRA, 1986). Since mining is active in this area, therefore there is need to evaluate the concentration of mercury in air in the vicinity of the mining area.

Langton Gully directly downstream from the Scottish Gympie Battery was found to contain high mercury levels. Of particular concern, however, is the proximity of Langton Gully to Mary River. Several mercury-contaminated sites including shafts, tailing sites and Old Victoria Battery exist downstream of the Scottish Gympie Battery. Langton Gully is a permanent watercourse, it could be expected to carry mercury downstream to Mary River, where it may be transformed into organic form and accumulate in fish. The consumption of fish from the river may indeed increase mercury levels in humans. Research into mercury levels in the river sediments and fish is, therefore, warranted.

Conclusions

About 1902 tons of mercury were released to the environment in the vicinity of Gympie during 60 years of gold mining and 65–87% of this ($<$1236 tons) was released into the air. High mercury levels were found in the soil around Scottish Gympie Battery. Mercury concentrations in contaminated soils in the area ranged from 0.11 to 104.72 μg g^{-1} (Mean 12.00 $\pm$ 21.14 μg g^{-1}). High mercury concentrations in samples from Victoria battery indicate that soils around all the 600 batteries could also expected to be contaminated with mercury. Analyses of sediment samples show that Langton Gully was also contaminated as result of considerable amount of washed from tailings into the Gully. As the Gully is a permanent water-course to

Mary River, the mercury contamination is expected to extend to the river via a race. Moreover, it is believed that tailings were also directly dumped into the river. Therefore, there is a strong need to investigate mercury levels in air around the mining areas as well as sediments and fish from the river.

References

ADELOJU, S.B., DHINDSA, H.S., and TANDON, R.K. (1994), *Evaluation of Some Wet Decomposition Methods for Mercury Determination in Biological and Environmental Materials by Cold Vapour Atomic Absorption Spectroscopy*, Anal. Chim. Acta *285*, 359–364.

BARNETT, M.O., HARRIS, L.A., TURNER, R.R., HENSON, T.J., MELTON, R.E., and STEVENSON, R.J. (1995), *Characterization of Mercury Species in Contaminated Floodplain Soils*, Water Air Soil Pollut. *1*, 643–646.

BEAUCHAMP, S. and TORDON, R. (1998), *Air-Surface Exchange of Gaseous Mercury from Gold Mine Tailing in the Atlantic Region*, In Preliminary Agenda and Abstracts for the Conference on Mercury in Eastern Canada and the Northeast States. Unpublished.

BYCROFT, B.M., COLLER, B.A.W., DEACON, G.B., COLEMAN, D.J., and LAKE, P.S. (1982), *Mercruy Contamination of the Lerderderg River, Victoria, Australia, from an Abandoned Gold Field*, Environ. Pollution (Series A) *28*, 137–147.

DAVIS, A., BLOOM, N.S., and HEE, S.S. (1997), *The Environmental Geochemistry and Bioaccessibility of Mercury in Soils and Sediments: A Review*, Risk Analysis *17*, 557–569.

D'ITRI, F.M., *Mercury pollution and cycling in aquatic system*. In *Proc. 5th Int. Conf. On Environmental Contamination* (CEP Consultants Ltd, Edinburgh 1992) pp. 391–402.

ENVIRONMENTAL PROTECTION AUTHORITY (Victoria), *An integrated report on mercury chemistry and biology of Raspberry Creek* (EPA (Victoria), Melbourne, 1975).

GLOVER, J.W., BACHER, G.J., and PEARCE, T.S., *Environmetal Studies Series Report* No. 279 (Ministry for Conservation, Victoria, Melbourne 1975).

GYMPIE GOLD MINING MUSEUM (GGMM), *Gold Facts: Gold in Gympie* (GGMM, Information Sheet, 1997).

GYMPIE and DISTRICT HISTORICAL SOCIETY Inc. (GDHSI), *Discovery of Gold by James Nash* (GDHSI, Information Sheet, 1997).

LACERDA, L.D. (1997), *Global Mercury Emission from Gold and Silver Mining*, Water, Air and Soil Pollution *97*, 209–221.

LACERDA, L.D. and MARINS, R. V. (1997). *Anthropogenic Mercury Emission to the Atmosphere in Brazil: The Impact of Gold Mining*, J. Geochem. Exploration *58*, 223–229.

LACERDA, L.D. and SOLOMONS, W., *Mercury in the Amazon. A Chemical Time Bomb?* (Dutch Ministry of Housing, Physical Planning and Environment, Haren, 1991).

MALLAS, J. and BENECDITO, N. (1986), *Mercury and Goldmining in the Brazillian Amazon*, Ambio. *15*, 248–249.

MASON, R.R. and MOREL, F.M.M., *An assessment of the principal pathways for oxidation of elemental mercury and production of methyl-mercury in Brazilian waters affected by gold-mining*, In *Proc. Int. Symp. Environ. Geochemistry in Tropical Countries*, (Niteroi 1993) pp. 413–416.

MELBOURNE AND METROPOLITAN BOARD OF WORKS (MMBW), *Report on environmental study into Thomson Dam and associated works*, vol. 1, Proposed Works and Existing Environment (MMBW, Melbourne 1975).

MELLOR, J.W., *A Comprehensive Treatise on Inorganic and Theoretical Chemistry* (Longman, Green and Co., London 1952).

MITRA, S., *Mercury in Ecosystem* (Trans. Tech. Publications, U.S.A., 1986).

NRIAGU, J.O. and PACYNA, J.M. (1988), *Quantitative Assessment of Worldwide Contamination of Air, Water and Soils by Trace Metals*, Nature *333*, 134–139.

PFEIFFER, W. C. and LACERDA, L. D. (1988), *Mercury inputs into the Amazon region, Brazil*. Environ. Technol. Lett. *1*, 325–330.

SMYTH, R.B., *The Gold Fields and Mineral Districts of Victoria* (Queensberry Hill Press, Melbourne, 1969, Facsimile Reprint, 1980).

WISE, E.M., *Gold and gold compounds*. In *Kirk-Othmer Encyclopedia of Chemical Technology* (John Wiley, New York, 1966).

(Received February 15, 2000, accepted December 14, 2000)

To access this journal online:
http://www.birkhauser.ch

Pure appl. geophys. 160 (2003) 157–187
0033–4553/03/020157–31

© Birkhäuser Verlag, Basel, 2003

█ **Pure and Applied Geophysics**

Chemistry of Forest Fires and Regional Haze
with Emphasis on Southeast Asia

Miroslav Radojevic[1]

Abstract — The current state of knowledge regarding the chemistry of forest fires and regional haze is reviewed. More than 100 compounds have been identified in wood smoke and many of these have also been observed in field studies. Products of biomass combustion can have different environmental effects: CO_2 and CH_4 may contribute to global warming, NO_x and SO_2 could contribute to rainwater acidity, whereas smoke particles and polynuclear aromatic hydrocarbons (PAHs) could affect human health. Also, photochemical reactions of primary emissions from biomass fires can lead to the production of secondary pollutants such as O_3. Regional haze episodes caused by forest fires have occurred in SE Asia on several occasions during the 1990s and the reported studies of these episodes are reviewed. Only total suspended particles (TSP) were determined in the earlier studies, and more comprehensive chemical investigations have only emerged during the more recent episodes, notably those of 1997 and 1998. To date, most of the measurements have centred on criteria pollutants (SO_2, NO_2, CO, O_3 and PM_{10}), however, other pollutants (e.g., VOCs, PAHs) have also been determined in certain studies. Rainwater analyses suggest that forest fires do not have a major acidifying effect because dissolved acidic gases (e.g., SO_2) are neutralised by alkaline substances (e.g., Ca, Mg, K) that are also emitted by forest fires. There is a need for further laboratory and field studies in order to investigate important pollutant transformation mechanisms.

Key words: Forest fires, haze, air pollution, atmospheric chemistry, acid rain, ozone.

1. Introduction

Forest fires in the tropics consume up to 80% of the total biomass burned on a global scale (CRUTZEN and ANDREA, 1990) and these have been of interest to researchers for more than a decade. More recently, the haze in SE Asia created by forest fires in Borneo and Sumatra has caused considerable concern amongst scientists, governments and members of the public alike. The so-called "haze" is a regional air pollution phenomenon caused by emissions from biomass fires. The pollutants emitted by forest fires can be transported thousands of kilometers and affect several countries for periods of weeks or months at a time (Fig. 1). The resulting haze can impact the lives of millions. The effects of this haze include impacts on: health, biogeochemical cycles (e.g., carbon cycle), weather and climate, tropospheric

[1] Department of Chemistry, University of Brunei Darussalam, Bandar Seri Begawan, BE1410, Brunei Darussalam. E-mail: miro@fos.ubd.edu.bn

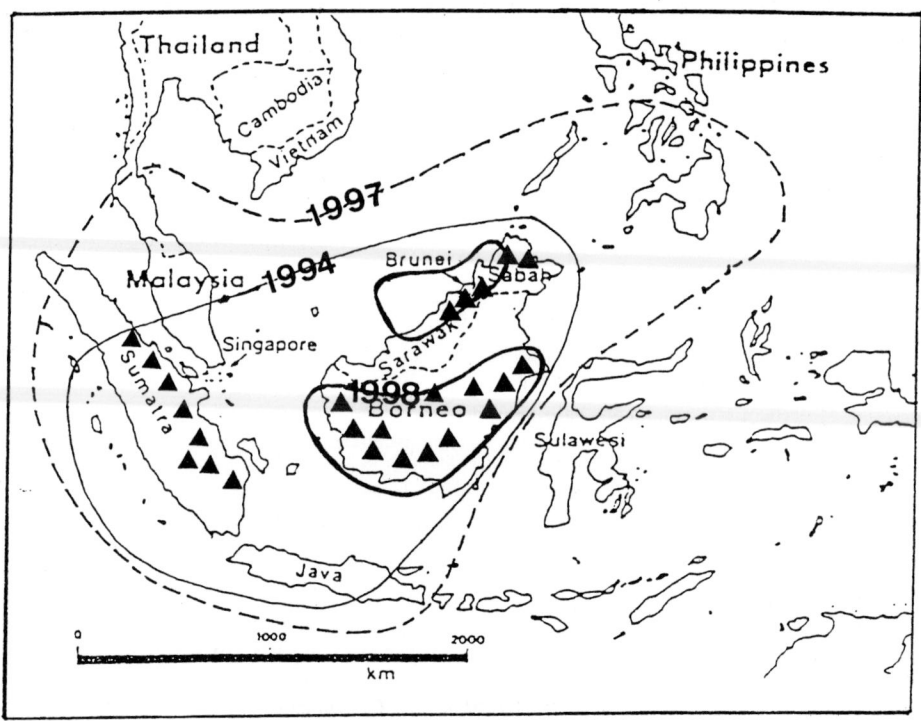

Figure 1
Location of forest fire hot-spots (▲) and extent of regional haze in SE Asia during August–October 1994,
August–October 1997, and February–May 1998.

ozone, atmospheric chemistry, and rainwater acidity. The worst haze episodes in SE
Asia occurred in 1997 and 1998, but forest fires in Mexico and the southern United
States caused a similar regional haze episode in Central America. During the 1997–
1998 period forest fires were also reported in Brazil, Spain, Greece, Australia,
Mongolia and Russia. In view of the growing incidence of forest fires and the resulting
haze throughout the world, there is need for a greater understanding of the various
aspects of these phenomena, including their chemistry. This paper reviews the current
state of knowledge regarding the chemistry of forest fires and haze.

The various environmental impacts of haze from biomass fires, and especially the
health effects, depend on the physicochemical properties of emissions from forest
fires and the resultant haze. Most of the information to date comes from air quality
stations in regions affected by haze. These stations routinely monitor the concentrations of *criteria pollutants*: fine particulate matter <10 μm in diameter (PM_{10}),
SO_2, NO_2, CO, and O_3 (RADOJEVIC, 1998). Although these pollutants are harmful to
health and ecosystems, the full environmental impact assessment of the haze requires
considerably more detailed chemical characterisation. Many potentially harmful

compounds may be present at trace levels in gases and smoke particles emitted by forest fires and these have generally not been determined during haze episodes resulting from biomass fires.

Most of our knowledge regarding forest fire chemistry originates from earlier studies conducted in various parts of the world (e.g., The United States, Australia, Brazil, Mexico, Africa). However, the results of these studies are of little use in assessing the health effects and other environmental impacts of the resulting pollution since their main objective was to quantify the flux to the atmosphere of various trace gases from biomass burning. Compounds commonly determined include CO_2, CO, CH_4, O_3, SO_2, N_2O, NO, NO_2, COS, and CH_3Cl (KUHLBUSH et al., 1991; CRUTZEN and ANDREA, 1990; SETZER and PEREIRA, 1991; WARD and HARDY, 1991; COFER et al., 1993; CRUTZEN and CARMICHAEL, 1993; ARTAXO et al., 1993) and most of these have little or no health implication, at least at the reported concentrations. Also, past studies of tropical forest fires have generally been confined to sparsely populated areas of Africa and the Amazon basin. Forest fires and resultant haze in SE Asia affect one of the most densely populated regions of the world, and it is the recent occurrence of haze in SE Asia that has brought the adverse impacts of this pollution to the attention of the experts and the public.

2. Biomass Combustion Chemistry

The fuel for biomass fires includes various types of vegetation (trees, cultivated plants, bushes, grass), peat and lignite coal. Forest fires in SE Asia generally occur in regions of peat swamps and underground lignite coal deposits. Vegetation is composed mainly of cellulose and hemicellulose (5–70% of dry matter). Other components of vegetation include: lignin (15–30%), nucleic acids, amino acids, volatile extractables (aldehydes, alcohols, terpenes), minerals (<10%) and water (<60% of fresh matter). Dry plant matter can be approximated to the empirical formula CH_2O but it also includes the following elemental constituents: N (0.3–3.8% w/w), K (0.5–3.4% w/w), S (0.1–0.9% w/w), and P (0.01–0.3% w/w).

Under ideal circumstances, combustion of biomass should lead only to CO_2 and H_2O:

$$CH_2O + O_2 \rightarrow CO_2 + H_2O.$$

However, incomplete combustion also releases CO and many unburned hydrocarbons. S and N present in the biomass are converted to oxides during the combustion process:

$$S + O_2 \rightarrow SO_2,$$

$$2N + O_2 \rightarrow 2NO.$$

N present in amino acids is mainly converted to NO during biomass combustion. At very high temperatures NO_x species may also result from the reaction of N_2 in the air, according to the Zeldovich chain reaction mechanism:

$$O_2 \Leftrightarrow 2O$$

$$O + N_2 \Leftrightarrow NO + N$$

$$N + O_2 \Leftrightarrow NO + O.$$

The chain reaction is initiated by the atomic oxygen which is formed from the dissociation of O_2 molecules at high temperatures reached during the flaming stage. However, high temperatures required for the Zeldovich mechanism are generally not expected in most biomass fires, except perhaps in very intense flames (LOBERT and WARNATZ, 1993). Although some NO_2 may be formed during combustion, most of it results from the oxidation of primary NO in the atmosphere.

A forest fire goes through several stages:

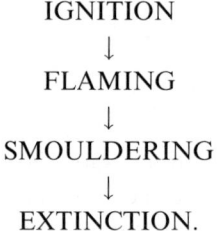

IGNITION
↓
FLAMING
↓
SMOULDERING
↓
EXTINCTION.

Most emissions take place during the flaming and smouldering stages. These two processes are quite different in appearance, types of chemical reactions involved, and products emitted. The flaming stage is characterised by high temperatures and visible flames, while temperatures are much lower during smouldering and the fire burns without any visible flames. During flaming, peak temperatures can be as high as 1800 K, producing emissions rich in H_2O, CO_2, NO, N_2O and N_2, along with particles high in elemental carbon. In general, reactions tend to go more to completion during the flaming stage and some of the more oxidised forms are released. Water and volatile extractables (e.g., alcohols, aldehydes, terpenes) are volatilised from the fuel material in an initial drying/distilling process. The subsequent pyrolytic step causes high temperature cracking of the fuel molecules. During this stage, components of high molecular mass are decomposed to compounds of low molecular mass. Char and tar products of intermediate molecular mass are formed and these serve as a primary energy source for the flame process, eventually decomposing to gaseous products. Gaseous hydrocarbons react with reactive atoms and radicals (e.g., OH·) formed in the hot regions of the flame. Unstable radicals formed during these reactions quickly decompose to more stable, smaller hydrocarbon radicals (e.g., CH_3, C_2H_5) which are slowly oxidised to H_2O, CO, and CO_2. CO_2 is formed by oxidation of CO by OH· radicals:

$$CO + OH\cdot \rightarrow CO_2 + H\cdot$$

CO can form in fuel-rich parts of the flame and at temperatures too low for the formation of OH· radicals. Flaming takes place under high oxygen concentrations ($\geq 15\%$).

Consumption of the biomass, buildup of charcoal layers on wet surfaces, increase in ash content, and the loss of volatile gases during flaming causes the flame to cease. This slows the rate of pyrolysis leading to the formation of less flammable products and lower temperatures characteristic of the smouldering stage, which can persist for several days. During smouldering a large quantity of incompletely oxidised compounds is emitted: CO, CH_4, non-methane hydrocarbons, polynuclear aromatic hydrocarbons (PAHs), CH_3Cl, H_2S, COS, $(CH_3)_2S$, $(CH_3)_2S_2$, amines, heterocycles, amino acids, and particles low in elemental carbon. Smouldering can take place at oxygen concentrations as low as 5% in densely packed fuel beds. CO is formed by low-temperature surface reaction of O_2 with carbon, and it cannot be oxidised to CO_2 to the same extent as during the flaming stage.

Other than the temperature, a number of other factors can influence biomass fires and the emissions produced: the type of vegetation, its water content, density and structure; weather (e.g., lightning can start fires while heavy rains can put them out, wind speed can determine whether the fire is static or moving); and topography (fires behave differently on slopes than on flat surfaces). The water content of the plant material is especially important as it can determine the burning efficiency. High water content may prevent a plant from igniting, or it may reduce flaming while enhancing smouldering combustion. Density and structure are also important; for example, high density stemwood is more difficult to ignite than low density grasses.

CO and CO_2 are the major carbon containing compounds released by biomass combustion, followed by hydrocarbons, carbon associated with particulate matter (soot), and other minor substances. Generally, more than 95% of carbon emitted by biomass fires is in the form of CO and CO_2. Smoke particles emitted during biomass combustion are generally < 10 µm in diameter. Studies of biomass fires reveal that the particles cover a broad size spectrum. Particles from 0.01 to 43 µm in diameter have been measured with a pronounced number concentration peak at 0.15 µm (WARD, 1990). Between 40 and 95% of the mass of particles consists of particles < 2.5 µm in diameter, while particles > 2.5 µm but smaller than 10 µm account for less than 10% of the particle mass (WARD, 1990). Numerous trace metals are also emitted during biomass combustion.

3. Biomass Combustion Studies

Field measurements and laboratory studies can both be used to evaluate the composition of emissions from biomass fires. These are widely used to estimate pollutant fluxes from biomass burning and predict global impacts of this pollution. Although forest fires in the United States contribute only 2–3% of global emissions, these have been the most widely investigated. Fairly comprehensive forest fire studies have also been carried out in Brazil and central Africa. Some field measurements have been conducted in Malaysia, Indonesia, Singapore and Brunei during the 1997 and 1998 haze episodes, however, there has been only one laboratory study utilising vegetation found in SE Asia (MURALEEDHARAN et al., 2000a).

Studies of emissions from forest fires have included:

- Laboratory studies:
    (i) microlaboratory (mg quantities of fuel)
    (ii) combustion laboratory (kg amounts of fuel)
- Field studies:
    (i) ground-based measurements
    (ii) airborne measurements
- Satellite measurements
- Modelling studies

Reported studies of biomass burning and forest fires are summarised in Table 1. Satellites can be used not only to detect forest fires and the extent and severity of the haze, but also to estimate pollutant emissions (e.g., CO, CH_4, NO_2, HCHO). Computer simulations involving many elementary combustion reactions have been used to model the chemistry of forest fires, and computer models have also been used to simulate the long-range transport of pollutants from areas of forest fires.

3.1. Laboratory Studies

In laboratory studies, weighed quantities (mg or kg) of biomass are burnt under controlled conditions and the combustion products analysed. In microcombustion experiments mg quantities of biomass are burnt. In one microcombustion experiment, 10 mg samples of ground-up pine needles were burnt and emission factors for total hydrocarbons, CO, CO_2, VOCs, and thermoparticulates were evaluated (CLEMENTS and MCMAHON, 1984). Experimental conditions were varied to include oxygen and non-oxygen (N_2) atmospheres and to simulate pyrolytic processes associated with smouldering combustion. An average emission factor of 85 g kg^{-1} was found for particulate matter in an oxygen atmosphere. In another study, employing thermo-gravimetric-evolved gas analysis, the production of NO_x was found to be proportional to the N content of the biomass fuel and to agree well with the emission factors determined in larger-scale fires of 1–10 g kg^{-1} (CLEMENTS and MCMAHON, 1980). The authors concluded that most of the NO emitted resulted from the prompt release of nitrogen bound up in the biomass fuel.

Combustion products in a test fire burning a sample of grass from Venezuela included: H_2, CH_3Cl, N_2 particulates, and various other minor components (CRUTZEN and ANDREA, 1990). Two distinct stages of combustion were identified:

Table 1

Studies into biomass combustion

A. Combustion studies

Microcombustion
Benner *et al.* (1997)
Clements & McMahon (1980)
Clements & McMahon (1984)
Edye & Richards (1991)
Kuhlbusch *et al.* (1991)
Muraleedharan *et al.* (2000a)

Combustion laboratory
Feldstein *et al.* (1963)
Darley *et al.* (1966)
Fritschen *et al.* (1970)
Vines *et al.* (1971)
Philpot *et al.* (1972)
Sandberg *et al.* (1975)
McMahon & Tsoulakas (1978)
Lobert *et al.* (1991)
Jenkins *et al.* (1991)
Griffith *et al.* (1991)

B. Field studies

Ground-based
Fritschen *et al.* (1970)
Ward (1989)
Crutzen *et al.* (1985)
White (1987)
Bonsang *et al.* (1991)
Ward & Hardy (1991)
Ward *et al.* (1992)
Sani *et al.* (1991)
Abas *et al.* (1995)
Zakaria *et al.* (1995)
Abas & Simoneit (1996)
Japony & Daliman (1999)
Muraleedharan & Radojevic (2000)
Radojevic & Hassan (1999)
Muraleedharan *et al.* (2000b)
de Jong & Hollander (1998)
Ponka *et al.* (2000)

Airborne
Vines *et al.* (1971)
Evans *et al.* (1977)
Radke *et al.* (1978)
Crutzen *et al.* (1979)
Stith *et al.* (1981)
Westberg *et al.* (1981)
Crutzen *et al.* (1985)
Radke *et al.* (1988)
Cofer *et al.* (1988)
Andrea *et al.* (1988)
Radke *et al.* (1990a,b)
Radke *et al.* (1991)
Kaufman *et al.* (1992)
Uthe *et al.* (1992)
Cofer *et al.* (1993)
Sawa *et al.* (1999)
Ward & Hardy (1991)

Combined ground & airborne studies
Ward *et al.* (1979)
Radke *et al.* (1990b)
Artaxo *et al.* (1993)
Tsutsumi *et al.* (1999)
Lacaux *et al.* (1993)

an early, flaming stage with high temperatures and bright flames; and a later, smouldering stage when the fire is cooler and flames are not visible. The flaming stage was characterised by more oxidised forms (e.g., CO_2, SO_2, NO_x), whereas the smouldering stage emitted incompletely oxidised compounds (e.g., CH_4, N_2O). The exact duration of each stage can vary depending on the conditions at the time of the fire. The ratio of CO to CO_2 is a good indicator of the degree of flaming combustion. Analysis of smoke samples from burning savanna grasslands in Northern Australia using Fourier Transform Infrared (FTIR) spectroscopy found CO/CO_2 ratios to vary between 2 and 20% depending on the combustion conditions (GRIFFITH, 1991). Low-temperature smouldering combustion and high moisture content gave rise to high CO/CO_2 ratios, whereas high temperature, well-oxygenated flaming combustion resulted in low CO production. The same study found that acetylene (C_2H_2) emissions were well correlated with CO_2. GRIFFITH (1991) also reported the use of FTIR to analyse emissions from biomass fires in a large-scale combustion laboratory. He found that NH_3 was the major nitrogen containing compound in the smoke; NH_3 constituted about 5 times ($NO + NO_2$). This is contrary to other studies which found NO_x emissions to dominate NH_3 (LOBERT et al., 1990). There is a good relationship between the fraction of N in the fuel emitted as N_2 and the smouldering/flaming (CO/CO_2) ratio indicating that emissions are favoured by flaming combustion, while emissions of CH_4 are favoured by smouldering combustion (KUHLBUSCH et al., 1991).

Approximately 90% w/w of N and ca. 95% w/w of C present in biomass is volatilised (LOBERT et al., 1990). About 11% w/w of the volatilised carbon is released as CO and about 1% w/w as CH_4, while about 13% w/w of the volatilised N is released as NO. On the basis of these results we can expect 1 kg of dry biomass to release: 0.5 g N as NO_x, 42 g of C as CO and 4 g C as CH_4.

Since temperatures in biomass fires are generally lower than those required for the formation of thermal NO_x from N_2 in air via the Zeldovich mechanism, most of the NO_x comes from nitrogen compounds in the biomass itself. It is possible to relate the NO_x emission rate to the nitrogen content of fuel using the following equation (DIGNON et al., 1991):

$$E_F(NO_x) = 3.9n_f - 1.5,$$

where $E_F(NO_x)$ is the emission rate of NO_x in gN kg^{-1} (dry biomass) and n_f is the percentage of nitrogen in the dry biomass.

As yet, few laboratory studies of emissions from biomass in SE Asia have been reported. In one microlaboratory study (MURALEEDHARAN et al., 2000a), samples of peat (30–50 mg) were combusted in closed pyrex (20 mL) and silica (30 mL) tubes at temperatures of 480 and 600°C, which were considered to be representative of the smouldering stage of combustion. Evolved gases were analysed for hydrocarbons (C_1 to C_4), CO, CO_2, aldehydes and polynuclear aromatic hydrocarbons (PAHs). Aldehydes and PAHs were present below the analytical detection limits, however, emission factors were estimated for the other gases. The most abundant combustion product was CO_2, followed by CO and CH_4. The concentration of alkanes in the evolved gases generally decreased with increasing molecular mass. The proportion of alkanes of lower molecular mass (methane, ethane) increased as the temperature increased, while the concentration of long chain alkanes decreased with increasing temperature. The emission factors for CH_4 in this study agree well with those reported by WARD and HARDY (1984) for smouldering combustion. Laboratory combustion experiments of different types of fuel wood from tropical and subtropical regions have been reported (KOPPMANN et al., 1999). Emissions of VOCs, CO, CO_2 and CH_4 were determined and found to vary considerably depending on the fuel type, water content of the fuel, the burning conditions, and the stages of the fires.

Experimental studies of wood combustion have revealed the presence of more than 100 compounds in the smoke, both organic and inorganic, from percentage levels for H_2O and CO_2, down to part per million levels for PAHs. Compounds that have been identified in wood smoke include: CO, CH_4, SO_2, NO, NO_2, CH_3Cl, benzene, various aldehydes, phenol, furans, dioxins, toluene, acetic acid, formic acid, cyclic di- and triterpenoids, various PAHs, normal alkanes (C_{24}–C_{30}), organic and elemental carbon, and numerous trace metals (LARSON and KOENIG, 1994). Products of wood combustion include compounds which are known carcinogens: benzene and benzo(a)pyrene (LARSON and KOENIG, 1994). Benzene has also been identified in flow reactor combustion chamber studies of biomass burning (WARD, 1979). Benzo(a) pyrene was measured in particulate matter emitted from pine needle combustion in controlled experiments at concentrations between 2 and 274 $\mu g g^{-1}$ of particulate matter (MCMAHON and TSOULAKAS, 1978). The products of wood combustion may have different environmental effects: NO_x, SO_2 and organic acids (e.g., formic, acetic) could contribute to rainwater acidity, CO_2 and CH_4 may contribute to global warming, while aldehydes and PAHs may affect human health.

3.2. Field Studies

While laboratory studies are convenient in that the experimental conditions can be carefully controlled, there is always the possibility that the conditions may not be fully representative of those under which real fires occur. Therefore, the results of laboratory experiments need to be substantiated with data from field studies of real biomass fires. Field studies can be either airborne or ground-based. In airborne studies, instrumented aircraft fly through forest fire plumes taking samples for chemical analysis, while in ground-based studies emissions are measured using either fixed monitoring stations or instrumented motor vehicles. Various configurations of sampling apparatus have been used to measure the flux of gases, particles and heat from near full-scale test fires in the different studies.

Most of the reported studies of biomass fires have been carried out in temperate and subtropical regions (e.g., EVANS et al., 1976; ROBERTS and ANDERSON, 1991; COFER et al., 1993; RADKE et al., 1995; CHENG et al., 1998). Emission factors of various pollutants measured over real biomass fires in the United States are summarised in Table 2. In addition to NO_x, CO and CH_4, non-methane hydrocarbons were also observed and these could contribute to secondary ozone formation via photochemical reactions. COFER et al. (1993) studied emissions of CO_2, CO, H_2, CH_4 and total non-methane hydrocarbons from prescribed biomass fires in Mexico using aircraft measurements. One interesting airborne study investigated the effect of aging on smoke from a forest fire (RADKE et al., 1995). A smoke plume from a large

Table 2

Emission factors (g kg^{-1} dry biomass) measured over real fires. Adapted from HEGG et al. (1990). n-C$_4$: Straight chain paraffin with carbon number 4

Fire	Vegetation	NO$_x$	CO	CH$_4$	C$_3$H$_6$	C$_2$H$_6$	C$_3$H$_8$	C$_2$H$_2$	n-C$_4$
Lodi I	Chapparal, Chamise	8.9 ± 3.5	74 ± 16	2.4 ± 0.15	0.58 ± 0.05	0.35 ± 0.12	0.21 ± 0.12	0.32 ± 0.05	0.11 ± 0.07
Lodi II	Chapparal, Chamise	3.3 ± 0.8	75 ± 14	3.6 ± 0.25	0.46 ± 0.03	0.55 ± 0.15	0.32 ± 0.12	0.21 ± 0.03	0.10 ± 0.05
Myrtle/Fall Creek	Pine, brush, Douglas Fir	2.54 ± 0.7	106 ± 20	3.0 ± 0.8	0.7 ± 0.04	0.60 ± 0.13	0.25 ± 0.05	0.22 ± 0.04	0.02 ± 0.04
Silver	Douglas Fir, True Fir, Hemlock	0.81 ± 0.69	89 ± 50	2.6 ± 1.6	0.08 ± 0.01	0.56 ± 0.33	0.42 ± 0.13	0.19 ± 0.09	0.02 ± 0.1
Hardiman	Jack Pine, Aspen, Birch	3.3 ± 2.3	82 ± 36	1.9 ± 0.5	0.58 ± 0.09	0.45 ± 0.26	0.18 ± 0.13	0.31 ± 0.35	0.02 ± 0.04
Eagle	Black Sage, Sumae, Chamise	7.2 ± 3.8	34 ± 6	0.9 ± 0.2	0.25 ± 0.06	0.18 ± 0.05	0.05 ± 0.02	0.08 ± 0.02	0.02 ± 0.08
Battersby	Jack Pine, White & Black Spruce	1.05 ± 1.33	175 ± 91	5.6 ± 1.7	0.9 ± 0.15	0.57 ± 0.45	0.27 ± 0.12	0.33 ± 0.06	0.07 ± 0.06
Overall average		3.9 ± 1.6	91 ± 21	2.9 ± 0.66	0.51 ± 0.14	0.47 ± 0.11	0.24 ± 0.06	0.24 ± 0.04	0.10 ± 0.05

fire in Oregon was sampled for three days as it travelled more than 1000 km over the Pacific Ocean. It was found that with increasing distance from the source, the dilution-normalised concentrations of particles in the nucleation mode (d $<$ 0.2 μm) in the plume decreased rapidly while the average size of the particles in the accumulation mode (d $=$ 0.2–2.0 μm) increased. The dilution-normalised optical scattering coefficient due to the particles in the smoke was found to decrease with increasing distance from the source. Open-path Fourier Transform Infrared (FTIR) spectroscopy has also been used to analyse *in-situ* emissions from biomass fires. Path lengths of 30 to 200 meters directly in the smoke from prescribed fires in The United States have been employed to identify and quantify emissions of CO, CO_2, CH_4, CH_2O, CH_3OH, HCOOH, CH_3COOH, NO, NO_2, N_2O, NH_3 and HCN (GRIFFITH, 1991).

Detailed studies of the composition of plumes from tropical forest fires have been previously carried out in the Amazon basin (ARTAXO *et al.*, 1993) and in central Africa (LACAUX *et al.*, 1993). These studies involved combined airborne and ground-based measurements, as well as additional data from satellite measurements. Compounds identified in these previous studies are summarised in Table 3 together with those from a recent study in Brunei Darussalam. While PAHs such as naphthalene, fluorene, phenanthrene, anthracene, fluoranthene, and pyrene have been observed in the African study, benzo(a)pyrene was not identified. Benzo(*a*)-pyrene was determined at concentrations between 7 and 58 μg g^{-1} of particulate matter in field experiments in southeastern United States (WHITE, 1987). In another field study in the United States, benzo(*a*)pyrene was found to be present at 13 $\pm$ 7 μg g^{-1} of particulate matter (WARD, 1989). Concentrations of benzo(a)py-rene of the order of 1 ng m^{-3} were determined in Brunei Darussalam during an intense haze episode caused by fierce forest fires in the vicinity (3–7 km) of the sampling site (DE JONG and HOLLANDER, 1998). ABAS *et al.* (1995) determined higher molecular weight organic compounds in aerosol particles emitted from controlled

Table 3

Chemical substances observed in plumes from tropical forest fires

---

Brazil (ARTAXO *et al.*, 1993):
Gases: O_3, CO_2, CO, CH_4
Particles: C black, Na, Mg, Al, Si, P, S, Cl, K, Ca, Ti, Mn, Fe, Zn, Rb, Sr, Zr, V, Pb, Cu, Ni, Br, Cr

Ivory Coast (LACAUX *et al.*, 1993):
Gases: CO_2, SO_2, CO, CH_4, N_2O, NO_x, NH_3, COS, O_3, non-methane hydrocarbons, organic acids (formic, acetic, propionic), PAHs
Particles: C, Na, Mg, Al, Si, P, S, Cl, K, Ca, Mn, Fe, Zn, PAHs

Brunei Darussalam (RADOJEVIC and HASSAN, 1999; MURALEEDHARAN *et al.*, 2000b):
Gases: SO_2, NO, NO_2, CO, O_3, hydrocarbons, aldehydes, benzene, toluene, ethylbenzene, xylene, phenol, acetic acid, PAHs
Particles: C, H, N, S, Cl, Na, K, Ca, Mg, Fe, Cu, Zn

---

biomass fires in Amazonia, Brazil. They found that the major organic compounds were straight-chain, oxygenated aliphatic compounds and triterpenoids from vegetation waxes, resins, gums, and bipolymers. They also observed PAHs and several other compounds (amyrones, friedelin, etc.) in the smoke samples.

3.3. Field Studies of Regional Haze in SE Asia

Although forest fires have caused regional haze episodes in various parts of the world, including Central and South America, it is the regional haze in SE Asia that has been most widely reported. The regional haze in SE Asia is a recurring environmental disaster due mainly to forest fires, and it is usually associated with dry weather and draughts caused by the El Niño Southern Oscillation phenomenon. Since 1982, regional haze episodes in SE Asia have been recorded during: August–September 1982, September 1983, September 1987, August 1990, August–September 1991, August–October 1994, August–October 1997 and February–May 1998 (RADOJEVIC, 1998; RADOJEVIC and HASSAN, 1999). Industrial pollution and volcanic eruptions (e.g., Mt. Pinatubo in 1991) have contributed to some of these episodes, but forest fires started deliberately by developers clearing land for settlements and plantations, and farmers engaged in traditional "slash and burn" practices are largely responsible for the regional haze in SE Asia (RADOJEVIC, 1997). The extent and severity of haze episodes has been increasing over time and the last three episodes have been studied more extensively than previous haze episodes. Reported studies of the chemistry of haze in SE Asia are summarised in Table 4.

There have been few detailed chemical studies of the haze in SE Asia; comprehensive forest fire studies in SE Asia have been conducted only during the 1997 and 1998 episodes. The few reported studies of earlier haze events have either been of a general or qualitative nature (NICHOL, 1997; TUSSIN, 1995; RINDAM, 1995; SANI et al., 1991) or they measured only a limited number of chemical components. Concentrations of criteria pollutants (PM_{10}, NO_2, CO, O_3, SO_2) are available for Kuala Lumpur and Singapore for the 1994, 1997 and 1998 haze episodes, and for Brunei for the 1998 episode. For earlier haze events, such as the 1991 episode, only suspended particulate matter (SPM) and CO measurements are available for some stations in SE Asia. For many sites, especially those in Indonesia, no chemical measurements of the haze are available and only visibility data have been reported. Particle concentrations and meteorological conditions during the earlier haze episodes in Malaysia (1990, 1991 and 1994) have been reviewed by SOLEIMAN et al. (2003).

With regard to the concentrations of organic micropollutants and the chemical composition of haze particles even less data are available. ZAKARIA et al. (1995) determined organic and elemental carbon, chloride, nitrate, sulphate, Ca, Mg, K, and several heavy metals (Fe, Mn, Cd, Cu, Pb, Cr, Ni) in aerosol particle in Kuala Lumpur on several days in September 1994 during a regional haze episode. ABAS and

SIMONEIT (1996) determined lipid fractions in solvent extracts of one aerosol sample collected on 29 September 1991 in Kuala Lumpur during a haze event. They observed *n*-alkanes, *n*-alkanoic acids and *n*-alkanols as the dominant components in the extracts, with minor amounts of *n*-alkanedioic acids, *n*-alkanones, terpenoids, and sterols. In a qualitative study using scanning electron microscopy, JAPONY and DALIMAN (1999) detected C, O, Si, Al, Fe, Mg, K, and Ca in several aerosol samples collected on 22–24 September 1997 in Kota Kinabalu, Sabah, during a severe regional haze episode.

Untill now, the only comprehensive study of haze composition in SE Asia was conducted in Brunei Darussalam during the 1998 haze episode (RADOJEVIC and HASSAN, 1999; MURALEEDHARAN *et al.*, 2000b; RADOJEVIC, 2003). This study involved the analysis of criteria pollutants (PM_{10}, SO_2, NO, NO_2, CO, O_3), total petroleum hydrocarbons (C_6–C_{36} excluding BTEX), BTEX (benzene, toluene, ethylbenzene, and xylene), aldehydes (formaldehyde, acetaldehyde, propionaldehyde, and butyraldehyde), polynuclear aromatic hydrocarbons (PAHs), acetic acid, phenol, and cresol. The suspended particles were also analysed for: C, H, N, S, Cl, Na, K, Mg, Ca, Fe, Cu, Zn, As, Cd, Ni, V, and Hg. Brunei is especially suited for studying the impact of biomass burning on air quality because of the absence of major industrial sources of air pollution. Of all the criteria pollutants, only PM_{10} was a significant contaminant during the haze episode. Concentrations of gaseous criteria pollutants were generally well below accepted air quality guidelines and standards. The 24-h average PM_{10} concentration at Bandar Seri Begawan exceeded the WHO guideline for PM_{10} on 60% of the days during the haze episode. Most of the observed effects of the haze (short-term health impacts, visibility reduction, soiling of materials, etc.) were ascribed to the high suspended particle concentrations. Analysis of suspended particles revealed that C was the major constituent, present at between 50 and 70% of the aerosol mass (Radojevic, unpublished data). Potassium, often considered a tracer of biomass burning, was present at between 10 and 20% of the aerosol mass, Ca was present at between 1 and 6% of the aerosol mass, and Zn was present at between 6 and 20% of the aerosol mass. Toxic metals (Cd, Ni, V, Hg, and As) and Mg were present at <0.01% of the aerosol mass (MURALEEDHARAN *et al.*, 2000b). A particle size analysis of haze particles revealed that >99% of the particles were <2.5 μm in diameter (MURALEEDHARAN *et al.*, 2000b), and these "respirable" particles could have greater impacts on human health than the "inhalable" PM_{10} particles routinely determined in air quality studies. The haze in Brunei during 1998 was mainly due to pollution from local fires in Brunei and the neighbouring Malaysian states of Sarawak and Sabah, rather than from the long-range transport of pollutants from forest fires in Indonesia (RADOJEVIC and HASSAN, 1999) which was responsible for earlier haze episodes in Brunei. There was significant spatial and temporal variation in the concentration of PM_{10} and other pollutants. Particle concentrations exhibited diurnal variations, generally rising to a peak in the early morning, and thereafter declining, usually by noon, due mainly to meteorological

Table 4

Field studies of regional haze in SE Asia

Reference	Location	Haze period	Comments
Sani et al. (1991)	Klang Valley Region, Malaysia	Aug 1990	Investigation of TSP, meteorological factors and pollution sources
Zakaria et al. (1995)	Kuala Lumpur, Malaysia	Sep 1994	Measurements of organic –C, Cl^-, NO_3^-, SO_4^{2-}, K, Ca, Mg, and heavy metals (Fe, Mn, Cd, Cu, Ni) in atmospheric aerosol
Hassan et al. (1995)	Kuala Lumpur, Malaysia	Sep–Oct 1994	Review of air quality data (CO, NO_2, O_3, SO_2, TSP, and PM_{10})
Rindam (1995)	Malaysia	Sep–Oct 1994 Aug 1990	Review of TSP and visibility data
Tussin (1995)	Malaysia	Sep–Oct 1991 Jul–Oct 1994	Review of TSP, visibility and meteorological data
Soleiman et al. (2003)	Kuala Lumpur, Malaysia	Aug 1990 Oct1991 Aug–Oct1994	Investigation of TSP and meteorological parameters
Abas & Simoneit (1996)	Kuala Lumpur, Malaysia	29 Sep 1991	Measurements of total-C, organic-C and some organic compounds (hydrocarbons, phenols, sterols, etc.) in atmospheric aerosol
Chee et al. (1997)	Singapore	Oct 1994	Measurements of PAHs in suspended particles
Nichol (1997)	Singapore	Sep–Oct 1994	Review of PSI and meteorological data
Kousa & Jantunen (1998)	Kuala Lumpur, Malaysia	17–23 Oct 1997	Measurements of fine particles in indoor and outdoor air
Japony & Dalimin (1999)	Kota Kinabalu, Sabah	22–24 Sep 1997	Scanning Electron Microscope (SEM) studies of atmospheric aerosol components (C, O, Si, Al, Fe, Mg, K, Ca)

Reference	Location	Date	Description
Radojevic & Hassan (1999)	Brunei Darussalam	Feb–May 1998	Review of air quality data (SO_2, NO_x, CO, O_3, PM_{10})
Muraleedharan et al. (2000b)	Brunei Darussalam	Apr 1998	Measurements of VOCs (hydrocarbons, benzene, toluene, xylene, phenol, aldehydes, etc.), PAHs, heavy metals (As, Ba, Cd, Ni, V, Hg), TSP, PM_{10}, O_3, SO_2, and NO_2.
Muraleedharan & Radojevic (2000)	Brunei Darussalam	Apr 1998	Measurements of fine particles in indoor and outdoor air
Radojevic (1999); Radojevic et al. (1999)	Brunei Darussalam	Aug–Oct 1994, Aug–Oct 1997, Feb–May 1998	Analysis of rainwater pH, conductivity, and composition (Cl^-, NO_3^-, SO_4^{2-}, Na^+, K^+, Ca^{2+}, Mg^{2+})
Balasubramanian et al. (1999)	Singapore	Aug–Oct 1997	Analysis of rainwater pH, conductivity, and composition (Cl^-, NO_3^-, SO_4^{2-}, Na^+, K^+, Ca^{2+}, Mg^{2+}, $HCOO^-$, CH_3COO^-)
de Jong & Hollander (1998)	Sarawak & Brunei Darussalam	Apr 1998	Measurement of PM_{10}, $PM_{3.5}$, VOCs, PAHs
Tsutsumi et al. (1999)	Kalimantan, Indonesia	Oct 1997	Aircraft measurements of O_3, NO_x, CO and aerosol
Fujiwara et al. (1999)	Watukosek, Indonesia	Oct 1997	O_3 measurements
Davies & Unam (1999)	Kuching, Sarawak, malaysia	Sept–Oct 1997	Review of air quality data (CO, SO_2, O_3, NO_x, CH_4, PM_{10})
Sawa et al. (1999)	Kalimantan, Indonesia	Oct 1997	Aircraft measurements of CO and H_2
Ponka et al. (2000)	Kuala Lumpur, Singapore & Jakarta	Oct 1997	Measurements of TSP, PM_{10}, VOCs

factors such as creation and dissipation of a low lying inversion, together with sea breeze and topographical effects.

Organic micropollutants were also determined in Kuala Lumpur, Singapore and Jakarta in October 1997 (PONKA et al., 2000) and in Sarawak during April 1998 (DE JONG and HOLLANDER, 1998). CHEE et al. (1997) observed 17 different PAHs in suspended particles sampled in Singapore during October 1994, including benzo(a)pyrene which was observed at concentrations between 0.07 and 0.80 ng m^{-3}. They compared their measurements with those of PAHs in other major cities and in Singapore in the absence of haze, and concluded that the marginally higher levels of PAHs observed in their study may be due to biomass fires during August–October 1994. Although some of the organic micropollutants observed in the haze are known or suspected carcinogens, mutagens and teratogens (e.g., benzene, toluene, xylene, PAHs) with the potential to cause significant long-term harm to human health, in the absence of epidemiological studies it is not possible to predict their health impacts.

3.4. Numerical Models of Haze in SE Asia

Forecasting of haze pollution required a three-fold approach: satellite observations, ground-based measurements and model simulations. Satellite observations can be used to locate the forest fires and quantify emissions from them, as well as quantifying the amount of smoke in air columns downwind of the fires. Satellite observations of haze can serve as input parameters in advanced dispersion models and these can be verified using ground-based observations.

To date, few modelling studies of forest fires and regional haze in SE Asia have been reported in the literature, however, several research groups are currently developing and applying numerical models in order to investigate the long-range transport of pollutants from source areas. These models could be useful for predicting the various impacts of forest fires in SE Asia.

PHANDIS et al. (2000) have applied the three-dimensional STEM-III regional-scale model to investigate the regional impact of gaseous and particulate emissions from forest fires in SE Asia. The model considers condensed state chemical mechanisms, and the numerical integration of the chemistry is done by means of a fully implicit integrator based on the Runge-Kutta-Rosenbrock type of solvers. Both anthropogenic and forest fire emissions of NO_x, SO_2, CO, and non-methane hydrocarbons were considered, as were emissions of particulate matter from forest fires. Meteorological parameters were included, however, no deep convections were considered. The decrease in the insolent solar radiation below the aerosol plume was also considered. Numerical model simulations were conducted for October 1997, when the haze in SE Asia was especially severe. Three cases were considered and these are summarised in Table 5. The results of the model showed that the region affected by forest fire emissions was mostly west of Indonesia towards the Indian

Table 5

Description of cases considered in the model simulations of Phandis et al. (2000)

Simulation Case	Description
Case 1	Chemistry included, effect of anthropogenic emissions included
Case 2	Effect of emissions from forest fires included
Case 3	Effect of the reduction of solar intensity below 3 km on photochemistry included

Ocean and that the impact of emissions decreased from the surface to the upper troposphere. Ozone concentrations were predicted to increase by more than 90 ppbv in the boundary layer over Sumatra and west of Kalimantan and by more than 20 ppbv over southern portions of India for case 2. For case 3, which included reduction of solar intensity, the model predicts a decrease in O_3 production and lower ambient levels of O_3. Model predictions were compared with ozonesondes launched over Malaysia and Indonesia during October 1997, and the results agreed with the vertical profiles of O_3 as measured by the ozonesondes. The model predicted extremely high levels of CO over the source regions of forest fires, however, the predicted CO concentrations were lower than the aircraft measurements of SAWA et al. (1999). The spatial and temporal distribution of particulate matter agreed well with surface measurements.

KOE et al. (1999) employed the DARLAM regional climate model to investigate the long-range transport of haze from regions of forest fires in East Kalimantan. Simulations were conducted on a 60-km grid resolution from February to April 1998. Wet and dry deposition of SO_2 and sulphate were accounted for in the model. Preliminary results suggested that the model is capable of simulating emissions of haze pollutants and that synoptic wind patterns play a major role in the long-range transport of haze. The majority of model images exhibited similarities in shapes and patterns with the NASA TOMS aerosol images and NOAA-AVHRR derived haze maps, although there were discrepancies in the location of plumes between model predictions and the satellite images. These discrepancies were ascribed to uncertainties in emission rates used as input parameters in the model. The model was able to represent the dynamics of transboundary haze pollution caused by forest fires and the effect of topography was evident. The vertical transport of haze was also investigated. A detailed sensitivity analysis of important model parameters is planned.

ROSWINTIARTI and RAMAN (1999) used the Fifth Generation of the Pennsylvania State University-National Center for Atmospheric Research (PSU-NCAR) Mesoscale Model (MM5) to simulate the circulation dynamics and the transport processes of aerosols and gases during the 1997 forest fire episode over Kalimantan. A three-dimensional non-hydrostatic version of the model was used to simulate the haze during September 1997, using double-nested grids with grid resolutions of 15 km and

45 km. Only small increases in forecast skill were gained by increasing the resolution from 45 km to 15 km. The authors concluded that the model could be a useful tool for trajectory analysis.

4. Tropospheric Chemistry and Secondary O_3 Formation

We may expect the physicochemical properties of the haze components to vary considerably during transport in the atmosphere. Close to the source of the fires, haze composition is dominated by primary pollutants and larger ash particles. With increasing distance from source areas, the relative proportion of secondary pollutants and fine particles will increase due to atmospheric reactions and deposition processes, respectively. Dry deposition is generally more significant closer to the source while wet deposition is more important further downwind. Primary pollutants emitted by forest fires are converted into secondary pollutants during long-range atmospheric transport as a result of chemical reactions in the troposphere. Due to the greater solar intensity and higher temperatures in the tropics, tropospheric chemistry is generally more active than in temperate regions.

Biomass fires could impact atmospheric chemistry in several ways:
- by emitting primary gaseous pollutants which could react in the atmosphere to produce O_3 and other secondary pollutants,
- by emitting suspended particles which could act as a medium for surface-catalysed reactions, and
- by affecting the intensity of solar radiation, which in turn would affect the photochemistry of the atmosphere.

LEVINE (1999, 2000) estimated the emission of gases (CO, CO_2, CH_4 and NO_x) and particles from Kalimantan and Sumatra during the 1997 haze episode and concluded that these emissions significantly exceeded those from the Kuwaiti oil fires of 1991. These emissions could have significant impact on tropospheric chemistry. The general chemistry of the atmosphere is dominated by photochemical reactions initiated by solar radiation. Free radicals (e.g., OH˙) play a fundamental role in many of the pollutant transformation mechanisms. The OH˙ radical is highly reactive and a strong oxidising agent for a variety of pollutant gases (e.g., SO_2, NO_x). Production of OH· radicals is initiated by the photolysis of O_3. Sunlight with wavelengths <315 nm splits O_3 into an O_2 molecule and singlet atomic oxygen O (^{1}D) which can react with water vapour to form OH˙ radical:

$$O_3 + h\nu \; (\lambda < 315 \text{ nm}) \rightarrow O_2 + O(^1D)$$

$$O(^1D) + H_2O \rightarrow 2OH˙$$

OH˙ radicals can also be formed by other atmospheric reactions (photolysis of HNO_2, photolysis of H_2O_2, etc.). The OH˙ radical is involved in the oxidation of SO_2

to H_2SO_4, oxidation of NO_x species to HNO_2 and HNO_3, and the oxidation of CO to CO_2.

The H atom can react with O_2 to produce the hydroperoxy radical ($HO_2\cdot$), another reactive species:

$$H + O_2 \rightarrow HO_2^\cdot$$

The $HO_2^\cdot$ radical is a strong oxidising agent and it is involved in several important atmospheric reactions, including the oxidation of NO to NO_2. Free radicals such as $OH^\cdot$ and $HO_2^\cdot$ also react with organic pollutants emitted by combustion processes (hydrocarbons, aldehydes, etc.). Aldehydes and PAHs may be emitted directly by combustion processes or they may be produced by atmospheric reactions of primary pollutants.

NO_2 is photolysed to produce $O(^3P)$ which reacts with O_2 to produce O_3:

$$NO_2 + h\nu\,(\lambda < 435\ \text{nm}) \rightarrow NO + O(^3P)$$

$$O(^3P) + O_2 + M \rightarrow O_3 + M,$$

where M is a third body reactant. O_3 reacts rapidly with NO to produce NO_2:

$$NO + O_3 \rightarrow NO_2 + O_2.$$

Assuming steady state, the O_3 concentration can be related to NO_x concentration according to the following equation:

$$[O_3] = J[NO_2]/k[NO],$$

where k is the reaction rate constant for the above NO oxidation reaction and J is the NO_2 photodissociation constant. In order for high concentrations of O_3 to build up, both high values of J and high NO_2/NO ratios are required. The value of J increases with increasing sunlight and we may expect lower values during haze periods than on clear days. This effect may be partly balanced by higher NO_2/NO ratios in haze if other NO oxidation reactions are effective. Multiphase chemical reactions are also important in the atmosphere. Pollutant gases may be adsorbed onto the surfaces of smoke particles where they may further react. They may also dissolve in cloud droplets, rain drops, and in deliquescent aqueous aerosols at high humidity and undergo liquid-phase reactions. For example SO_2 may be oxidised to H_2SO_4 on soot particles, in cloud droplets, and in deliquescent aerosols (RADOJEVIC, 1992). The very high concentration of smoke particles during haze episodes may act as a medium for pollutant transformation reactions.

Extensive measurements of SO_2, NO_2, and O_3 concentrations during periods of intense forest fires and haze in SE Asia failed to reveal major impact on ground level concentrations (RADOJEVIC, 1998; 1999; RADOJEVIC and HASSAN, 1999; RADOJEVIC et al., 1999; MURALEEDHARAN et al., 2000b). Although concentrations of SO_2, NO, NO_2, O_3, and CO were somewhat higher during haze episodes, the measured concentrations are unlikely to have a major impact on human health, natural ecosystems, or atmospheric chemistry. While these pollutants are produced in various amounts by combustion of biomass, they are readily dispersed and diluted to levels which are of little importance in the local and regional atmosphere (RADOJEVIC, 1999).

Particles tend to settle and remain close to the ground while gaseous pollutants may diffuse upward out of the smoke plumes. While higher concentrations of O_3 may result from photochemical reactions of pollutants diffusing out of the top of the plumes, mean ground-level concentrations of O_3 during forest fires in Brunei were around 60 µg m^{-3} and the highest O_3 concentration was around 100 µg m^{-3}. According to the U.S. NAAQS, photochemical smog episodes are defined by O_3 concentrations >0.12 ppmv (i.e., >235 µg m^{-3}). Ground-level O_3 concentrations observed in other studies of forest fires are generally between 5 and 45 ppbv (CRUTZEN and ANDREA, 1990; LACAUX et al., 1993; CRUTZEN and CARMICHAEL, 1993; ANDREA et al., 1990), well below the standard value. Even the elevated levels of O_3 observed at higher altitudes (1–10 km) in these studies are generally in the 40–60 ppbv range and sometimes in the 90–100 ppbv range. Since sunlight intensity is dramatically reduced within thick plumes of smoke, we may expect significant reduction in photochemical activity in areas blanketed by haze from forest fires. TSUTSUMI et al. (1999) also measured low O_3 concentrations (ca. 20 ppbv) near the surface, however concentrations as high as 80.5 ppbv in the middle layer of smoke haze over Kalimantan in October 1997. They concluded that low O_3 concentrations near the ground may be due to the reduction in solar flux because of high particle concentrations in the haze and loss of O_3 on particle surfaces. On the other hand, FUJIWARA et al. (1999) observed O_3 concentrations greater than 100 ppbv at several altitudes over Indonesia in October 1997. KITA et al. (2000) analysed O_3 observations from the TOMS satellite and a MKIV-type Brewer spectrophotometer during the 1994 and 1997 haze episodes over SE Asia, and found that the two observations were consistent with each other. Significant increases in total O_3 were observed in Indonesia during both episodes and these were attributed to increases in tropospheric O_3. Horizontal and temporal variations in O_3 increases were similar during both years. BALASUBRAMANIAN (2000) observed several photochemical air pollution episodes with elevated O_3 levels (>70 ppbv) in Singapore due to the transport of pollutants from forest fires in Sumatra. These studies may be compared to those in subtropical and temperate regions. Using airborne measurements, EVANS et al. (1976) observed significant increases in O_3 concentration in the top few hundred metres of plumes from prescribed biomass fires in Australia, although little effect on the concentrations of SO_2 and NO_2. Model calculations suggest substantial formation of O_3 in aging plumes from biomass fires and that this may have consequences for the global O_3 budget (KOPPMANN et al., 1999).

FORBERICH et al. (1996) observed significant increases in the concentrations of NO, NO_2, CH_4, peroxyacetyl nitrate (PAN), formaldehyde and naphthalene following a forest fire at a subtropical site (Tenerife). They also determined the diurnal variation in OH˙ radical concentrations using multipass optical absorption spectrometry (MOAS). They compared their measurements with results of a box model and found that formaldehyde plays a major role in the OH˙ balance.

In addition to affecting the chemistry of the troposphere, forest fires may also affect stratospheric O_3. Methyl chloride and methyl bromide, although released in trace amounts by biomass burning, may have a negative impact on stratospheric O_3 concentrations (LEVINE, 2000). The contribution of these compounds to strato-spheric O_3 depletion will become more important as other sources of chlorine and bromine are phased out as a result of the Montreal Protocol.

5. Impacts of Forest Fires on Rainwater Chemistry

Over the last ten years a number of reports have suggested that biomass burning in the tropics may have an acidifying effect on precipitation with consequent effects on tropical ecosystems (CRUTZEN and ANDREA, 1990; LACAUX et al., 1991–1993; CACHIER and DUCRET, 1991). Impacts of forest fires on rainwater chemistry and acidity have previously been investigated in South America and Africa. In one study in Venezuela (SANHUEZA et al., 1989) higher pH values were observed in rainwater during periods of biomass burning, while in another study (SANHUEZA et al., 1992) it was concluded that biomass burning had no significant effects on rainwater pH. LACAUX et al. (1987) observed higher pH values in rainwater collected in the Ivory Coast during the dry season when there were extensive biomass fires as compared to the rainy season, however, in later studies they claimed a significant contribution of biomass burning during the dry season to rainwater acidity (LACAUX et al., 1991–1993). However, most reported studies (e.g., LACAUX et al., 1987, 1991–1993) compared the pH and composition of individual rainwater samples rather than the deposition rates. Biomass fires are inversely related to precipitation; they generally occur during dry periods when precipitation amounts are very low or non-existent since heavy rains tend to extinguish the fires. Most deposition in the tropics takes place during the wet season when precipitation amounts are exceedingly large, and it is the deposition rates that should be related to ecological impacts.

Only two studies reported the impacts of forest fires and haze on the acidity of rainwater in SE Asia; one in Singapore (BALASUBRAMANIAN et al., 1999) and one in Brunei (RADOJEVIC et al., 1999). Radojevic et al., (1999) analysed precipitation samples collected in Brunei Darussalam during three haze episodes in 1994, 1997, and 1998. Comparison of the pH of individual rainwater samples during haze episodes with the pH of rainwater collected during periods when forest fires and haze were absent revealed no significant acidification. Similar conclusions were drawn by BALASUBRAMANIAN et al. (1999) who analysed rainwater samples collected in

Singapore during the 1997 haze episode. Although biomass burning produces many acidifying compounds such as SO_2, NO_x and organic acids (e.g., formic acid, acetic acid), it also produces ions such as K, Mg, and Ca (WARD and HARDY, 1991), which may act to neutralise the acidity. It has been shown that alkaline substances (Ca, Mg) can be leached from particles of industrial origin (e.g., fly ash) into cloud-water and rainwater neutralising the acidity (WILLIAMS et al., 1988), and similar leaching may take place from smoke particles originating from biomass burning. Heavy metals (Fe, Cu, Mn, Zn) can also be leached from atmospheric particles (WILLIAMS et al., 1988). In their study, BALASUBRAMANIAN et al. (1999) noted the neutralisation of acidity by NH_3 and $CaCO_3$. The conductivity and the ionic content of individual rainwater samples were greater during haze episodes than during periods when haze was absent (BALASUBRAMANIAN et al., 1999; RADOJEVIC, 1999; 2003; RADOJEVIC et al., 1999), however, deposition rates were generally higher during non-haze periods due to the greater precipitation amounts (RADOJEVIC, 1999; 2003; RADOJEVIC et al., 1999). The monthly contributions during periods of forest fires and haze to the total annual deposition of H^+ in Brunei varied between 0.0 and 5.35%, indicating that forest fires are a minor source of acidity in wet deposition (RADOJEVIC, 1999; 2003; RADOJEVIC et al., 1999).

WU et al. (1995) detected three carcinogenic heterocyclic aromatic amines: 3-amino-1-methyl-5H-pyrido[4,3-b]indole (Trp-P-2), 2-amino-1-methyl-6-phenylimidazo[4,5-b]pyridine (PhIP), and 2-amino-3,4,8-trimethylimidazo[4,5-f]quinoxaline (4,8-DiMeIQx) by capillary zone electrophoresis in rainwater and suspended particles collected in October 1994 in Singapore. On 9 October, rainwater concentrations were 84.65, 34.51, and 16.83 ng L^{-1} for PhIP, Trp-P-2, and DiMeIQx, respectively. These compounds were not detectable in rainwater samples collected on 28 November 1994, well after the haze dissipated from the region.

6. Impacts of Forest Fires on Weather and Climate

Air pollutants from forest fires could conceivably affect weather and climate in a number of ways:
- by emitting greenhouse gases (e.g., CO_2, CH_4, N_2O) which could contribute to global warming,
- by emitting smoke particles which could affect the earth's radiation balance by scattering and absorbing radiation, and
- by emitting smoke particles which could act as cloud condensation nuclei (CCN) and affect precipitation patterns.

Biomass burning is a significant global source of CO_2 and CH_4, both of which are greenhouse gases that can lead to global warming. After fossil fuel combustion in industrialised countries, biomass burning in the tropics is the largest contributor to global carbon emissions. Recent forest fires in SE Asia released significant quantities

DAVIES, S.J. and UNAM, L. (1999), *Smoke-haze from the 1997 Indonesian Forest Fires: Effects on Pollution Levels, Local Climate, Atmospheric CO₂ Concentrations, and Tree Photosynthesis*, Forest Ecology and Management *124*, 137–144.

DAVIES, S.J., *Fires and smoke: effects on tropical rain forests in Southeast Asia*. In *Forest Fires and Regional Haze in Southeast Asia* (eds. Eaton P. and Radojevic M.) (Nova Science Publishers, Inc., New York, 2000).

DE JONG, G. and HOLLANDER, K. (1998), *Air Quality Monitoring and Risk Assessment Carried out During Haze Episode in Malaysia and Brunei (April 1998)*, Shell International Exploration and Production B.V., The Hague, 1998.

DIGNON, J., ATHERTON, C.S., PENNER, J.E., and WALTON, J.J. (1991), *NOₓ pollution from biomass burning: A global study*, Paper presented at the 11th Conference on Fire and Forest Meteorology, April 1991, Missoula, MT.

EDYE, L.A. and RICHARDS, G.N. (1991), *Analysis of Condensates from Wood Smoke: Components Derived from Polysaccarides and Lignins*, Environ. Sci. and Techn. *25*(6), 1133–1137.

EVANS, L.F., KING, N.K., MACARTHUR, D.A., PACKHAM, D.R., and STEPHENS E.T. (1976), *Further Studies of the Nature of Bushfire Smoke*, CSIRO Aust. Div. Appl. Organic Chem. Tech. Pap. No. 2, 1–12.

EVANS, L.F., WEEKS, I.A., ECCLESTON, A.J., and PACKHAM, D.R. (1977), *Photochemical Ozone in Smoke from Prescribed Burning of Forests*, Environ. Sci. and Techn. *11*(9), 896–900.

FELDSTEIN, M., DUCKWORTH, S., WOHLERS, H.C., and LINSKY, B. (1963), *The Contribution of the Open Burning of Land Clearing Debris to Air Pollution*, J. Air Pollution Control Assoc. *13*, 896–900.

FORBERICH, O., WALTER, J., and COMES, F.J. (1996), *Measurement of OH Concentration During a Forest Fire Episode: Atmospheric Implications for Biomass Burning*, Chem. Phys. Lett. *259*, 408–414.

FRITSCHEN, L., BOVEE, H., BUETTNER, K., CHARLSON, R., MONTEITH, L., PICKFORD, S., MURPHY, J., and DARLEY, E. (1970), *Slash Fire Atmospheric Pollution*, U.S. Department of Agricultrure, Forest Service, Research Paper PNW-97, Pacific Northwest Research Station, Portland, OR, 1970.

FUJIWARA, M., KITA, K., KAWAKAMI, S., OGAWA, T., KOMALA, N., SARASPRIYA, S., and SURIPTO, A. (1999), *Tropospheric Ozone Enhancement During the Indonesian Forest Fire Events in 1994 and in 1997 as Revealed by Ground-based Observations*, Geophy. Res. Lett. *26*, 2417–2420.

GRIFFITH, D.W.T. (1991), *FTIR, bushfires and atmospheric chemistry*, In *8th International Conference on Fourier Transform Spectroscopy, SPIE Vol. 1575*, pp. 59–69.

GRIFFITH, D.W.T., MANKIN, W.G., COFFEY, M.T., WARD, D.E., and RIEBAU, A., *FTIR remote sensing of biomass burning emissions of CO₂, CO, CH₄, CH₂O, NO, NO₂, NH₃ and N₂O*. In *Global Biomass Burning: Atmospheric Climate and Biospheric Implications* (ed. Levine J.S.) (Cambridge MA, MIT Press, 1991) pp. 230–239.

HARGER, J.R.E., *Climate, El Niño, drought and fire*. In *Forest Fires and Regional Haze in Southeast Asia* (eds. Eaton P. and Radojevic M.) (Nova Science Publishers, Inc., New York, 2000).

HASSAN, M.N., ABIDIN, A.Z., YUSOFF, M.K., GHAZALI, A.W., MUDA, A., and ZAKARIA, M.P. (1995), *Damage costs of the 1991 and 1994 haze episodes in Malaysia*. In *Proceedings of the International Symposium on Climate and Life in the Asia Pacific* (ed. Sirinanda K.U.) 10–13 April 1995, University of Brunei Darussalam, Bandar Seri Begawan, Brunei Darussalam, pp. 105–124.

HEGG, D.A., RADKE, L.F., HOBBS, P.V., RASMUSSEN, R.A., and RIGGON, R.J. (1990), *Emissions of Some Trace Gases from Biomass Fires*, J. Geophy. Res. *95*, 5669–5675.

HORVATH, H., *Effects on visibility, weather and climate*. In *Atmospheric Acidity: Sources, Consequences and Abatement* (eds. Radojevic M. and Harrison R.M.) (Elsevier Applied Science, London, 1992) pp. 435–466.

JAPONY, M. and DALIMAN, M.N. (1999), *Analysis of haze particles using scanning electron microscope*. Paper presented at the *International Workshop on Biomass Burning and its Transport*, 1–3 March 1999, Kuala Lumpur, Malaysia.

JENKINS, B.M., TURN, S.Q., WILLIAMS, R.B., CHANG, D.P.Y., RAABE, O.G., PASKIND, J., and TEAGUE, S., *Quantitative assessment of gaseous and condensed phase emissions from open burning of biomass in a combustion wind tunnel*. In *Global Biomass Burning: Atmospheric Climate and Biospheric Implications* (ed. Levine J.S.) (Cambridge, MA, MIT Press, 1991) pp. 305–320.

KAUFMAN, Y.J., SETZER, A., WARD, D., TANRE, D., HOLBEN, B.N., KIRCHHOFF, V.W.J.H., MENZEL, P., PEREIRA, M.C., and RASMUSSEN, R. (1992), *Biomass Burning Airborne and Spaceborne Experiment in the Amazonas (BASE-A) 1991*, J. Geophy. Res. *97*, 14,581–14,599.

KITA, K., FUJIWARA, M., and KAWAKAMI, S. (2000), *Total Ozone Increase Associated with Forest Fires over the Indonesian Region and its Relation to the El-Niño-Southern Oscillation*, Atmos. Environ. *34*, 2681–2690.

KOE, L.C.C., McGREGOR, J.L., and ARELLANO, Jr. A.F. (1999), *Application of DARLAM to regional haze modelling*. Paper presented at the *International Conference with Workshop on Air Quality Management*, 15–19 November, University of Brunei Darussalam.

KOPPMANN, R., Von CZAPIEWSKI, K., and KOMENDA, M. (1999), *Emissions of Volatile Organic Compounds from Biomass Burning*, Volatile Organic Compounds in the Environment *16*, 91–107.

KOUSA, A. and JANTUNEN, M. (1998), *Centrally Monitored, Home Indoor and Outdoor and Personal Fine PM Levels during the Haze, 17–23.10.1997, in Kuala Lumpur, Malaysia*, Abstracts of paper presented at the 1998 Annual Conference of ISEE and ISEA. Epidemiology *9*(4), S103.

KUHLBUSCH, T.A., LOBERT, J.M., CRUTZEN, P.J., and WARNECK, P. (1991). *Molecular Nitrogen Emissions from Denitrification During Biomass Burning*, Nature *351*, 135–137.

LACAUX, J.-P., CACHIER, H., and DELMAS, R., *Biomass burning in Africa: an overview of its impacts on atmospheric chemistry*. In *Fire in the Environment: The Ecological, Atmospheric, and Climatic Importance of Vegetation Fires* (eds. Crutzen, J.P. and Goldammer, J.G.) (John Wiley and Sons, Chichester, 1993) pp. 159–191.

LACAUX, J.-P., DELMAS, R.A., CROS, B., LEFEIVRE, B., and ANDREAE, M.O., *Influence of biomass burning emissions on precipitation chemistry in the equatorial forests of Africa*. In *Global Biomass Burning* (ed. Levine, J.S.) (MIT Press, Cambridge, MA, 1991) pp. 167–174.

LACAUX, J.P., LOEMBA-NDEMBI, J., LEFEIVRE, B., CROS, B., and DELMAS, R. (1992), *Biogenic Emissions and Biomass Burning Influences on the Chemistry of the Fogwater and Stratiform Precipitations in the African Equatorial Forest*, Atmos. Environ. *26A*, 541–551.

LACAUX, J.-P., SERVANT, J., and BAUDET, J.G.R., *Acid rain in the tropical forests of western Africa*. In *Acid Rain- Scientific and Technical Advances* (eds. Perry, R., Harrison, R.M., Bell, J.N.B., and Lester, J.N.) (Selper Ltd., London, 1987) pp. 264–269.

LARSON, T.V. and KOENIG, J.Q. (1994), *Wood Smoke: Emissions and Noncancer Effects*, Annu. Rev. Public Health *15*, 133–156.

LEVINE, J.S. (1999), *The 1997 Fires in Kalimantan and Sumatra, Indonesia: Gaseous and Particulate Emissions*, Geophy. Res. Lett. *26*, 815–818.

LEVINE, J.S., *Global impacts and climate change*. In *Forest Fires and Regional Haze in Southeast Asia* (eds. Eaton P. and Radojevic M.) (Nova Science Publishers, Inc., New York, 2000).

LOBERT, J.M., SCHARFFE, D.H., HAO, W.M., and CRUTZEN, P.J. (1990), *Importance of Biomass Burning in the Atmospheric Budgets of Nitrogen-containing Gases*, Nature *346*, 552–554.

LOBERT, J.M., SCHARFFE, D.H., HAO, W.M., KUHLBUSCH, T.A., SEUWEN, R., WARNECK, P., and CRUTZEN, P.J., *Experimental evaluation of biomass burning emissions: Nitrogen and carbon containing compounds*. In *Global Biomass Burning: Atmospheric Climate and Biospheric Implications* (ed. Levine J.S.) (Cambridge, MA, MIT Press, 1991) pp. 289–304.

LOBERT, J.M. and WARNATZ, J., *Emissions from the combustion process in vegetation*. In *Fire in the Environment: The Ecological, Atmospheric, and Climatic Implications of Vegetation Fires* (eds. Crutzen P.J. and Goldammer J.G.) (John Wiley & Sons, Chichester, 1993) pp. 15–37.

McMAHON, C.K. and TSOULAKAS, S.N., *Polynuclear aromatic hydrocarbons in forest fire smoke*. In *Polynuclear Aromatic Hydrocarbons* (eds. Jones P.W. and Freudenthal R.I.) Carcinogenesis *3*, (New York, Raven Press, 1978) pp. 61–73.

MURALEEDHARAN, T.R. and RADOJEVIC, M. (2000), *Personal Particle Exposure Monitoring Using Nephelometry During Haze in Brunei*, Atmos. Environ. *34*, 2733–2738.

MURALEEDHARAN, T.R., RADOJEVIC, M., WAUGH, A., and CARUANA, A. (2000a), *Emissions from the Combustion of Peat: An Experimental Study*, Atmos. Environ. *34*, 3033–3035.

MURALEEDHARAN, T.R., RADOJEVIC, M., WAUGH, A., and CARUANA, A. (2000b), *Chemical Characterisation of the Haze in Brunei Darussalam During the 1998 Episode*, Atmos. Environ. *34*, 2725–2731.

NICHOL, J. (1997), *Bioclimatic Impacts of the 1994 Smoke Haze Event in Southeast Asia*, Atmos. Environ. *31*, 1209–1219.

PHANDIS, M.J., CARMICHAEL, G.R., and LIM, S.F., *Air pollution measurements and modelling in Southeast Asia during the El Niño of 1997*. In *Forest Fires and Regional Haze in Southeast Asia* (eds. Eaton P. and Radojevic M.) (Nova Science Publishers, Inc., New York, 2000).

PHILPOT, C.W., GEORGE, C.W., BLAKELY, A.D., JOHNSON, G.M., and WALLACE, Jr. W.H. (1972), *The Effect of Two Flame Retardants on Particulate and Residue Production*. Res. Paper SE-231. U.S. Dept. of Agriculture, Forest Service, Intermountain Research Station, Asheville, NC, 1972.

PONKA, A., VARTIALA, T., and RADOJEVIC, M. (2000), *Measurements of organic micropollutants during regional haze in South-East Asia*. Proc. Sixth Eurasia Conference on Chemical Sciences, 27 February–2 March 2000, University of Brunei Darussalam.

RAHN, K.A. (1984), *Who's polluting the Arctic?* Natural History 5, 30–38.

RADKE, L.F., STITH, J.L., HEGG, D.A., and HOBBS, P.V. (1978), *Airborne Studies of Particles and Gases from Forest Fires*, J. Air Pollution Control Assoc. *28*, 30–34

RADKE, L.F., HEGG, D.A., HOBBS, P.V., NANCE, J.H., LYONS, J.H., LAURSEN, K.K., WEISE, R.E., RIGGAN, P.J., and WARD, D.E., *Particulate and trace gas emissions from large biomass fires in North America*. In *Global Biomass Burning: Atmospheric Climate and Biospheric Implications* (ed. Levine J.S.) (Cambridge MA, MIT Press, 1991) pp. 209–224.

RADKE, L.F., HEGG, D.A., LYONS, J.H., BROCK, C.A., HOBBS, P.V., WEISS, R.E., and RASMUSSEN, R., *Airborne measurements on smokes from biomass burning*. In *Aerosols and Climate* (eds. Hobbs P.V. and McCormick M.P., Hampton, VA) (A.Depak Publishing, 1988) pp. 411–422.

RADKE, L.F., HOBBS, P.V., and BROCK, C.A. (1990a), *Airborne lidar studies of the smokes from large biomass fires*. Paper presented at the International Laser Radar Conference, Tomsk, USSR, 23–27 July 1990.

RADKE, L.F., LYONS, J.H., HOBBS, P.V., HEGG, D.A., SANDBERG, D.V., and WARD, D.E. (1990b), *Airborne Monitoring and Smoke Characterisation of Prescribed Fires on Forest Lands in Western Washington and Oregon*. Gen. Tech. Rep. PNW-GTR-251, U.S. Dept. of Agriculture, Forest Service, Pacific Northwest Research Station, Portland OR, 1990.

RADKE, L.F., HEGG, A.S., HOBBS, P.V., and PENNER, J.E. (1995), *Effects of Aging on the Smoke from a Large Forest Fire*, Atmos. Res. *38*, 315–332.

RADOJEVIC, M., *SO$_2$ and NO$_x$ oxidation mechanisms in the atmosphere*. In *Atmospheric Acidity: Sources, Consequences and Abatement* (eds. Radojevic M. and Harrison R.M.) (Elsevier Applied Science, London, 1992) pp. 73–137.

RADOJEVIC, M. (1997), *The Haze in Southeast Asia*, Environ. Scientist 6(6), 1–3.

RADOJEVIC, M. (1998), *Burning Issues*. Chemistry in Britain *34*(12) 38–42.

RADOJEVIC (1999), *Impacts of biomass burning and regional haze on rainwater and atmospheric chemistry in the tropics*. In *Proc. of the Fifth International Joint Seminar on Regional Deposition Processes in the Atmosphere*, 12–16 October 1999, Department of Atmospheric Sciences, Seoul National University, pp. 211–220.

RADOJEVIC, M. (2003), H*aze research in Brunei Darussalam during the 1998 episode*, Pure appl. Geophys., this issue.

RADOJEVIC, M. and HASSAN, H. (1999), *Air Quality in Brunei Darussalam During the 1998 Haze Episode*, Atmos. Environ. *33*, 3651–3658.

RADOJEVIC, M. and TAN, K.S. (1999), *Impacts of Biomass Burning and Regional Haze on the pH of Rainwater in Brunei Darussalam*, Atmos. Environ. *34*, 2739–2744.

RADOJEVIC, M., TAN, K.S., and RAMALINGAM, P. (1999), *Potential Health Impacts of Criteria Pollutants Present in the Haze*. Brunei Internat. Med. J. *1*, 299–303.

RINDAM, M. (1995), *The haze phenomenon in Malaysia: problems and challenges*. In *Proc. International Symposium on Climate and Life in the Asia Pacific*, (ed. Sirinanda K.U.) 10–13 April 1995, University of Brunei Darussalam, Bandar Seri Begawan, Brunei Darussalam, pp. 74–84.

ROBERTS, P.T. and ANDERSON, J.A. (1991), *Pollutants measured in forest fires and agricultural burn plumes and their potential influence on exceedances of the ozone standard*. In *Proc. 84th Annual Meeting of the Air Waste Management Assoc. vol. 5*, Paper no. 91/67.11.

ROSWINTIARTI, O. and RAMAN, S. (1999), *Three-dimensional simulations of mean transport during the 1997 forest fires in Kalimantan, Indonesia using a mesoscale numerical model*. Paper presented at the International Conference cum Workshop on Air Quality Management, 15–19 November, University of Brunei Darussalam.

SANDBERG, D.V., PICKFORD, S.G., and DARLEY, E.F. (1975), *Emissions from Slash Burning and the Influence of Flame Retardant Chemicals*, J. Air Pollution Control Assoc. *25*(3), 278–281.

SANHUEZA, E., ARIAS, M.C., DONOSO, L., GRATEROL, N., HERMOSO, M., MARTI, I., ROMERO, J., RONDON, A., and SANTANA, M. (1992), *Chemical Composition of Acid Rains in the Venezuelan Savannah Region*, Tellus *44B*, 54–62.

SANHUEZA, E., ELBERT, W., RONDON, A., ARIAS, M.C., and HERMOSO, M. (1989), *Organic and Inorganic Acids in Rain from a Remote Site of the Venezuelan Savannah*, Tellus *41B*, 170–176.

SANI, S., KHAN, C.B., PENG, L.C., and FOOK, L.S. (1991), *The August 1990 Haze in Malaysia with Special Reference to the Klang Valley Region*, Technical Note No. 49, Malaysian Meteorological Service, Petaling Jaya, Malaysia.

SAWA, Y., MATSUEDA, H., TSUTSUMI, Y., JENSEN, J.B., INOUE, H.Y., and MAKINO, Y. (1999), *Tropospheric Carbon Monoxide and Hydrogen Measurements over Kalimantan in Indonesia and Northern Australia During October 1997*, Geophys. Res. Lett. *26*, 1389–1392.

SETZER, A. and PEREIRA, M.C. (1991), *Amazonia Biomass Burning in 1987 and an Estimate of their Tropospheric Emissions*, Ambio *20* (1), 19–22.

SOLEIMAN, A., OTHMAN, M., SHAMAH, A.A., SULEIMAN, N.M., and RADOJEVIC, M. (2003), *The Occurrence of Haze in Malaysia: A Case Study in an Urban Industrial Area, Klang Valley*, Pure appl. Geophys., this issue.

SOROOS, M.S. (1992), *The Odyssey of Arctic Haze: Toward a Global Atmospheric Regime*, Environment *34*(10), 6–27.

STITH, J.L., RADKE, L.F., and HOBBS, P.V. (1981), *Particle Emissions and the Production of Ozone and Nitrogen Oxides from the Burning of Forest Slash*. Atmos. Environ. *15*, 73–82.

TSUTSUMI, Y., SAWA, Y., MAKINO, Y., JENSEN, J.B., GRAS, J.L., RYAN, B.F., DIHARTO, S., and HARJANTO, H. (1999), *Aircraft Measurements of Ozone, NO_x, CO, and Aerosol Concentrations in Biomass Burning Smoke over Indonesia and Australia in October 1997: Depleted Ozone Layer at Lower Altitude over Indonesia*, Geophys. Res. Lett. *26*, 595–598.

TUSSIN, A.M. (1995), *Thick haze in Malaysia in the 90's*. In *Proc. International Symposium on Climate and Life in the Asia Pacific* (ed. Sirinanda K.U.) 10–13 April 1995, University of Brunei Darussalam, Bandar Seri Begawan, Brunei Darussalam, pp. 56–73.

UTHE, E.E., MORLEY, B.M., and NIELSEN, N.B. (1982), *Airborne Lidar Measurements of Smoke Plume Distribution, Vertical Transmission and Particle Size*, Appl. Optics *21*, 460–463.

VINES, R.G., GIBSON, L., HATCH, A.B., KING, N.K., MACARTHUR, D.A., PACKHAM, D.R., and TAYLOR, R.J. (1971), *On the Nature, Properties and Behaviour of Bush-fire Smoke*. CSIRO Australia, Div. Appl. Chem., Tech. Paper, 1971.

WARD, D.E. (1979), *Particulate Matter and Aromatic Hydrocarbon Emissions from the Controlled Combustion of α-pinene*. Dissertation. University of Washington, Seattle, WA.

WARD, D.E. (1989), *Air toxics and fireline exposure*. Paper presented at the *10th Conference on Fire and Forest Meteorology*, 17–21 April 1989, Ottawa, Canada.

WARD, D.E., *Factors influencing the emissions of gases and particulate matter from biomass burning*. In *Fire in the Tropical Biota: Ecosystem Processes and Global Challenges* (ed. Goldammer J.G.) (Springer-Verlag, Berlin, 1990) pp. 418–436.

WARD, D.E. and HARDY, C.C. (1984), *Advances in the characterization and control of emissions from prescribed fires*. In *Proc. 78th Annual Meeting of the Air Pollution Control Association*, 24–29 June 1984, San Francisco, *CA*, Air and Waste Management Association, Pittsburgh, PA, Paper No. 84–363.

WARD, D.E. and HARDY, C.C. (1991), *Smoke Emissions from Wildland Fires*, Environ. International *17*, 117–134.

WARD, D.E. NELSON, R.M, and ADAMS, D. (1979), *Forest fire documentation*. In *Proc. 72nd Annual Meeting of the Air Pollution Control Association*, 21–29 June 1979, Cincinnati, OH, Air and Waste Management Association, Pittsburgh, PA, paper No. 79–6.3.

WARD, D.E., SUSOTT, R.A., KAUFMANN, J.B., BABBITT, R.E., CUMMINGS, D.L., DIAS, B., HOLBEN, B.N., KAUFMAN, Y.J., RASMUSSEN, R.A., and SETZER, A.W. (1992), *Smoke and Fire Characteristics for Cerrado and Deforestation Burns in Brazil-BASE-B Experiment*, J. Geophys. Res. *97*, 14,601–14,619.

WESTBERG, H., SEXTON, K., and FLYCKT, D. (1981), *Hydrocarbon Production and Photochemical Ozone Formation in Forest Burn Plumes*, J. Air Pollution Control Assoc. *31*, 661–664.

WHITE, J.D. (1987), *Emission Rates of Carbon Monoxide, Particulate Matter, and Benzo(a)pyrene from Prescribed Burning of Fine Southern Fuels*. Res. Note SE-346, U.S. Dept. of Agriculture, Forest Service, Southeastern Forest Experiment Station, Asheville, NC, 1987.

WILLIAMS, P.T., RADOJEVIC, M., and CLARKE, A.G. (1988), *Dissolution of Trace Metals from Particles of Industrial Origin and its Influence on the Composition of Rainwater*, Atmos. Environ. *22*, 1433–1442.

WU, J., WONG, M. K., LEE, H.K., and ONG, C.N. (1995), *Capillary Zone Electrophoretic Determination of Heterocyclic Aromatic Amines in Rain*, J. Chromatographic Science *33*, 712–716.

YOKELSON, R.J., GRIFFITH, D.W.T., and WARD, D.E. (1996), *Open-path Fourier Transform Infrared Studies of Large-scale Laboratory Biomass Fires*, J. Geophys. Res. *101*, 21067–21080.

YOKELSON, R.J., SUSOTT, R., WARD, D.E., REARDON, J., and GRIFFITH, W.T. (1997), *Emissions from Smouldering Combustion of Biomass Measured by Open-path Fourier Transform Infrared Spectroscopy*, J. Geophys. Res. *102*, 18,865–18,877.

ZAKARIA, M.P., AWANG, M., KASIM, A.R.H.A., KUANG, H.S.Z., HARON, M.J., IBRAHIM, R., JARGONA, M.K., KASIM, Z.A., PHENG, T.K., and SIMAN, N.M. (1995), *Organic compounds, sulphate, nitrate and heavy metals in the atmosphere of the city of Kuala Lumpur*. In *Proc. International Symposium on Climate and Life in the Asia Pacific* (ed. Sirinanda K.U.) 10–13 April 1995, University of Brunei Darussalam, Bandar Seri Begawan, Brunei Darussalam, pp. 85–88.

(Received February 15, 2000, accepted December 14, 2000)

To access this journal online:
http://www.birkhauser.ch

Pure appl. geophys. 160 (2003) 189–204
0033–4553/03/020189–16

© Birkhäuser Verlag, Basel, 2003

❙ Pure and Applied Geophysics

Application of DARLAM to Regional Haze Modeling

Lawrence C.C. Koe,[1][*] Avelino F. Arellano, Jr.,[1,2]
and John L. McGregor [3]

Abstract — The CSIRO Division of Atmospheric Research limited area model (DARLAM) is applied to atmospheric transport modeling of haze in southeast Asia. The 1998 haze episode is simulated using an emission inventory derived from hotspot information and adopting removal processes based on SO_2.

Results show that the model is able to simulate the transport of haze in the region. The model images closely resemble the plumes of NASA Total Ozone Mapping Spectrometer and Meteorological Service Singapore haze maps. Despite the limitation of input data, particularly for haze emissions, the three-month average pattern correlation obtained for the whole episode is 0.61. The model has also been able to reproduce the general features of transboundary air pollution over a long period of time. Predicted total particulate matter concentration also agrees reasonably well with observation.

The difference in the model results from the satellite images may be attributed to the large uncertainties of emission, simplification of haze deposition and transformation mechanisms and the relatively coarse horizontal and vertical resolution adopted for this particular simulation.

Key words: Atmospheric transport modeling, biomass burning, haze, southeast Asia.

1. Introduction

The immense public health and economic impact of transboundary haze in southeast Asia (SEA) has generated a lot of scientific and political attention in recent years. Several regional and international programs have been initiated to minimize uncontrolled forest/land fires in the region as well as to mitigate their effects on the environment. In the area of regional haze monitoring and mitigation, atmospheric transport models are important in providing scientific support to related policy issues and their implementation. They can be used in the analysis and understanding of the evolution and transport of haze over an extended period of time. The ability of these models to predict pollution concentration downwind of emission sources and to

---

[1] Department of Civil Engineering, National University of Singapore, 10 Kent Ridge Crescent, 119620, Singapore.

[2] Now at Nicholas School of the Environment, Duke University, Durham, NC, U.S.A. 27708.

[3] CSIRO Atmospheric Research, PB1 Aspendale, Vic. 3195, Australia.

[*] Corresponding Author, now at Aromatrix Pte Ltd., 1 Butik Batok St 22, GRP Indl Bldg #03-02, Singapore 659592. E-mail: ceo@aromatrix.com

forecast the local and regional areas most likely to be affected by haze makes them valuable tools in air quality management.

The CSIRO Division of Atmospheric Research limited area model, DARLAM, has been used to simulate long-range atmospheric transport of trace gases and aerosols. It has been successfully applied to studies on CO_2, radon and sulfate aerosol transport as well as in the representation of their sources and sinks. Recently, collaboration between National University of Singapore and CSIRO Atmospheric Research was initiated to investigate the transport and behavior of haze from the biomass burning in southeast Asia using DARLAM. The main focus of the research is the study of 1997–1998 Indonesia forest fires.

This paper describes some results of the study. It presents a plausible system for regional haze modeling using CSIRO's DARLAM as an atmospheric transport model. It also shows the performance of the model on the simulation of 1998 regional haze by comparing the output to satellite images of NASA Total Ozone Mapping Spectrometer (TOMS) and Meteorological Service Singapore (MSS) haze maps and to a few observed total particulate matter (TPM) concentrations. Finally, the merits and limitations of the model are evaluated for future impact assessment and regional haze modeling studies.

2. Model Description

2.1 Regional Climate Model

DARLAM is a regional climate model. It is a two-time level semi-implicit hydrostatic primitive equations model, as described originally by MCGREGOR (1987). The horizontal advection is carried out by a semi-Lagrangian scheme where the departure points for each time step are determined by the scheme of MCGREGOR (1993). Bicubic Lagrange interpolation is employed for the associated updating of variables. For moisture and trace gas advection, the positive definite enforcement of BERMEJO and STANIFORTH (1992) is also incorporated to preserve the shape of the advected quantities where strong gradients and discontinuities may be present. Vertical advection is performed as a split process using the VAN LEER (1974) total-variation scheme advocated by THUBURN (1993); this is a simple limited scheme with small numerical diffusion.

DARLAM has extensive parameterizations to describe key physical processes. For longwave and shortwave radiation, a recent version of the scheme by FELS and SCHWARTZKOPF (1975) is adopted; cloud cover is diagnosed from the relative humidity. Cumulus cloud convection uses a version of the ARAKAWA (1972) mass flux scheme modified by CSIRO's H.B. Gordon (personal communication). The vertical mixing and surface transfer processes are represented by the LOUIS (1979) approach in conjunction with the GELEYN (1987) shallow convection scheme. Surface

temperatures and moisture are parameterized by the canopy scheme of KOWALCZYK
et al. (1994). The vegetation types are based on DORMAN and SELLERS (1989) while
the soil characteristics are classified according to ZOBLER (1988). Soil temperatures
over three layers are prognostically updated in the model.

The updating of the fields near the lateral boundaries is carried out by a
procedure similar to DAVIES (1976). The quantities in the outermost rows take the
values from a set of data interpolated through time and space from 12-hourly
analyses. Certain weights, which decrease exponentially away from the boundaries,
are incorporated in the calculations such that large-scale forcing is quite insignificant
after about six rows.

A more detailed description of the physical parameterizations is given in
McGREGOR et al. (1993).

2.2 Aerosol Model

The aerosol model follows similar dynamics and physical parameterization as for
moisture. The horizontal and vertical advection and diffusion schemes described
earlier are adopted for haze transport.

Removal Processes

The model uses simple parameterizations for both dry and wet deposition. The
schemes are based on removal processes of airborne sulfur (SO_2 and sulfate). For dry
deposition, the deposition velocities are assumed to be similar to the values for SO_2.
Table 1 shows the model dry deposition velocities for five seasonally varying surface
types. The values are modified from those suggested by ROBERTSON et al. (1995) and
a factor of three smaller than for SO_2 is adopted.

The parameterization of wet removal is based on the SO_2 schemes of FEICHTER
et al. (1996) and BERGE (1993). For in-cloud scavenging, all haze within the cloudy
part of a grid box is assumed to be dissolved in cloud droplets. The average change in
haze concentration, $\Delta\mu$, within the model time step is given by

$$\Delta\mu = \mu \cdot f \cdot \left(e^{-\beta\Delta t} - 1\right) . \tag{1}$$

Table 1

Dry deposition velocities as a function of surface type

Category	Description	Dry deposition velocity* (m s^{-1})
1	Urban and bare ground	0.0007
2	Low vegetation	0.0010
3	Medium height vegetation	0.0017
4	High vegetation	0.0023
5	Ocean	0.0033

*Sulfate deposition velocity divided by three

where f is an empirically calculated fraction of the grid box that is covered by precipitating clouds, Δt is equal to the model time step (15 minutes) and β represents the frequency of cloud rainwater conversion. For a given grid point, the concentration decreases by an amount equal to that lost from the cloud region at an exponential rate β for a period of time Δt. As pointed out by GIORGI and CHAMEIDES (1986), the expression also ensures numerical stability since no more than the amount of material present in the grid box can be removed during a time step.

Between cloud layers or below the lowest clouds, haze particles are scavenged by the falling raindrops. The removal is calculated in the same way as in-cloud scavenging replacing cloud water by rainwater. This is given by

$$\Delta\mu = \mu \cdot \left(e^{-(C \cdot P_F \cdot E_m)\Delta t} - 1\right) , \tag{2}$$

where $C = 5.2$ m^3 kg^{-1} s^{-1}, P_F is the precipitation mass in kg m^{-3} and $E_m = 0.1$, the mean collection efficiency.

The effect of the release of haze to the air from evaporating precipitation is also incorporated. When raindrops or cloud water evaporate completely, part or all of the haze removed from higher levels is returned. The evaporated fraction is calculated by taking the difference between the amount of rainfall (cloud water) passing through the k and $k + 1$ vertical level divided by the amount of rainfall (cloud water) passing through $k + 1$ vertical level.

Although secondary aerosols play a critical role in the study of long-range transport of the haze, as indicated by their presence in observations in Kuala Lumpur in 1997 (VON HOYNINGEN-HUENE *et al.*, 1999), chemical transformations of particulate matter and other products of biomass burning are not included in the model. The simulations focus only on the transport of primary total particulate matter with no associated aqueous and gas phase chemistry involved.

3. Emissions

The emission inventory is taken from Integrated Forest Fire Management (IFFM)/German Technical Cooperation (GTZ) hotspot data. The data consist of daily hotspots derived from NOAA AVHRR and European Remote Sensing Synthetic Aperture Radar (ERS-2 SAR) satellite images for the whole haze episode of 1998. The hotspot information includes the time of detection and its corresponding geographical location in latitude and longitude. The hotspot data are a product of the IFFM/GTZ project in Samarinda, East Kalimantan. The project investigated the effectiveness of the multi-temporal ERS-2 SAR images to complement and expand the existing NOAA-AVHRR fire detection system (SIEGERT and HOFFMANN, 1998). The data have the advantage of providing spatial and temporal distribution of fires; however, the data which came from a more detailed analysis of hotspots, focused only on East Kalimantan.

Table 2

Biomass loading and burning efficiency assumptions

	Forest	Peat Swamps
Biomass loading (tonnes km^{-2})	23,000	97,500
Burning efficiency	20%	50%

The duration of each hotspot is taken to be one whole day; the aerosol is injected uniformly for 24 hours. The NOAA-AVHRR image has a resolution of 1.1 km^2. To allow for the possibility that the burnt area may occupy a smaller area, and that the duration of burning may be less than 24 hours, it is assumed that each hotspot has an effective burnt area of 0.9 km^2. This is only arbitrary as the actual burnt area per hotspot cannot be exactly determined from the data. The computation of particulate emissions was based on UNEP (1999) estimates of biomass burnt; the biomass burnt M is a product of the area burnt A, biomass loading B and its associated burning efficiency E. The calculation includes the contributions of forest and peat swamp fires with corresponding biomass loading and burning efficiencies shown in Table 2.

$$M = A \cdot B \cdot E \ . \tag{3}$$

The resulting biomass burnt can be converted to the amount of total particulate matter by multiplying an emission factor of 30 kg TPM ton^{-1} C with the biomass having a 45% carbon content. These assumptions are based from the works of WARD (1990), ANDREAE (1991) and LEVINE (1999).

The height of injection is assumed in the simulation, and the rationale for its choice is based on the type of fire for the associated hotspot. Aerosols from smoldering fires tend to be emitted at a lower vertical level while aerosols from flaming fires rise up to higher levels (higher than 1000 m above ground) before undergoing normal advection.

4. 1998 South East Asia Haze Simulation

The simulation is conducted over southeast Asia at a horizontal resolution of 60 km. The domain is centered on the island of Borneo and covers the neighboring countries of Thailand, Malaysia, Singapore, Sulawesi, the southern part of Philippines and the northern tip of Australia. This domain is chosen to include all the areas most affected by the haze episodes of 1997 and 1998. The 1998 Indonesia forest fire episode is also selected to simplify the simulations, since the scope of the fires was not very large and was only localized in central eastern Kalimantan. The simulation is also carried out at 18 vertical levels with the lowest four levels comprising the boundary layer.

MODEL DOMAIN (Lon 90.4,138.1;Lat -15.1,11.2)

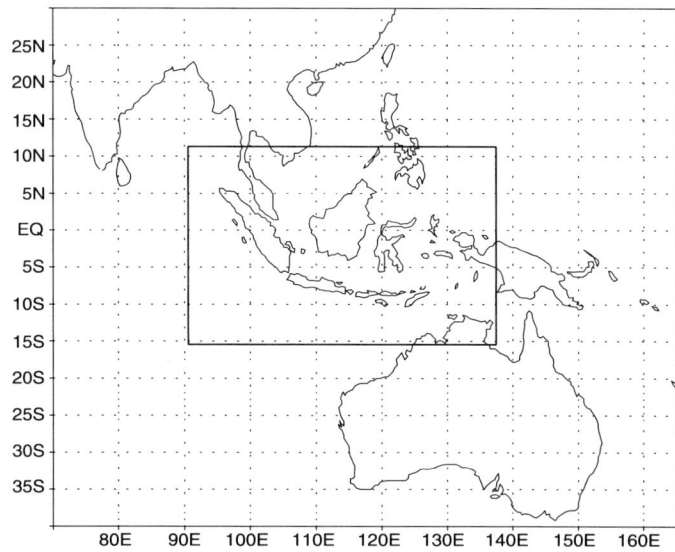

Figure 1
Spatial domain of the simulation.

The boundary data and the initial data were taken from the National Center for Environmental Prediction (NCEP) reanalysis for February to April 1998 at 1° resolution. In order to ensure that the dynamic variables of the simulation continued to resemble the observations throughout the 3-month simulation, the model temperatures, wind and surface pressure were nudged each time step towards the interpolated reanalysis values using a 12-hour e-folding time.

Initial qualitative calibration of model parameters with satellite images and observed TPM concentration for this period shows that a reasonable height of injection is the third vertical level (about 500 m above ground); the emissions originated from smoldering fires in this area.

5. Results

5.1 Comparison of Model Results with Satellite Images

The model output is compared to the satellite images of NASA TOMS and NOAA AVHRR-derived MSS haze maps. The similarity of the plume shape, relative intensity, spatial and temporal patterns are evaluated for each daily image. The daily model output in this case is an average of the TPM concentration ($\mu g\ m^{-3}$) from ground level to the sixth vertical level (about 2100 m). It is observed that no appreciable haze was transported above this vertical level. NASA TOMS daily

aerosol index (taken as a composite of a two-day data) is the main verification data for this study while the haze maps are used as supportive information. It has been found that the haze maps are especially useful in providing extra details that are not shown in the NASA TOMS such as hotspot location and extent of haze. The following images are chosen to represent each major haze episode of the entire simulation.

Qualitative Comparison

The orientation of the plumes from the model is similar to the satellite images, particularly with the MSS haze maps. From direct observation, the daily model images in Figures 2 to 4 closely resemble NASA TOMS and MSS images. It can also be seen that the plumes follow the general pattern of the wind streamlines. As the orientation of the plumes is associated with the regional wind flow, the results verify that the synoptic wind pattern is a major driving force in the transport of haze and this has been fairly captured by the model.

The shapes of the plumes from the model output agree reasonably with the satellite images. These shapes are defined by the particulate matter concentrations. While it is noted that the scale of the NASA TOMS (aerosol index) cannot be directly compared with the predicted particle concentrations ($\mu g\ m^{-3}$), the similarity of the shape was evaluated by the relative concentration within the plumes. This means that the highest aerosol index should correspond to the highest particle concentration.

The similarity of the NASA TOMS with the model concentrations in terms of the extent of the plume is rather low. The reason for this is that the satellite images only exhibit the higher concentration of the plume. Low particle concentrations are not usually captured by NASA TOMS. The plume extent of the model concentration however, is better identified in the MSS haze maps. In some instances, the large plume is due to the gross emission estimates. This type of error affects not only a particular daily image but may influence the succeeding images. The plume extent on March 18, for example, is partly influenced by larger emissions during the previous week.

Some of the plumes have offsets in the location of the plume center. This may be due to the uncertainty of the emission inventory and the low horizontal resolution of the model. The location of the hotspots was assumed to cover a 60-km grid but it may be smaller in reality. Also, the plumes are not easily transported in the low-wind areas, particularly in southern Kalimantan. When compared to the satellite images, the plumes are actually moving at a faster rate than that predicted. This observation suggests the presence of smaller scale circulations (local wind systems) that may affect the transport of the particulate in these areas other than the synoptic wind pattern.

Lastly, the haze in the northern part of Borneo (Miri) and Sumatra is not depicted in the model since the hotspot data used are only limited to east Kalimantan.

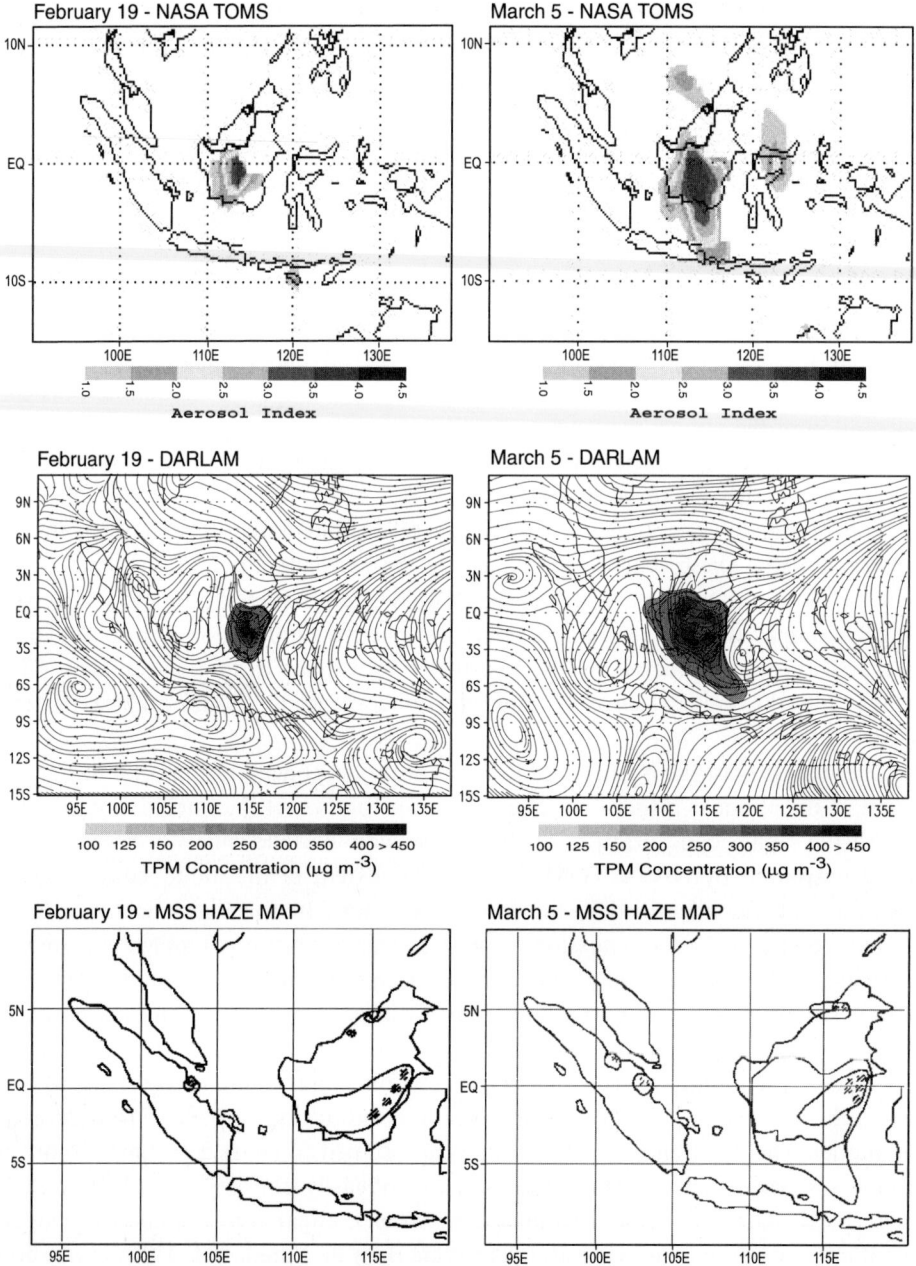

Figure 2
Comparison of model output with NASA TOMS and MSS haze map for February 19 and March 5 over
Indonesia (model output in TPM concentration with wind (u, v) streamlines at $\sigma = 0.90$).

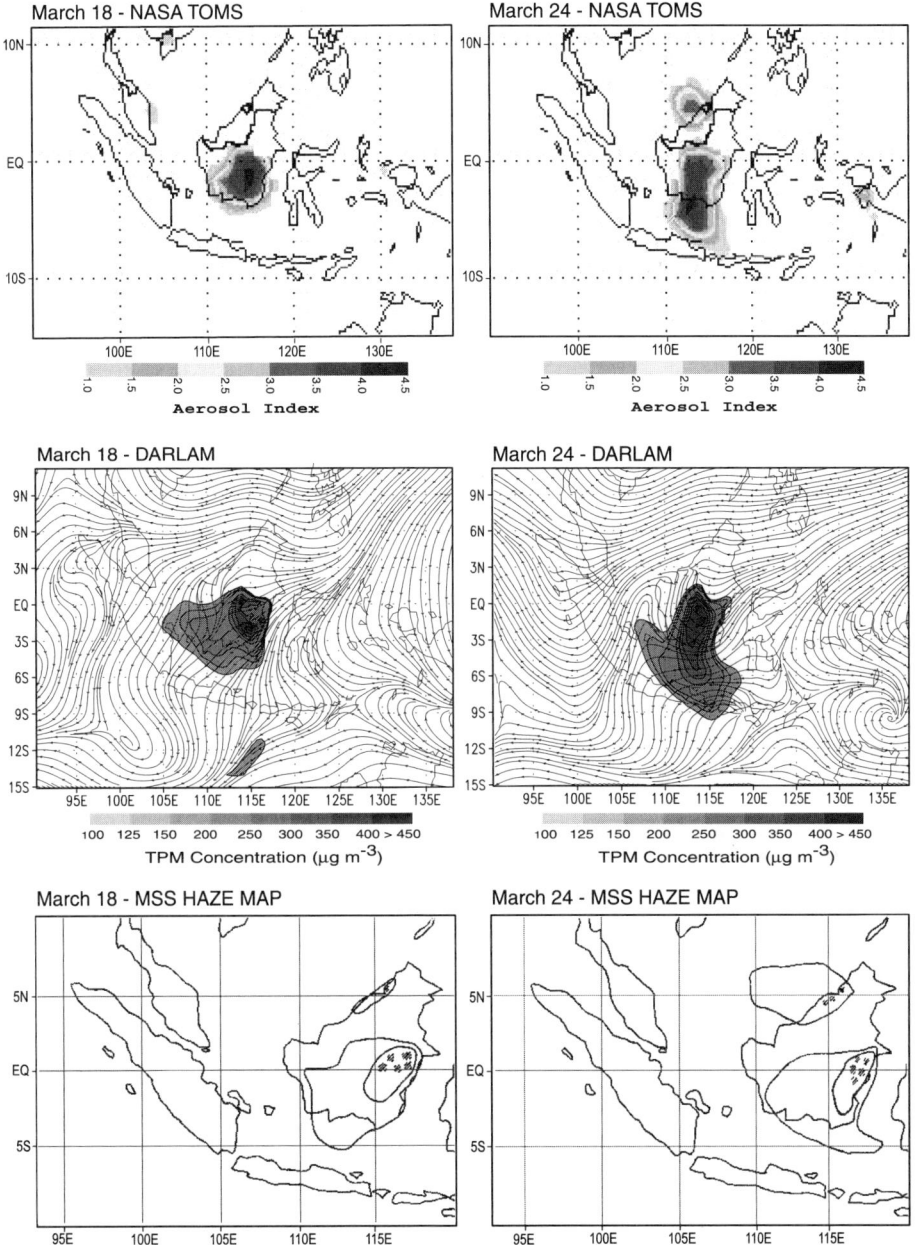

Figure 3
Comparison of model output with NASA TOMS and MSS haze map for March 18 and March 24 over
Indonesia (model output in TPM concentration with wind (u, v) streamlines at $\sigma = 0.90$).

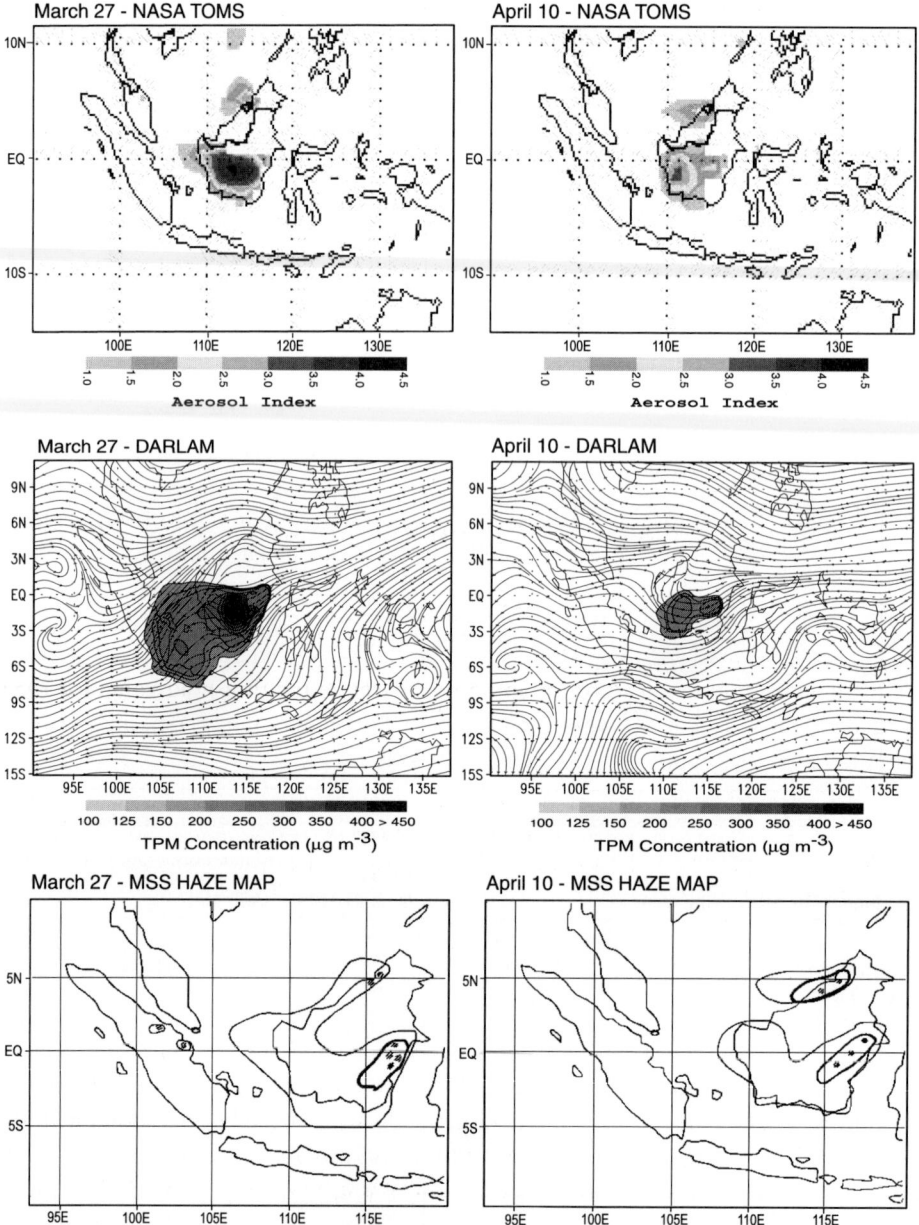

Figure 4
Comparison of model output with NASA TOMS and MSS haze map for March 27 and April 10 over
Indonesia (model output in TPM concentration with wind (u, v) streamlines at $\sigma = 0.90$).

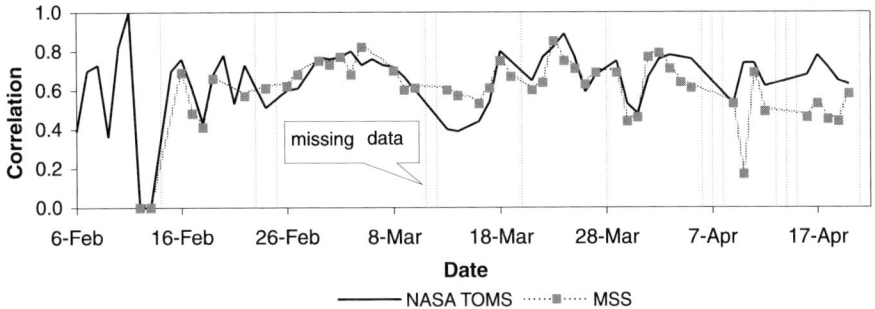

Figure 5
Time Series of Model haze pattern correlation with NASA TOMS and MSS haze maps.

Pattern Correlation

In order to quantify the performance of the model, the results are statistically compared to digitized satellite images. Each model grid point is assigned a value of either 1 or 0 to correspond to a haze event or no-haze event, respectively. For the NASA TOMS comparison, a haze event corresponds to a minimum haze concentration (averaged over the lowest six vertical sigma levels) of 125 µg m^{-3}. On the other hand, a concentration of 75 µg m^{-3} is set as the threshold value for MSS comparison. These values are assigned arbitrarily after fine-tuning. Upon assigning a value to each grid point, a spatial pattern correlation matrix is calculated.

The three-month average pattern correlation from NASA TOMS comparison is 0.61 with the highest monthly average correlation obtained in March (0.68). Figure 5 shows the time series of pattern correlation for the entire period of simulation. The model results compare favorably with the satellite images.

5.2 Comparison of Model Results with Observed TPM Concentrations

The model results are also compared to the TPM observations in southern Borneo. Figure 6 shows the comparison of predicted ground TPM concentration to available measurements. The model TPM ground concentrations in Pontianak (southwest Borneo) and Samarinda (southeast Borneo) strongly correspond with the observed data. The predicted concentrations are realistic considering that the observation measurements may be in error by almost a factor of 2 (A. Heil, personal communication, 1999). This is based on 1997 measurements done by two different agencies at a similar location. The discrepancy of the measurements between the two agencies is almost a factor of two. This is probably due to procedural differences. On the other hand, the Palangkaraya (south Borneo) TPM observed concentrations are significantly low compared to the model. The observed data in Palangkaraya may be anomalous since satellite images show a high concentration of haze in the area. Unfortunately, there are only three station measurements available for comparison.

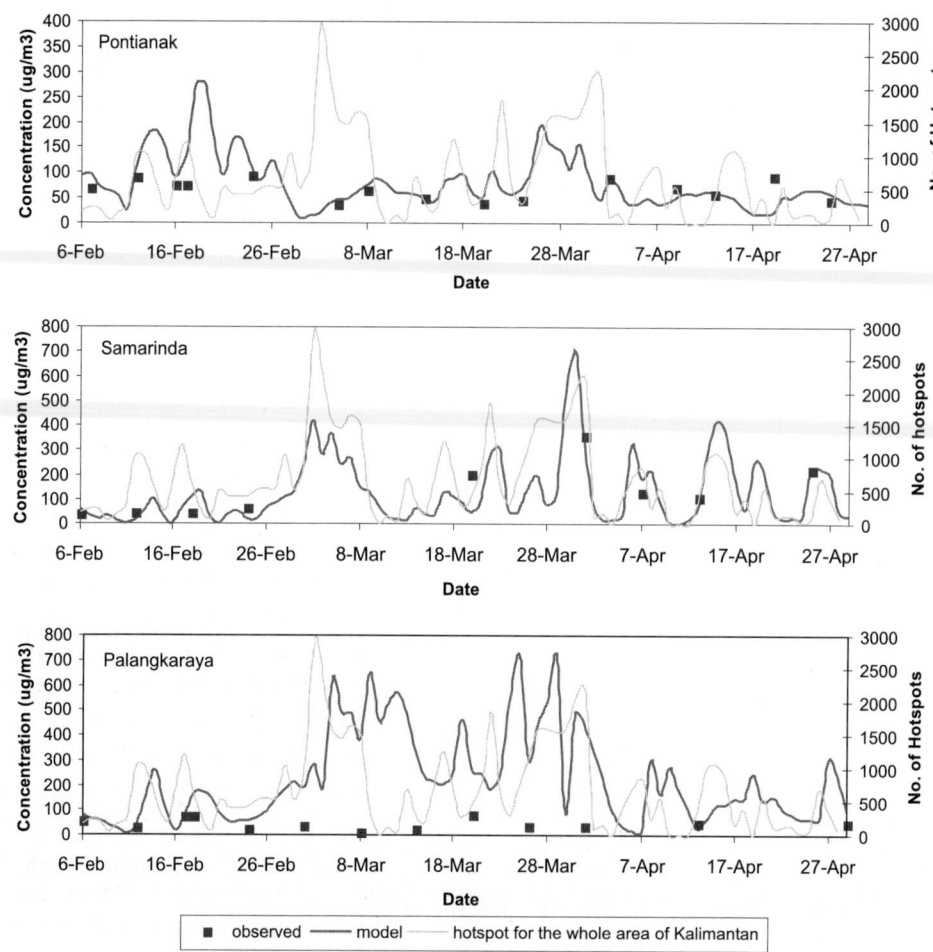

Figure 6
Predicted TPM concentration versus observed TPM* concentration.
(* Original Source: BMG Meteorological and Geophysical Agency, Indonesia)

Hence, these comparisons can only be used to suggest that the model prediction is realistic.

5.3 Discussion on Haze Transport

The February to April 1998 haze episode was the second fire season following the immense forest fires in 1997. The fires resumed and burned out of control after an abnormally short wet season. March 1998 was the peak of the haze event. Unlike in 1997 where fires are scattered all over Sumatra and Borneo, the peat fires in Kalimantan primarily caused the haze event. It has immensely affected the southern part of Borneo and some parts of Java, Sumatra and northern Borneo. Most of the fires originated from East Kalimantan and some areas in northern Borneo. The area burnt in Kalimantan for 1998 (based on UNEP 1999 report) is estimated to be 8,000 km^2.

The transport of haze in the region is primarily driven by the wind system. The El Niño phenomenon still continued to cause low rainfall in the region such that the wet deposition is relatively low at this period. Unlike the 1997 haze events, the low-level wind is predominantly from the northeast. The wind over the island of Borneo is very light owing to its large land area. Strong winds rarely occur there and land breezes are very light relative to other parts of the archipelago. The monsoon is also channeled through the South China Sea and the Java Sea and through the straits of Makasser. These cause the haze to be localized in the southern part of Borneo and the Java Sea. The February and early March haze was eventually transported to the southeast of Borneo and to the east of Java Sea, while the late March and April haze was transported to the southeast of Sumatra due to the transition of the monsoon from northeast to southeast at this time.

The influence of land and sea breezes is apparent in Kalimantan. The observed surface winds in the coastal plains for the whole of Kalimantan (particularly on the south) exhibit a diurnal cycle. This behavior gradually develops during the start of the inter-monsoon, when synoptic winds begin to weaken. This also supports the observation on the offsets of plume location in the model images. In Samarinda, where most hotpots originated, the actual wind speed dramatically changes from being low at nighttime to high in early afternoon. Figure 7 shows a comparison of observed and predicted wind speeds in Samarinda. While the model was able to simulate the diurnal variation, the simulation should be made at finer resolution to fully capture the wind speeds.

The vertical transport of the haze is also seen to be limited up to the sixth vertical layer (layer top about 2500 m above ground) and with its plume center lying between about 900 m to 2100 m. Above the sixth vertical layer, there is no appreciable amount of haze advected (Fig. 8). This result is similar to the airborne observations of VON HOYNINGEN-HUENE et al. (1999) which show the upper level of haze is

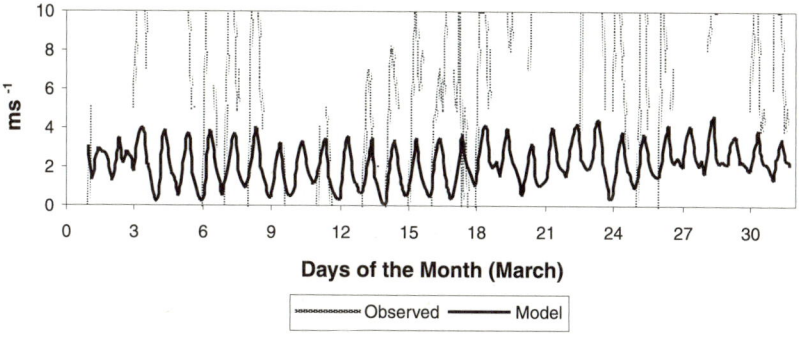

Figure 7
Surface wind speeds at Samarinda, Kalimantan (model versus observation).

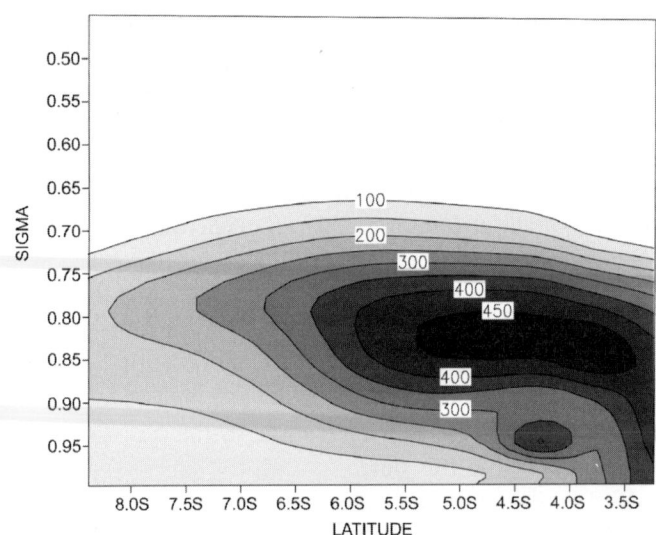

Figure 8
Vertical profile of haze on March 23, 1998 1500 UTC (in μg m $^{-3}$). The profile is a cross section along the longitude 115.3E.

between 2000 m to 2500 m and of BRAAK (1929) who observed that the haze is most dense between 1000 and 2000 m. This also confirms the observation that the boundary layer and the trade wind inversion within these levels largely influence the haze transport in the tropics. The trade wind inversion serves as a lid or capping for the haze (HEIL *et al.*, 1998; BRAAK, 1929).

6. Conclusion

Despite the limited scientific information on southeast Asia regional haze and biomass burning in general, the 1998 haze episode has been successfully simulated using the regional climate model DARLAM in conjunction with NCEP data. A simple system is created for this particular simulation. The system mainly includes an emission inventory with hotspot information as the basis for its spatial and temporal distribution. It also utilizes simplified removal processes similar to those for SO_2. It has been shown that such a simple system can work effectively in simulating haze transport from biomass burning in the region.

This simulation reproduces reasonably the mesoscale-β meteorology. This is very important in atmospheric pollution studies as it greatly affects the transport behavior of the pollutant. DARLAM effectively captures the meteorological variability in regional scale.

However, atmospheric transport modeling should also account for well-represented emissions as well as include realistic mechanisms for atmospheric chemistry and removal processes. Only simple approaches have been used in this study. More sophisticated schemes should be included in future transport modeling research.

Clearly, the main limitation of the simulation is the accuracy of emission data and its representation in the model. Finer horizontal and vertical resolution should improve the latter. Currently, the horizontal resolution is not sufficient in representing local area sources or for fully capturing the effects of local circulations. The boundary-layer processes can be better described by increasing the number of vertical levels within the boundary layer.

There is a need to improve the information on input data, particularly for emission rates from biomass burning and its associated deposition rates. The lack of reliable scientific data is still the most critical issue to be resolved in the area of haze transport modeling in southeast Asia.

Acknowledgments

The authors are grateful to Dr. Jack Katzfey of CSIRO Atmospheric Research, for preparing most of the model input files, to Meteorological Service Singapore for the regional meteorological observations and haze maps, to Angelika Heil of GTZ for the TPM and meteorological data, to Dr. Anja Hoffmann of IFFM for the hotspot data, to the Laboratory for Atmospheres, NASA/Goddard Space Flight Center for the NASA TOMS images and to NCEP for the reanalyses.

REFERENCES

ANDREAE, M.O., *Biomass burning: Its history, use and distribution and its impact on environmental quality and global climate.* In *Global Biomass Burning: Atmospheric, Climatic, and Biospheric Implications* (ed. Levine, J.) (The MIT Press, Cambridge, Mass. 1991) pp. 3–21.

ARAKAWA, A. (1972), *Design of the UCLA General Circulation Model. Numerical Simulation of Weather and Climate,* Technical Report No. 7, Dept. of Meteorology, University of California, Los Angeles.

BERGE, E. (1993), *Coupling of Wet Scavenging of Sulfur to Clouds in a Numerical Weather Prediction Model,* Tellus *45B,* 1–22.

BERMEJO, R. and STANIFORTH, A. (1992), *The Conversion of Semi-Lagrangian Advection Schemes to Quasi-Monotone Schemes,* Mon. Wea. Rev. *120,* 2622–2632.

BRAAK, C. (1929), *Transparency of the air: Haziness.* In *The Climate of the Netherland Indies; Volume 1. Verhandelingen No. 8,* Koninklijk Magnetisch en Meteorologisch Observatorium te Batavia, 190–198.

DAVIES, H.C. (1976), *A Lateral Boundary Formulation for Multi-Level Prediction Models,* Quart. J. Roy. Meteor. Soc. *102,* 405–418.

DORMAN, J.L. and SELLERS, P.J. (1989), *A Global Climatology of Albedo, Roughness Length and Stomatal Resistance for Atmospheric General Circulation Models as Represented by the Simple Biosphere Model (SiB),* J. Appl. Meteor. *28,* 833–855.

FEICHTER, J., KJELLSTROM, E., RODHE, H., DENTENER, F., LELIEVELD, J. and ROELOFFS, G.J. (1996), *Simulation of the Tropospheric Sulfur Cycle in a Global Climate Model,* Atmos. Environ. *30,* 1693–1707.

FELS, S.B. and SCHWARTZKOPF, M.D. (1975), *The Simplified Exchange Approximation: A New Method For Radiative Transfer Calculations,* J. Atmos. Sci. *32,* 1475–1488.

GELEYN, J.F. (1987), *Use of a modified Richardson number for parameterizing the effect of shallow convection*. In *Short-and Medium-Range Numerical Weather Prediction* (ed. Matsuno T.), Special Volume of the J. Meteor. Soc. Japan, 233–246.

GIORGI, F. and CHAMEIDES, W. (1986), *Rainout Lifetimes of Highly Soluble Aerosols and Gases as Inferred from Simulations with a General Circulation Model*, J. Geophys. Res. *91*, D13, 14,367–14,376.

HEIL, A., STOLLE, F., MAHMUD, M., and ELFIAN, E. (1998), *Air Pollution from Large Scale Forest and Land Fires in Indonesia 1997/98: Development and Impacts*. Draft document for the SEA-SPAN Electronic Conference on Transboundary Air Pollution, Oct. 1998, http://www.icsea.or.id/sea-span/scipol2/study42.htm.

KOWALCZYK, E.A., GARRATT, J.R., and KRUMMEL, P.B. (1994), *Implementation of a Soil Canopy Scheme into the CSIRO GCM – Regional Aspects of the Model Response*, CSIRO Atmospheric Research Tech. Paper No. 32, 59 pp.

LEVINE, J.S. (1999), *The 1997 fires in Kalimantan and Sumatra, Indonesia: Gaseous and Particulate Emissions*, Geophys. Res. Lett. *26*, 815–818.

LOUIS, J.F. (1979), *A Parametric Model of Vertical Eddy Fluxes in the Atmosphere*, Boundary-Layer Meteor. *17*, 187–202.

McGREGOR, J.L. (1987), *Accuracy and Initialization of a Two-time-level Split Semi-Lagrangian Model*. Short-and -medium-range Numerical Weather Prediction (ed. T. Matsuno), Special Volume of the J. Meteor. Soc. Japan, 233–246.

McGREGOR, J.L. (1993), *Economical Determination of Departure Points for Semi-Lagrangian Models*, Mon. Wea. Rev. *121*, 221–230.

McGREGOR, J.L., GORDON, H.B., WATTERSON, I.G., DIX, M.R., and ROTSTAYN, L.D. (1993), *The CSIRO 9-level Atmospheric General Circulation Model*, CSIRO Atmos. Res. Tech. Paper No. *26*, 89 pp.

ROBERTSON, L., RODHE, H., and GRANAT, L. (1995), *Modelling of Sulfur Deposition in the Southern Asian Region*, Water, Air and Soil Pollution *85*, 2337–2343.

SIEGERT, F. and HOFFMANN, A. (1998), *Evaluation of the 1998 Forest Fires in East-Kalimantan (Indonesia) Using multi-temporal ERS-2 SAR Images and NOAA-AVHRR Data*, as presented at the International Conference on Data Management and Modelling Using Remote Sensing and GIS for Tropical Forest Land Inventory, Jakarta, Indonesia, 26–29 October 1998.

THUBURN, J. (1993), *Use of a Flux-limited Scheme for Vertical Advection in a GCM*, Quart. J. Roy. Meteor. Soc. *119*, 469–487.

UNEP (1999), LEVINE, J.S., BOBBE, T., RAY, N., SINGH, A., and WITT, R.G. *Wildland Fires and The Environment: A Global Synthesis*, UNEP/DEIAEW/TR.99-1.

VAN LEER, B. (1974), *Towards the Ultimate Conservative Difference Scheme. II: Monotonocity and Conservation Combined in a Second Order Scheme*, J. Comput. Phys. *14*, 361–370.

VON HOYNINGEN-HUENE, W., SCHMIDT, T., SCHIENBEIN, S., KEE, C.A., and TICK, L. J. (1999), *Climate-relevant Aerosol Parameters of South-East Asian Forest Fire Haze*, Atmos. Environ. *33*, 3183–3190.

WARD, D., *Factors influencing the emissions of gases and particulate matter from biomass burning*. In *Fire in the Tropical Biota: Ecosystem Processes and Global Challenges* (ed. J.G. Goldammer) (Springer-Verlag, Berlin 1990), Ecological Studies, vol. 84, pp. 418–436.

ZOBLER, L. (1988), *A World Soil File for Global Climate Modeling*. NASA Tech. Memo. 87802 [available from NASA Goddard Institute for Space Studies, 2880 Broadway, New York, NY 10025].

(Received February 15, 2000, accepted January 12, 2001)

To access this journal online:
http://www.birkhauser.ch

Pure appl. geophys. 160 (2003) 205–220
0033–4553/03/020205–16

❙ Pure and Applied Geophysics

Haze in Southeast Asia: Needed Local Actions for a Regional Problem

JOHN ONU ODIHI[1]

Abstract—Southeast Asia experienced disastrous haze episodes in 1997 and 1998 as a result of natural factors such as the El-Niño occurrence and its attendant drought, which aided desiccation and facilitated high combustibility of forest resources, and human activities such as widespread burning of forests in the lush tropical environment. The human factors that caused the haze are presented in this paper, and the conclusion is that Nature is not guilty. A comprehensive approach that tackles the root causes of the haze problem is suggested. Particularly, socioeconomic problems of the poor who use fire as part of their land management, promulgation and effective enforcement of environmental laws are needed on the one hand to reduce the frequency and magnitude of haze episodes in the future. On the other hand, education of the public on 'safe' behavior during haze episodes is necessary to avert large scale haze-related disasters or to prevent haze from becoming instantly synonymous with disaster in the region.

Key words: Haze, environmental laws, education, economic empowerment, Southeast Asia.

1. Introduction

Haze episodes, which have become familiar environmental hazards in Southeast Asia, are the result of a combination of human activities such as forest burning in the lush tropical environment and the cyclic El-Niño-spurned drought, which aids desiccation and facilitates high combustibility of forest resources. El Niño and its attendant drought in Southeast Asia, which are endemic natural phenomena, will likely continue to occur in the region. This is in line with the views expressed in reviews of El Niño and Southern Oscillation (ENSO) by many scholars including RASMUSSON and WALLACE (1983), CANE (1983), and YARNAL and KILADIS (1985). However, there is no need for these natural events to become instantly synonymous with haze or haze hazard, as was the case with their recent occurrences in the region. Preliminary studies of the recent haze episodes show that their adverse impacts were both serious and multidimensional, including ecological, economic, social, and even political (ODIHI, 1998; ANAMAN and IBRAHIM, 1999; RADOJEVIC

[1] Department of Geography, University Brunei Darussalam, Bandar Seri Begawan BE1410.
E-mail: odihi@fass.ubd.edu.bn

et al., 1999; ANAMAN and LOOI, 2000). The impacts were also widespread, affecting not only the ASEAN region but also the Indian subcontinent (JIM, 1999). Primarily, forest fires that devastated close to two million hectares of forests in the region caused these episodes of haze. One estimate put the area of forests that was burned in Indonesia alone at 1.5 million ha (JIM, 1999). Studies also show that some of the adverse impacts were not inevitable consequences of haze occurrence as they were caused by human actions that could be minimized if not totally avoided (ODIHI, 1998, 2001). At other times, hazards in haze episodes resulted from inaction on the part of the victims of haze and/or the villains of the piece (i.e., those responsible for the haze episodes).

The fact that human actions are involved or implicated in haze hazards in the region means that such hazards are not inevitable because actions that make haze episodes hazardous can be avoided or minimized. For example, recent experience indicated that though events in Kalimantan Borneo and Sumatra were mostly responsible and rightly blamed for the haze episodes and their attendant woes, few, if any local communities were totally innocent. The case for local responsibility for haze or the hazards that attended it is in the burning of local forests. For example, forests were burned in Brunei causing additional problems such as pungent smell and soot that covered surfaces in homes and offices, which was not the case with Kalimantan fires. Local firefighters were busy round the clock especially in March and April 1998 fighting fires on the ground and from the air. Also, refuse was continuously burned in spite of the law and/or appeal to not do so. Exposure to haze was largely a voluntary action on the part of individuals who played or carried on with recreational activities in the haze. For entrepreneurs such exposure was an economic choice between losses in sales or production that would result from closure of business during dangerous levels of haze-related air pollution and hazarding personal health or that of employees. For the risk-group employees who worked outside mainly in the construction industry, it was a dilemma of risking personal health or job when employers ignored the hazard and carried on with business as usual. But exposure, for whatever reason, was nearly if not always a choice made by, not forced on individuals.

Related to all this is the role of governments as policymaking institutions and as enforcers of policies. Are there laws that protect the environment from being 'raped' or pillaged in the many ways in which modern humans have done? Are such laws being adequately enforced to deter environmental offenders? There is also much that can be done in the area of environmental education to prevent the privatization of profits and socialization of costs arising from environmental offenses. This paper presents the results of local haze-related studies showing various dimensions of haze episodes and suggests local actions in a country like Brunei Darussalam (hereafter Brunei) and other ASEAN members that may be subsequently useful in reducing haze episodes and/or mitigating their attendant hazards in the region.

Hypotheses

The hypotheses formulated to guide the study and subsequent suggestions that would be needed for the reduction of haze hazard damage in the region are as follows:

1. The number of people that experienced health problems during the 1978/98 haze episodes was not significantly more than the number that experienced health problems at a corresponding period when there was no haze.
2. The proportion of outdoor workers (including players) who experienced haze-related health problems was not significantly more than the proportion of indoor people who experienced haze related health problems.

These hypotheses were verified using the *t* statistic. The model has the following form

$$t = \frac{|\bar{x}_1 - \bar{x}_2|}{\sqrt{\frac{s_1^2}{n_1} + \frac{s_2^2}{n_2}}} \; ,$$

where in Hypothesis 1: $\bar{x}_1$ = the average monthly number of people experiencing health problems before the onset of haze, $\bar{x}_2$ = the average monthly number of people experiencing health problems during the haze period; s_1^2 = standard deviation of the monthly average number of people that experienced health problems before the onset of haze, s_2^2 = standard deviation of the monthly average number of people that experienced health problems during the haze, n_1 and n_2 denote the sample sizes of the periods in question. In Hypothesis 2: $\bar{x}_1$ = proportion of indoor workers that experienced haze-related health problems; $\bar{x}_2$ = proportion of outdoor workers that experienced haze-related health problems; s_1^2, s_2^2 =standard deviations of the monthly average of indoor workers and outdoor workers that experienced health problems respectively; n_1 and n_2 denote the sample size of indoor and outdoor workers, respectively. It would be rational to reject any of the above null hypotheses and accept the alternative hypothesis that there was significant difference if the corresponding computed *t* is greater than the critical (table) *t* value.

2. Study Methods and Field Activities

Several methods which were used in this study may, for convenience, be divided into three groups of survey, environmental and archival components. The survey component comprised interviews, questionnaire administration and focus group discussions. The environmental component comprised mainly physical field activities carried out by the author at forest fire sites. The third component was literature survey of studies of the recent (1990–98) haze episodes in the Southeast Asian region to understand where and why they occurred, their duration and impact. The respective methods are presented and explained more fully below.

2.1 Survey Component

The survey component of interviews, questionnaire administration and focus group discussions described below took place at different times at different places in Brunei. While surveys in the Brunei-Muara District involved urban residents, surveys in Kuala Belait and Temburong Districts involved mostly rural communities in longhouses except for about 80 respondents in Bangar, the headquarters of Temburong District. These surveys involved approximately 300 randomly selected households in the case of the Brunei-Muara District. Entire households in Iban, Dusun and Penan longhouse communities (e.g., Iban in Amo A and Amo B, Batan Duri and others in Temburong District, Iban in Melilas and Biadong in Kuala Belait, the Dusun and the Penan in Kampong Sukang) were surveyed.

2.1.1 Interviews and questionnaire administration

From September 1997 to September 1999, five surveys were conducted to address different aspects of the haze and related problems in three of the four districts in Brunei Darussalam. Both interview schedules and questionnaire administration were used to sample targets that were randomly selected or whole communities as was the case with longhouses in Kuala Belait and Temburong districts. The content of the interviews and the questionnaire were the same and use depended on the suitability and convenience of the respondent. The household was the unit or target in most of the interviews and questionnaire administration. However, one survey code named "Why the Forest Burns" had different age groups as targets as it aimed to understand the level of comprehension of episodes of forest fires in the region and what the different groups of respondents perceived to be their solutions. This questionnaire was also, therefore, administered to students/pupils in primary and lower secondary schools. The purpose was to discern what these young people thought were the causes and cures of the forest fires (i.e., actions necessary to lessen their occurrence and forestall their devastation in the future). Knowledge of children's perception of forest fires would be used to tailor suggestions to the needs of children, which environmental education teachers or policy makers can employ in the quest to invest the younger generation with a better sense of responsibility for the environment.

2.1.2 Focus group discussions

The uniqueness of longhouse communities proved advantageous to group discussions. The packing of an entire community or village into one superstructure which has a common area for the community as well as private areas for individual families or households (KALOKO, 1998) facilitated gathering people for group discussion. Targets such as age groups or women were easy to interview as they stayed together and consent of participation and cooperation was always given prior to the visits to the longhouses.

2.2 Field Activities

During the haze episodes of 1997 and 1998, the author visited several sites where forests were burning in the Brunei-Muara, Kuala Belait and Tutong Districts in Brunei. One of these visits covered a large area (approximately 1000 square kilometers) in Tutong and Kuala Belait Districts where many hotspots or burning sites including peat fires were carefully observed and studied. The understanding of the parameters of local fires, which the author discerned from observing them, enabled him to distinguish between smoke/haze from local fires and those from distant parts of Borneo. Activities carried out at burned/burning sites included taking photographs and, making observations and qualitative assessment of the nature of the burning at different sites. Such qualitative assessment attempted to fit forest burning into categories of "burned well," "slightly burned well," "did not burn well" depending on the estimated 'green area' within a burned site. Additionally, refuse burning in the Brunei-Muara District was monitored and the profile of the areas and people involved was determined in an attempt to understand their actions. For example, the neighbourhood characteristics (i.e. type of houses and the socioeconomic class) and cultural background of those burning refuse was determined. This was done to determine whether refuse burners were recent immigrants; had access to mass media and understood the language in which the regulations and policies were communicated.

2.2.1 Awareness test at burning sites

The author determined four important qualitative parameters that could be used to recognize whether a haze episode had a local component (i.e., whether a local rather than distant fire in Kalimantan was responsible for, or aggravated the haze condition). These parameters included soot production, pungency of burning smell, uniformity of haze cover and color of smoke/haze. Certain of these parameters correlated with distance. Some had a distance decay effect (i.e., the closer the burning area, the more prominent would parameters appear). For example, the closer the burning forest, the more abundant the soot that fell in a place; the more pungent was the burning smell; the less uniform would be the smoke and visibility. While smoke increased towards the burning site, visibility was usually better away from the thick smoke and the flames. It was observed that wind dually influenced soot production. While it was necessary for entraining dispersion of soot, wind could destroy soot when it blew swiftly. Rainfall was another factor which affected soot production by dissolving it. Respondents were asked if they could tell whether haze on a given day resulted from distant forest fires, local fires (in Brunei) or both. The question posed to respondents was "How can you differentiate between local (Brunei) and distant forest fires if you do not see the flames?" The purpose of asking the question was to sensitize observers to the fact that the haze problem had a local dimension and to make them think about what they could do to mitigate it. The author explained the ways respondents could qualitatively determine that fires were burning locally though they might not see them.

3. Results

Interesting results were obtained from the different surveys and other methods of the study. These results or findings, which inform the suggestions given in this paper are presented and explained as found below.

3.1 Perception of the Haze Problem in Brunei

Many respondents (over 70 percent) attributed health problems, business loses, social dislocations such as the separation of families as a result of evacuation of members to safe areas (e.g., home countries) and ecological destruction visited on the region by the 1997/98 haze to several factors. The most frequently cited factor was forest burning by peasant farmers, concession loggers and plantation farmers in Indonesia. Less frequently cited were activities of local environmental offenders such as those who burned forests or refuse that contributed to the air pollution problem. These perceived factors, however, were assigned varying degrees of culpability. Some times respondents blamed haze rather than the people whose activities caused haze and related damages. The multidimensional nature of the impact of the 1997/98 haze is evident in Table 1.

3.2 Location of Forest Fires

It was found that over 100 cases of forest fires occurred locally (within Brunei) during the 1997/98 period. Estimates of the areas burned in each case ranged from less than a hectare to scores of hectares. Over 90 percent of respondents observed that distant forest fires in Indonesia (Kalimantan and Sumatra) were the primary cause of haze in Brunei Darussalam. The two types of fires manifested several different conditions which could be used to distinguish them. Some conditions, which were more or less symptomatic of local fires, included relative abundance of soot production, pungency of the burning smell, non-uniform visibility and irregular pattern of soot distribution. Respondents regardless of their age or status were all aware that the burning of forests in Kalimantan (Indonesia) was the primary cause of the haze problem in Brunei and the region in general. The surveys indicated that people were aware of local forest fires and the burning of refuse within compounds, which exacerbated the haze problems. However, they did not always identify these as a cause of the haze problem. Also, most respondents (over 80 percent) failed to mention exposure to haze, a voluntary action which was a necessary condition for haze-related health problems.

Table 1

Adverse impact of the 1997/98 haze in Brunei Darussalam

IMPACT	IMPACT DESCRIPTION
Health	*Many adverse health conditions occurred during the haze episodes. Some of these were new cases while others, which were preceded the haze episodes were triggered by them. Outdoor workers especially in the construction industry, seniors and under-5 year olds were the most adversely affected segments of the population (*ODIHI, 1999).*
COPDs	Chronic obstructive pulmonary diseases (disorders) increased significantly during the haze period in 1998 (ANAMAN and IBRAHIM, 1999; ODIHI, 1998,1999).
Environmental	*Thousands of hectares of forests were lost in the Brunei forest fires. Many species were consequently lost because the forests were home to many life forms. While no studies have yet established biodiversity impact of the fires, there is no doubt that many life forms were lost in the fires.*
Loss of forests	Estimated to be above 100,000 ha
Species reduction	Number of species to which burnt forest was habitat was reduced directly in the fires and consequently through deteriorating habitat-related conditions such as overpopulation and increased competition.
Increased flooding	Preliminary examinations have revealed that flood problems in Brunei have increased since the haze without significant increase in rainfall amount and other parameters.
Increased erosion problems	Soil erosion has increased due to loss of protection of land surface as a result of vegetal cover. This is evident in increased sedimentation in drains, rivers and roads.
Communication	*Generally communication between Brunei and the outside world was adversely affected by the 1997–98 haze episodes due to the perception of increased risks which lowered the attraction of the region blanketed by haze.*
Haze-related accidents	About 10 road accidents were attributed to the haze problem.
Cancelled flights	Flights from the International Airport in Bandar Seri Begawan were delayed or cancelled in a number of days during the 1998 haze period.
Reduced intra- and interregional interaction	There was as much as 50% reduction in the intra-Borneo weekend flows of people between Brunei and Sarawak during some weekends in April 1998. Tourist visits Brunei from extra-ASEAN region were similarly affected.
Economic	*Economic adversities of haze resulted from reduced production, sale or distribution and from costs borne to treat haze-related health conditions or to prevent them.*
Production	Surveys of rural areas in the Temburong District showed that farmers and fishermen could not carry out their activities for a number of days. Also many sick workers did not work because they were too sick to do so or given sick leave.
Distribution	The fresh food markets (tamu) could not receive produce from farms in Malaysian Borneo due to bad (haze) weather.
Sales	Sales of goods and commodities were adversely affected as less people visited markets and shops.
Services	
Salaries	Both public and private sectors lost money through paying employees for days that work was not done due to cancellation of schools or sickness.

Table 1

Continued

IMPACT	IMPACT DESCRIPTION
Educational	*The haze episodes of 1998 disrupted the school year in the country as schools were prematurely closed due to hazardous haze levels.*
Closure of schools	The School Year was disrupted when schools were closed between April and May and students had to be in school from June to end of November without the usual break.
Lateness and absenteeism	Many students were late to or absent from school between February and April when the decision to allow children go to school was left to the discretion of parents. Schools closed arbitrarily when headmasters/ principals were empowered to close schools when they judged haze levels in their school area hazardous.
Tourism	*The tourism industry suffered in Brunei as in the rest of the region because of the reduction in the attractiveness of the region during the haze episodes. It had been estimated that about* *B$3 million may have been lost by the tourism sector during the haze.*
Hospitality industries Depressed sales	Hotels registered lower levels of check-ins as well as length of stay/visit. Souvenirs especially in the Kianggeh Tamu usually patronized by visitors from outside Southeast Asia witnessed lower levels of sales (ODIHI, 1998, 1999)
Social Disruption	Families had to be separated through evacuation to home countries/ safer places for holiday. The disruption of the school year affected holiday and other family plans.

3.3 Differences between Local and Distant Fires

A sizeable proportion of respondents (about 35%) could identify at least one way to determine if a forest fire occurred locally when the flames were not visible. The criteria used for the identification of the factors varied somewhat amongst respondents, with thick smoke or fumes being the most cited criterion. Respondents were taught ways (criteria) to qualitatively determine if haze was caused or exacerbated by local fire(s). Following this, about 27 percent of respondents could offer an acceptable explanation of why the criteria might apply. People could relate the pungency of the smell of local fires to distance-mediated diffusion effects of smell and soot. Smell decreased from the source of burning, and soot would dissipate, break and fall down sooner or later. Haze from distant fires always appeared more uniform over an area whereas local fires generally resulted in patchy haze with the severity of haze always increasing, near the burning forests.

3.4 Causes of Local Forest Fires

The author's observation and results of the surveys indicated that forest fires occurring outside Brunei (in Kalimantan), which were the primary causes of the haze problem, preceded local cases. Such distant forest fires existed nearly one month, with haze conditions well established in the region before local forest fire problems became serious. Though it was clear that people started local fires, it was not clear why they did so. The contribution of local fires to the haze problem might have been comparatively small. However, their overall environmental and economic damage was not. They caused the decimation of several thousands of hectares of natural environment directly and indirectly through the actions of property owners who cleared and burned forests around their houses under 'guided' supervision that failed in most cases, resulting in uncontrolled burning and loss of resources or biodiversity. Initially, these local fires were usually along major roads however later they started to manifest in remote areas of lush pristine environment. It puzzled many respondents that after people had seen that forest fires caused haze, they would set forests on fire when their livelihood did not necessitate it, like farmers and loggers in Indonesia.

3.5 Haze-related Hazards

The study revealed from both primary and secondary sources of information that the haze problem in the region had several dimensions as Table 1 shows. In Brunei the most serious problem associated with the 1997/98 haze was health. Such health problems stemmed mainly from exposure to air pollution, which caused new cases of diseases or triggered the attack (recurrence) of existing ones. Health problems associated with the 1997/98 haze episodes included chronic obstructive pulmonary diseases (COPD), with asthma and bronchitis among the serious conditions. The different health problems associated with the haze problem are presented in Table 2. The financial losses from these health problems resulted from treatment, declining production and loss of business.

The result of the testing of the hypothesis that there was no significant increase in the incidence of diseases associated with haze showed that the increase was significant. To accept the null hypothesis of no difference between conditions during haze episodes and non-haze periods at 99% confidence level ($\alpha = 0.01$) at various respective degrees of freedom (df), the calculated t had to be less than 3 in all the cases. This was, however, not the case in any of the testings. In the case of asthma (calculated t for $\alpha = 0.01$ was 7.52, at $df = 13$ for new cases; and $t = 15.02$, $df = 34$ for pre-1997/98 existing cases). The results for other health problems were similarly statistically significant. For example, emphysema had a calculated $t = 11.54$; at $df = 25$). Headache, conjunctivitis, skin rashes and eye irritation also had significant increases during the haze episodes.

3.6 Environmental Vigilance

Respondents were divided in their opinions on the issue of what people would do if they found environmental offenders. While the majority of respondents (over 85

Table 2

Health problems associated with the 1997/98 haze in Brunei Darussalam

Health Problem	Those Most Affected (i.e., group(s) accounting for 45% or more of cases)	Number of cases as % of pre-haze (January –May 1997) level
COPDs/Respiratory tract infections	The worst affected were under-5 and seniors above 60 years of age.	
Asthma	Under-5 children were the most vulnerable group showing highest levels of new cases. Amongst this group there were differences in the level of male-child and female-child affected. The male child accounted for over 50% of the cases in this group. Seniors are the second group that indicated high vulnerability.	Under-5 (133.47%) Old people (117.21%)
Bronchitis	Under-5 and outdoor Workers.	Under-5 (122.73%) Outdoor workers (119.97%)
Pneumonia	Outdoor workers, especially those involved in road and estate construction. Many construction workers have more or less continuous exposure to haze. On and off work those who occupied ramshackle or uncompleted buildings without windows or doors were exposed to the polluted air during haze. Children were the next most vulnerable group.	Outdoor workers (122.31%) Under-5 (119.63%)
Emphysema	Outdoor workers	233.17%
Adenoid	Under-5 children and those pupils between the age of 6-10 years	247.44%
Cough	Outdoor workers	A general problem for all groups
Running nose (catarrh)	Though this was a rather general problem, outdoor workers were the most affected.	1871.82%
Dizziness	Outdoor workers, especially those involved in outdoor cooking in the night food markets.	213.81%
Headache	Outdoor workers	3
Conjunctivitis	Outdoor workers	2241.53%
Skin rashes/irritation	Over-60 years, outdoor workers	2724.15%

percent) would feel bad, only about 23 percent of the respondents would confront environmental offenders about their action. Those who would report such people to the authorities were even fewer (less than 20 percent). The majority (over 80 percent) of those who would confront people burning forests or refuse in violation of environmental laws/policies in Brunei were mostly whites from Western and Southern Pacific countries. Over 60 percent of Brunei citizens and permanent residents would not confront or report environmental offenders. Other parameters that appeared important in the behavior of people were income and/or type of job.

Close to 80 percent of low-income nonwhite immigrant workers in the construction sector and of Asian origin were not likely to report environmental offenders.

Those who would report environmental offenders would do so in order that they might be "apprehended," "fined" or "made examples of" to deter other offenders, or to "avert environmental and other calamities." Others would do so because they wanted "to breathe clean air," and because "haze is a menace" that should be stopped because it was costly both socially and economically. Those who would not confront or report environmental offenders who burned forests or refuse in their compounds gave reasons for their option. Such reasons included they: "did not think it was good to report neighbors;" "did not want neighbors (other people) to suffer," or "did not think that it was their business to do so."

Although many people did not favor punitive actions against local environmental offenders, there was near unanimity (over 98 percent) in the view that those who burned forest in Indonesia should be punished. The most popular reasons for this opinion included the fact that they were "responsible for the haze," "haze resulting from their activities caused numerous problems to people," "destroying habitats for wildlife" such as orang-gutang.

3.7 Awareness of Consequences of Haze

The range of consequences of haze that respondents identified was wide, including environmental, social, economic and political. This included health, economic, social, ecological and recreational problems. All respondents, including the pupils surveyed in schools, were aware of one or another of the many consequences of haze. Even primary school pupils were able to identify health and school related problems of the haze episodes. They observed that it led to closedown of schools, ban on their playing outdoors and contraction of diseases.

4. Discussion

The hovering spectre of haze existed in the region for a very long time. One record, which is more than a century old, indicated the occurrence of a serious haze hazard in the late 1880s (BRASSEY, 1889). The hazard will in all probability be present in the region in the foreseeable future due to the apparent cyclic nature of the ENSO phenomenon and droughts with which it is associated. The drought will enhance combustibility of forests due to drought-induced desiccation. The expected long-term nature of haze episodes in the region is also based on what JIM (1999) called the "deeply-ingrained" human actions and their 'inertia' that has worked against accepting drought as part of the natural way of things in the humid tropical environment of Southeast Asia.

Respondents appear readily willing to blame people and events outside their region for the haze problem and are blind to or deny local responsibility. Perceiving

haze problems as a result of environmental (natural) conditions such as the ENSO phenomenon or the activities of distant people is a hazard teleconnection syndrome. Human culpability at the local level is, however, not in question. Surprisingly, this teleconnection syndrome is not limited to the general public. Scholars studying recent haze episodes have maintained silence on local dimensions of the problem and have blamed the episodes on teleconnections such as El Niño, forest fires and other human activities elsewhere. However, as this study shows, local factors were involved in haze and related hazard causation.

The results of the hypotheses tested in this paper are in some ways similar and dissimilar to other studies done in Brunei. For example, differences in the occurrence of certain conditions such as asthma, emphysema, bronchitis and influenza were similarly statistically significant. However, others such as pneumonia, conjunctivitis and different types of acute respiratory infections, which recorded statistically significant increases in the present study did not appear so in, for example, ANAMAN and IBRAHIM (1999). Possible causes of the difference may be found in the sources of data used in the two studies. While the ANAMAN and IBRAHIM (1999) used aggregate data obtained from the hospital, the present study used primary data. Many respondents apart from self-medication or using traditional methods for treating certain conditions did not go to hospitals or fee-paying health centers if they did not consider their health condition 'serious'. Several motives could be responsible for this course of action. Patients might not want to or could not afford the time or money that might be involved. For the low income immigrants working in the construction industry, seeking treatment for ailments was not only avoided for immediate financial reasons but also for its implication for job security. As daily paid manual workers, it was very important for them to appear healthy as sickness resulted in their being sacked from the job. Thus underreporting was a severe problem at least among this group of workers. Studies based on data from hospitals would therefore underestimate haze-related health problems.

The lack of understanding of the motives behind the local fires, which were not for precautionary purposes to safeguard property, is a challenge for authorities and environmental education teachers. It will be difficult to arrest the problem without proper understanding of the motive(s) for these actions, which clearly are inimical to environmental and also human interests. Peasant farmers and concession loggers in Kalimantan may need the use of fire in their environmental resources management or exploitation while people who are not farmers or loggers do not need to set forests ablaze for such purposes in Brunei.

Is nature really the cause of haze and related problems in Southeast Asia in the 1990s? We need to only ask whether the natural conditions which might be necessary for haze occurrence were indeed sufficient for haze episodes and related hazards to occur on the scale and frequency that they have occurred in the region since the 1990s? Or we need to ask whether El Niño, which caused droughts and

desiccation that enhanced the combustibility of forests, was occurring for the first time only in the 1990s when haze problems became rather prominent in the region? Importantly, the correct answer to these questions is "no" in both cases. Drought is not necessarily an invitation to forest burning by humans. Also, the occurrence of El Niño in the Eastern Pacific did not start in the 1990s – it has always had its cyclical occurrence. These facts absolve nature of the blame for haze both here and elsewhere and underpin recent conclusions by GARCIA (1981) that "nature is not guilty" of human problems arising from its workings. In the same vein, TIMBERLAKE (1990) observed concerning Africa that accepting nature's responsibility for the so-called natural disasters can only 'tempt leaders to apathy'. What could leaders do if nature, which is largely beyond human control, causes problems? GARCIA (1981) also observed that it is 'pseudo-science' to stop the inquest at the doors of nature when a natural event is involved in human misery or death. He encourages us to go beyond the snowstorm that ends the misery-filled life of the homeless to find the true killers in principles and policies in society that create homelessness.

There is no consensus in the literature regarding the culpability of poor farmers practising slash and burn agriculture in the haze problem in Southeast Asia (see JIM, 1999). However, ODIHI (1993) and Le Dien Duc (quoted in JACOBS, 1995) showed that poverty causes environmental degradation universally. RAVINDRANATHAN and HANAFI (1999: 465) observed that "The greatest of evil[s] and worst of crimes is poverty." Following the Garcian principle we must examine the societies in which haze-causing activities occur to ascertain if there are human connections such as adverse socioeconomic conditions whose alleviation would mitigate haze hazards. If such connections exist, there is need to rescue desperate people taking desperate actions that compromise our corporate security both in the short- and long-terms, to make the environment safer for all.

Solution to the problems of haze in Southeast Asia requires a bold appraisal of the political, economic, social and cultural pillars upon which a particular society is erected. Are these just and capable of providing for the basic needs of the people or do they create conditions of destitution and desperation? In the light of the link between poverty and environmental degradation or hazards, we must ask the following questions to lead us to the appropriate local actions for reducing haze hazards and their related multidimensional costs.

1. Are the peasants that have been blamed for burning forests villainous? (i.e., are they very bad people willing to harm other people or break the law in order to get what they want or are they caught in the serious survival instinct that desperation and destitution demand)?

2. Are there principles and/or actions within the larger society which adversely impact those burning forests, and which when modified or changed will meaningfully reduce their forest-burning actions?

As GARCIA (1981) reminds us, to stop short of diagnosing problems to their root cause is 'pseudo-science', which would at best result in symptom treatment that would not provide a cure for the problem.

4.1 Suggestions

The surveys, which involved all socioeconomic strata of the Brunei population and covered a wide range of age classes including primary school pupils, showed that generally people were aware of both the causes and adverse consequences of haze in the region. These conditions are favorable to effective environmental education aimed at reducing haze hazards in the future. Environmental educators should enthusiastically seize the opportunity to inculcate in the public responsibility for the environment, which will bring about what CHAN (1998) called "pro-environment" behaviour. The following suggestions are put forward in the belief that their implementation may make the future more haze-free and/or reduce haze damage when haze occurs in the region.

1. Governments in the region should ensure equitable distribution of resources and access to viable means of livelihood to reduce undue dependence on, and overexploitation of natural resources such as forests. Supporting development programmes in poor communities so that they would not pillage or interact with the environment in ways that cause widespread serious environmental problems can achieve this. Such development will be an insurance against haze and haze-induced problems costing millions of dollars to mitigate.

2. Relevant environmental laws should be made, widely popularized and strictly enforced to prevent environmental abuse that compromises the wellbeing of the environment and humans that depend on it.

3. Environmental education should be popularized through sustained programmes using mass media. The goals of such education should be clearly thought out and the programmes should be regularly monitored to ensure that all segments of society (citizens and immigrants, old and young, literate and illiterate) possess necessary knowledge to become environmentally responsible citizens.

4. Training of adequate environmental education teachers to teach environmental education. Religious elite and other respected groups in society can be trained or involved in the environmental education of the public to ensure widespread acceptance of the environmental ethic.

5. The ASEAN Environment Ministers conference or meeting, which has arisen as a response to recent environmental problems, should be strengthened to enable the ministers to discuss environmental issues in the region more regularly with the possibility of arresting problems before they become serious.

6. A central institution should be established for training nationals of ASEAN countries in environmental education and for information networking. This will reduce costs arising from duplication of institutions in different countries and

arrest the problem of not having skilled people in nations that may not be able to afford training institutions of their own.

5. Conclusion

This paper has shown that haze in Southeast Asia is a serious environmental problem because of the severity and multidimensional nature of its impacts. The haze problem, which is not new in the region, is likely to continue because of the enduring nature of the natural and human dimensions of the problem. The study has shown that haze-related problems are largely human-made and because they are so, such problems are preventable or can be meaningfully reduced. The suggested mechanisms for the reduction of haze problems in the region include paying attention to the social, cultural and economic problems of communities engaged in haze-causing activities. Particularly, improved access to resources by the poor should be encouraged. One recurring finding of the survey indicates that some groups in society were unwilling to report environmental offenders. Appropriate environmental education programmes should be promoted with the intent that individuals in society, irrespective of their backgrounds, are environmentally informed.

Acknowledgement

The author gratefully acknowledges the logistical support in the form of transport provision by the University Brunei Darussalam (UBD) during the fieldwork. He is also grateful to Dr. Pangiras Michael and Dr. Abdul Aziz Kaloko both of the Department of Geography, UBD for their encouragement. He wishes to acknowledge the assistance of third-year students in the department who over the years helped him administer questionnaires or served as his interpreters during interviews. The efforts of teachers and students and indeed all respondents are also gratefully acknowledged.

REFERENCES

ANAMAN, K. A. and IBRAHIM, N. (1999), *Economic Analysis of Human Health Impact of the 1998 Haze-related Air Pollution Episode in Brunei Darussalam*. In ICB-ICU Book of Abstracts, *Biometeorology and Urban Climatology at the Turn of the Millennium* Sydney, Australia: ICB-ICUC. p 139.

ANAMAN, K. A. and LOOI, C. N. (2000), *Economic Impact of Haze-related Air Pollution on the Tourism Industry in Brunei Darussalam*, J. Econ. Analysis and Policy 30(2), 133–143.

BRASSEY, LADY A., *The Last Voyage (1886–87)* (1889) (London, Longman and Co. 1889), 212 pp.

CANE, M. A. (1983), *Oceanographic Events during El Niño*, Science 222, 1189–1195.

CHAN, K. (1998), *Mass Communication and Pro-environmental Behaviour: Waste Recycling in Hong Kong*, J. Environ. Management 52(4), 317–325.

GARCIA, V. I., *Drought and Man: Nature Pleads Not Guilty*, vol. 1 (New York: Pergamon Press. 1981)
    455 pp.
JACOBS, J. W. (1995), *Mekong Committee History and Lessons for River Basin Development*, Geograph. J.
    161(2), 135–148.
JIM, C. Y. (1999), *Forest Fires in Indonesia 1997-98: Possible Causes and Pervasive Consequences*,
    Geography *84*(364), part 3, 251–260.
KALOKO, A. A. (1998), *Longhouse Communities in Ulu Belait in Brunei Darussalam*. Bandar Seri Begawan:
    Educational Technology Centre, University of Brunei Darussalam, 69 pp.
ODIHI, J. O. (1993), *Afforestation and deforestation propaganda: Government ambiguity and environmental
    resources development and sustainability problems in NE Nigeria*. In *Resource Management for
    Sustainable Development in Nigeria* (J. M. Baba, D. O. Adefolalu and G. E. Uwaya, eds.) (Minna:
    Geography Department, Federal University of Minna. 1993) 243 pp.
ODIHI, J. O. (1998), *Gains in the Pains: Using the Southeast Asian 1997/98 Drought and Haze Experience for
    Better Environmental Management*. In: South East Asian Geography Association, *Geography and
    Geography Education in the 21st Century: Directions and Challenges* (Singapore: SEAGA. 1998), p 66.
ODIHI, J. O. (2001), *Environmental Education: Beckoning Roads for Brunei Darussalam*. In *Internat. Res. in
    Geograph. and Environ. Education,* accepted
RADOJEVIC, M., TAN, K. S., and RAMALINGAM, P. (1999), *Potential Health Impacts of Criteria Pollutants
    Present in Haze*, Brunei Internat. Medical J. *1*, 299–303.
RASMUSSON, E. M. and WALLACE, J. M. (1983), *Meteorological Aspects of El Niño/Southern Oscillation*,
    Science *222*, 1195–1202.
RAVINDRANATHAN, N. and HANAFI, Z. (1999), *Improving Health Care Services in BIMP-EAGA: A
    Collaborative Opportunity among Neighbours*, Brunei Internat. Medical J. *1*(1), 465–466.
TIMBERLAKE, L., *Africa in Crisis: The Causes, the Cures of Environmental Bankruptcy* (London: Earthscan
    1990), 203 pp.
YARNAL, B. and KILADIS, G. (1985), *Tropical Teleconnections Associated with El Niño/Southern
    Oscillation*, Progress in Phys. Geography *9*(4), 524–558.

(Received March 31, 2000, accepted November 12, 2000)

To access this journal online:
http://www.birkhauser.ch

Pure appl. geophys. 160 (2003) 221–238
0033–4553/03/020221–18

© Birkhäuser Verlag, Basel, 2003

❘Pure and Applied Geophysics

The Occurrence of Haze in Malaysia: A Case Study in an Urban Industrial Area

Aiman Soleiman,[1,2] Mazlan Othman,[1] Azizan Abu Samah,[2]
Nik Meriam Sulaiman,[2] and Miroslav Radojevic[3]

Abstract — Klang Valley, a heavily industrialized urban area in Malaysia, has experienced severe haze episodes since the early 1980s. Total Suspended Particulate matter (TSP) is used in studying this phenomenon. Three severe haze episodes during the early 1990s are reviewed; August 1990, October 1991, and August–October 1994. The nature of these episodes, their possible causes, and their major features are discussed. Meteorological conditions associated with these episodes were analyzed. Results of the study indicate that stability and trapping of particles are the main factors affecting the pollution during haze periods. Maximum total suspended matter (TSP) was recorded in October 1991. The August–October 1994 episode was the most persistent and least affected by meteorological variables. Analysis of wind direction data showed that southerly and southwesterly winds coincided with the worst haze periods.

Key words: Haze, TSP, temperature, wind, rainfall, Malaysia.

Introduction

Haze is defined as the presence of fine particles (0.1–1.0 µm in diameter) dispersed at a high concentration through a portion of the atmosphere that diminishes the horizontal visibility, giving the atmosphere a characteristic opalescent appearance (MMS, 1995). Particles, which are of respirable sizes, are of concern because of their negative effect on health, as well as their other environmental impacts. Particles < 10 µm can affect meteorological processes (visibility and solar radiation), and they can be involved in chemical reactions in the atmosphere producing secondary pollutants.

The haze phenomenon in the Klang Valley region is an important and serious problem. Since the early 1980s Klang Valley was reported to experience high concentrations of particulates. Unusually thick haze, which occurred during September 1982 (Sham, 1984), was the first to attract a great deal of public

---

[1] Department of Physics, National University of Malaysia, 43600, Bangi, Malaysia.
[2] Air Pollution Research Unit (APRU), Department of Chemical Engineering, University of Malaya, 50603, Kuala Lumpur, Malaysia. E-mail: oklat@hotmail.com
[3] Department of Chemistry, University of Brunei Darussalam, B.S.B. BE1410, Brunei Darussalam.

attention and to be extensively reported in the local media. Subsequently, Klang Valley experienced episodes of severe haze conditions lasting for varying periods of time. The Malaysian Meteorological Service (MMS) reported these haze episodes which occurred during April 1983 (CHOW and LIM, 1984), August 1990 (SHAM *et al.*, 1991), June 1991, October 1991 (CHEANG *et al.*, 1991), August to October 1994 (MMS, 1995), and August to October 1997.

Haze episodes in the Klang Valley were usually associated with dry weather which suppressed convection. A dry layer in the lower troposphere during haze episodes was observed, and it was speculated that the dry layer could act as a lid for vertical mixing and hence contribute to the trapping of the emitted pollutants (SAMAH, 1995; SHAM *et al.*, 1991).

In the Klang Valley region there are two kinds of haze; shallow localized haze and dense haze. The former, which usually occurs in urbanized areas, arises from trapping of pollutants from anthropogenic emissions in response to stabilization of the atmosphere, while the latter is thought to be due to different reasons (MMS, 1995) listed below:

• The advection by prevailing winds of suspended ash particles from large-scale forest fires and open burning in Indonesia (April 1983, October 1991, August to October 1994, and August to October 1997),

• the haze conditions combined with local open-burning (August 1990), and

• the injection of suspended ash particulates from volcanic eruptions such as Mount Pinatubo in June 1991 (MMS, 1995).

The aim of our work is to review the three severe haze episodes experienced in the Klang Valley during the early 1990s, namely: August 1990, October 1991, and August to October 1994. Sources and concentrations of particulate matter, as well as meteorological conditions associated with each of the haze episodes are investigated. The pollutant trapping mechanism is discussed, and a comparison between the severe haze episodes is presented on the basis of meteorological variables.

Description of the Study Area

Occupying a central location on the West Coast of Peninsular Malaysia, the Klang Valley region is a highly populated and industrialized urban area. This region includes the following: Kuala Lumpur, the capital city of Malaysia, Petaling Jaya, a densely populated and industrialized area, Shah Alam, the state capital, Klang and Port Klang (Fig. 1). The entire area is fully developed, with many manufacturing industries producing a range of products from electrical appliances to cars and motorcycles. Power generation, iron smelting, chemical industry and various construction activities are also located in the region.

This region has been shown to have a high potential for pollution because of the increase in pollution sources following rapid urbanization and industrial expansion

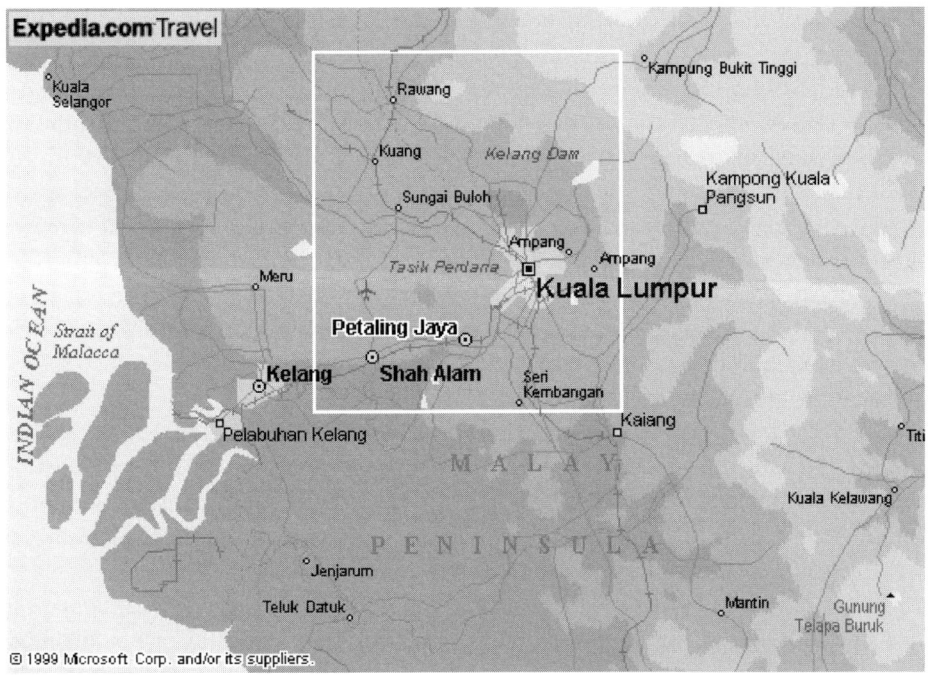

Figure 1
Geographic location of experimental site, Petaling Jaya, Malaysia.

(SHAM, 1987). The Klang Valley is surrounded by hills to the north, east and south, and this impedes the horizontal transport of surface generated pollutants. Hence, meteorological variables have a strong influence on the concentration and transport of TSP. These factors have resulted in the Klang Valley becoming the most polluted area in Malaysia.

SHAM *et al.* (1991) reported that the main sources of pollutants in this area are motor vehicle and industrial emissions. Moreover, the development in this area during the last two decades has resulted in a stream of construction activities including the clearing of agricultural land for building roads, housing estates, industrial parks, airport expansion, etc. In the early 1990s the number of the pollutant sources increased due to the economic boom in the area.

Petaling Jaya station, the MMS headquarters, was chosen to represent the pollution levels in the Klang Valley. At this station, where the only complete record of TSP levels during the haze period was available, a more realistic interpretation of air quality in the area is possible. The colocation with the climatological station allowed simultaneous and continuous observation of both the meteorological variables and TSP concentrations. This enables a comprehensive assessment and interpretation of the air quality data (MMS, 1995) to be carried out.

TSP Measurement and Sampling

In its headquarters station in Petaling Jaya, the MMS uses a High Volume Air Sampler (HVAS) for sampling suspended particulate, which agrees with the standard set by the Environmental Protection Agency (EPA). The exposure of the instrument meets the guidelines as required by (EPA) and is placed approximately 50 feet above ground level. The sampler is described as follows: it consists of a suction motor blower which draws air through a glass fiber filter paper which is placed on a filter holder consisting of a cone-shaped filter support screen. The filter paper is held in position by the cast aluminium face plate with a neoprene sponge rubber gasket to prevent leakage. All these parts are enclosed in an aluminium housing which allows sufficient air movement and access, yet keeps the weather out.

All filter papers, exposed and unexposed, are dried in a dessicator at temperatures of around 25 degree Celsius and a relative humidity of about 45% for at least 24 hours before being weighed. The TSP load is obtained by subtracting the weight of the unexposed from the exposed filters and dividing by the volume of air that has penetrated the instrument during the sampling period.

This sampler was fitted with a constant flow rate controller that maintained the flow rate throughout the sampling period. The flow rate was set at one cubic meter per minute. To avoid great stress on the motor blower due to overloading of the filter medium, the sampling period was fixed at 12 hours, hence, sampling was done twice a day (CHOW and LIM, 1984). According to them, this practice is advantageous in that a comparison could be made between day and night-time particulate concentrations. In this work we make use of another advantage, that is morning values (at 7:00 am) of total suspended particulate concentration were used in the investigation of the relationship between the concentration and meteorological variables. It was decided that the 12-hour wind speed averages prior to the sampling time at 7:00 am were most appropriate and they appeared to have the most bearing on TSP concentration.

One important note on TSP sampling was the one by SHAM *et al.* (1991), who pointed out that the readings at the Petaling Jaya station could be an underestimate of the actual ground level pollution. This is due to the fact that the sampler was located on the roof of a three-storey building, which means that a certain amount of attenuation in pollution levels should be expected with height.

Review of Severe Haze Episodes

August 1990

Klang Valley was reported to have hazy conditions during the period from the 15th to the 30th of August 1990, with the worst conditions persisting from the 20th to the 30th of August. The high concentration of suspended particles in the atmosphere was found to persist during a dry spell in August 1990, and there were numerous reports of open burning in Selangor, Negri Sembilan and Malacca (SHAM *et al.*, 1991).

Figures 2 and 3 illustrate the wind speed and rainfall amount together with the concentration of TSP for the period: August 1st–September 8th. TSP levels were already high from the first week of August and fluctuated until the start of an increasing trend on August 16th which continued until August 30th, with a peak of 516 g m^{-3} over a 12-h period on August 27th.

Low wind speed and low mixing depth, together with the absence of rainfall, persisted during the worst period of the haze episode which continued from the 20th to the 30th of August (Figs. 2 and 3). Over the next few days the onset of widespread rain in Selangor cleansed the atmosphere.

SHAM *et al.* (1991) concluded that a persistent supply of haze particles from sources coupled with the surface inversion associated with the dry spell, light surface winds, and the nature of the topography resulted in the haze particles being trapped in the Klang Valley. The mixing depths and the wind speeds are relatively low, giving rise to inefficient ventilation. This condition is exacerbated by the occurrence of a relatively "long" dry spell, which, during the August 1990 haze episode was associated with the active phase of the southwest monsoon. The latter resulted in the development of meteorological conditions that were unfavorable for efficient transport and dispersion of pollutants from the surface (SHAM *et al.*, 1991). The lingering presence of deep subsidence in the atmosphere associated with the dry weather brought about persistent surface inversion throughout the day during the haze episode. In the early morning, the surface inversion traps the haze particles near

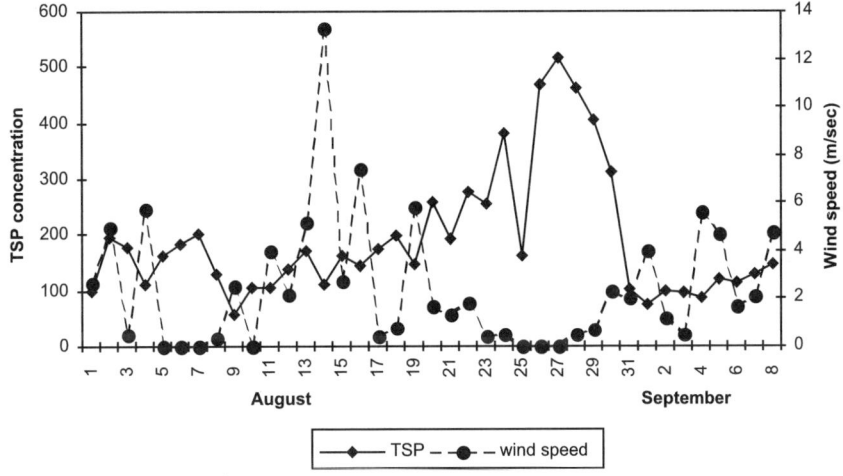

Figure 2
The daily variation of TSP (μg m^{-3}) and wind speed during August 1990 haze episode.

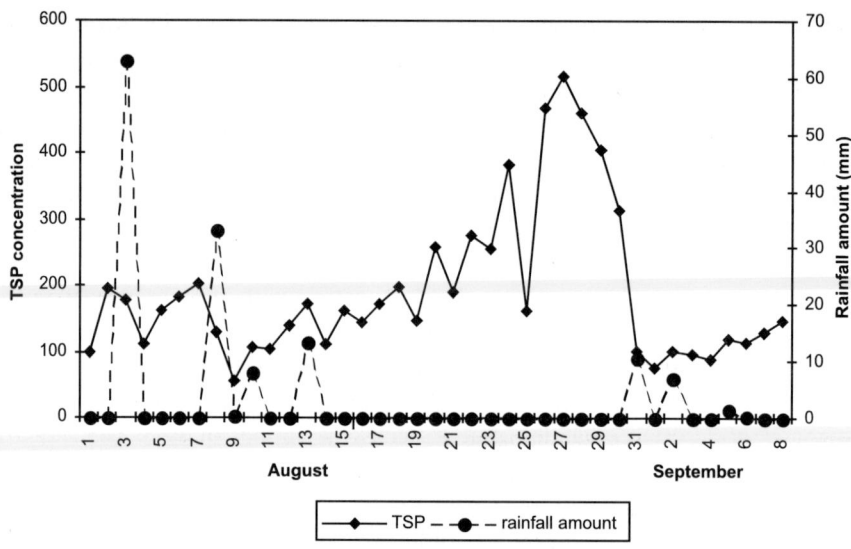

Figure 3
The daily variation of TSP (μg m^{-3}) and rainfall amount during the August 1990 haze episode.

the earth's surface. During the day, when the earth's surface is heated, the uprising convective current tends to carry the haze particles upward. Due to the presence of the second inversion during the dry spell, these particles could not be transported very high. The mountain range in Peninsular Malaysia acts as a barrier preventing the dispersion of haze particles from the Klang Valley. Also, the sinking air currents from the return flow (from the west) at levels between 1500 m and 3000 m of the land-breeze circulation in the early morning acted as another factor contributing to the entrapment of haze particles in the Klang Valley.

The hazy condition ended on August 31st because of widespread rain and particle dispersion due to the occurrence of wind. SHAM *et al.* (1991) found that the clearing of haze was caused by the change in the weather. A stronger convective current in the afternoon from the 1st to the 8th of September may have dispersed the haze by turbulent mixing and other processes listed below:

• The upward convective currents carried the light particles to great heights where they were transported horizontally by strong winds,
• heavy rain and particle entrainment washed the heavy particles down to the earth's surface, and
• strong gusts at the surface dispersed the haze particles.

October 1991

The Klang Valley was reported to experience haze conditions during much of October 1991. The haze episode occurred during a relatively dry period of the year resulting in poor visibility and disruption of flight schedules; this haze was considered to be the most severe and persistent to date in the Klang Valley. The reason for the high concentration of suspended particles was reported to be intrusion of particulates arising from forest fires in Indonesia brought over by the southerly winds. The continual influx of particulates had made it very difficult for the atmosphere to effectively disperse and dilute the particles.

Unlike gaseous pollutants, which exhibited the usual urban diurnal variation patterns, consistently high levels of particulates were observed throughout the day during the haze episode. A probable cause for this observation was the continual influx of particles from outside sources, making it very difficult for the atmosphere to effectively disperse and dilute them, even during the day. This could also be further evidence that the sources of particles were not local. CHEANG et al. (1991) also reported the October haze episode and presented evidence of long-range transport of pollutants from forest fires to Malaysia. According to them, the severe haze condition mainly resulted from smoke particles transported from forest fires in Sumatra and Kalimantan, and aggravated by local emissions from vehicles and industry. Observations that support this suggestion were reported to be:

- The haze affected the whole of Peninsular Malaysia and western Sarawak, including places such as Pulau Langkawi, Kelantan and Terengganu, where haze development was considerably lower than in some west coast states.
- The haze dispersed when low-level winds changed direction from southeasterlies (blowing from Sumatra and Kalimantan) to easterlies (from the South China Sea) on October 11th. The haze reappeared as soon as the southeasterlies and southwesterlies (from Sumatra) returned on October 22nd.
- Satellite images (visible channel) taken by the Japanese Geostationary Meteorological Satellite.

Two spells of hazy conditions were observed to occur during this haze episode; the first was from September 27th to October 11th, while the other was from October 22nd onwards. TSP levels during the haze episode are presented in Figure 4 for the period of September 15th to October 31st, together with the windspeed and rainfall amount.

TSP concentrations recorded in Petaling Jaya during this period were above 200 $\mu g\ m^{-3}$ (about three times the long-term average value). A maximum of 570 $\mu g\ m^{-3}$ was recorded over a 12-h period on October 7th. This value was higher than the highest value recorded during the August 1990 haze episode in Peninsular Malaysia. The fluctuation in TSP can be summarized as follows: From October 3rd to the 10th there was a dramatic increase in TSP levels, followed by a period of heavy and widespread rain from October 11th to the 19th which significantly lowered the concentrations of particulate matter. Haze intensity increased again from the 21st to the 29th of October because of the break in rain. By October 31st the TSP levels began to drop quickly because the rain resumed.

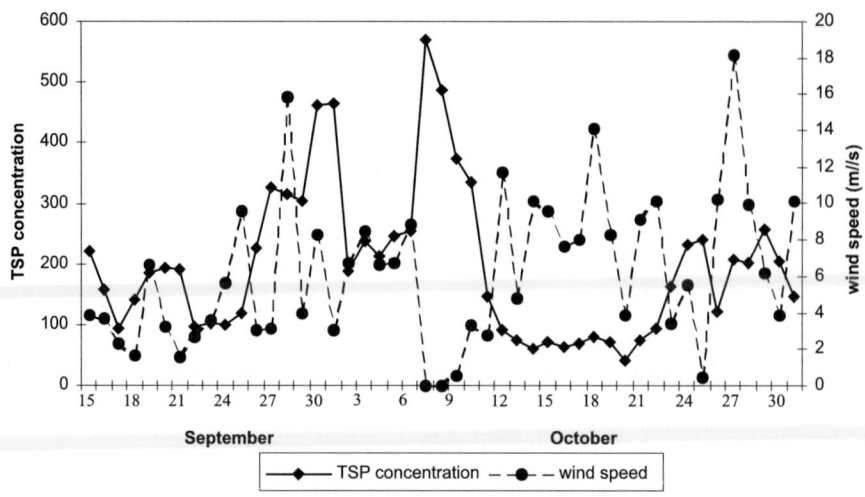

Figure 4
The daily variation of TSP (μg m^{-3}) and wind speed during October 1991 haze episode.

Meteorological conditions during the haze episode indicated widespread subsidence of air over the Malaysian area, resulting in the frequent occurrence of temperature inversions. This further prevented the particles from being lifted into the higher atmosphere. The presence of a thick inversion layer, about 1 km above the surface, was also believed to have aggravated the condition, as it provided an effective lid in preventing the particles from being dispersed more effectively. SHAM *et al.* (1991) hypothesized that the development of such an inversion layer would have prevented the interaction between the lower surface region and the region above the inversion layer.

The advent of heavy and widespread rain accompanied by thunderstorm activities over the region cleared the haze. Figure 5 indicates the rainfall amount coupled with the TSP concentration. The widespread and heavy rainfall (most western stations recorded 24-hour rainfall exceeding 20 mm) on the 11th of October helped to clear up the haze. Of particular significance is the heavy rain on October 11th in Kuala Lumpur and Petaling Jaya, resulting in severe flooding. Petaling Jaya was exceedingly wet during October 1991 when compared to the long-term October mean. This was triggered by several occurrences of heavy rain from the 11th to the 21st of October when the haze temporarily cleared up (CHEANG *et al.*, 1991). From October 31st onwards, when the rain resumed, the synoptic patterns changed to more unstable conditions, augmented by the appearance of unstable easterlies flow.

August–October 1994

During the August–October 1994 period, Klang Valley was reported to have experienced a haze episode. Hazy conditions were first observed over the

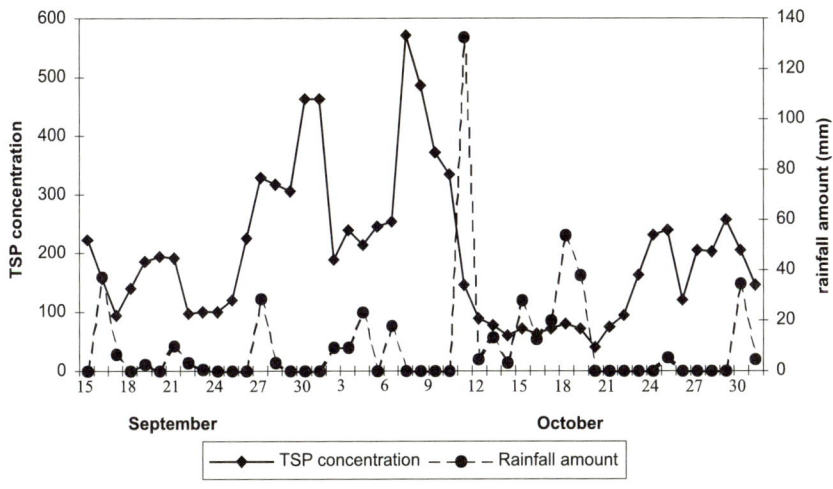

Figure 5
The daily variation of TSP (μg m^{-3}) and rainfall amount during the October 1991 haze episode.

northwestern states of Peninsular Malaysia and over the Klang Valley in early July. These conditions persisted throughout July with hazy conditions spreading across the central states in early August. The haze conditions worsened towards the middle of August in the central and southern states, and persisted with further deterioration into a dense haze condition in September and early October. Haze recurred in southern Peninsular Malaysia from October 21st and cleared towards the end of the month (MMS, 1995).

The widespread haze observed in the region was believed to have been due to both eddy diffusion and advection by the prevailing wind of the smoke particles from forest and plantation fires in Indonesia; the latter being the more efficient in spreading haze pollution (MMS, 1995). The evidence for the advection of particles from Indonesia included:

- Evidence of large-scale burning in the southern part of Sumatra and Kalimantan provided by enhanced satellite images,
- daily meteorological reports of smoke conditions over these areas indicating large-scale open burning and
- enhanced visible NOAA-11 satellite imagery indicating the presence of an extensive haze covering Malaysia, Singapore, parts of Indonesia and the South China Sea.

In its report on air quality, MMS (1995) reported that the haze occurred, as usual, during the relatively dry period, i.e., during the months of July to September of the southwest monsoon season.

TSP levels for Petaling Jaya were ca. 50% higher than the long-term mean (1977–1993) for most of July, and only slightly higher than the mean during the early half of August. However, from the latter half of August to the latter half of October, a drastic increase in the TSP values was recorded (2–4 times higher than the long-term mean). The highest TSP morning value was recorded on September 29 with a value well over 463.5 µg m^{-3}.

TSP levels during this episode are shown in Figures 6 and 7 accompanied by the wind speed and rainfall amount. According to the MMS report, winds played an

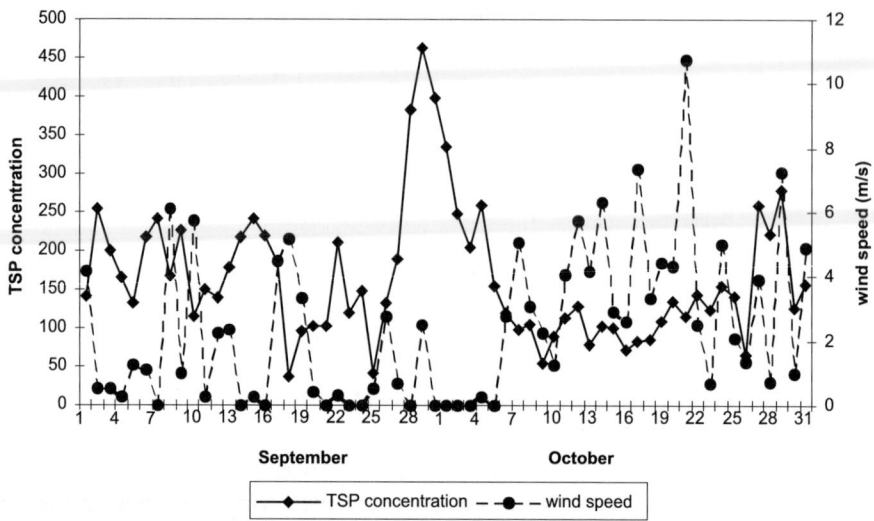

Figure 6
The daily variation of TSP (µg m^{-3}) and wind speed during August–October 1994 haze episode.

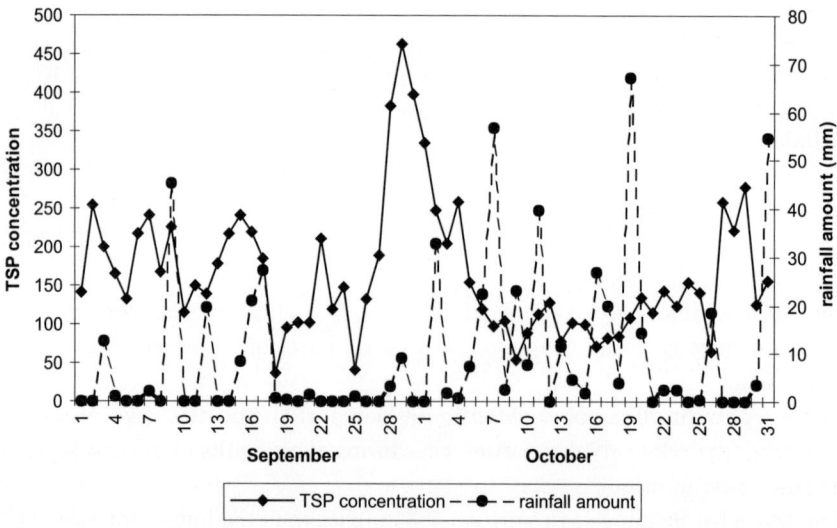

Figure 7
The daily variation of TSP (µg m^{-3}) and rainfall amount during the August–October 1994 haze episode.

important role during this severe haze episode. The importance of the advection mechanism in bringing the smoke particles into the Malaysian region is illustrated by the presence of intense haze over parts of Peninsular Malaysia when low-level southeasterly winds prevailed over southern Sumatra and southwesterlies prevailed over central and northern Peninsular Malaysia. However, when the winds over Peninsular Malaysia changed to northwesterlies, from September 18th to the 22nd, temporary improvement in the haze condition over areas of Peninsular Malaysia was observed. The haze returned when the winds shifted to south-southeasterlies again. During this period the confluence of wind flow led to even denser haze conditions over the peninsular region (MMS, 1995).

The haze began clearing when the low-level winds changed to easterlies and southeasterlies again. In addition, the rainfall contributed to the clearing of the haze. The rainfall for Malaysia during July 1994 was considerably less than normal (when compared to the long-term average). For August the rainfall was well above normal, while for September the rainfall over Peninsular Malaysia was normal (MMS, 1995).

Analysis of Atmospheric Conditions During Haze Episodes

The meteorological conditions during the haze episode are presented in this section in order to investigate the extent to which these variables affected the TSP concentrations. Moreover, this will give the dominant variable that controls the dispersion of pollutants during severe haze episodes experienced in the Klang Valley.

Data Analysis Technique

The least median of squares technique (ROUSSEEUW, 1984) was used in the analysis. It was used in investigating the extent and strength of the relationships between the TSP concentration and meteorological variables during each of the examined severe haze episodes. Daily morning values (at 7:00 am) of TSP concentration, and daily average values for temperature and relative humidity were used. In the case of the wind speed, the daily 12-hour average prior to the sampling time at 7:00 am was considered most appropriate and it appeared to have the most bearing on the TSP concentration (ABDULLAH and SHAM, 1991).

Episode Days and Related Meteorological Variables

The occurrence of severe haze episodes associated with high concentrations of TSP in the Klang Valley when weather conditions are favorable has been verified by measurements carried out by the MMS and by other researchers. The data on TSP concentration obtained by the MMS, as well as the meteorological variables which were also measured for Klang Valley at the same station in Petaling Jaya, have been analyzed statistically.

The influence of selected and available meteorological variables which are known to affect the pollutant concentration; wind speed, wind direction, rainfall amount, relative humidity and temperature, has been investigated on a weekly basis for the years 1990–1994. The results show a relatively strong correlation.

General Investigation

Severe haze episodes with high TSP concentrations during this period have occurred on numerous days, especially in certain months of dry periods as shown in the previous section. In this section, the most intense episodes selected from the 1990–1994 period will be used to investigate some of the above-mentioned relationships between the TSP concentration and meteorological variables on a daily basis.

Results of the least median squares technique are shown in Table 1. The strong correlations indicate two important points. The first can be seen from the negatively strong correlation between the TSP concentration and each of wind speed, rainfall amount, and the relative humidity. This demonstrates the importance of these variables in removing the particles from the atmosphere and facilitating clearance of the haze condition from the Klang Valley by different processes: dispersion, washout and rainout.

The second, which was the strongest correlation among the investigated variables, was the positive correlation between the TSP and temperature. This result is very interesting because it expresses the important role that temperature plays in influencing the pollutant concentration during the intense and severe haze conditions in this region. This indicates that the atmosphere tended to be very stable, preventing deep convection and dispersion of pollutants during all the severe haze periods.

Correlations found during the first two haze episodes were stronger than those for the general period of 1990–1994. While the August–October 1994-haze episode, correlated similar to the general period. This implies that the meteorological variables play a major role in creating conditions for pollutant levels to remain high

Table 1

Correlation coefficients between TSP concentrations and meteorological variables for the years 1990–1994 and during the severe haze episodes

Haze episode	Meteorological variable			
	Wind speed	Temperature	Relative humidity	Rainfall amount
Aug.1990	−0.679	+0.807	−0.764	−0.7337
Oct. 1991	−0.74199	+0.8447	−0.686491	−0.64
Aug.–Oct. 1994	−0.396913	+0.51784	−0.52488	−0.399
General (1990–1994)	−0.362	+0.50	−0.583	−0.597

and to remain in the same area. Correlations for the general period are also presented in Table 1.

Wind Direction

It is generally known that the prevailing winds over Southeast Asia in the lower troposphcrc are mainly southwesterlies during the southwest monsoon, i.e., during the period from the end of May to September. From the end of September to the end of October low level winds begin to vary frequently, with winds sometimes blowing from the west and sometimes from the east, while northeastern monsoonal winds begin to prevail over Malaysia from early November (CHEANG et al., 1991).

It has been always reported that the haze phenomenon in the Klang Valley is correlated with the injection of pollutants from biomass burning from Sumatra and Kalimantan. This particle injection can only occur if wind direction at the surface layer is favorable, i.e., either southerly or southwesterly (SAMAH, 1995).

In the present research an investigation of the surface wind direction record in Petaling Jaya was carried out. The three haze episodes were examined to identify the wind direction, which coincided with the highest TSP concentrations. For this purpose, wind direction with the highest frequency occurrence was adopted. If two or more directions were found to have the same frequency of occurrence, the one with the higher mean wind speed was chosen. This is because winds with the higher mean wind speed were believed to have the most influence on particulates either by bringing them in or dispersing them. Furthermore, the periods with the worst hazy conditions were chosen for the investigation of each haze episode. Daily morning values of TSP were averaged for each period according to the wind directions. The investigated periods, dominant wind direction, and the average TSP concentrations for this direction are presented in Table 2.

A general observation in the wind record for these periods reveals that mean surface wind speeds were light and varied little during the hazy days. Results in Table 2 showed that, generally, the highest TSP concentrations coincide with the winds prevailing from the south and southwest. This indicates that the particles may have been brought from external sources by these winds. This agreed with earlier studies that reported these haze episodes and related them to the external sources brought by southerly and southwesterly winds (CHEANG et al., 1991; MMS, 1995).

Unlike other haze periods, dominant winds for the second spell of the August–October 1994 haze episode were found to be northeast, west and southeast. This indicates that hazy conditions could be due only to the local build-up of sources.

This leads to two questions; one concerns the origin of the particles, while the other concerns the trapping mechanism of pollutants in this area. As shown by the

Table 2

Results of the wind direction investigation during the haze episodes

Time period	Dominant wind direction	Average TSP concentration ($\mu g/m^3$) for the wind direction
20–30 Sept. (1990)	S	516.2
	SW	461.2
	W	360.43
27 Sept.–11 Oct. (1991)	SW	418.4
	S	391.35
	NE	321.63
22–31 Oct. (1991)	W	231.7
	NE	222.85
	S	202.8
27 Sept.–4 Oct. (1994)	SW	333.95
	W	310
	S	295.53
25–31 Oct. (1994)	NE	209.5
	W	188.65
	SE	161.95

wind direction analysis, particles seem to have different origins; the first is external sources when the wind direction is favorable (i.e., southerly and southwesterly), while the second is the build-up of pollution from local sources. In his study of the August–October 1994-haze episode, SAMAH (1995) analyzed the wind direction in combination with an analysis of the particle size. He suggested that there seemed to be two different origins of the haze: local pollutants during the periods of suppressed low-level convection, and finer aerosols from forest fires advected into the boundary layer when wind direction was favorable.

Trapping Mechanism

Temperature, among the investigated meteorological variables, appeared to have the strongest correlation with TSP concentrations during the severe haze episodes. Dominance of temperature in all the severe haze episodes is a strong indication that stability and trapping are the main factors affecting pollution during these periods.

According to a number of researchers, the existence of an inversion layer in the lower troposphere during the haze episodes was reported to be the reason behind the trapping of pollutants in the Klang Valley (SHAM *et al.*, 1991). This dry layer could act as a lid for vertical mixing and hence contribute to the trapping of the emitted pollutants (SAMAH, 1995). In addition, in the October 1991 haze episode, it was reported that the meteorological conditions indicated a widespread subsidence of air over the Malaysian area, resulting in frequent occurrences of temperature inversions. This further prevented the particles from being lifted into the higher atmosphere.

With the aim of studying the possible mechanism of trapping and atmospheric mixing capacity, SAMAH (1995) carried out a detailed analysis of the radiosonde ascents during the haze episode of 1994. He used two conservative thermodynamic parameters to determine the atmospheric stratification and subsequently vertical mixing. His analysis demonstrated the important role of moisture in the process of fumigation and trapping of the haze and not of inversion layers as in the case of mid-latitude areas. He mentioned that vertical mixing between the surface layer and the well mixed layer above could not take place due to the observed stable stratification and the suppression of convection. He added that this condition would trap pollutants in the surface layer and contribute to the build-up of pollutants observed during the haze period. Accordingly, the build-up of the haze was associated with a dry layer capping convectional mixing and not an inversion layer as suggested by many workers.

Comparison between Haze Episodes

Although haze episodes had been shown to occur during the same period of different years, each haze has been associated with different meteorological conditions and has been related to different sources. Consequently, a comparison between haze episodes could be of interest. To this end a simple comparison between the selected haze episodes has been carried out. Elements of this comparison were maximum-recorded TSP concentration, period of the worst hazy conditions, dominant meteorological variable, and the role of wind and rainfall during the peaks of each episode.

In fact, all the severe episodes of air pollution over the Klang Valley were associated with very high values of TSP. On the basis of morning values, the maximum concentration was $570~\mu g~m^{-3}$, for the October 1991 episode, $516~\mu g~m^{-3}$ for August 1990, and $463.5~\mu g~m^{-3}$ for the August–October 1994 episode.

According to the duration of dense hazy conditions, it is obvious that the August–October 1994 episode was the most persistent with dense haze conditions in September, early October, and the last week of October. During the 1991-haze episode, a dense haze condition persisted for two spells: September 27th–October 11th and 22nd–31st October. The 1990 episode had the minimum duration compared to the other two episodes, with dense haze persisting from the 15th to the 30th of August.

Among the meteorological variables, temperature was the dominant variable with the strongest effect on TSP concentration, for all the haze episodes. Its effect was to, locally, trap particles in the polluted area and suppress vertical mixing. The highest correlation was for the October 1991 episode ($r = +0.84$) followed by August 1990 ($r = +0.81$). The lowest, although still strong correlation was for the August–October 1994 episode ($+0.52$). On the other hand, analysis of the wind direction indicated its importance in identifying the sources of the particles during the worst hazy days.

Wind speed and rainfall amount were found to strongly affect TSP, using the correlation analysis. This reflects the efficiency of these variables in cleansing particles from the atmosphere by dispersion and precipitation, respectively. It can be noted that during the TSP peaks, both wind speed and rainfall amounts were extremely low (Figs. 2–7). Hence, calm wind and dry days coincided with the maximum concentrations of TSP for all the haze episodes.

NICHOL (1997) also observed higher temperature and lower precipitation during the 1994-haze episode in Singapore. She considered the possibility that this may have been due to the ENSO phenomenon, but concluded that it was mainly attributable to haze-induced stability in the lower atmosphere restricting the amount of outgoing radiation.

Summary

Three severe haze episodes during the early 1990s in the Klang Valley were reviewed: August 1990, October 1991 and August–October 1994. The main conclusions can be summarized as follows:

- The increasing numbers of sources, especially in the urban-industrial areas, and the generally restrictive nature of the atmosphere to disperse and transport pollutants in this part of the world are the reasons behind the haze phenomenon.
- In the Klang Valley region there are two kinds of haze: shallow localized haze and dense haze. The former, which usually occurs in urbanized areas, arises from trapping of pollutants from anthropogenic emissions, in response to stabilization of the atmosphere. The latter is due either to the injection of suspended ash particles from large-scale forest fires and open burning in Indonesia (October 1991 and August to October 1994), or these external sources combined with local open-burning (August 1990).
- Haze usually occurs during the southwest monsoon season. Low-level winds over Malaysia are generally southwesterlies and the weather is generally dry during the active phase of this season. In addition, meteorological conditions are unfavor-able for efficient dispersion and dilution of pollutants through convective mixing or for their washout by widespread heavy rain.
- An investigation of the meteorological variables during haze episodes revealed a strong correlation between the TSP concentrations and all the investigated meteorological variables. Therefore, haze episodes are strongly affected and influenced by meteorological variables. Meteorological conditions would trap the pollutants and contribute to the build-up of pollutants observed during the haze.
- Of the meteorological variables, temperature was found to have the strongest correlation with TSP for the considered haze episodes. Dominance of temper-

ature in all the severe haze episodes is a strong indication that stability and
trapping are the main factors affecting pollution during these periods. We can
conclude that atmospheric convection and dispersion were suppressed during all
the severe haze episodes and the atmosphere tended to be very stable, leading to
the suspension of the pollutants in the valley.

- A simple comparison between the investigated haze episodes revealed that the
  maximum TSP concentration was recorded during the October 1991 episode.
  The August–October 1994 episode was the most persistent and the least affected
  by the meteorological variables.

- To investigate the sources of the particles during the hazy days, analysis of the
  wind direction data for the periods with the worst hazy conditions showed that
  southerly and southwesterly winds coincide with these periods. This indicates
  that the particles may have been brought from external sources by these
  favorable winds.

Acknowledgements

The authors would like to thank the Malaysian Meteorological Service for
providing the data used in the analysis.

References

ABDULLAH, MOKHTAR and SHAM, SANI (1991), *Use of Robust Methods in the Analysis of Suspended
Particulate Air Pollution: A Case Study in Malaysia*, Environmetrics. *2*, 201–215.

CHEANG, B. K., LEONG, C. P., OOI, S. H., and TUSIN, A. M. (1991), *Haze episode October 1991.*
Information Paper no. 2, 10 pp., Malaysian Meteorological Service, Jalan Sultan, 46667 Petaling Jaya,
Selangor, Malaysia.

CHOW, K. K. and LIM, J. T., *Monitoring of suspended particulate in Petaling Jaya.* In *Urbanization
and Ecodevelopment* (Yip, Y. H. and Low, K. S., eds.) (University of Malaya Publications, 1984)
pp. 178–185.

MMS (1995), *Report on Air Quality in Malaysia as Monitored by the Malaysian Meteorological Service
1994.* Technical Note No. 55, Malaysian Meteorological Service. Jalan Sultan, 46667 Petaling Jaya,
Selangor, Malaysia.

NICHOL, J. (1997), *Bioclimatic Impacts of the 1994 Smoke Haze Event in Southeast Asia*, Atmospheric
Environment *31* (8), 1209–1219.

ROUSSEEUW, P. J. (1984), *Least Median of Squares Regression*, J. Am. Statistical Assoc. *79*, 871–879.

SAMAH, AZIZAN, A. (1995), *The Necessity of a Good Scientific Foundation in Tackling Environmental
Issues.* National Review of Environmental Quality Management In Malaysia. 10–12 October, Kuala
Lumpur, Malaysia. (can be obtained from LESTARI, National University of Malaysia, 43600, Bangi,
Malaysia).

SHAM, S. (1984), *Suspended Particulate Air Pollution over Petaling Jaya during the September 1982 Haze*,
Ilmu Alam (in English). (12 and 13), 83–90.

SHAM, S. (1987), *Urbanization and the Atmospheric Environment in the Low Tropics: Experiences for the
Klang Valley Region, Malaysia* (UKM Press, 1987).

SHAM, S., CHEANG, BOON KHEAN, CHOW PENG, and LIM, SZE FOOK (1991), *The August 1990 Haze in Malaysia with Special Reference to Klang Valley Region*, Technical Note 49, 60 pp., Malaysian Meteorological Service, Jalan Sultan, 46667 Petaling Jaya, Selangor, Malaysia.

(Received April 1, 2000, accepted February 2, 2001)

 To access this journal online:
http://www.birkhauser.ch

Pure appl. geophys. 160 (2003) 239–250
0033–4553/03/020239–12

▌Pure and Applied Geophysics

A Review of some Recent Radiation Fog Prediction Studies and the Results of Integrating a Simple Numerical Model to Predict Radiation Fog over Brunei

GANDIKOTA V. RAO[1] and JAMES O'SULLIVAN[1]

Abstract—Several radiation fog studies with emphasis on numerical simulation and prediction are reviewed. One of the earliest attempts started with a given surface diurnal variation of temperature and water vapor, and concluded by forecasting the onset of saturation at various levels; thus fog, by examining the spread of temperature and moisture in the vertical. The one-dimensional (1-D) models are still popular. Some of the recent numerical simulations use more than 100 levels in the vertical and treat various kinds of vegetation, aerosols, and soils with moisture contents. Some also employ a mesoscale model in conjunction with a 1-D model to consider the advective effects. In the following a simple 1-D numerical model was used to predict the onset of fog at Brunei, based on a desktop computer and routine surface observations of dry bulb temperature (T), dewpoint temperature (T_d), and wind speed at 1800 Local Time (LT). Optimism exists in improved predictions of fog and stratus as 1-D models incorporate many physical processes, and mesoscale models continue to improve in predicting advection and cloud cover.

Key words: 1-d fog model, radiation fog, fog prediction, fog over Brunei, numerical modeling of fog, review of fog.

1. Introduction

Fog is important to study because it occurs close to the ground and obstructs the free flow of transportation. Fog occurs by definition when the horizontal visibility falls below 1 km. When the visibility is still poor but extends to a few kms, the phenomenon goes by the name, mist or haze. While fog may occur because of a variety of factors, such as: frontal, advection, mixing, upslope, or radiation; the one due to radiation is relatively simple to discuss. PETTERSSEN (1940, 1956) was one of the first to examine the conditions leading to the formation of radiation fog. According to him, the nocturnal air temperature in contact with the ground falls because of the outgoing longwave radiation. The surface air layers subsequently are cooled because of conduction and the eddy diffusion of heat. Consequently, an inversion develops near the ground. This inversion may deepen with time if eddy activity continues. With

---

[1] Department of Earth and Atmospheric Sciences, Saint Louis University, St. Louis, Missouri, U.S.A.
E-mail: rao@eas.slu.edu

the progression of cooling, the capacity of air to hold moisture decreases while the relative humidity increases. At a certain relative humidity, depending on the composition of the condensation nuclei and their concentration, mist forms and subsequently fog develops. GEORGE (1951) hypothesized, that dew and fog form depending on the initial vertical variation of humidity and on the balance between radiative cooling and turbulent diffusions of heat and moisture. Additional factors, such as: Nocturnal cloud cover with upslope and downslope motion also were found to influence the occurrence of radiation fog. Differences among various kinds of fog were elaborated. Later, GEORGE (1960) discussed in depth the synoptic forecasting, climatology, and movement of fog over the United States.

2. Some Basic Microphysics

ROACH (1994, 1995) presented the definitions and basic microphysics of fog. According to him, fog consists of water droplets typically 10–20 μm in diameter. PODZIMEK (1997) has given droplet concentration and size distribution in haze and fog. Fog droplets contain insoluble dust or smoke particles or traces of dissolved salts. Occasionally, fog is reported in very polluted atmospheres before a significant number of drops have grown (PODZIMEK et al., 1992). According to the measurements of PODZIMEK and HOPKINS (1989) near Naples, Italy, a polluted city, approximately 7.3% of the mean haze elemental volume was formed by insoluble substances, such as, minerals and ash; and the percentage grew to 11.7% during a low relative humidity situation, when the winds blew from an industrial center. This information is important, because the rate of growth of drops depends on the temperature and the degree of supersaturation, the nature of nucleus, and the droplet curvature. Direct loss of infrared radiation by a drop can cool and influence its growth. When there is an abundance of condensation nuclei, a large number of droplets is likely to form. These droplets experience slow gravitational settling, so that fog may persist. According to FITZJARRALD and LALA (1989), fog that persists for more than 30 minutes is likely to be more than 50 meters deep. A thick fog with a visibility of about 100 meters measures 0.1 to 0.2 g m^{-3} liquid water content (LWC).

3. Some Early Numerical Models

With the availability of electronic computers in the 1950s, numerical modeling of fog was pursued. FISHER and CAPLAN (1963) attempted fog prediction by solving equations for temperature, water vapor, and liquid water. These equations were of the form

$$\frac{\partial q}{\partial t} = \frac{\partial}{\partial z} K_q \left(\frac{\partial q}{\partial z} \right) - \vec{V} \cdot \vec{\nabla}_q + S \ , \tag{1}$$

where q is the particular variable (temperature, vapor, or liquid water) under consideration, K_q is the eddy diffusivity of that variable, $\vec{v} \cdot \vec{\nabla} q$ is the horizontal advection, and S represents the resultant of sources and sinks of q. For detailed examination, a 1-D model having variation only in the vertical was considered. Air temperature and water vapor pressure at the surface were prescribed resembling a typical diurnal variation. The variables: temperature and humidity were integrated with height and time. The time of onset of saturation near the surface and subsequent generation of liquid water were regarded as the beginning of fog. This study suffers, among other things, from the lack of an energy budget equation, explicit formulas of radiative cooling, and shortwave radiation. Nevertheless, it paved the way for numerical modeling of fog.

ZDUNKOWSKI and NIELSON (1969) predicted the diurnal changes in temperature and specific humidity through a 1-D model. A budget equation for water vapor was set up so that the transformation of vapor into liquid and *vice versa* was specified. Longwave radiation emanating from water vapor and liquid water was calculated. Specific heat and thermal conductivity of soil were so chosen that they yielded reasonable cooling rates of soil and the adjacent air layers at the time of fog onset. Their model integration started at the local sunset with the prescribed temperature and humidity values. Eddy diffusivity values were assumed so that a realistic nocturnal boundary layer had developed. Nevertheless, fog was predicted unrealistically early. The authors realized that the eddy diffusivities must be internally computed (from the local wind profile) rather than assumed. To facilitate an internal computation the wind field must be predicted. This was done by ZDUNKOWSKI and BARR (1972) by assuming a geostrophic wind and a simple momentum equation.

BROWN and ROACH (1976) used essentially ZDUNKOWSKI and NIELSON's (1969) model but added gravitational settling of droplets. They experimented with eddy diffusivities of varying magnitudes. Fog seemed to have developed within 7 hours with smaller eddy coefficients of the order of 0.25 m^2 s^{-1}. However, verification was not feasible because no synoptic surface data were used. The investigators realized that the nocturnal surface (soil) temperatures predicted by the model differed from the temperature at the level of vegetation such as grass. This temperature difference added an extra degree of complexity to the fog prediction problem.

4. More Recent Numerical Models

The more recent numerical models incorporated better turbulence formulations. For example, TURTON and BROWN (1987) improved the BROWN and ROACH (1976) model by reformulating the mixing length and allowing the eddy diffusivities to vary with stability. They specified a gamma size distribution of droplets so that the gravitational settling of fog and its longwave transmissivity can be computed. They also employed a parameterized five-band radiation scheme to fit the calculations of molecular transmissivity as a function of absorber path length. Solar radiation was parameterized. The authors did not specifically compute the solar radiation because forecasting the dissipation of fog was not considered a priority. Surface temperature was calculated by incorporating basic soil physics. Model integration started with the midday observations. For the first time, it was discovered that integrations performed with a homogeneous clay soil overestimated the soil heat flux and consequently the minimum surface temperature. By using the physical properties of an inhomogeneous mixture of clay and peat soil such as those given in MONTEITH (1973), the authors were able to obtain reasonable nocturnal minimum temperatures. The results were verified with special tethered balloon and mast observations. In one simulation, fog formed at 40 m and descended to the surface, while the field observations showed no fog development, apparently because of warm air advection in the middle boundary layer. On another day, fluctuating fog was observed in the evening and fog was simulated only with molecular diffusion. Critical analysis of the discrepancy between the observed structures of the atmosphere and the simulated structures was made. Obviously, the model wind and its vertical shear played a key role in generating turbulence, which in turn influenced the fog simulation.

5. Numerical Models Based on Energy Balance Method

MUSSON-GENNON (1987) added considerable sophistication to the 1-D modeling of fog. Noteworthy additions were:
(a) Providing for liquid water content distribution in the grid volume.
(b) Deriving the diffusivities from a turbulent kinetic energy equation (TKE).
(c) Obtaining the soil temperature through an energy balance equation.
(d) Formulating visibility in terms of the computed liquid water.
Visibility has a higher practical value and the results helped to distinguish the development of a moist boundary layer from one that impaired visibility. Special field observations from the Netherlands were used to verify the results. This study emphasized the necessity of obtaining the detailed observations of temperature and humidity in the proximity of the surface and also obtaining soil moisture values. The lack of horizontal advection in 1-D models was recognized but it was hoped that its effect could be incorporated with a synoptic model. In general, routine synoptic models do not handle well the horizontal advection of boundary-layer moisture and clouds. The optimism expressed may be realized with mesoscale models.

6. Numerical Models Emphasizing Vegetation

The role of surface vegetation (in particular grass) in the onset of fog was examined by DUYNKERKE (1991), using a 1-D model similar to the ones presented earlier. It was deduced that the resistance of the grass and the stagnant air against heat transport would cause a temperature difference of 10 K between the grass and the soil, and that a model lacking a parameterization of vegetation would underestimate the nocturnal cooling to therefore mistime the onset of fog. As was the case with the earlier models, TKE was used to compute eddy diffusion. From the measured upward longwave radiation near the surface and the soil measurements of temperature, the local Monin-Obukhov parameters were estimated. The model boundary layer developed a three-layer structure before the formation of fog. Within the surface layer the total cooling rate was a slight difference between longwave cooling and turbulent warming. The midregion cooled by the divergence of turbulent heat flux, while the uppermost region cooled by radiative flux divergence. Thus, the radiative and turbulent heat fluxes are thereby to be accurately formulated at least in the lower two-thirds of the boundary layer.

DUYNKERKE (1991, 1997) used the special 200-m tower data collected in the Netherlands to verify his model. The model run was initialized at 1800 LT. In 1991 and 1997 simulations for a day in August and in April were discussed. Verification consisted of examining the predicted and observed profiles of temperature and humidity. It was concluded that the parameterization of gravitational settling of fog drops should be further improved.

The above models pointed out that there was a temperature difference between the tip of a grass blade and the soil, and the fact that fog could possibly occur on the basis of the lower temperature at the grass level. VON GLASOW and BOTT (1999) thoroughly examined the evolution of radiation fog based on certain initial conditions at 1000 LT. Their model consisted of various components: a soil module, a microphysical module that starts from a given aerosol and fog droplet radii, radiation, and turbulence modules. While their model had the standard differential equations governing u, v, q, potential temperature, and the frictional stresses; they had included a form drag in terms of the shielding factor to simply describe the coverage of soil with foliage. They observed that fog developed in four stages.

(a) Soon after sunset, fog appeared on the dense portion of 20-meter high vegetation.
(b) By about 0200 LT, fog spread to all the lower leaves.
(c) Because of the enhanced cooling caused by the droplets, fog propagated to levels as high as 70 meters.
(d) Finally, sunrise started the dissipation stage.

The model extended up to 300 meters with a 2-meter resolution. Turbulence and radiation were parameterized consistent with the state of the art. It was realized by the authors that the development of fog was sensitive to the initial conditions of humidity and the shielding factor. The dependence on aerosol type, whether it was urban or maritime seemed to influence the LWC of fog. The urban aerosols produced almost ten times more LWC than the maritime aerosol. A similar sensitivity was found with the shielding factor. A shielding factor of 0.5

produced double the LWC and double the depth of fog compared to a factor of zero. Thus the presence of vegetation prominently contributes to the production of fog.

7. A 1-D Simulation of Fog Coupled with an Operational Model

Following the earlier advice of MUSSON-GENNON (1987), BERGOT and GUEDALIA (1994) employed a 1-D model along with a mesoscale three-dimensional operational model for predicting fog. This model has features common to many of the models previously discussed (e.g., TURTON and BROWN, 1987). Thirty vertical levels were distributed in a log-linear manner between ground and 1400 meters. It was deduced that visibility was dependent on surface dew deposition and fog-layer height which were critically affected by geostrophic wind. The warm and cool air advection, as well as the low-level cloud cover seemed to influence the time of fog formation and its vertical development. Vegetative components were not considered. GUEDALIA and BERGOT (1994) compared the model results with two cases of detailed observations of wind, temperature, and humidity at 0.7 m and 5 m in France. Sensitivity tests concluded that the choice of initial conditions played an important role in the prediction of fog.

8. A 1-D Fog Prediction Model

The above review shows that many sophisticated processes were incorporated into fog modeling and that several levels were designed to resolve the fine structure of fog. However, in routine radiation fog forecasting the data required by these detailed models is lacking. It is believed that radiation fog is generally a local problem and as such is best handled by a resident forecaster, who is familiar with the unique conditions of that site. MEYER and RAO's (1999) simple model was directed to predicting fog, using the routinely available synoptic surface observations by the military meteorologists, utilizing desk top computers, and some local parameters carefully estimated. The 24-hour prediction is based on the 1800 LT observation of dry bulb, dewpoint temperature, and wind. Only the highlights of the model are presented here.

(a) Four levels: One at the surface, the second at 10 meters, the third at 200 meters, and the fourth at 1000 meters constitute the model; the most important of these levels being the one at 10 meters where the routinely available 1800 LT observations were placed. Data at other levels were extrapolated on the basis of the 10 m observations and certain assumptions about lapse rates from the surface to 10 meters, and in the middle and upper boundary layers, and relative humidity at 200 meters and 1000 meters. Surface roughness parameter was assigned 0.2 m.

(b) An analytical formula of radiative cooling due to HALTINER and MARTIN (1956) was modified by MEYER and RAO (1999), for use in the model. The physical properties of the soil, such as density and thermal conductivity, were obtained from climatology (GARRATT, 1992).

(c) The turbulent diffusion of heat and moisture formulation was essentially due to MAHRT *et al.* (1991).

(d) Shortwave radiation was explicitly computed and a radiative energy balance condition applied at the soil-atmosphere interface. Sensible heat flux was explicitly computed.

(e) Evaporation from the surface was calculated from the radiative energy balance condition and Bucket model (GARRATT, 1992), from a knowledge of the climatological moisture content of the soil appropriate to the geographical area (MATTHEWS, 1985), and 24-hour rainfall during the previous day at Brunei.

The 1-D model of MEYER and RAO (1999) does have limitations. For example, there is no means of accounting for horizontal advection. MEYER and RAO's (1999) model can be combined with a larger 3-D model which will generate synoptic conditions. Although the 1800 LT wind speed is specified, it is strictly to calculate the diffusion coefficients of heat and moisture. The model has no explicit calculations of longwave radiation due to fog droplets. Thus no cooling occurs at the top of the fog level. Because the calculation is done at only one level, fog depth is not calculated in the model.

9. Purpose of this Research

The purpose of this research is to apply the MEYER and RAO's (1999) model to the prediction of mist, haze, or fog at Brunei, based on the 1800 LT routine observations of dry bulb, dewpoint temperature, and wind. Two distinct time periods were selected: one in December 1997, when the Kalimantan region was relatively dry because of the strong El Niño in the eastern Pacific; the second in December 1998, when the same region was relatively wet. One day in December 1997 and similarly another day in December 1998 were chosen from the routine surface observations of Brunei which reported a drop in the visibility during the night of forecast, indicating the onset of mist, haze, and fog. The purpose of this exercise was to test whether the MEYER and RAO (1999) model, based on the initial routine surface observations of T, T_d, and wind at 1800 LT at Brunei, would predict saturation conditions (liquid water, if supersaturation arose) at night or early morning times. Because the surface observations were available to us only at 3-hour intervals, our estimation of the beginning of saturation within the model to the observed drop in visibility was only approximate. A more systematic skill score of the success of fog prediction at Brunei by the model can be undertaken later, when a complete set of hourly observations becomes available. Thus, the present exercise should be regarded as preliminary.

10. Data

Because of the strong El Niño in late 1997, Southeastern Asia was subjected to large subsidence (CLIMATE PREDICTION CENTER (CPC), 1997) in December 1997, while the same region experienced large-scale ascent and rainfall in December 1998. These two months were selected for calibrating the MEYER and RAO's (1999) model, originally developed for extratropics at Brunei's tropical station. It turned out that only minor changes in radiation energy at sunset and sky radiation were needed to apply the model to Brunei. The surface data for Brunei were available at the National Center for Atmospheric Research only at 3-hour intervals, necessitating an interpolation. This made verification of the results difficult. Three days were selected in December 1997 and four in December 1998, initially. From this sample; 26 December 1997 (Table 1) and 22 December 1998 (Table 2) were chosen for discussion here, because they underwent no significant weather and precipitation during the previous day.

11. Results

Figure 1 shows T and T_d for 26 December 1997. The dewpoint depression became less than 1°C at 2106 LT and further became less than 0.1°C at 2356 LT. Synoptic observations available at 3-hour intervals reported reduced (from 20 km to 15 km) visibility at 2300 LT, and light fog at 0500 LT 27 December. This indicates that the model has developed near saturation conditions slightly ahead of the observation, but indicates agreement between simulation and observation. The dewpoint

Table 1

Data for model run at 1800 LT 26 December 1997 Brunei (4.93°N, 114.93°E)

Dry-bulb temperature	28.0°C
Dewpoint temperature	23.2°C
Wind speed	7.3 knots
Precipitation	None
Assumed soil type	Semiwet, sandy
Assumed lapse rate	0.9 of dry adiabatic

Table 2

Data for model run at 1800 LT 22 December 1998 Brunei (4.93°N, 114.93°E)

Dry-bulb temperature	27.8°C
Dewpoint temperature	23.8°C
Wind speed	4.3 knots
Precipitation	None
Assumed soil type	Semiwet, sandy
Assumed lapse rate	0.9 of dry adiabatic

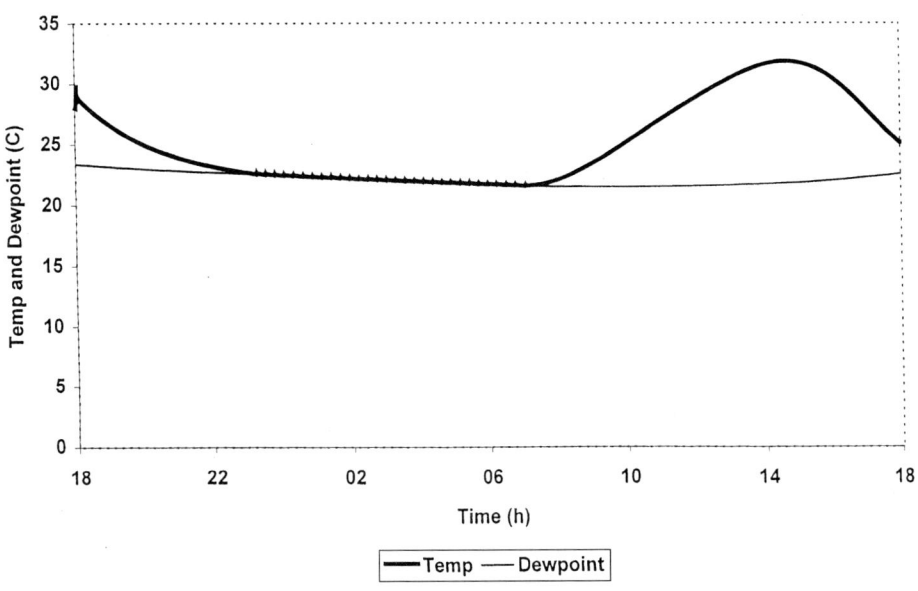

Figure 1
Shows the temperature (T) and dewpoint temperature (T_d) for 26 December 1997. Note the onset of saturation signifying the likelihood of fog beginning at 2106 Local Time.

depression in Figure 1 exceeded 1.0°C at 0824 LT 27 December 1997, while the synoptic observation showed a slightly better visibility at 0800 LT. Consequently, simulation is supported by observation.

Figure 2 shows T and T_d for 22 December 1998. The dewpoint depression became less than 1°C at 2149 LT and further became less than 0.1°C at 2317 LT. Synoptic observation reported reduced (from 15 km to 10 km) visibility at 2300 LT and light fog at 0500 LT 23 December. As in the previous example, the model developed saturation conditions ahead of the observation. The dewpoint depression in Figure 2 exceeded 0.1°C at 0742 LT 23 December, while the synoptic reports indicated visibility improving at 0800 LT. Thus the simulation is considered favorable. Perhaps a forecaster stationed in Brunei can further improve the results by inserting site-relevant quantities such as, lapse rate in the boundary layer, albedo, and radiation energy at sunset, etc.

12. Conclusions

One-dimensional fog modeling research has become sophisticated through improved calculations of radiative, turbulence, microphysical processes, and soil

22 Dec 1998

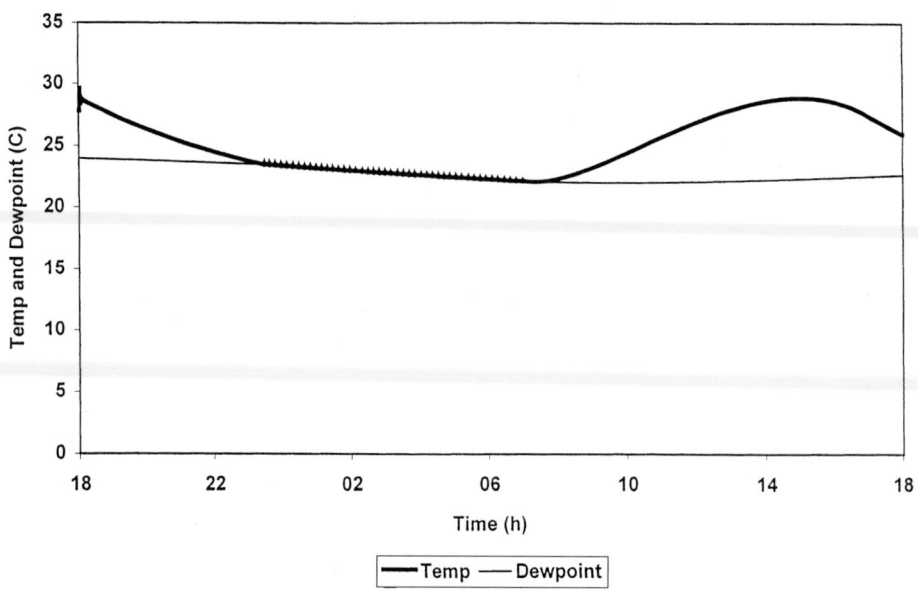

Figure 2

Shows the temperature (*T*) and dewpoint temperature (*T_d*) for 22 December 1997. Note the onset of saturation signifying the likelihood of fog beginning at 2149 Local Time.

moisture incorporation. However, more research is needed to model wind and its vertical structure in the boundary layer. Special observations are needed at various locations to understand the roles of vegetation and different kinds of aerosols in the fog development.

On the forecasting side, site relevant climatological data should be made available to forecasters such as: the physical characteristics of soil, radiative energy at the time of sunset, and sky radiation. The forecasters, with the aid of 1-D models such as the one presented here, climatological data, and a mesoscale model should be able to arrive at better predictions.

Acknowledgments

The authors wish to thank: Professor M. P. Singh for suggesting the composition of this article, Dennis Joseph for supplying the Brunei meteorological data, and Dr. W. Dale Meyer for his discussions. The research relating to fog modeling was sponsored by the University Corporation for Atmospheric Research (UCAR) Cooperative Program for Operational Meteorology Education and Training (COMET) under S94-43838.

REFERENCES

BERGOT, T. and GUEDALIA, D. (1994), *Numerical Forecasting of Radiation Fog. Part I: Numerical Model and Sensitivity Tests*, Mon. Wea. Rev. *112*, 1218–1230.

BROWN, R. and ROACH, W. T. (1976), *The Physics of Radiation Fog. Part II: A Numerical Study*, Quart. J. Roy. Meteor. Soc. *102*, 335–354.

CLIMATE PREDICTION CENTER (CPC) (1997), *Climate Diagnostics Bulletin*, 97 pp. [Available from U.S. Dept. of Commerce, NOAA/NWS, W/NCEP, 5200 Auth Rd., Washington, D.C. 20233.]

DUYNKERKE, P. G. (1991), *Radiation Fog: A Comparison of Model Simulation with Detailed Observations*, Mon. Wea. Rev. *119*, 324–341.

DUYNKERKE, P. G. (1999), *Turbulence, Radiation, and Fog in Dutch Stable Boundary Layers*, Bound. Layer Meteor. *90*, 447–477.

FISHER, E. L. and CAPLAN, P. (1963), *An Experiment in Numerical Prediction of Fog and Stratus*, J. Atmos. Sci. *20*, 425–437.

FITZJARRALD, D. R. and LALA, G. L. (1989), *Hudson Valley Fog Environments*, J. Appl. Meteor. *28*, 1303–1328.

GARRATT, J. R., *The Atmospheric Boundary Layer* (Cambridge University Press, New York 1992), 316 pp.

GEORGE, J. J., *Weather Forecasting for Aeronautics* (Academic Press, New York and London 1960), 673 pp.

GEORGE, J. J., *Fog, Compendium of Meteorology* (American Meteorological Society, Boston, MA 1951), 1334 pp.

GUEDALIA, D. and BERGOT, T. (1994), *Numerical Forecasting of Radiation fog. Part II: A Comparison of Model Simulation with Several Observed Fog Events*, Mon. Wea. Rev. *122*, 1231–1246.

HALTINER, G. J. and MARTIN, F. L., *Dynamical and Physical Meteorology* (McGraw-Hill Book Co. 1957), 470 pp.

MAHRT, L., EK, M., KIM, J., and HOLTSLAG, A. A. M. (1991), *Boundary-Layer Parameterization of a Global Spectral Model*, U.S. Air Force Phillips Laboratory Tech. Report 91-2031, 204 pp.

MATTHEWS, E. (1985), *Atlas of Archived Vegetation, Land-use, and Seasonal Albedo Datasets*, Tech. Memo 86199 NASA, 53 pp. [Available from the National Technical Information Service, Springfield, VA 22161.]

MEYER, W. D. and RAO, G. V. (1999), *Radiation Fog using a Simple Numerical Model*, Pure Appl. Geophys. *155*, 57–80.

MONTEITH, J. L., *Principles of Environmental Physics* (American Elsevier Publishing Co., Inc., New York 1973), 241 pp.

MUSSON-GENON, L. (1987), *Numerical Simulation of a Fog Event with a One-Dimensional Boundary-Layer Model*, Mon. Wea. Rev. *115*, 592–607.

PETTERSSEN, S., *Weather Analysis and Forecasting* (McGraw-Hill 1940), 503 pp.

PETTERSSEN, S., *Weather Analysis and Forecasting. Vol. II: Weather and Weather Systems*, 2nd Ed. (McGraw-Hill 1956), 266 pp.

PODZIMEK, J. and HOPKINS, R. A. (1989), *Surface Wind Characteristics at the Bay of Naples and the Effect of Fluctuating Supersaturation on the Haze Element Evolution*, Estrato Annali dell Istituto Universatario Navale di Napoli *57*, 146–161.

PODZIMEK, J., SPEZIE, G., and IANNIRUBERTO, M. (1992), *An Effective Cleaning of the Napoli Polluted Atmosphere by the Activated Haze Elements*, Annali IUNN *59*, 111–125.

PODZIMEK, J. (1997), *Droplet Concentration and Size Distribution in Haze and Fog*, Studia Geophys. et Geodaet. *41*, 277–296.

ROACH, W. T. (1994), *Back to Basics, Fog. Part I: Definitions and Basic Physics*, Weather *49*, 411–415.

ROACH, W. T. (1995), *Back to Basics, Fog. Part 2: The Formation and Dissipation of Land Fog*, Weather *50*, 7–10.

TURTON, J. D. and BROWN, R. (1987), *A Comparison of a Numerical Model of Radiation Fog with Detailed Observations*, Quart. J. Roy. Meteor. Soc. *113*, 34–54.

VON GLASOW, R. and BOTT, A. (1999), *Interaction of Radiation Fog with Tall Vegetation*, Atmos. Environ. *33*, 1333–1346.

ZDUNKOWSKI, W. and NIELSON, B. (1969), *A Preliminary Prediction Analysis of Radiation Fog*, Pure Appl. Geophys. *19*, 45–66.

ZDUNKOWSKI, W. and BARR, A. (1972), *A Radiative-Conductive Model for the prediction of Radiation Fog*, Bound. Layer Meteor. *3*, 152–157.

(Received July 31, 2000, accepted January 22, 2001)

To access this journal online:
http://www.birkhauser.ch

Pure appl. geophys. 160 (2003) 251–264
0033–4553/03/020251–14

© Birkhäuser Verlag, Basel, 2003

Haze Research in Brunei Darussalam During the 1998 Episode

Miroslav Radojevic[1]

Abstract — Brunei Darussalam experienced a severe haze episode between the beginning of February and the end of April 1998 due mainly to local peat and forest fires in Brunei and in neighbouring Sabah and Sarawak. The extensive research studies of the haze carried out in Brunei are outlined together with selected results. Particulate matter (PM_{10}) was the only significant criteria pollutant and it exceeded WHO guidelines and accepted air quality standards on most days during the haze episode. Gaseous criteria pollutants (CO, SO_2, NO_2, O_3) were generally well below WHO guidelines and at these concentrations they are expected to have no significant health or environmental effects. Measurements of volatile organic compounds (VOCs) revealed the presence of benzene, toluene, ethylbenzene, and xylenes (BTEX), aldehydes, phenol, and polynuclear aromatic hydrocarbons (PAHs). Personal exposure monitoring of PM_{10} revealed significant differences in exposure patterns between different individuals depending on the location, time and activity. Data on outpatient visits showed an increase for some illnesses (e.g., acute respiratory infection) during the months of haze. No significant impacts of haze on rainwater acidity or deposition were noted. Emission factors for some volatile compounds were determined in combustion experiments in which peat was burned at temperatures typical of smouldering.

Key words: Biomass fires, haze, air pollution, particulate matter (PM_{10}), volatile organic compounds (VOCs), acid deposition.

1. Introduction

The regional haze in SE Asia during 1997 and 1998 was a major pollution disaster affecting the lives of millions of people in the region and vitally impacting their health and economic development (Radojevic, 1997; 1998). The forest and peat fires, together with the associated haze produced numerous potential impacts, including: health effects, reduced visibility resulting in transportation accidents and closures of airports, loss of biodiversity, lower productivity (agricultural, industrial, commercial), downturn in tourism, economic costs, effects on rainwater acidity, effects on weather and climate, effects on global cycles (e.g., carbon), and impacts on people's lifestyle.

The present paper summarises research into forest fires and associated haze carried out in Brunei Darussalam during the 1998 episode. The haze event of February–April 1998 was fundamentally different from those of 1994 and 1997.

---

[1] Department of Chemistry, University of Brunei Darussalam, Bandar Seri Begawan, BE1410, Brunei Darussalam. E-mail: miro@ubd.edu.bn

These haze episodes were largely due to the long-range transport of air pollutants from forest fires in Kalimantan and Sumatra, while the 1998 event was mainly due to local fires in Brunei, Sarawak and Sabah. During the 1994 and 1997 episodes pollutants were transported over distances of between 800 and 1600 km before reaching Brunei, while during the 1998 episode the sources of pollution were between 5 to 100 km from the sampling stations. While forest fires were also burning in Kalimantan in 1998, the plumes from these fires did not reach northern Borneo due to the prevailing wind direction which was generally northerly and northeasterly. The 1994 and 1997 fires occurred during July–October when the prevailing wind direct is southwesterly.

2. Methodology

Gases, airborne particles and rainwater were sampled during recent haze episodes in Brunei Darussalam. Instrumental methods were used to determine major gaseous air pollutants:

- SO_2 by pulsed fluorescence,
- NO and NO_2 by chemiluminescence,
- CO by infrared spectrometry,
- O_3 by ultra-violet absorption spectrometry.

SO_2, NO_2 and O_3 were also determined by means of diffusion tubes and absorption trains. Particulate matter less than 10 μm in diameter (PM_{10}) was determined using several different techniques:

- High volume (Hi-Vol) sampler,
- β-gauge automated particle sampler,
- Tapered Element Oscillating Microbalance (TEOM®),
- *Personal*DataRAM® based on nephelometry.

Standard NIOSH methods (NIOSH, 1994) were employed to determine a number of organic pollutants:

- Total petroleum hydrocarbons (TPH), C_6-C_{36}, excluding BTEX,
- Benzene, toluene, ethylbenzene, and xylene (BTEX),
- Polynuclear aromatic hydrocarbons (PAHs),
- Aldehydes including formaldehyde, acetaldehyde, propionaldehyde and butyraldehyde,
- Acetic acid,
- Cresol,
- Phenol.

Suspended particles were analysed for several components:

- C, H and N using a CHN analyser,
- S, Cl, K, Ca, Fe, Cu, and Zn using energy dispersive X-ray fluorescence spectroscopy (XRF),

- Mg, Na, As, Cd, Ni, V and Hg using inductively coupled plasma spectrometry (ICP).

Gases and particles were determined extensively at various sites throughout Brunei during the 1998 haze episode.

Rainwater was collected by means of bulk and automatic wet-only samplers during haze episodes of 1994, 1997 and 1998. Samples were immediately analysed for pH and conductivity and stored at 5°C for subsequent chemical analysis. Concentrations of Cl^-, NO_3^- and SO_4^{2-} were determined by ion chromatography (IC). Flame atomic absorption spectrometry (AAS) was employed to determine Na, Ca and Mg, while flame atomic emission spectrometry (AES) was used to determine K.

Location of monitoring stations is illustrated in Figure 1. A more detailed study of micropollutants was carried out in the Seria/Kuala Belait area, close to the location of major fires as indicated in the figure.

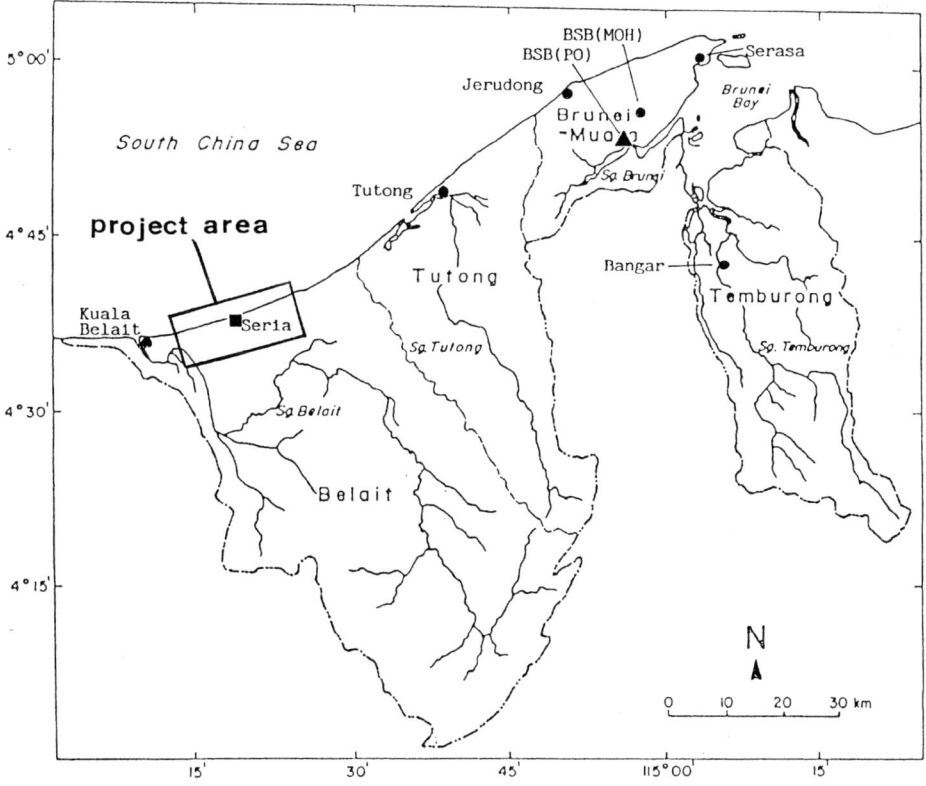

Figure 1

Map of Brunei Darussalam indicating location of monitoring stations. ▲ Fully automated AQMS. ■ BSP AQMS. ● TEOM® PM_{10} monitors. Location of project area where more detailed studies were carried out is indicated.

3. Results

Concentrations of PM_{10} and major gaseous pollutants (SO_2, NO, NO_2, CO and O_3) determined using instrumental methods in Brunei during the 1998 haze episode are summarised in Table 1. Similar concentrations of SO_2, NO_2 and O_3 were determined using diffusion tubes and absorption trains.

Concentration ranges of organic micropollutants are summarised in Table 2.

Analysis of suspended particles revealed that C was the major component, generally present at between 40 and 70% of the aerosol mass. The 24-h average PM_{10} concentrations at BSB during the course of the 1998 episode is illustrated in Figure 2. Also shown are the concentrations determined in Seria during the month of April 1998, when the haze was especially severe.

Table 1

Range of hourly concentrations of major air pollutants in Brunei Darussalam determined using instrumental methods. All concentrations in µg m^{-3} except CO in mg m^{-3}. Adapted from RADOJEVIC and HASSAN (1999)

Pollutant	PM_{10}	SO_2	NO	NO_2	O_3	CO
(a) Haze						
Range	1.2–999	0.78–87.3	6.8–97.2	5.8–99.1	5.1–99.9	1.2–21.9
Mean	109.9	7.27	19.9	41.5	63.2	4.20
(b) Non-haze						
Range	9.0–393	3.6–23.3	6.6–90.4	3.6–99.4	1.4–97.0	0.6–2.77
Mean	20.4	5.24	20.8	28.6	48.5	1.29

Table 2

Concentration ranges of organic micropollutants in haze during the 1998 haze episode in Brunei and recommended guideline values. Adapted from MURALEEDHARAN et al. (2000a)

Pollutant	Concentration (µg m^{-3})	Recommendation/ guideline (µg m^{-3})	Authority
TPHs	<0.1–307		–
Acetic acid	<0.1–19.4	500	Australia (Victoria)
Cresol	<0.1		
Formaldehyde	<1.0–21.6	120	WHO
Acetaldehyde	<1.0–15.9	76	Australia (Victoria)
Propionaldehyde	2.0–17.0	–	
Butyraldehyde	2.8–71.6	–	
Benzene	<0.1–24.8	1.2	US EPA (long-term goal)
Toluene	<0.1–15.5	260	EU (proposed limit)
Ethyl benzene	<0.1–2.01	20	Russia
Xylenes	<0.1–28.7	200	Russia
Phenol	<0.1–0.41	100	US EPA
PAHs (total)	1.0–33.8	0–0.1	USA
Naphthalene	0.5–29.8	1000	Australia (Victoria)

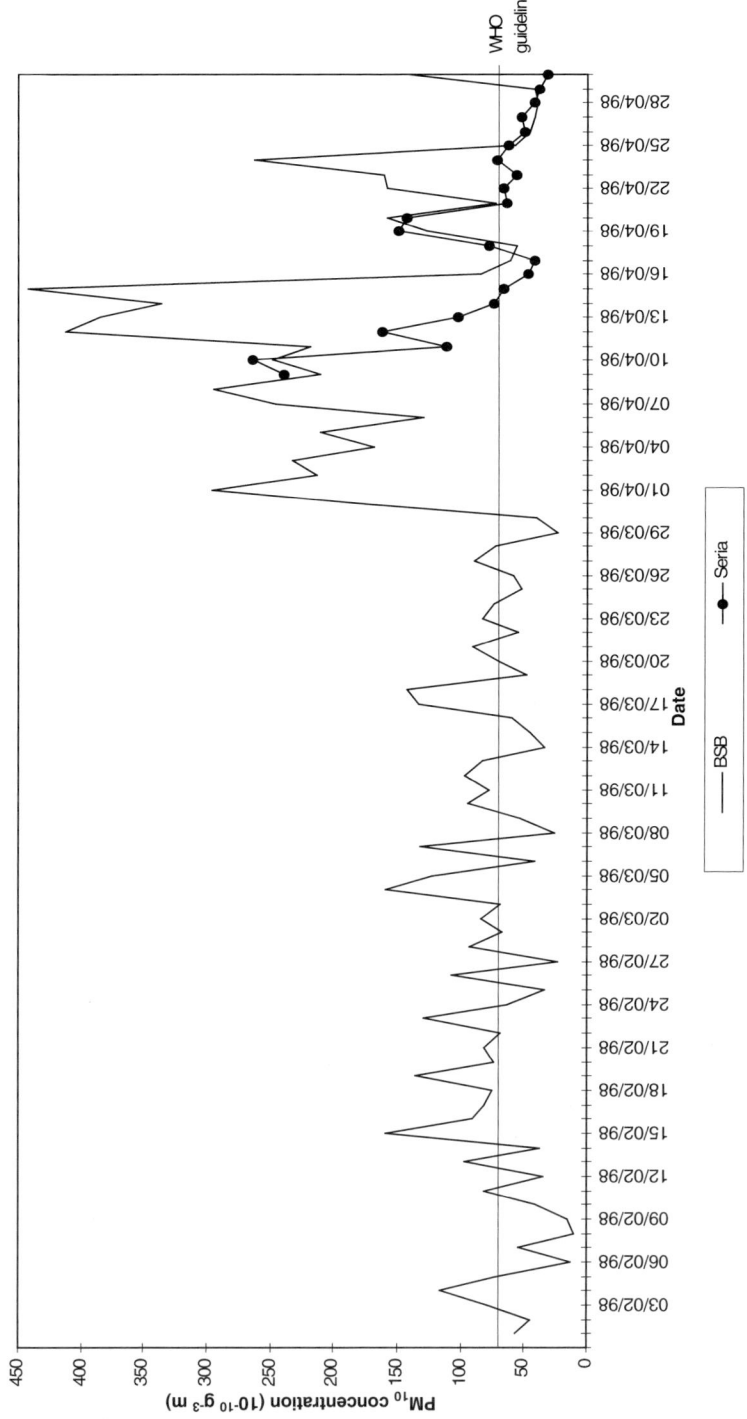

Figure 2

Concentrations of PM$_{10}$ (24-h average) during the haze episode of 1998 at Bandar Seri Begawan (BSB) and Seria.

There was significant diurnal variation in haze intensity, with concentrations of PM_{10} typically rising through the night to very high levels in the morning and thereafter decreasing due largely to meteorological factors (Fig. 3).

4. Discussion

4.1 Health Impacts

Symptoms and illnesses experienced by those exposed to the haze include: upper respiratory tract infection, asthma, conjunctivitis, bronchitis, eye and throat irritation, coughing, breathlessness, blocked and runny noses, skin rashes, and cardiovascular disorders. The criteria gaseous air pollutants were generally below accepted air quality standards (RADOJEVIC and HASSAN, 1999) and at the observed concentrations these are expected to have no harmful health effects. However, the WHO guideline for PM_{10} (24-hour average) was exceeded on 60% of the days during the entire haze episode (1 February–30 April 1998) at Bandar Seri Begawan (RADOJEVIC and HASSAN, 1999). The observed short-term health effects of the haze are due mainly to particulate matter. Particle size analysis of haze particles in Brunei revealed that more than 99% of particles were < 2.5 μm in diameter (MURALEE-DHARAN et al., 2000a) and these could significantly impact the respiratory system because they can penetrate deeper into the lungs than PM_{10}. No deaths to date have been attributed to the haze in Brunei but we will be examining mortality statistics.

Concentration ranges of organic micropollutants are summarised in Table 2, together with recommended guideline values where these are available. It should be

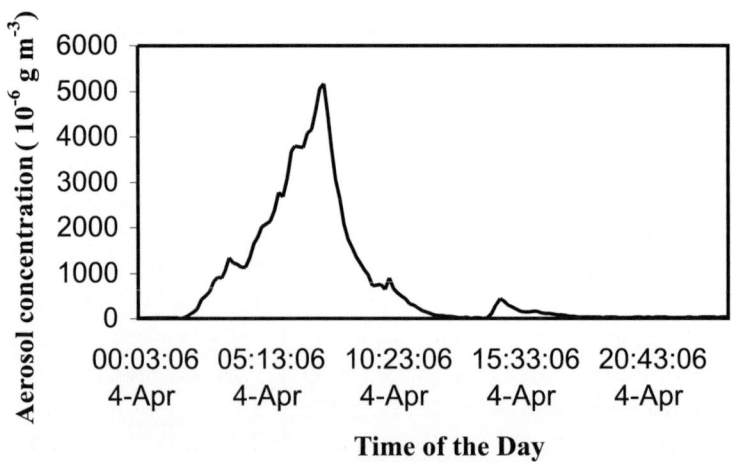

Figure 3
Diurnal variation in PM_{10} at Jerudong Park on 4 April 1998 determined with a *personal*DataRAM®.

noted that air quality standards do not exist for many of these compounds and the recommended values are based mainly on occupational exposure studies. There may be considerable uncertainty in extrapolating from occupational studies to the general population which also includes sensitive groups such as pregnant women, children, the elderly, and the sick. Although some of the organic substances identified in the haze are known or suspected carcinogens, mutagens, and teratogens (e.g., benzene, toluene, xylene, PAHs) with the potential to cause harm to human health, in the absence of epidemiological studies it is not possible to predict their long-term health effects.

Monthly outpatient data for some illnesses from the government hospital in BSB are summarised in Figure 4 during the haze period of February, March and April, together with data for January and May when haze was absent. The data were classified by age. The peak in the incidence of some of these diseases, notably acute respiratory infection, was during the months of haze: February, March and April. With the possible exception of acute respiratory infection, the hospital data show no clear trends of any effects. However, to date only limited data have been examined and it is necessary to consider longer term for all the diseases.

The psychological aspects of the haze, although often ignored, are also an important medical consideration. The haze dramatically affects people's lifestyle by severely restricting normal everyday activities. These effects, in combination with the physiological symptoms and discomfort, can cause considerable stress. The psychological impacts could act to exacerbate the physiological effects in certain individuals.

It is not clear to what extent publicity of the haze may have influenced hospital attendance. During the haze period, citizens were advised by the health authorities to attend outpatient clinics if they experienced any symptoms. Individuals suffering from symptoms unrelated to haze may have attended clinics during the haze period, when normally they would not have done so.

4.2 Impacts on Acid Deposition

We have previously noted the occurrence of acidic precipitation in Borneo similar to that observed in other tropical regions (RADOJEVIC and LIM, 1995). Although intense fires burn during dry periods when rainfall amounts are unusually low, precipitation samples were collected and analysed during the 1994, 1997, and 1998 episodes. All samples were analysed for pH and conductivity, but only certain samples were analysed for ionic content. We have complete data on the pH of rainwater during the three haze episodes and these showed no significant acidifying effect (RADOJEVIC and TAN, 1999). Results for samples in which ionic concentrations were determined are shown in Table 3. Also given in Table 3 is the total monthly deposition for September 1994 and February 1998, periods when Brunei was affected by haze. Although rain fell on 25 February 1998, this was only a trace (0.2 mm) and could not be analysed. No rain fell during the severe forest fire and haze events in March 1998.

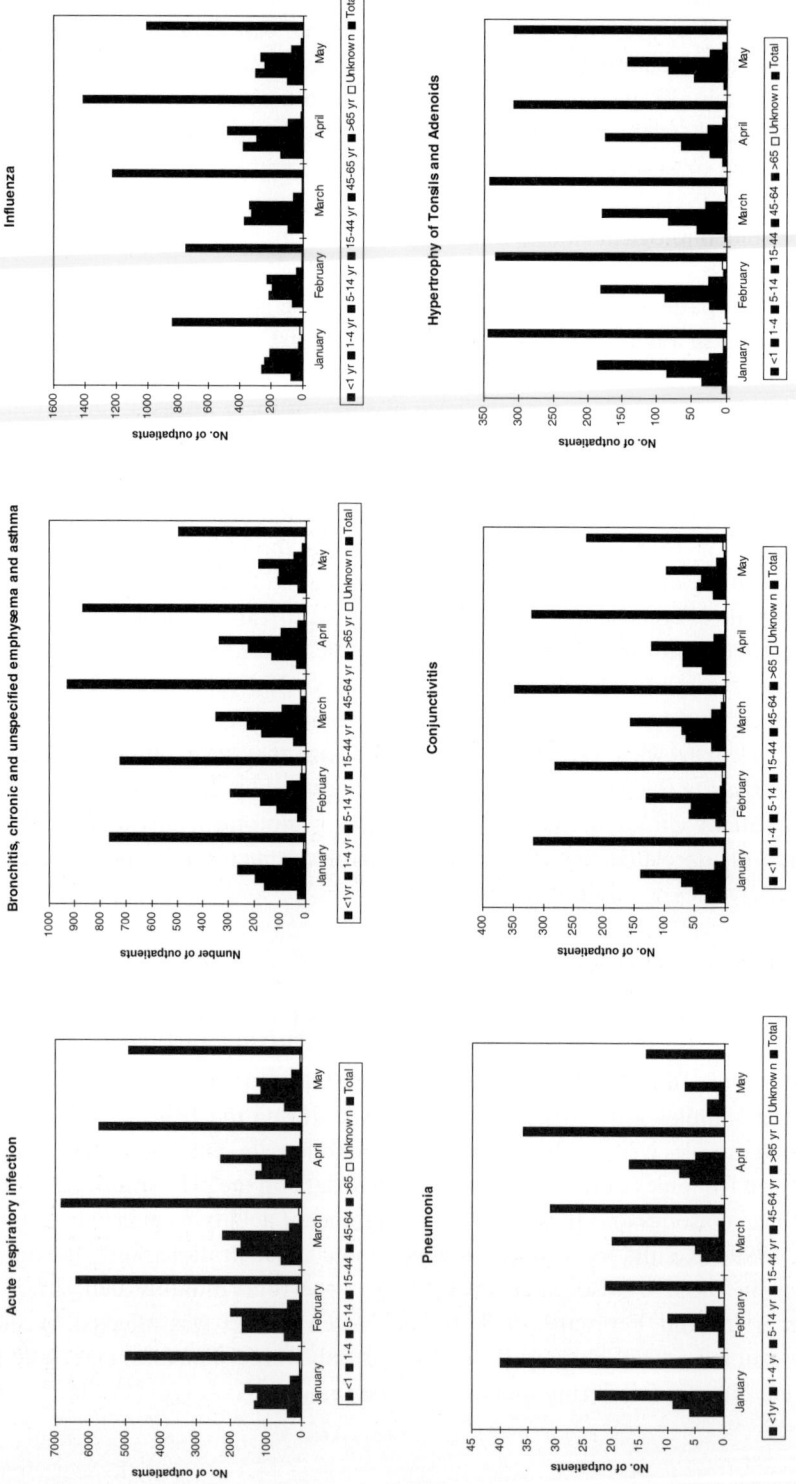

Figure 4

Monthly outpatient morbidity by age. Data for the main government hospital (RIPAS) in BSB, 1998.

Table 3

Rainwater composition during episodes of biomass fires and haze in Brunei Darussalam in 1994 and February 1998. Concentrations in μeq L⁻¹. Conductivity in μmhos cm⁻¹. Deposition in μeq m⁻². A–E = different sites. VWM = volume weighted mean. Mean and VWM of pH calculated from mean and VWM of H^+. (a) Non-haze pH: 228 samples, other measurements: 28–44 samples. (b) Arithmetic average of conductivity. * August 1994 deposition excluding haze days

Date	mm rain	pH	Cond.	Cl⁻	NO₃⁻	SO₄²⁻	Na⁺	K⁺	Mg²⁺	Ca²⁺	H⁺
10–13/8/94	32.5	5.03	3.4	7.62	2.23	5.58	3.70	0.69	1.07	4.14	9.33
23–27/8/94	2.8	4.96	14.4	21.72	32.68	31.75	20.87	5.32	15.64	61.10	10.96
3–7/9/94	34.4	4.48	8.8	9.59	15.84	24.27	10.00	3.63	4.94	22.94	33.11
7–13/9/94	8.3	4.35	22.9	114.19	41.53	58.77	2.52	6.14	10.70	43.14	44.67
14–26/9/94	32.2	4.81	17.0	24.26	18.21	36.33	23.70	4.35	7.16	17.46	15.49
26–28/9/94	9.7	4.85	16.5	18.05	23.39	53.02	22.83	3.52	5.25	31.17	14.12
28/9–1/10/94	19.7	4.64	20.0	20.12	22.51	18.04	19.35	1.08	2.95	17.71	22.91
1–4/10/94	4.4	4.88	16.0	16.92	31.16	40.38	14.15	4.51	5.52	45.88	13.18
Mean	–	4.69	16.0	29.06	23.44	33.52	14.64	3.65	6.78	30.44	20.47
VWM	–	4.68	14.9	20.92	16.99	26.46	14.12	3.89	6.40	19.92	20.76
22/2/98 A	20.6	5.52	19.5	14.92	13.31	2.21	35.43	9.36	20.00	21.80	3.02
22/2/98 B	32.4	5.44	16.1	10.29	6.47	5.38	12.56	26.19	29.38	12.82	3.63
22/2/98 C	4.2	5.81	55.4	23.22	3.31	6.44	10.26	2.53	66.17	8.88	1.55
22/2/98 D	18.0	5.13	6.7	3.81	32.44	12.00	18.30	13.53	2.55	23.64	7.41
22/2/98 E	16.0	5.01	17.2	10.94	32.48	8.02	64.04	39.13	23.62	127.88	9.77
Mean	–	5.29	22.98	12.64	17.6	6.81	28.12	18.15	28.35	39.00	5.08
VWM	–	5.28	–	10.77	17.56	6.48	27.79	21.07	22.65	36.99	5.22
Non-haze VWM (a)	–	4.93	9.65 (b)	14.2	7.6	11.5	10.3	5.2	3.4	14.14	11.70
Haze 94 VWM/non-haze VWM	–	0.95	1.54	1.47	2.24	2.30	1.37	0.75	1.88	1.41	1.77
Haze 98 VWM/non-haze VWM	–	1.07	2.38	0.76	2.32	0.56	2.70	4.06	6.60	2.62	0.45
Deposition Sept. 94 (haze)	–	–	–	2630	2146	3362	1730	371	598	2360	2597
Deposition Feb. 98 (haze)	–	–	–	196	320	118	507	384	413	674	95
Deposition May 94 (non-haze)	–	–	–	731	861	1256	959	261	389	1712	475
Deposition June 94 (non-haze)	–	–	–	5477	1017	2260	3278	336	991	2295	2025
Deposition July 94 (non-haze)	–	–	–	1402	2044	2894	1566	157	727	2039	2252
Deposition August 94 (non-haze) *	–	–	–	843	1256	1550	1627	275	359	991.6	3679

In addition, samples were collected in the town of Seria, closer to the area of major fires, on several days in April 1998. Only the pH was determined in these samples, with the exception of one sample collected on 12 April 1998 with a very high pH of 6.4 which was analysed for several parameters (conductivity = 58.5 μmhos cm^{-1}; other measured ions in μeq L^{-1}: Cl$^-$ = 180.5, SO$_4^{2-}$ = 10.4, HCO$_3^-$ = 280.3, and Ca^{2+} = 269.7). The Ca^{2+} concentration was considerably higher than those shown in Table 3. The high Ca^{2+} and HCO$_3^-$ concentrations are not unusual given the high pH of this sample.

From the data in Table 3 it is apparent that the conductivity and concentrations of most ions in rainwater were higher during periods of intense biomass fires and haze than during periods when there were no significant forest fires. Monthly deposition of various ions for some non-haze periods is shown in Table 3 for comparison. More H$^+$ was deposited during the non-haze period of August 1994 than during the haze-affected period of September 1994. None of these results indicates significant impacts of biomass burning on precipitation pH or the deposition of H$^+$ or other ions in rainwater. Also shown in Table 3 are the enrichment factors (i.e., ratio of concentrations in rainwater during haze/concentrations in the absence of haze) in rain during the 1994 and 1998 haze episodes (February 1998 data only). An interesting observation is that during the 1994 episode there was no enrichment of K, which is often considered to be an indicator of biomass burning, unlike the February 1998 period when significant enrichment of K was noted. Since K is present in particulate matter, much of it could have been deposited out of the plumes before the air mass reached Brunei from regions of forest fires in Indonesia in 1994. The high enrichment of Na, K, Mg and Ca during 1998 is indicative of biomass burning (WARD and HARDY, 1991) and this is expected in view of the dominance of local fires. Comparison of other ionic ratios and correlations also revealed significant differences between rainwater composition during September 1994, February 1998, and in the absence of haze, however, the number of data points is insufficiently high to justify a detailed statistical analysis.

We have a complete annual data set of pH measurements from July 1995 to July 1996 when there was no significant biomass burning or haze in Borneo. The total annual deposition of H$^+$ was 0.046 eq m^{-2}. Taking this value as a reference, the contribution of H$^+$ deposition during periods of biomass burning to the annual deposition of H$^+$ was calculated (Table 4). The monthly contributions during episodes of biomass fires and haze to the total annual deposition varied between 0.0 and 5.35%. Even assuming that all of the H$^+$ during haze episodes was due to biomass fires (which it obviously is not) the results suggest that forest fires are a minor source of acidity in wet deposition.

Although biomass burning produces many acidifying compounds, for example SO$_2$, NO$_2$, and organic acids such as formic and acetic acid, it also produces ions such as Ca^{2+}, Mg^{2+} and K$^+$, which act to neutralise the acidity. Furthermore,

Table 4

*Deposition of H^+ in Brunei Darussalam during episodes of forest fires and haze and percentage of contribution to total annual deposition of H^+. * mean of 5 samples collected at different sites on the same date. (a) reference annual H^+ deposition taken as 0.046 eq m^{-2} determined during a one year period (1995– 1996) in the absence of forest fires and haze. Adapted from RADOJEVIC and TAN (1999)*

Haze period	Deposition of H^+ (μeq m^{-2})	% of total annual deposition of H^+ (a)
August 1994	334	0.73
September 1994	2460	5.35
October 1994	58	0.13
Total 1994 episode	2852	6.21
August 1997	459	1.00
September 1997	1830	3.98
October 1997	853	1.85
Total 1997 episode	3142	6.83
February 1998*	95	0.21
March 1998	0	0.00
April 1998	141	0.31
Total 1998 episode	236	0.52

biomass fires and haze generally occur during dry periods since heavy rains tend to extinguish the fires. Hence, even though the ionic content of individual rainwater samples may be high during haze periods, the net deposition rates tend to be low simply because rainfall amounts are so low. Most deposition takes place during the wet season when precipitation amounts are exceedingly substantial.

4.3 PM_{10} Personal Exposure Monitoring

The traditional approach to assessing the health impacts of air pollution has been to compare measurements from fixed outdoor ambient air quality monitoring stations (AQMSs) with air quality standards and guidelines. While this is quite a convenient and basic approach it is not representative of personal exposure because most people spend more time indoors than outdoors, and even more so during haze episodes when normal everyday activity is severely restricted. During the haze episode of 1998 we carried out extensive PM_{10} personal exposure monitoring using portable PM_{10} monitors (MURALEEDHARAN and RADOJEVIC, 2000). Exposure of different individuals (office workers, drivers, etc.) was monitored inside homes, offices, classrooms, motor vehicles, helicopters and outdoors during the course of various activities (sleep, work, fire-fighting, seismic surveying, oil spill inspection, etc.).

A PM_{10} personal exposure profile of a driver is shown in Figure 5. Considerable differences were observed between individual exposure patterns depending on the activity, location and time. Outdoor concentrations were considerably higher than indoor concentrations. Concentrations of PM_{10} inside motor vehicles were generally

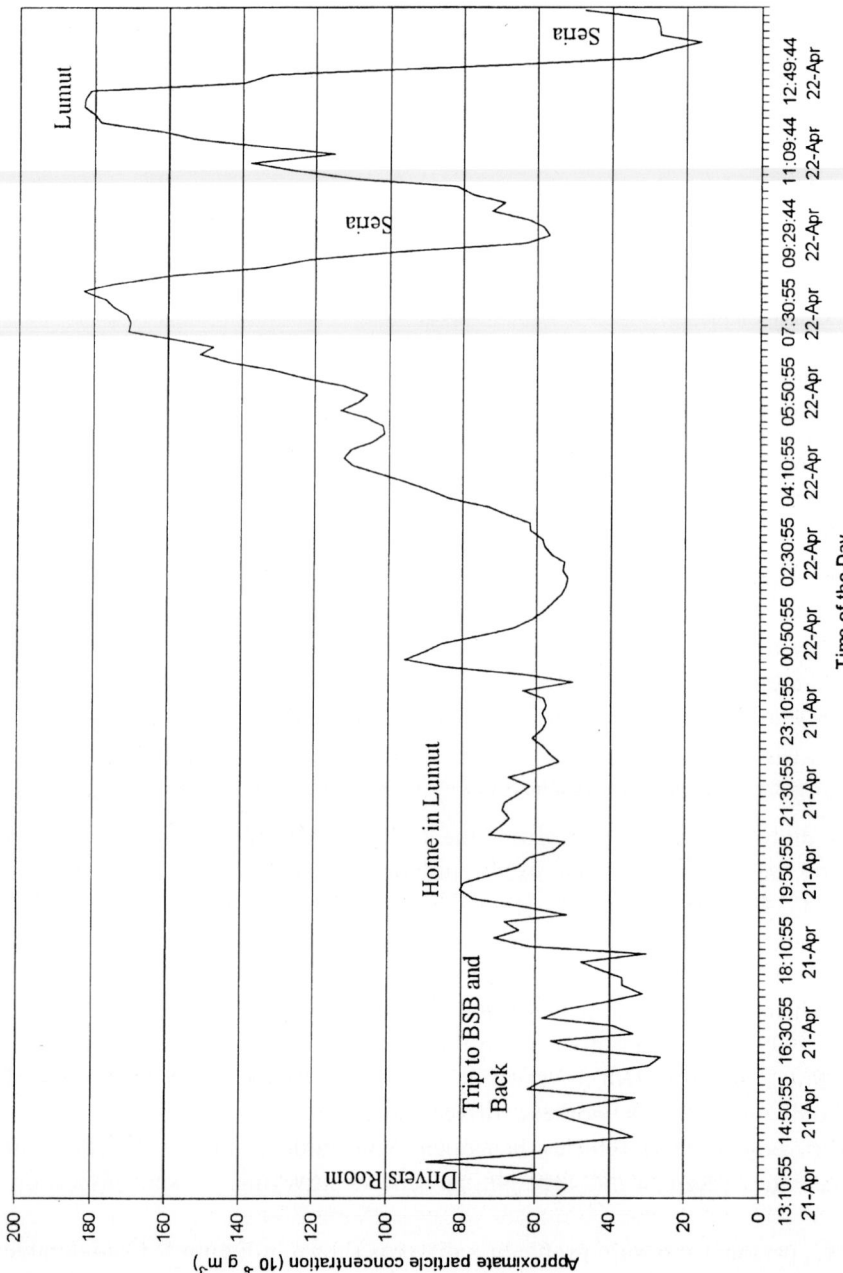

Figure 5
Personal exposure profiles of a driver for PM₁₀.

Table 5

Emission factors (in g kg⁻¹ peat) based on peat combustion experiments. Adapted from MURALEEDHARAN et al. (2000b)

Substance	Temperature (°C)	
	480	600
Moisture	514	514
Loss-on-ignition	750	770
Ash	250	230
CH_4	5.78	11.34
C_2H_6	1.32	3.12
C_3H_8	1.96	0.011
C_4H_{10}	0.875	0.203
C_2H_4	0.721	0.113
CO	37.13	15.28
CO_2	185.00	149.59

about one half of outdoor levels (MURALEEDHARAN and RADOJEVIC, 2000). Monitoring of PM_{10} during helicopter flights revealed higher concentrations while in flight than on the ground, probably due to higher PM_{10} levels aloft. Personal PM_{10} monitoring also revealed the effectiveness of air purifiers and air conditioners in minimising indoor particle concentrations.

4.4 Emissions from Peat Combustion

Emissions of volatile compounds from peat combustion were studied in laboratory experiments (MURALEEDHARAN et al., 2000b). Milligram quantities of peat were burned inside silica or pyrex tubes at 480 and 600°C, temperatures typical of smouldering combustion. The emission factors determined in these experiments are summarised in Table 5. Concentrations of aldehydes and PAHs in the emission products were below the instrumental detection limits.

5. Conclusions

This paper presents data collected during the 1998 haze episode in Brunei Darussalam. The main conclusions of the research are:
- The only criteria pollutant of significance was PM_{10},
- The PM_{10} mass was dominated by the $PM_{2.5}$ fraction,
- particulate matter was dominated by the carbon component,
- there was no significant impact of the haze episode on rainwater acidity or the deposition of H^+ ions,
- personal exposure to PM_{10} was dominated by an individual's location and activity, and

- the impacts on public health are inconclusive due to the limited data that were examined.

Acknowledgements

Thanks are extended to the Meteorological Service, Department of Civil Aviation, Ministry of Communications, Government of Brunei; The Ministry of Health, Government of Brunei; Dr. T.R. Muraleedharan, formerly Head of Environmental Studies at Brunei Shell Petroleum Co. Sdn Bhd; Allan Waugh, Watson Hawksley Asia; Anthony Caruana, formerly of Consil-Global Environment Group, Australia; and the staff and students of the University of Brunei Darussalam who contributed to the research.

References

MURALEEDHARAN, T. R. and RADOJEVIC, M. (2000), *Personal Particle Exposure Monitoring Using Nephelometry during Haze in Brunei*, Atmos. Environ. *34*, 2733–2738.
MURALEEDHARAN, T. R., RADOJEVIC, M., WAUGH, A., and CARUANA, A. (2000a), *Chemical characterisation of the haze in Brunei Darussalam during the 1998 episode*, Atmos. Environ. *34*, 2725–2731.
MURALEEDHARAN, T. R., RADOJEVIC, M., WAUGH, A., and CARUANA, A. (2000b), *Emissions from the Combustion of Peat: An Experimental Study*, Atmos. Environ. *34*, 3033–3035.
NIOSH (1994), *NIOSH Manual of Analytical Methods (NMAM)*, 4th edition, US Department of Health, Education, and Welfare.
RADOJEVIC, M. (1997), *The Haze in Southeast Asia*, Environ. Scientist *6* (6), 1–3.
RADOJEVIC, M. (1998), *Burning issues*, Chemistry in Britain *34*(12), 38–42.
RADOJEVIC, M. and HASSAN, H. (1999), *Air Quality in Brunei Darussalam during the 1998 Haze Episode*, Atmos. Environ. *33*, 3651–3658.
RADOJEVIC, M. and LIM, L. H. (1995), *A Rain Acidity Study in Brunei Darussalam*, Water, Air and Soil Pollution *85*, 2369–2374.
RADOJEVIC, M. and TAN, K. S. (1999), *Impacts of Biomass Burning and Regional Haze on the pH of Rainwater in Brunei Darussalam*, Atmos. Environ. *34*, 2739–2744.
WARD, D. E. and HARDY, C. C. (1991), *Smoke Emissions from Wildland Fires*, Environ. Internat. *17*, 117–134.

(Received March 1, 2000, accepted August 15, 2000)

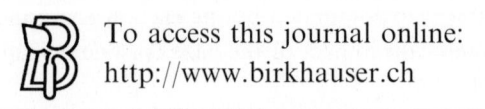

To access this journal online:
http://www.birkhauser.ch

Pure appl. geophys. 160 (2003) 265–277
0033–4553/03/020265–13

© Birkhäuser Verlag, Basel, 2003

Visibility and Incidence of Respiratory Diseases During the 1998 Haze Episode in Brunei Darussalam

ANIL KUMAR YADAV,[1,*] KRISHAN KUMAR,[1] AWG MAKARIMI BIN HJ AWG KASIM,[2]
M. P. SINGH,[3,*] S. K. PARIDA,[4] and MAITHILI SHARAN[5]

Abstract — Air pollution episodes as a result of forest fires in Brunei Darussalam and neighbouring regions have reached hazardous levels in recent years. Such episodes are generally associated with poor visibility and air quality conditions. In the present study, data on PM_{10} (particulate matter of size less than 10 microns) and CO in Brunei Darussalam have been considered to study the incidence of respiratory diseases whereas data on relative humidity (RH) in addition to PM_{10} have been used to explain the visibility with a particular emphasis on haze episode during 1998.

Initial exploratory analysis indicates significant correlation of visibility with PM_{10} and RH. An attempt has been made to explain visibility on the basis of PM_{10} and RH using multiple linear regression analysis. The regression model shows that PM_{10} and RH are two significant factors affecting the visibility at a given site. Further, canonical correlation, a multivariate method of analysis, has been used to explain the incidence of respiratory diseases as a function of air quality during the haze period. The results indicate that PM_{10} and CO levels during the haze period have a significant bearing on the incidence of respiratory diseases (Asthma, Acute Respiratory Infections and Influenza (ARII)).

Key words: Haze, air pollution, visibility, respiratory diseases, PM_{10}, canonical correlation.

Introduction

Intense forest fires during February–April 1998 in Brunei Darussalam located on the NE part of Borneo Island; 5°N, 115°E and neighbouring regions led to the emission of unusually high concentration of particulates, resulting in prolonged haze formation over the entire southeast Asian region (see Fig. 1). The regional haze in SE Asia is a recurring air pollution problem (NICHOL, 1997; RADOJEVIC, 1997; RADOJEVIC and HASSAN, 1999). The resultant haze reduced visibility to hazardous

[1] Guru Jambheshwar University, Hisar-125 001, Haryana, India. E-mail: aky68@yahoo.com
[*] Current affiliation: Ansal Institute of Technology, Gurgaon-122003 India
[2] University Brunei Darussalam, Brunei Darussalam.
[3] Visiting Consultant, University Brunei Darussalam, Brunei Darussalam.
[4] Ministry of Health, Brunei Darussalam.
[5] Centre for Atmospheric Sciences, Indian Institute of Technology, Delhi-110 016, India.

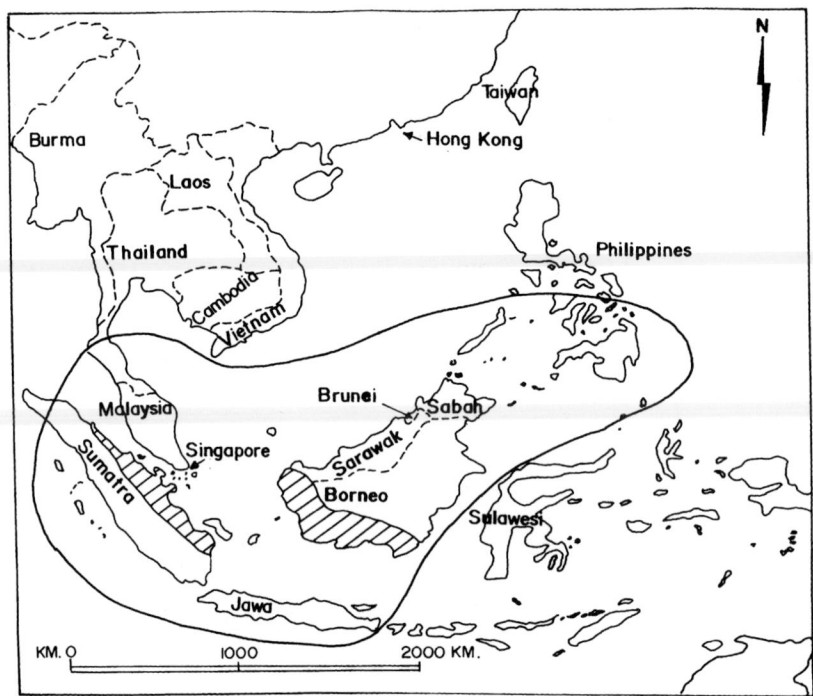

Figure 1
Region in Southeast Asia affected by haze in 1998.

levels and affected the health of the exposed population adversely. A reasonably good background of earlier studies on haze-related aspects such as location of sources, effects on health and general day-to-day life has been provided in a recent study by RADOJEVIC and HASSAN (1999). They have analysed the air quality data of Brunei Darussalam collected in the recent past. Their analysis included overall air quality (SO_2, NO, NO_2, CO, O_3 and PM_{10}), diurnal variation of these pollutants and PSI (Pollution Standards Index) vis-à-vis haze. PSI is an index developed by the US Environmental Protection Agency on the basis of epidemiological studies carried out in several US cities. It provides a simplified system of reporting air quality status of a place to a layman. PSI of a pollutant converts daily measured concentrations to a number on a scale of 0 to 500 through an empirical formulation. Using this technique a nomogram (Fig. 2) is prepared which classifies PSI into five different categories, namely Good, Moderate, Unhealthy, Very Unhealthy and Hazardous. Further information on PSI may be found in SINGH and HASSAN (1998). On the basis of air quality data analysis during the 1998 haze episode, RADOJEVIC and HASSAN (1999) have reported that only particulate matter is a significant pollutant. They have looked at 1 h, 8 h and 24 h average concentrations of PM_{10} and reported a number of exceedances with reference to WHO guidelines.

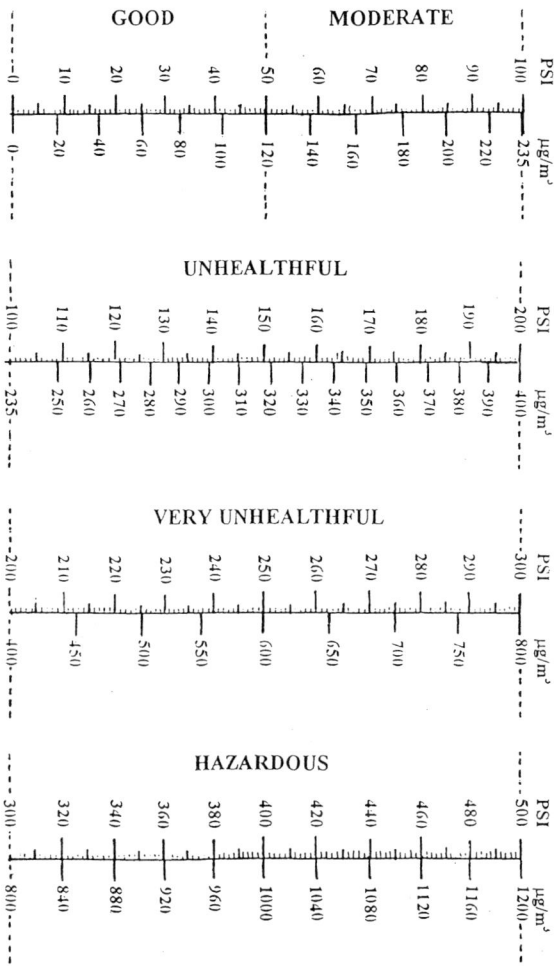

Figure 2
Pollution Standards Index (PSI) Nomogram for PM$_{10}$ (24-hour average) [Based on US EPA Guidelines]
(SINGH and HASSAN, 1998).

Though most studies on haze, to the best of our knowledge, have followed a somewhat qualitative approach regarding its effects on visibility and human health, no attempt has been made relative to statistical verification of some of their conclusions. Further, the studies mentioned above have focused only on PM$_{10}$ as the sole pollutant affecting the general air quality. The present study attempts to bridge this gap by doing a quantitative analysis of haze related hazards. Two vital statistical techniques – multiple linear regression analysis and canonical correlation analysis have been applied to explain the visibility and the incidence of respiratory diseases, respectively.

Data Description

In the present study, hourly data on PM_{10}, RH and visibility from the airport in Brunei Darussalam have been considered for the period April 7 to December 31, 1998, which includes a part of the haze episode of 1998. Though the haze period extended from Feb. to May 1998, PM_{10} data for the airport were available only from April 7 onwards. Data on PM_{10} from another site (BSB Post Office) were also available but have certain drawbacks such as missing values and a value repeating consecutively on some occasions. Therefore, the data from this site have not been considered for the present study. Hourly data of CO were also considered during the haze period. Data on the number of cases reported for various respiratory diseases, namely Asthma (ASTH), Acute Respiratory Infections and Influenza (ARII) in Brunei during this period were also collected from OPD records of Civil Hospitals of Brunei.

Time Behaviour of PM_{10}

PM_{10} being the most significant pollutant contributing to haze formation, its time traces must be examined before any further analysis. Figure 3 gives the time series plot of PM_{10} for the entire period considered in this study. The trends of PM_{10} values

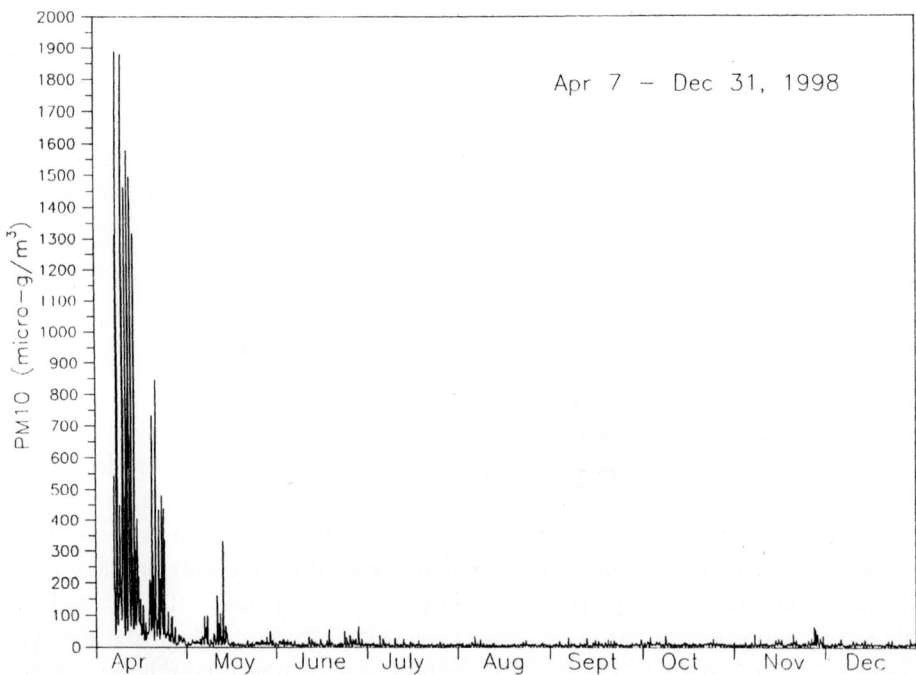

Figure 3
Time series plot of PM_{10} at the airport in Brunei Darussalam during April–Dec. 1998.

from Figure 3 indicate clearly the occurrence of unusually high PM_{10} concentration during the period April–May 1998. Based on the PSI nomogram for PM_{10} (Fig. 2), it appears pertinent to focus on PM_{10} concentrations above 50 µg/m^3. Therefore, in our analysis, we have identified the interval from April 7 to May 15 as a sample of the "haze" period and the remaining portion as "non-haze" period. It may be relevant here to mention that the diurnal behaviour of PM_{10} concentration for the haze period (Fig. 4a) (i.e., April 7th to May 15th 1998) is quite different from that during the non-haze period (Fig. 4b). In the haze period, hourly PM_{10} concentration starts building up at about 2 am onwards and peaks at about 8 am. Thereafter, it starts declining and reaches a steady minimum at about 3 pm. The build-up of PM_{10} during night in the non-haze months is never so pronounced (very small). The reduced levels of PM_{10} during the day can be explained on the basis of diurnal plots of horizontal wind speed and mixing height. A typical plot of diurnal variation of the ventilation coefficient, a product of wind speed and mixing height, shows behaviour complementary to Figure 4b (MAKARIMI et al., 1999). This offers a plausible explanation for the observed diurnal behaviour of PM_{10}.

Data Analysis

Visibility

Initial exploratory analysis indicates significant negative correlation of visibility (km) with PM_{10} (µg/m^3) and relative humidity (RH in percentage) (Table 1). The values of the Pearson's correlation coefficient are −0.31 and −0.58 with PM_{10} and RH, respectively. This provides the basis to explain visibility (VIS) in terms of PM_{10} and RH using a multiple linear regression model of the following form:

$$VIS = \beta_0 + \beta_1 PM_{10} + \beta_2 RH + e , \tag{1}$$

where e is the error (residual) term and β_0, β_1 and β_2 are the regression coefficients.

The regression model (Table 2) shows that PM_{10} and RH are two significant factors affecting visibility at a given site.

The equation representing the regression model may be expressed as

$$VIS = 42.28 - .0195\,PM_{10} - .323\,RH + e . \tag{2}$$

The results of ANalysis Of VAriance (ANOVA*) (Table 3) for the regression model indicate the goodness of fit for the overall model. The R^2 (adj.) value obtained reveals

---

* Analysis of variance may be defined as a technique whereby the total variation present in a set of data is partitioned into several components (DANIEL, 1991; MOORE and COBBY, 1998). In regression analysis, ANOVA partitions the total variation present in the data into two components: variation explained by the model and the residual variation.

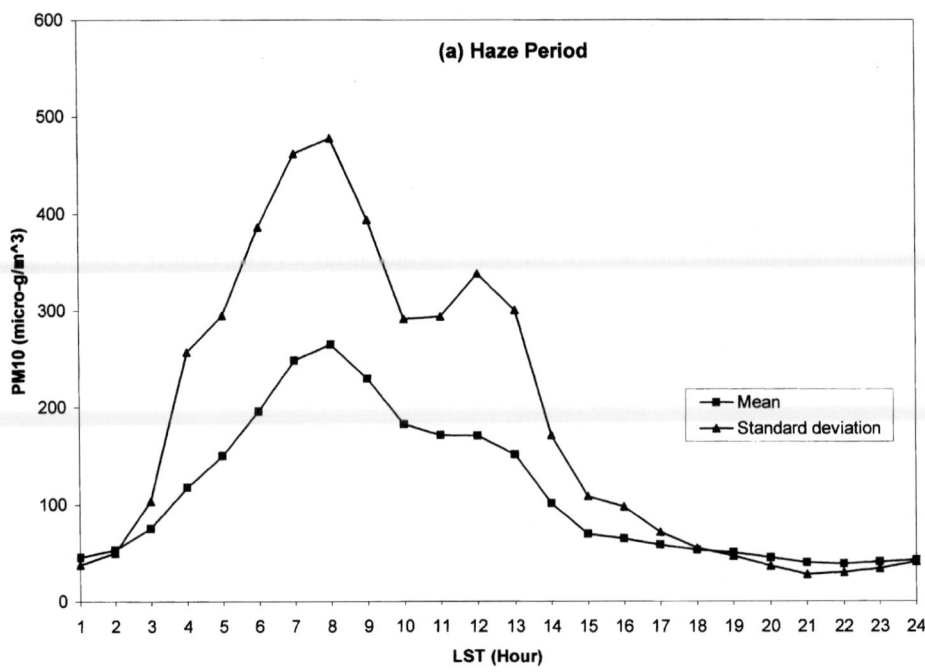

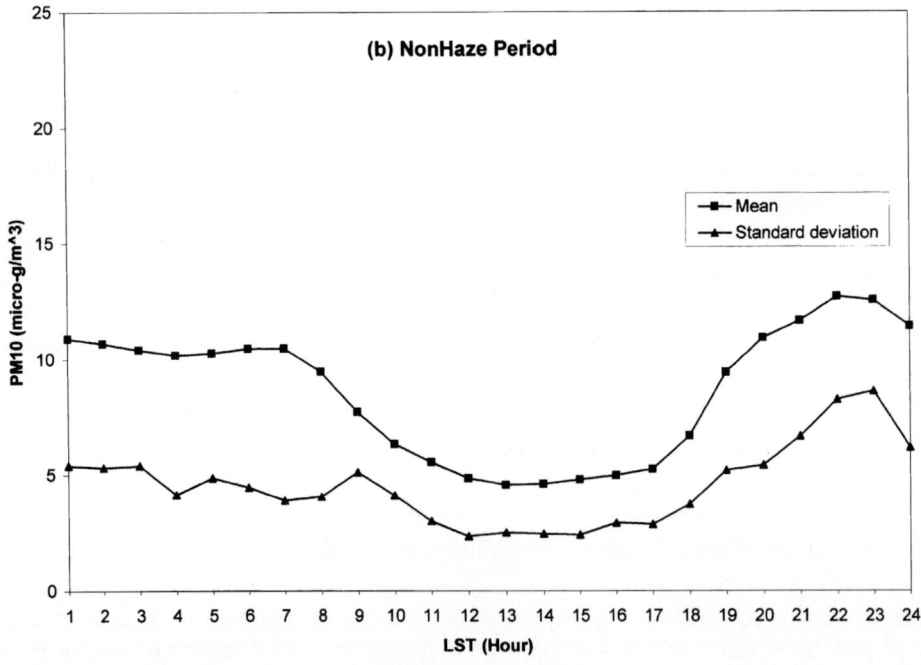

Figure 4
Diurnal variation of mean PM_{10} values and the standard deviations regarding these during (a) haze, and
(b) non-haze period in 1998.

Table 1

Correlation matrix (bracketed values indicate significance levels[#])

	Visibility	PM_{10}	RH
Visibility	1.000	−0.31 (.000)	−0.58 (.000)
PM_{10}		1.000	−0.007 (.598)
RH			1.000

[#] Significance levels are values that help accept/reject a hypothesis about a statistical parameter. A value close to zero (generally less than 0.05) indicates that the computed statistical parameter is significant.

Table 2

Results of the regression model

Variable	Coefficient	Standard error	t-value	Significance level
Constant	42.28	0.465	90.898	0.0000
PM_{10}	−0.0195	0.001	−31.500	0.0000
RH	−0.323	0.006	−58.521	0.0000

Table 3

ANOVA for the model above

Model	Sum of squares	Degrees of freedom	Mean sum of squares	F-ratio	Significance level
Regression	120468.0	2	60233.994	2198.224	0.0000
Residual	156159.5	5699	27.401		
Total	276627.5	5701			

R^2 *(adj.)* = .435

that the model is able to explain about 44 percent variation in the visibility. This performance of the model, though reasonably good, points to the fact that PM_{10} and RH may not be the sole factors affecting the visibility of a place. Factors such as the concentration of coarser particles, mist, sun reflection etc. may also be influential in determining the visibility of a place. However, it was not possible to include these factors in the present study due to lack of information about them.

Incidence of Diseases

Significantly high correlations of PM_{10} (Table 4) are obtained with the data regarding the number of cases reported for most of the respiratory diseases during the haze period at a time lag of one day. This only confirms the time period of 24 hour taken into account for obtaining PSI values of PM_{10} (SINGH and HASSAN, 1998).

Table 4

Correlations of PM$_{10}$ with respiratory diseases at various lags during haze period. Significance levels are indicated in the brackets

Disease	Lag 0 day	Lag 1 day	Lag 2 days
ASTH	.489 (.002)	.788 (.000)	.654 (.000)
ARII	.521 (.001)	.816 (.000)	.445 (.005)

A look at the scatter plot of PM$_{10}$ with various diseases (Figs. 5a and b) at lag of one day for the entire study period, on the other hand, indicates that PM$_{10}$'s role in affecting the incidence of various respiratory diseases becomes significant as its concentrations rise above the background levels. Low PM$_{10}$ values (< 50 µg/m^3) are associated with substantial variation (scatter) in the incidence of diseases, thus indicating the insignificance of PM$_{10}$ and involvement of other factors in causing these diseases. In other words, PM$_{10}$ is an inappropriate variable to explain the incidence of diseases when its concentration is at background levels. Its correlation with the diseases improves as its concentration increases. In econometric parlance, such behaviour is termed as the presence of heteroscedasticity in the data. PM$_{10}$ values more than 50 µg/m^3 are associated with relatively much less variation (scatter) in the incidence of diseases, thus indicating that as PM$_{10}$ values increase beyond 50 µg/m^3, it becomes a dominating factor in explaining the occurrence of diseases like asthma.

However, PM$_{10}$ is not the only pollutant that affects human health (respiratory illness) adversely. RADOJEVIC and HASSAN (1999) reported occasions when even CO concentrations exceeded WHO standards during the haze period. Looking at the monthly mean of maximum rolling PSI values of pollutants (see Table 5) at post office BSB from Oct. 1997 to April 1998 in Brunei Darussalam as reported by SINGH and HASSAN (1998), we can see a sudden rise after Jan 1998 only in PM$_{10}$ and CO while others show no significant change. Recalling that haze in 1998 in SE Asia extended from Feb. to May, it is pertinent to consider CO in addition to PM$_{10}$ during haze period. Since CO concentration reduces the oxygen carrying capacity of blood, it affects the respiratory health adversely. To ascertain the impact of CO in addition to PM$_{10}$ in causing respiratory illness, the technique of canonical correlation, a multivariate method of analysis, has been applied. This is particularly relevant to the present study since the number of dependent variables exceeds one. The variables involved can be broadly grouped into two categories: a set of dependent variables (ASTH and ARII) and a set of independent variables (PM$_{10}$ and CO).

An attempt has been made to explain the set of dependent variables ASTH and ARII (Asthma and Acute Respiratory Infections and Influenza, respectively) in terms of the set of independent variables (PM$_{10}$ and CO) using canonical

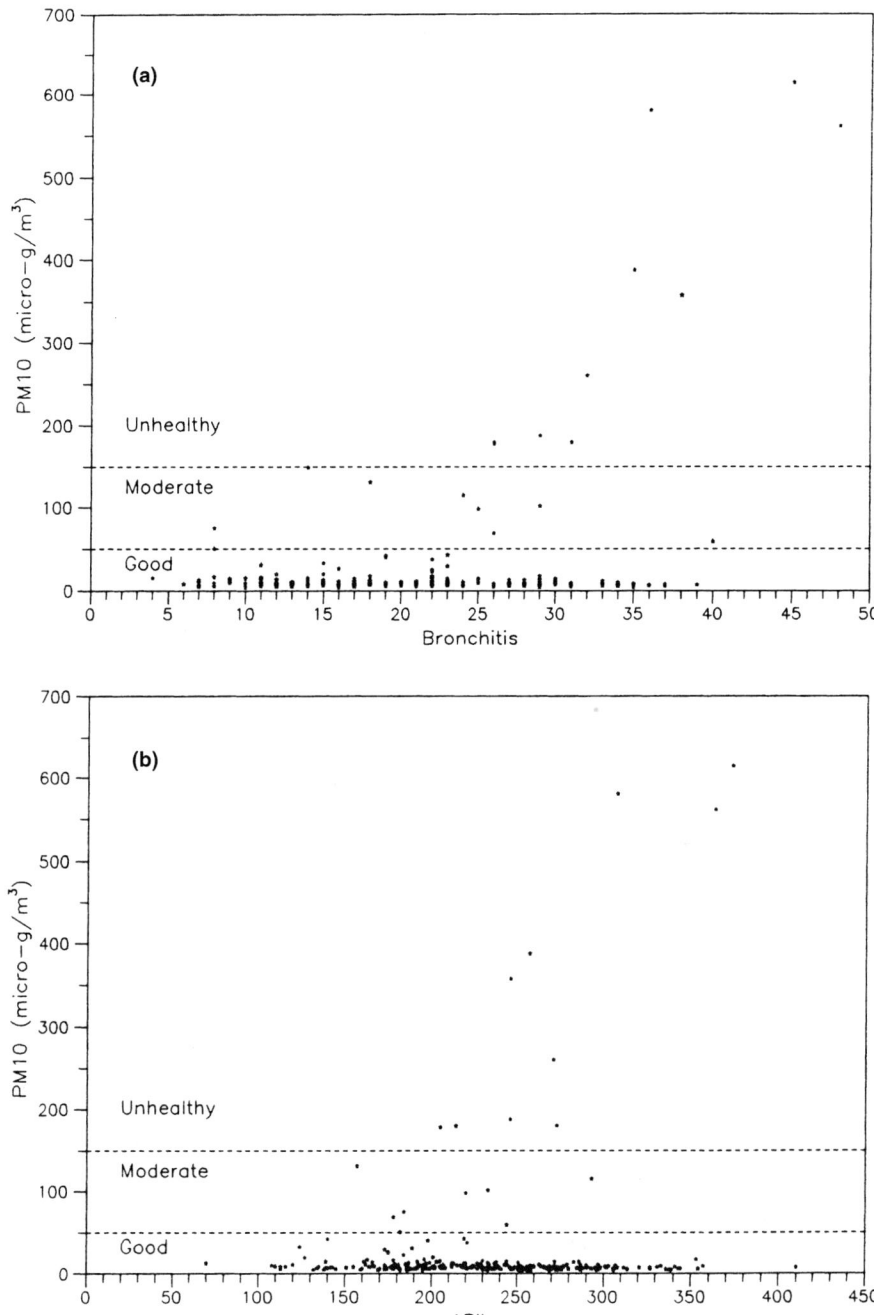

Figure 5
Scatter plot of daily averaged PM$_{10}$ concentration versus the number of cases reported for Bronchitis and ARII (Acute Respiratory Infections and Influenza) during the period April 7–Dec. 31, 1998 in Brunei Darussalam.

Table 5

Monthly mean of maximum rolling PSI values of pollutants at Post Office BSB (SINGH and HASSAN, 1998)

Month	PM_{10} (24-hr average)	SO_2 (24-hr average)	NO_2 (1-hr average)	O_3 (1-hr average)	CO (8-hr average)
Oct. 97	18	2	22	25	10
Nov. 97	16	4	24	33	13
Dec. 97	23	4	19	32	15
Jan. 98	21	4	19	32	20
Feb. 98	75	2	23	35	38
March 98	108	8	23	40	73
April 98	159	7	26	37	64

correlations. Canonical correlation analysis can be viewed as a logical extension of multiple regression analysis. Whereas multiple regression involves a single dependent variable, canonical correlation deals with multiple dependent variables. The underlying principle is to develop a linear combination of each set of variables (both dependent and independent) in a manner that maximizes the correlation between the two sets. Procedural details of the method may be found in HAIR *et al.* (1992), PRESS (1972) and RAO (1973). The results of canonical correlation analysis have been summarized in Tables 6–8. Two canonical functions orthogonal to each other have been fitted using STATGRAPHICS (ver 3.0) and are indicated in Tables 6 and 7. It is evident from Table 6 that the first canonical function displays a very high value of canonical correlation, i.e., 0.85 (eigenvalue = .72) as compared to second function which yields a highly insignificant (significance level = .9546) value of canonical correlation (= .0097). Therefore, only the first function is retained to explain the set of dependent variables. The canonical coefficients (weights) corresponding to the first canonical function are shown in the first column of Table 7. The values of canonical weights indicate the relative importance of PM_{10} and CO in the canonical relationship. However, a better method to assess the relative importance of individual variables in the canonical relationship is to examine the canonical loadings and cross-loadings (HAIR *et al.*, 1992).

Canonical loadings reflect the variance that the observed variable shares with its canonical variate. In other words, canonical loadings indicate the relative contribu-

Table 6

Canonical correlations

Function	Eigenvalue	Canonical correlation	Significance level
1	.7366	.8583	.0000
2	.0000	.0020	.9906

Table 7

Canonical weights for the two canonical functions

	Variables	Function 1	Function 2
Predictor	PM_{10}	1.02379	−1.54955
set	CO	−0.02837	1.85700
Criterion	ASTH	0.46730	1.44104
set	ARII	0.60021	−1.39094

Table 8

Canonical loadings and cross-loadings for the first canonical function

	Variables	Loadings	Cross-loadings
Predictor	PM_{10}	1.000	0.835
Set	CO	0.843	0.697
Criterion	ASTH	0.803	0.788
Set	ARII	0.997	0.816

tion of each variable to the canonical variate representing them. Canonical cross-loadings, on the other hand, provide a measure of the contribution of each variable in explaining the variance of the canonical variate of its opposite set (HAIR *et al.*, 1992).

Table 8 shows the canonical loadings and cross-loadings computed for the first canonical function/relationship. It may be seen that the canonical loadings and cross-loadings pertaining to PM_{10} are greater in magnitude in comparison to those for CO. Nevertheless, the canonical loadings and cross-loadings pertaining to CO are substantial in magnitude and indicate that CO concentration is also one of the important factors explaining the occurrence of respiratory diseases. This justifies the hypothesis of the authors that CO concentration should not be ignored while explaining the incidence of diseases. Finally, it is noted that the set of independent variables is able to explain satisfactorily (eigenvalue = .74) the set of dependent variables. The study also demonstrates that in conditions when the number of dependent variables is more than one, canonical correlation may prove to be a useful technique for explaining the relationship between a set of dependent variables and a set of independent variables.

Conclusions

In this study the visibility and the incidence of respiratory diseases in Brunei Darussalam have been studied on the basis of data on PM_{10}, carbon monoxide and relative humidity during haze episode of 1998. The results indicate a significant correlation of visibility with PM_{10} and relative humidity. Here, an endeavour is made to explain visibility on the basis of PM_{10} and RH using multiple linear regression analysis. The model fitted establishes that PM_{10} and RH are two significant factors affecting the visibility at a given site. Furthermore, it is expected that the results would further improve if other factors such as the concentration of coarser particulates and the effect of mist and sun reflection, etc. could also be taken into account.

During the haze period (April 7 to May 15, 1998), significantly high correlations of PM_{10} with the data on the number of cases reported for Asthma, Acute Respiratory Infections and Influenza (ARII) have been found at a time lag of one day; the values of correlation coefficients being .788 and .816, respectively. The non-haze period did not show any significant correlation of PM_{10} with these diseases. It is inferred that good correlations for April–May 1998, in contrast to poor correlations in the remaining months, may be due to the direct impact of PM_{10} levels on respiratory health when PM_{10} values are above certain safe limits (TLV/PSI). The overall impact of pollutants such as PM_{10} and CO in producing the incidence of respiratory diseases like Asthma and Acute Respiratory Infections and Influenza has been studied using canonical correlations, by classifying these diseases as the dependent variables and pollutants as the independent variables. The canonical correlation between these two sets of variables has been found to be good, thus underlining the importance of both PM_{10} and CO in explaining the incidence of respiratory diseases.

REFERENCES

DANIEL, W. W., *Biostatistics: A Foundation for Analysis in the Health Sciences* (John Wiley & Sons, 1991), 274 pp.
HAIR, J. F. Jr., ANDERSON, R. E., TATHAM, R. L., and BLACK, W. C., *Multivariate Data Analysis* (Maxwell Macmillan Publications, 1992), 544 pp.
MAKARIMI, HJ AWG KASIM, SINGH, M. P., and HJ SIDUP BIN HJ SIRABAHA (1999), *Mixing depth, wind speed and air pollution potential in Bandar Seri Begawan*, Presented at *International Conference on Air Quality Management (University Brunei Darussalam, Nov 15–17)*.
MOORE, P. and COBBY, J., *Introductory Statistics for Environmentalists* (Prentice-Hall Europe, 1998), 250 pp.
NICHOL, J. (1997), *Bioclimatic Impacts of the 1994 Smoke Haze Event in Southeast Asia*, Atmos. Environ. *31*, 1209–1219.
PRESS, S. J., *Applied Multivariate Analysis* (Holt, Rinehart and Winston INC, 1972), 521 pp.
RADOJEVIC, M. (1997), *The Haze in Southeast Asia*, Environ. Scientist 6(6), 1–3.
RADOJEVIC, M. and HASSAN, H. (1999), *Air Quality in Brunei Darussalam during the 1998 Haze Episode*, Atmos. Environ. *33*, 3651–3658.
RAO, C. R., *Linear Statistical Inference and Its Applications* (Wiley Eastern Private Limited, New Delhi, 1973), 625 pp.

SINGH, M. P. and HASSAN, H. (1998), *Air Quality of Brunei Darussalam and its Measurement*, Technical Report, University Brunei Darussalam, Available at The Educational Technology Centre, University Brunei Darussalam, 56 pp.

(Received July 1, 2000, accepted February 16, 2001)

To access this journal online:
http://www.birkhauser.ch

Pure appl. geophys. 160 (2003) 279–293
0033–4553/03/020279–15

© Birkhäuser Verlag, Basel, 2003

Statistical Estimation of Dose-response Functions of Respiratory Diseases and Societal Costs of Haze-related Air Pollution in Brunei Darussalam

Kwabena A. Anaman[1] and Norsinah Ibrahim[1]

Abstract — The effects on human health resulting from the January to April 1998 haze-related air pollution episode in Brunei Darussalam were analysed for five groups of diseases of the respiratory system. The analysis concentrated on the statistical estimation of dose-response functions which related the number of cases of respiratory diseases to the level of quality of ambient environment as measured by the pollutants standards index (PSI) and other environmental variables. The total number of cases of the five groups of diseases was shown to be significantly related to PSI and temperature. Societal costs were also estimated. The results showed that societal costs were significantly related to PSI, temperature and relative humidity. Societal costs increased with higher PSI and relative humidity but decreased with increasing temperature.

Key words: Air pollution, Brunei Darussalam, dose-response functions, economic damages function, environmental economics, haze health economics, time-series models.

1. Introduction and Problem Statement

Haze-related air pollution originating from forest fires in Indonesia engulfed much of the Southeast Asian region from July to September 1997. The pollution subsided by the end of September 1997 as the forest fires were quenched by monsoon rains. Brunei Darussalam was mildly affected by the July to September 1997 haze-related pollution episode. However from January to April 1998, major forest fires, once again from Indonesia, spread haze-related air pollution around the Southeast Asian Region. Brunei Darussalam was seriously affected by the January to April 1998 haze-related episode leading to the closure of schools and the rescheduling of work hours in offices of both Government and the private sector. Very dry weather conditions associated with the El-Niño-Southern Oscillation weather phenomenon contributed to the severe haze-related air pollution problem which led to widespread increases in respiratory diseases (Anon, 1998).

[1] Department of Economics, Faculty of Business, Economics and Policy Studies, Universiti Brunei Darussalam, Bandar Seri Begawan BE1410, Brunei Darussalam, Southeast Asia.
E-mail: kanaman@fbeps.ubd.edu.bn

Increased air pollution related to haze is known to increase the levels of respiratory diseases such as asthma, bronchitis, emphysema and pneumonia, and also death rates from these diseases. However the relationships between the concentration of air pollutants, for example, as measured by the PSI, and the levels of increases in respiratory diseases and related deaths vary from region to region. Most of the comprehensive studies addressing these relationships have been conducted in North America and Western Europe (RAUFER, 1997). These studies have often been the basis for adaptation to estimate dose-response functions and related impacts of air pollution for developing countries as was done by RAUFER (1997) for Cairo, Egypt and DIXON et al. (1994) for Indonesia. Because of the highly motorised population in Brunei and the frequent occurrence of haze-related air pollution in the Southeast Asian Region over the last 15 years, which have affected the country (SINGH and HASSAN, 1998; SIRINANDA, 1997), there is a growing need to determine authentic locally-based dose-response functions and societal economic damages for air pollution. This is necessary to better gauge the extent of the air pollution problem from an economic perspective and to determine the extent of costs and benefits from economy-wide measures that would be necessary to control levels of air pollution in emergency cases as happened in 1998. The objectives of this study were to estimate the relationships between the ambient concentration of air pollutants and other environmental variables and the number of cases of respiratory diseases and to establish societal costs in terms of increases in respiratory diseases based on data from the January–April 1998 haze-related pollution episode. This paper is organised as follows: the next section discusses the theoretical framework of the study followed by the methods and procedures used. The results and the conclusions follow.

2. Theoretical Framework

The study of the effects on human health due to pollutants is often done through an economic damage function approach using dose-response relationships (FIELD, 1994). Damages indicate the negative impacts that users of environmental amenities such as air and water experience as a result of pollution of these amenities. Dose-response relationships relate the dose of various pollutants to the response in terms of the level of human illnesses. The change in air quality with regards to certain pollutants is assumed to be statistically linked to changes in the level of human illnesses. With regards to haze-related air pollution, the degree of pollution is measured by the PSI. The PSI therefore represents the degree of the dose of air pollutants. The human health response is the number of cases of respiratory diseases induced by the pollutants. Hence establishing a dose-response relationship relating the level of PSI of the environment to the number of respiratory diseases establishes a damage function. This forms the basis for estimating the direct costs of haze-related air pollution. Monetary values can then be placed on health effects of the haze such as increased morbidity (illness), pain and suffering and deaths.

Figure 1 shows hypothetical marginal damage and marginal cost of preventing damages resulting from haze-related air pollution. It is assumed that marginal damage to society increases with increasing ambient concentration of pollutants as measured by the PSI. The 100 PSI level is considered the equilibrium level of ambient concentration of pollutants with the societal marginal damage being 0 Co with respect to the case of no pollution. Above 100 PSI ambient air concentration, marginal damage to society exceeds marginal cost of preventing damages and hence society will institute measures to curtail the level of pollutants. However, below the equilibrium level of pollutants, the marginal cost of preventing damages exceeds the marginal damage inflicted on society and hence recommended societal action is likely

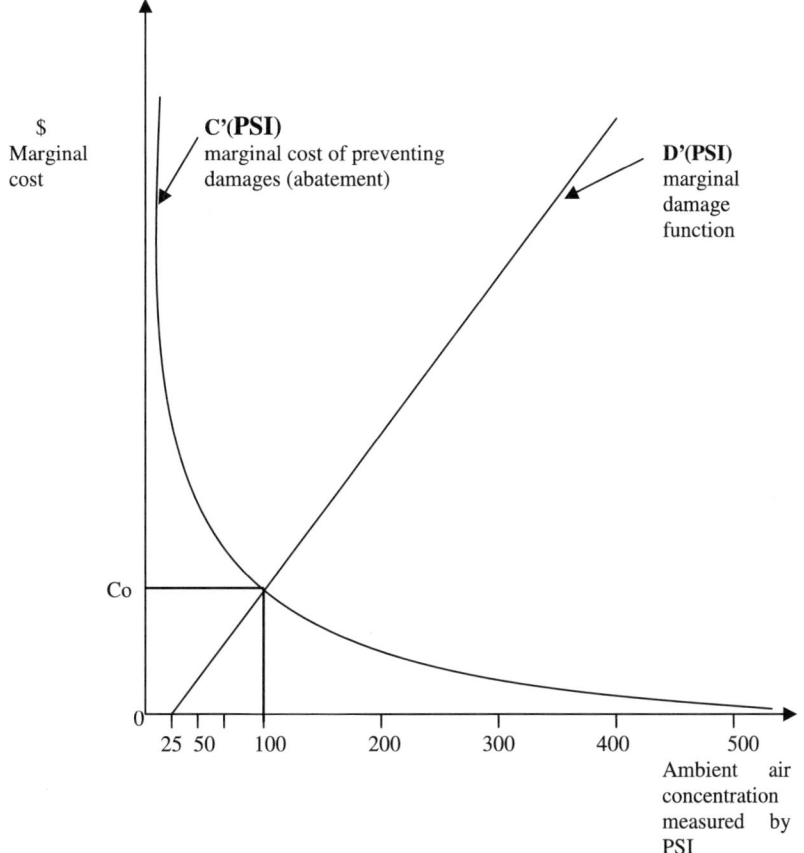

Figure 1

Representation of societal marginal damage and marginal abatement functions for haze-related air pollution.

to be minimal. Knowledge of the marginal damage function is important in order to derive the benefits (cost savings) from reducing high and unhealthy levels of PSI to low and healthy levels. The Brunei Darussalam National Committee on Haze (BNCOH) has outlined several measures that must be undertaken by businesses and households in the case of moderate to severe air pollution (i.e., for conditions where the 24-hour average PSI reading exceeds 100). The measures indicated by BNCOH entail societal costs in order to reduce societal damages from haze-related air pollution. The benefits of these measures can be estimated if knowledge of the cost savings that accrue to society from reduction of the levels of ambient concentration of pollutants to healthy levels is known. This study attempts to establish such benchmark information based on data gathered during the January to April 1998 haze-related air pollution episode.

As noted earlier, dose-response relationships often established using United States data are adapted to conditions of other countries, especially developing countries. The 1998 haze-related air pollution episode provided a form of an "unwelcome" natural experiment in the form described by BABBIE (1998, pp. 249–251), whereby wide variation in the ambient environmental air quality measured by the PSI was observed in Brunei Darussalam. Widespread increases in respiratory illnesses were observed and reported by local newspapers such as the Borneo Bulletin (ANON, 1998). As such, the 1998 haze-related air pollution episode provided data for the determination of local dose-response relationships with many potential uses including environmental impact assessment of air pollution reduction projects funded by national Governments in East Asia and international lending agencies.

The PSI is the most commonly used index to measure the level of air pollution in the ambient environment. This index was developed by the United States Environment Protection Agency (USEPA) to provide reliable and accurate data on the daily levels of air pollution that can allow the public at large to determine whether the quality of air in a particular location is good (USEPA, 1994). The PSI measures five main pollutants. These include particulate matter such as soot, dust and particles and very fine particulate matter of size 10 microns (ug/m^3) or less (PM_{10}). The other four pollutants are sulphur dioxide measured in parts per million, nitrogen dioxide, carbon monoxide and ozone. For each pollutant, air quality standards over a few hours and up to 24 hours are established and explicitly incorporated in the PSI-related warning information (USEPA, 1994). A PSI reading is measured for each of the five pollutants. The highest of the five figures is reported each day. Because of the nature of haze-related pollution in Brunei Darussalam arising from forest fires, the actual recorded PSI figures reported are heavily biased towards PM_{10}. The human body mechanisms can remove inhaled dusts or particulate matter in excess of 10 microns. Particulate matter less than 10 microns is invisible to the naked eyes and can enter the respiratory system and cause diseases, especially if people are extendedly exposed to the pollutant.

3. Methods and Procedures

Time-series econometric models can be classified into three groups based on the driving force of the underlying stochastic process. These are trend, seasonal and irregular models (ENDERS, 1995). Seasonal models are often expressed with sine or cosine time components and can be estimated with data over a period of at least 12 months to capture the seasonality. Irregular models are those in which the dependent variable is a function of its lagged values. For this study, due to the limited available data, over a period of six months (from January to June 1998), linear and log-linear dose-response relationships and societal cost (damage) functions were estimated, based on a trend type model using SAS (1996) regression software. These functions are described below.

$$TRD = A_0 + A_1\,PSI + A_2\,LAGTEMP + A_3\,HUMIDITY + A_4\,TREND + U_1$$
$$LTRD = B_0 + B_1\,LPSI + B_2\,LLAGTEMP + B_3\,LHUMIDITY + B_4\,LTREND + U_2$$
$$TCOSTS = C_0 + C_1\,PSI + C_2\,LAGTEMP + C_3\,HUMIDITY + C_4\,TREND + U_3$$
$$LTCOSTS = D_0 + D_1\,LPSI + D_2\,LLAGTEMP + D_3\,LHUMIDITY + D_4\,LTREND + U_4,$$

where TRD is the total daily number of all five groups of respiratory diseases recorded for public and private hospitals, health centres and clinics in the Brunei-Muara district (Bandar Seri Begawan and Muara). The five groups of respiratory diseases are: (1) asthma, bronchitis and emphysema group (2) influenza, (3) pneumonia, (4) acute upper respiratory infections and (5) conjunctivitis;

TCOSTS is the estimated total daily societal costs of the five groups of respiratory diseases in terms of cost of treatment of outpatients, cost of hospital admissions, cost of self-medication, cost of labour productivity losses and cost of pain, discomfort and suffering;

PSI is the previous day's average 24-hour daily reading (an average of 24 hourly records) of air quality measured by the pollutant standards index at the Ministry of Health Headquarters in Bandar Seri Begawan (BSB);

LAGTEMP is the previous day's average daily temperature in BSB;

HUMIDITY is the previous day's average relative humidity in BSB;

TREND is a trend variable with a value of 1 for February 7th 1998 and increasing every day by 1 to 144 for the 30th of June 1998;

$LTRD_i$, LPSI, LLAGTEMP, LHUMIDITY and LTREND are the natural logarithm of the variables TRD, PSI, LAGTEMP, HUMIDITY and TREND respectively.

A_i, B_i, C_i and D_i are the parameters to be estimated and

U_1, U_2, U_3, and U_4 are the error terms initially assumed to have zero mean and constant variance.

PSI data were collected from the Environmental Unit, Ministry of Development. Data on respiratory diseases were collected from all public health centres in the Brunei-Muara District which covered the capital city of Bandar Seri Begawan and contained about two-thirds of the entire population of Brunei Darussalam (Government of Brunei Darussalam, 1999). These public health centres included the RIPAS General Hospital in Bandar Seri Begawan, several government clinics and Jerudong Park Medical Centre in Bandar Seri Begawan. Data for the 16 private clinics were derived from the detailed records kept by the Borneo Clinic, a typical small private clinic which offered its detailed outpatient diseases and expenses data over the January to June 1998 period for our research work. Data on weather variables such as temperature and relative humidity likely to affect the incidence of respiratory diseases were collected from the Meteorological Services Unit of the Department of Civil Aviation. The PSI data were only available on a daily basis from the 7th of February 1998 to June 30th 1998. Data on cases of the five groups of respiratory diseases were available on a daily basis from the 1st of January 1998 to the 30th of June 1998. The dose-response functions were therefore based on data from the 7th of February to 30th of June 1998.

Short-term human health costs estimated in the study comprised of five components. These were (a) direct outpatient treatment costs consisting of the costs of medicine and doctor consultation costs (b), costs of increased hospital admissions and emergency cases at health centres and hospitals (c), increased self-medication costs of those who chose not to see a doctor or visit health centres; (d) short-term human productivity losses based on lost work days and (e) the cost of human suffering, discomfort and pain of the diseases.

Figure 2 provides a schematic representation of the societal costs from human respiratory illnesses arising from the worsening air quality related to air pollution based on natural/human causes. Natural and human induced causes lead to worsening air quality measured by an increase in PSI. This then leads to increases in human respiratory diseases. The increases in human respiratory diseases impose six sets of costs on society including the five components discussed above. The sixth cost component is the increased expenditure by the public on masks, air filters and others and increased use of energy. The establishment of these costs requires a survey of the general public. This has not yet been undertaken and hence the societal costs established in this study are only for the first five components. The details of the costs figures derived for the first five components are available from ANAMAN and IBRAHIM (1999). The assumptions involved in the derivation of these cost figures are briefly summarised in the following paragraphs.

Direct outpatient treatment costs based on an incremental increase in the number of respiratory diseases were derived by comparing the data from January to April 1998 with those obtained during the same period from January to April 1997 and adjusted downwards by the 3% natural growth rate of human population. The societal costs for patients staying overnight at hospitals were based on figures provided by Jerudong Park Medical Centre, a private medical centre in Bandar Seri Begawan. The average cost per bed per day was B$350 in 1998. The length of stay in

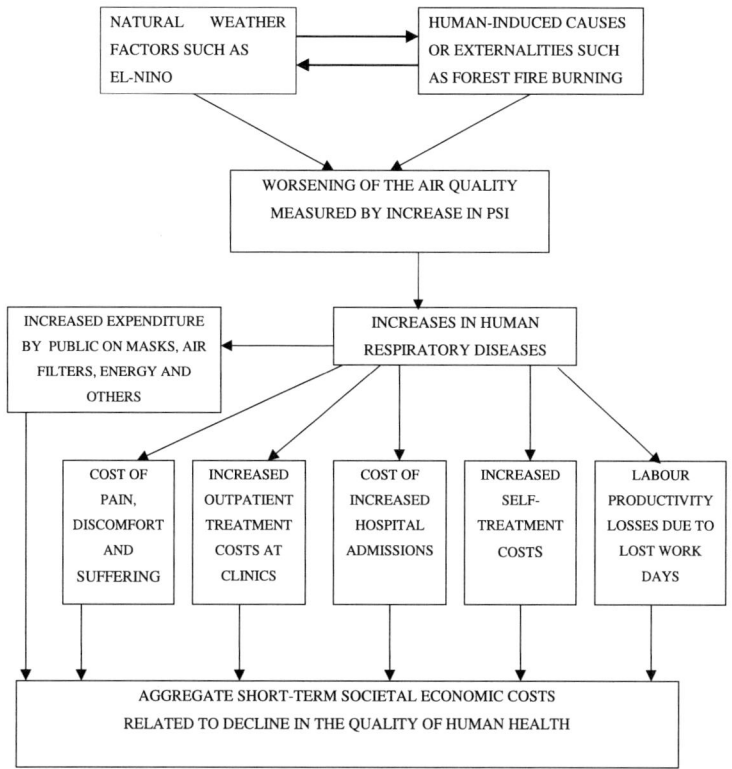

Figure 2
Schematic representation of the societal costs from human illness arising from the worsening of air quality
due to haze-related air pollution.

the hospitals was assumed to be five days. Patients who chose self-medication were assumed to be two-thirds of those (outpatients) who chose to visit clinics for influenza and conjunctivitis. It was also assumed that patients who chose self-medication were 10% of those (outpatients) who chose to visit clinics for asthma, bronchitis and emphysema group, pneumonia and AURI.

Human productivity costs were estimated as number of days absent from work by adults multiplied by the average wage rate. Medical leave was assumed to be on average three days for outpatients plus additional five days hospital stay for those hospitalised. The proportion of working adults to children was approximately 1:1 based on data provided by the Borneo Clinic and information from other clinics. The average real daily wage rate for Brunei Darussalam was assumed to be B$100 based on information from the Labour Department and other sources (Government of Brunei Darussalam, 1999, p. 125–128). The cost of human discomfort, suffering and pain was based on the willingness to pay (WTP) amounts by individuals to avoid respiratory diseases. The cost of discomfort, suffering and pain used in this study was assumed to be a 1:1 ratio with the cost of treatment plus costs of productivity losses based on similar work on Singapore (HON, 1998).

4. Results

Linear and log-linear dose-response equations were estimated for the total number of all five groups of respiratory diseases collectively and also for each of the five groups of diseases: asthma, bronchitis and emphysema as one group, influenza, pneumonia, AURI and conjunctivitis. The choice of whether to use the linear or log-linear functional form as the appropriate model was decided, based on several factors: (a) the MWD test developed by MACKINNON *et al.* (1983) and summarised by GUJARATI (1995, pp. 265–266); (b) relative prominence of these two econometric problems: heteroscedasticity and multicollinearity. Based on the MWD test analysis, the linear dose response functional form was preferred to the log-linear form for the asthma, bronchitis and emphysema group of diseases, and acute upper respiratory infections. However the log-linear functional form was preferred to the linear form for the other three groups of diseases: influenza, pneumonia and conjunctivitis. The estimated dose-response functions for each of the five groups of diseases are contained in the Appendices.

For the estimated dose-response functions for the total number of cases of all five groups of diseases, the linear functional form was superior to the log-linear model based on the MWD test. However the former model had a severe heteroscedasticity problem as measured by the WHITE (1980) general heteroscedasticity test. Hence the log-linear model was preferred. Table 1 presents the results of the estimated log-linear dose response function for the total number of all five groups of diseases. Due to the initial first-order autocorrelation, the dose-response function was estimated by the generalised least squares method (GLS) whereby all the variables in the model were transformed based on the first-order autocorrelation coefficient derived from the first stage ordinary least-squares (OLS) estimation. The results showed that the total number of cases of all five groups of diseases was quite significantly related to the level of PSI as indicated by the parameter estimates (p value of 0.0001). The total number of cases was also significantly related to temperature and time trend variable. Humidity was negatively related to the total number of cases of diseases. However, this relationship was not statistically significant even though the humidity variable had a t value of over 1.0, which meant it had to be kept in the model to avoid misspecification bias as outlined by GUJARATI (1995, 253–254).

Standardized regression coefficients measure the relative influence of various factors in influencing a dependent variable. These were computed by dividing the parameter estimate by the ratio of the sample standard deviation of the dependent variable to the sample standard deviation of the independent variable (STATISTICAL

Appendix 3

Results of the preferred log-linear dose-response equation with the natural logarithm of the number of pneumonia cases (LPNEUMO) recorded in the Brunei-Muara district (Bandar Seri Begawan and Muara) from January to April 1998 as the dependent variable and LPSI, LLAGTEMP, LHUMIDITY and LTREND as independent variables estimated by the ordinary least-squares method

Explanatory Variable	Parameter Estimate	T Statistic	P Value	Standardised Estimate
INTERCEPT	−235.392	−6.824	0.0001**	0.000
LPSI	1.0251	2.585	0.0108**	0.189
LLAGTEMP	68.421	6.934	0.0001**	0.500
LHUMIDITY	2.701	1.334	0.184	0.082
LTREND	−2.428	−6.223	0.0001**	−0.426
Adjusted R^2				0.485
F value				34.417
Durbin-Watson statistic				1.541
First-order autocorrelation coefficient				0.225
Significance level of White general heteroscedasticity test				0.001

Note:
** denotes statistical significance at 10% level

Appendix 4

Results of the preferred linear dose-response equation with the number of acute upper respiratory infections, cases (AURI) recorded in the Brunei-Muara district (Bandar Seri Begawan and Muara) from January to April 1998 as the dependent variable and PSI, LAGTEMP, HUMDITY and TREND as independent variables estimated by the ordinary least-squares method

Explanatory Variable	Parameter Estimate	T Statistic	P Value	Standardised Estimate
INTERCEPT	645.328	3.851	0.0002**	0.000
PSI	0.603	5.551	0.0001**	0.374
LAGTEMP	−10.486	−2.051	0.0421**	−0.131
HUMIDITY	−0.339	−0.537	0.5922	−0.033
TREND	−0.965	−7.360	0.0001**	−0.477
Adjusted R^2				0.501
F value				36.588
Durbin-Watson Statistic				1.865
First-order autocorrelation coefficient				0.055
Significance level of White general heteroscedasticity test				0.033

Note:
** denotes statistical significance at 10% level

Appendix 5

Results of the preferred log-linear dose-response equation with the natural logarithm of the number of conjunctivitis cases (LCONJUN) recorded in the Brunei-Muara district (Bandar Seri Begawan and Muara) from January to April 1998 as the dependent variable and LPSI, LLAGTEMP, LHUMIDITY and LTREND as independent variables estimated by the ordinary least-squares method

Explanatory Variable	Parameter Estimate	T Statistic	P Value	Standardised Estimate
INTERCEPT	3.330	0.866	0.3882	0.000
LPSI	0.255	5.759	0.0001**	0.489
LLAGTEMP	0.024	0.022	0.9827	0.002
LHUMIDITY	−0.371	−1.642	0.1029	−0.117
LTREND	−0.083	−1.900	0.0595**	−0.151
Adjusted R^2				0.305
F value				16.552
Durbin-Watson statistic				1.574
First-order autocorrelation coefficient				0.211
Significance level of White general heteroscedasticity test				0.751

Note:
** denotes statistical significance at 10% level

REFERENCES

ANON (1998), *Hazy Days Are Here Again*, Borneo Bulletin (28 February).

ANAMAN, K. A. and IBRAHIM, N. (1999), *Economic Analysis of Human Health Impact of the 1998 Haze-related Air Pollution Episode in Brunei Darussalam*, Proceedings, International Congress of Biometerology and International Conference on Urban Climatology, Sydney, 8–12 November, 6 pp.

BABBIE, E., *The Practice of Social Research*, Eight Edition (Wadsworth, New York 1998).

Brunei Darussalam National Committee on Haze, Master Plan of Action Based on 24-Hour Averaged PSI Readings (Ministry of Health, Bandar Seri Begawan 1997).

DIXON, J. A., SCURA, L. F., CARPENTER, R. A., and SHERMAN, P. B., *Economic Analysis of Environmental Impacts*, Second Edition (Earthscan, London 1994) 210 pp.

DURBIN, J. and WATSON, G.S. (1951), *Testing for Serial Correlation in Least Squares Regressor*, Biometrika *38*, 159–171.

ENDERS, W., *Applied Econometric Time Series* (John Wiley and Sons, New York 1995) 433 pp.

FIELD, B. C., *Environmental Economics: An Introduction* (McGraw-Hill Inc., Sydney 1994) 482 pp.

Government of Brunei Darussalam, Brunei Darussalam Statistical Yearbook 1998 (Department of Economic Planning and Development, Bandar Seri Begawan 1999).

GUJARATI, D. N., *Basic Econometrics*, Third Edition (McGraw-Hill, Singapore 1995) 838 pp.

HON, P. *Economic Value of the 1997 Haze Damages to Singapore* (Institute of Southeast Asian Studies, Economy and Environment Program for Southeast Asia, Singapore 1998) 34 pp.

MACKINNON, J., WHITE, H., and DAVIDSON, R. (1983), *Tests for Model Specifications in the Presence of Alternative Hypothesis: Some Further Results*, J. Econometrics *21*, 53–70.

RAUFER, R. K. (1997), *Particulate and Lead Pollution Control in Cairo: Benefit Valuation and Cost-effective Control Strategies*, Natural Resources Forum *21*, 209–219.

SINGH, M. P. and HASSAN, H., *Air Quality of Brunei Darussalam and its Measurement* (Department of Mathematics, Universiti Brunei Darussalam, Bandar Seri Begawan 1998) 56 pp.

SIRINANDA, K. U., *Climate and Life in the Asia Pacific: Proceedings* (Universiti Brunei Darussalam, Bandar Seri Begawan 1997) 438 pp.

STATISTICAL ANALYSIS SYSTEM INSTITUTE, *SAS/STAT User's Guide*, Volume 2, Version 6, Fourth Edition (SAS Institute, Cary, North Carolina 1990).

STATISTICAL ANALYSIS SYSTEM INSTITUTE, *SAS Procedures Guide for Personal Computers with Windows* (SAS Institute, Cary, North Carolina 1996).

UNITED STATES ENVIRONMENTAL PROTECTION AGENCY, *Measuring Air Quality: The Pollutant Standards Index* (Office of Air Quality Planning and Standards, Washington, D.C. 1997).

WHITE, H. (1980), *A Heteroscedasticity Consistent Covariance Matrix Estimator and a Direct Test of Heteroscedasticity*, Econometrica *48*, 817–818.

(Received March 1, 2000, accepted October 2, 2000)

To access this journal online:
http://www.birkhauser.ch

C. Air Quality Modeling

Pure appl. geophys. 160 (2003) 297–316
0033–4553/03/020297–20

© Birkhäuser Verlag, Basel, 2003

▌ **Pure and Applied Geophysics**

Recent Advances in Air–Sea Interaction Studies Applied to Overwater Air Quality Modeling: A Review

S. A. Hsu[1] and B. W. Blanchard[1]

Abstract — Air quality modeling in the marine environment requires a better understanding of air–sea interaction. The behavior of the sea is different from that of the land, particularly with respect to the aerodynamic roughness parameter which changes temporally and spatially as a function of wind-generated wave characteristics. Recent improved understanding in air–sea interaction as applied to topics in air quality modeling are discussed. They include stability characteristics, variation of the wind speed with height, mixing height, standard deviation of crosswind and vertical wind, and the eddy diffusion. These topics are synthesized for scientists and engineers who need to improve the air quality modeling for overwater applications.

Key words: Overwater dispersion, stability length, friction velocity, mixing height, drag coefficient.

1. Introduction

According to Lyons and Scott (1990), the atmosphere contains a mixture of gases and solid particles ranging in size from less than one thousandth of a millimeter to a millimeter or more. This mixture is affected by continuous, dynamic change with vast quantities being added to or removed from the atmosphere by various natural and industrial processes. Air pollution meteorology is concerned with the fate of these pollutants once they are emitted into the atmosphere and addresses natural releases, deliberate industrial emissions, and accidental spills. According to Zannetti (1990), air quality modeling is an essential tool for most air pollution studies. For more detail regarding air pollution meteorology and air quality modeling, see, e.g. Hanna *et al.* (1982), Zannetti (1990), and Arya (1999).

Overwater dispersion characteristics required in air quality modeling include stability classification, turbulence intensity, eddy diffusion, and mixing height. The purpose of this review is to synthesize recent advances in these parameterizations. The most important difference between sea and land is that the sea surface is in

---

[1] Coastal Studies Institute, Louisiana State University, Baton Rouge, Louisiana 70803, U.S.A.
E-mail: sahsu@antares.esl.lsu.edu

motion while the land surface is fixed. Sea waves and currents will affect the aerodynamic roughness so that rapid changes in both space and time depend upon the wind-generated wave characteristics. Hence the drag coefficient, which is fixed on land, cannot be treated as a constant. This in turn affects a host of other overwater characteristics.

All pertinent parameter symbols are identified in the text; however, since we present over 50 equations incorporating many variables, a list of symbols is summarized as the Appendix for the reader's convenience. All units used are in meter –kilogram–second (MKS) or the Systeme International (SI).

2. Determining the Stability Characteristics

a. Parameterization of the Stability Length, L

For overwater air quality modeling, the Offshore and Coastal Dispersion Model (OCD) (see HANNA et al., 1985) has been recommended for regulatory use by both the U.S. Environmental Protection Agency (EPA) and the U.S. Minerals Management Service for emissions located on the outer continental shelf (see ZANNETTI, 1990). In this EPA preferred model, the stability parameter z/L is required, where z is the height above the surface and L is the Monin-Obukhov stability length, which is defined as (see, e.g., GARRATT, 1992, p. 38)

$$L = -\frac{u_*^3 \theta_v}{g \kappa \overline{w'\theta_v'}} \tag{1a}$$

or as (e.g., PANOFSKY and DUTTON, 1984, p. 132)

$$L = -\frac{u_*^3 \rho C_{\mathrm{ph}} T_{\mathrm{air}}}{g \kappa H \left(1 + \frac{0.07}{B}\right)} \tag{1b}$$

where u_* is the friction velocity, θ_v is the virtual potential temperature, g is the gravitational acceleration, κ is the von Karman constant, $\overline{w'\theta_v'}$ is the surface buoyancy flux, ρ is the air density, C_{ph} is the specific heat of air at constant pressure, T_{air} is the air temperature, H is the surface layer sensible heat flux, and B is the Bowen ratio (the ratio of sensible to latent heat flux). Note that T_{air} should be T_v, the virtual temperature. However, since $T_v = T_{\mathrm{air}} (1 + 0.68\ q)$ and q is at most 5% (see, e.g., KOMEN et al., 1994, p. 59), we use $T_v \cong T_{\mathrm{air}}$.

Since direct measurements of u_* and H are not available routinely, the following parameterization is employed.

From Eq. (1b) and SMITH (1980, Eqs. (3) and (4))

$$\frac{H}{\rho C_{\mathrm{ph}}} = C_T U_{10}(T_{\mathrm{sea}} - T_{\mathrm{air}}) \tag{2}$$

and

$$C_d = \left(\frac{u_*}{U_{10}}\right)^2 \tag{3}$$

or $u_*^3 = U_{10}^3 C_d^{3/2}$, we have for the unstable condition ($T_{\text{sea}} > T_{\text{air}}$)

$$L = -\frac{T_{\text{air}} C_d^{3/2} U_{10}^2}{g\kappa C_T (T_{\text{sea}} - T_{\text{air}})\left(1 + \frac{0.07}{B}\right)} \tag{4a}$$

and for the stable condition ($T_{\text{sea}} < T_{\text{air}}$)

$$L = -\frac{T_{\text{air}} C_d^{3/2} U_{10}^2}{g\kappa C_T (T_{\text{sea}} - T_{\text{air}})} \tag{4b}$$

where C_T is the heat flux coefficient ($= 1.1 \times 10^{-3}$ for unstable, 1.0×10^{-3} for neutral and 0.83×10^{-3} for stable conditions (see SMITH, 1980 and 1988)), C_d is the drag coefficient, U_{10} is the wind speed at 10 m above the sea surface, and T_{sea} is the sea-surface temperature.

On the basis of thermodynamic conditions, a relationship between B and ($T_{\text{sea}} - T_{\text{air}}$) under unstable conditions (i.e., $T_{\text{sea}} > T_{\text{air}}$) has been proposed by HSU (1998) that

$$B = a(T_{\text{sea}} - T_{\text{air}})^b \tag{5a}$$

where a and b are to be determined by field experiments. Based on the availability of additional data sets from tropical oceans and coastal seas, HSU (1999) found

$$B = 0.146(T_{\text{sea}} - T_{\text{air}})^{0.49} \tag{5b}$$

with a high correlation coefficient of 0.94 between B and ($T_{\text{sea}} - T_{\text{air}}$). It is the purpose of this paper to verify Eq. (4) by employing this B parameterization along with a proven C_d formulation used successfully in the third generation wave model (see WAMDI, 1988, p. 1784) that

$$C_d = \begin{bmatrix} 1.2875 \times 10^{-3}, & U_{10} < 7.5\,\text{m s}^{-1} \\ (0.8 + 0.065\,U_{10}) \times 10^{-3}, & U_{10} \geq 7.5\,\text{m s}^{-1} \end{bmatrix} \tag{6}$$

b. Defining the "Near-neutral" Condition

When the atmospheric surface layer is dominated by purely mechanical turbulence, $z/L = 0$, which is said to be neutral stability (PANOFSKY and DUTTON, 1984, p. 114, Table 5.1). This means that L is approaching infinity. Since the "absolute zero" for z/L is not likely to be encountered in field experiments, some accurate digits apart from zero, i.e., significant digits, are usually associated with this "absolute neutral." Therefore, near-neutral or almost neutral conditions are normally designated.

In the atmospheric boundary layer, two stability parameters are usually used. One is z/L as defined in Eq. (1) and the other is the Richardson number defined as (see, e.g. BUSINGER et al., 1971)

$$\text{Ri} = \frac{g\frac{\partial \bar{\theta}}{\partial Z}}{\bar{\theta}\left(\frac{\partial \bar{u}}{\partial Z}\right)^2} \tag{7}$$

where $\partial\bar{\theta}/\partial Z$ represents the potential temperature profile and $\partial\bar{u}/\partial Z$ the wind profile in the vertical.

The effect of stability on the wind profile is shown by KARLSSON (1986, p. 339) that near-neutral stability can be classified as being

$$-0.02 \leq \text{Ri} \leq 0.02 \tag{8}$$

According to BUSINGER et al. (1971), the relationship between z/L and Ri is as follows: under unstable conditions when the sea surface is warmer then the air, we have

$$\text{Ri} = \frac{0.74\xi(1 - 15\xi)^{1/2}}{(1 - 9\xi)^{1/2}} \tag{9}$$

and under stable conditions when the air is warmer than the sea

$$\text{Ri} = \frac{\xi(0.74 + 4.7\xi)}{(1 + 4.7\xi)^2} \tag{10}$$

where $\xi = z/L$.

From Eqs. (8) through (10), we find that under near-neutral conditions

$$-0.03 \leq z/L \leq 0.03 \ . \tag{11}$$

This is also supported by SORBJAN (1986, Fig. 19) in that z/L is approximately symmetrical on both sides of neutral ($z/L = 0$) for stable and unstable. Therefore, we propose $|z/L| < 0.03$ as our definition of "near-neutral" conditions from routine wind measurements (such as from NDBC buoys).

c. Verification of the L Formulation

In order to verify Eq. (4), simultaneous measurements of T_{air}, T_{sea}, U_{10}, and L are necessary. The data sets provided in DONELAN et al. (1997) supply all of these parameters. Based on the previous discussion, we classify $z/L \leq -0.03$ as unstable, $|z/L| < 0.03$ as near-neutral, and $z/L \geq 0.03$ as stable. In this verification, we set $z = 10$ m to correspond to U_{10} and the C_d formulation used in Eq. (6). Our results are shown in Figures 1 and 2 for unstable and stable conditions, respectively. If one accepts the small RMSE values (≈ 8 to 11 m) for the range of L between zero and

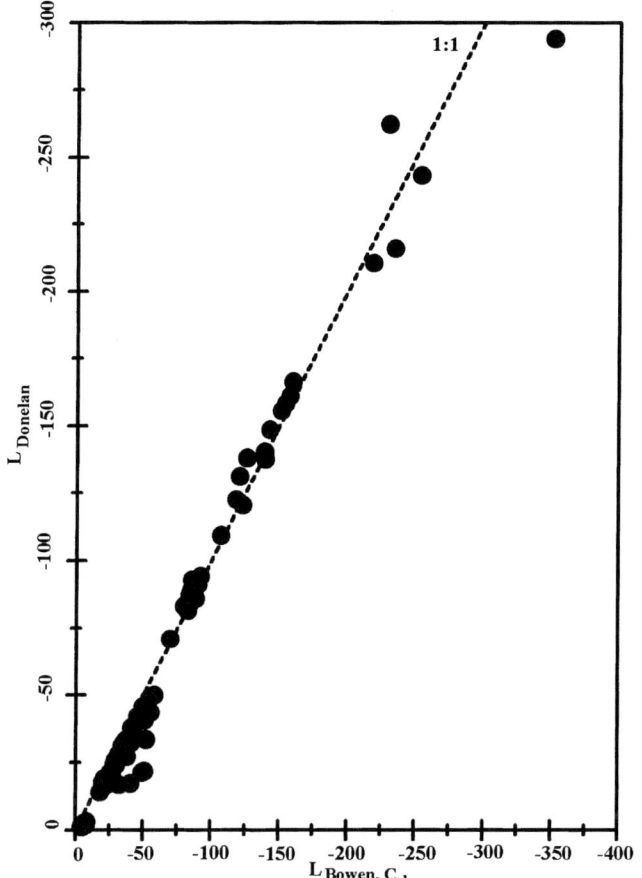

Figure 1

A verification of Eq. (4a) for unstable conditions ($-7.69 \leq z/L \leq -0.03$) using data from DONELAN *et al.* (1997). Root-mean-square error for 90 samples is 10.5.

approximately 350 m, Eq. (4) is verified. Note that the term of B, i.e. $(1 + 0.07/B)$ in the denominator of Eq. (4a) was dropped for stable conditions in Eq. (4b) when T_{air} is larger than T_{sea} and therefore no sensible heat is directed from the sea to the air.

3. Variation of the Wind Speed With Height

In the parameterization of z/L, the wind speed at 10 m above the sea surface, U_{10}, is used (see Eqs. 2, 3, 4a, 4b, and 6). However, not all wind measurements are located at 10 m, e.g., the anemometers on smaller buoys used by the National Data Buoy Center are located at 5 m height. Certainly, the wind measurements from offshore oil platforms as well as from ships are located at much higher elevations (see HSU *et al.*,

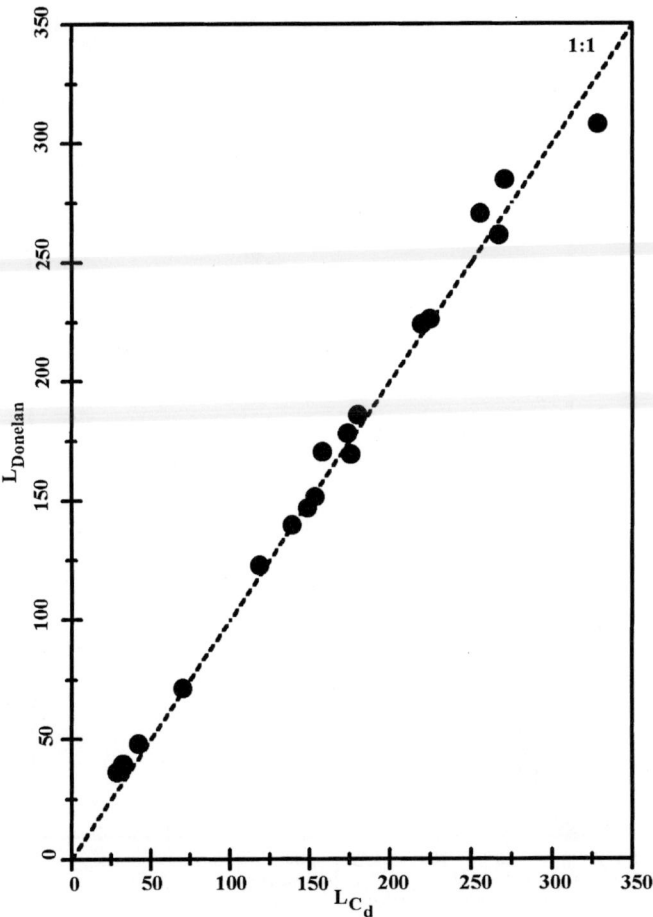

Figure 2

A verification of Eq. (4b) for stable conditions ($0.03 \leq z/L \leq 0.28$) using data from DONELAN, *et al.*
(1997). Root-mean-square error for 21 samples is 8.2.

1994). Therefore, in order to determine the L value, it is prudent to adjust all wind
speeds to the reference height of 10 m. This is done operationally as follows:

In the atmospheric boundary layer, the power-law wind profile states that

$$\frac{u_2}{u_1} = \left(\frac{z_2}{z_1}\right)^p . \tag{12}$$

From HSU (1988)

$$p = \frac{\sqrt{C_d}}{\kappa} \Phi_m\left(\frac{z}{L}\right) . \tag{13}$$

According to SMITH (1980), when $z/L \geq 0$

$$10^3 C_d = 1.46 - 0.97\frac{z}{L} \tag{14}$$

and when $z/L < 0$

$$10^3 C_d = 1.56 - 0.33\frac{z}{L} \ . \tag{15}$$

Therefore, setting $z/L = 0$ and averaging Eqs. (14) and (15) one gets

$$C_d = 1.51 \times 10^3 \ . \tag{16}$$

Thus,

$$p = 0.10\Phi_m\left(\frac{z}{L}\right) \ . \tag{17}$$

According to PANOFSKY and DUTTON (1984), when $z/L < 0$

$$\Phi_m\left(\frac{z}{L}\right) = \left(1 - 16\frac{z}{L}\right)^{-1/4} \ . \tag{18}$$

When $z/L = 0$

$$\Phi_m\left(\frac{z}{L}\right) = 1 \tag{19}$$

and when $z/L > 0$

$$\Phi_m\left(\frac{z}{L}\right) = 1 + 5\frac{z}{L} \ . \tag{20}$$

Thus, when $z/L < 0$

$$p = 0.10\left(1 - 16\frac{z}{L}\right)^{-1/4} \ . \tag{21}$$

When $z/L > 0$

$$p = 0.10\left(1 + 5\frac{z}{L}\right) \tag{22}$$

and when $z/L = 0$

$$p = 0.10 \ . \tag{23}$$

Further verifications of this $p(= 0.10)$ value under near-neutral conditions are provided in SIMPSON and RIEHL (1981), HSU (1988), and HSU et al. (1994).

On the basis of pertinent data presented in BANNER et al. (1999), under unstable conditions, Eq. (21) is verified as shown in Figure 3. It can be seen that p approaches 0.10 as z/L approaches zero. Therefore, as a first approximation we recommend that Eq. (23) be used for Eq. (12), so that U_{10} may be obtained from any height within the atmospheric boundary layer. It is interesting to note that for offshore applications HSU (1988, Table 8.5, p. 202) has recommended that $p = 0.06$ for unstable, 0.10 for

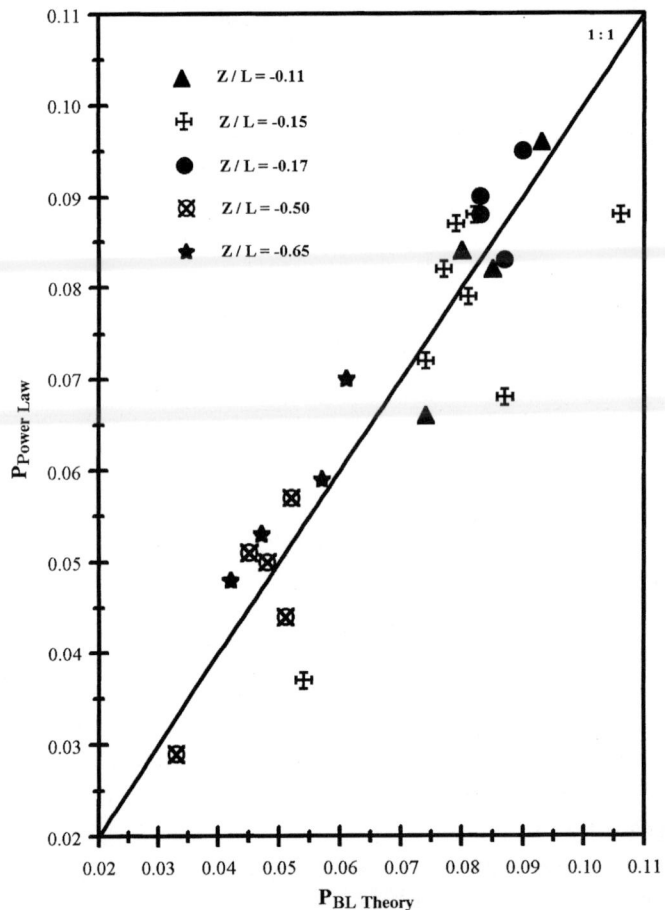

Figure 3

A verification of Eq. (21) under unstable conditions ($z/L \leq -0.11$) using data from BANNER et al. (1999). Root-mean-square error is 0.008.

near-neutral, and 0.27 for stable conditions. If one substitutes $z/L = +0.4$ and -0.4 into Eqs. (21) and (22), respectively, one gets $p = 0.06$ for unstable and 0.30 for stable. Note that $z/L = +0.4$ and -0.4 are the demarcation lines between stable, neutral, and unstable as used in the OCD model (HSU, 1992).

4. Determining Overwater Friction Velocity, u_*

In the atmospheric boundary layer the friction (or shear) velocity, u_*, is related to the wind speed, aerodynamic roughness, and the stability parameter as (see, e.g., PANOFSKY and DUTTON, 1984)

$$U_z = \frac{u_*}{\kappa} \left[\ln \frac{Z}{Z_0} - \psi_m \left(\frac{z}{L} \right) \right] , \tag{24}$$

where U_Z is the wind speed at height Z, $\kappa (= 0.4)$ is the von Karman constant, and Z_0 is the aerodynamic roughness length. For aerodynamically rough flow, when $z/L < 0$, i.e., under unstable conditions (from HSU et al., 1999):

$$u_* = \kappa U_Z \left[\ln \frac{Z}{Z_0} - 1.05 \left(-\frac{z}{L} \right)^{0.46} \right]^{-1} . \tag{25}$$

The effect of wave characteristics on Z_0 needs to be elaborated. According to PANOFSKY and DUTTON (1984), at the surface, eddy viscosity is taken to represent the product of eddy size and eddy velocity, and we see now that Z_0 represents eddy size at the surface. Clearly, the rougher the ground, the larger these eddies can be. Thus Z_0 is a measure of surface roughness, hence it is called the roughness length. Measurements show that Z_0 varies from about 0.01 cm over ice or water to several meters over cities or irregular woods, illustrating that Z_0 is a measure of how efficiently momentum can be transferred to the ground at a given wind speed. Furthermore, since Z_0 represents eddy size at the surface, it depends not only on height but also on shape and spacing of the surface features.

Parameterization of Z_0 over the water surface has been summarized in HSU (1988, p. 116). Briefly, it is related to the wave height and wave age, C_p/u_*, where C_p is the wave speed at the spectral peak. The wave age parameter can be further classified depending on the stage of wave development. Operationally, C_p/u_* is replaced by C_p/U_{10} such that when the waves are not fully-developed, i.e., when $C_p/U_{10} < 1.29$ (from DONELAN et al., 1993):

$$\frac{Z_0}{\sigma} = 6.7 \times 10^{-4} \left(\frac{U_{10}}{C_p} \right)^{2.6} . \tag{26}$$

Substituting $\sigma = H_s/4$ into Eq. (26) and rearranging one gets

$$\frac{Z}{Z_0} = \left(\frac{5.97 \times 10^4}{H_s} \right) \left(\frac{C_p}{U_{10}} \right)^{2.6} , \tag{27}$$

where σ is the rms wave height and H_s is the significant wave height.

For fully-developed seas, when $C_p/U_{10} \geq 1.29$, set $C_p/U_{10} = 1.29$ and $gH_s/U_{10} = 0.2433$,

$$\frac{Z}{Z_0} = \frac{11.57 \times 10^4}{H_s} = \frac{4.66 \times 10^6}{U_{10}^2} . \tag{28}$$

Now, from Eq. (24)

$$u_* = \kappa U_Z \left[\ln \frac{Z}{Z_0} - \psi_m \left(\frac{z}{L} \right) \right]^{-1} \tag{29}$$

or

$$u_* = \kappa U_Z \left[\frac{\kappa}{\sqrt{C_{dn}}} - \psi_m \left(\frac{z}{L} \right) \right]^{-1} \tag{30}$$

$$\therefore \ln \frac{Z}{Z_0} = \frac{\kappa}{\sqrt{C_{dn}}} \;, \tag{31}$$

where C_{dn} is the drag coefficient under near-neutral conditions.

From GARRATT (1992), for $4 < U_{10} < 20$ m s^{-1}

$$C_{dn} = (0.75 + 0.067 U_{10}) 10^{-3} \;. \tag{32}$$

From BANNER et al. (1999), for $5.1 < U_{10} < 19.5$ m s^{-1}

$$C_{dn} = (0.7549 + 0.0713 U_{10}) 10^{-3} \;. \tag{33}$$

Notice that the difference between Eqs. (32) and (33) is small. u_* can also be obtained from Eqs. (6) and (3).

Our results are shown in Figures 4 through 6. Since the RMSE for all three figures is small and compatible, it is concluded that the WAMDI's C_d formulation for u_* can be explained by the wave characteristics. The inclusion of z/L into the u_* determination shown in Eq. (25) does not improve the RMSE much (≈ 8.01 cm s^{-1} in Figure 5 with z/L and 8.09 cm s^{-1} without z/L), hence the mechanical contribution due to wind waves is more significant in u_* determination than the thermal stratification.

Under stable conditions, i.e., when $z/L > 0$, and $\Psi_m(z/L) = -5z/L$ (see PANOFSKY and DUTTON, 1984), Eq. (30) becomes

$$u_* = \kappa U_Z \left[\frac{\kappa}{\sqrt{C_{dn}}} + 5 \frac{z}{L} \right]^{-1} \;. \tag{34}$$

Since wind speeds are generally light under stable conditions, the aerodynamic roughness is said to be in smooth flow (see, e.g., GARRATT, 1992, p. 101); the wind-wave interaction involves capillary waves where surface tension plays an important role. Under these conditions, the C_d formulation shown in Eq. (6) must be modified. According to YELLAND and TAYLOR (1996) and YELLAND et al. (1998)

$$10^3 C_{dn} = \begin{bmatrix} 0.29 + \frac{3.1}{U_{10n}} + \frac{7.7}{U_{10n}^2} & 3 \le U_{10n} \le 6 \, \text{m s}^{-1} \\ 0.50 + 0.071 U_{10n} & 6 < U_{10n} \le 26 \, \text{m s}^{-1} \end{bmatrix} \tag{35}$$

where U_{10n} is the wind speed at 10 m under near-neutral conditions. For operational use, $U_{10n} \approx U_{10}$. Thus, using z/L values as determined from Sec. 2 and C_{dn} from Eq. (35), u_* can be obtained from Eq. (34) under stable conditions. However, more field experiments are needed to further verify Eq. (35), particularly when $U_{10} < 6$ m s^{-1}.

In summary, u_* can be estimated based on Eqs. (25) for unstable and (34) for stable with the incorporation of wave characteristics provided in Eqs. (27) for not

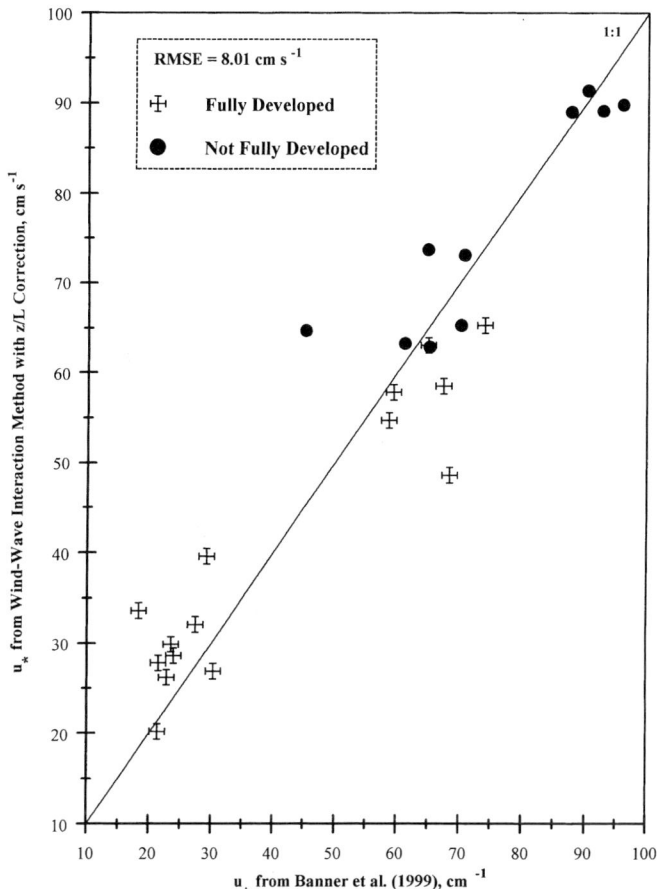

Figure 4

A comparison between Eqs. (25) through (28) and the measurements made by BANNER *et al.* (1999) to determine u_*.

fully developed seas and (28) for fully developed. An example is shown in Figure 7. If one accepts that the composite RMSE (<6 cm s^{-1}) is small as compared to the u_* range from 10 to 60 m s^{-1}, the formulas as recommended for u_* estimation are useful operationally if wave information is available. Otherwise, Eqs. (3) and (6) may be used as a first approximation.

5. The Mixing Height

Due to potential evaporation, the air over the water is usually moister than that over land, and the top of the marine boundary layer is often times capped by clouds (see, e.g., GARRATT, 1992). On the basis of analysis of vertical soundings by research

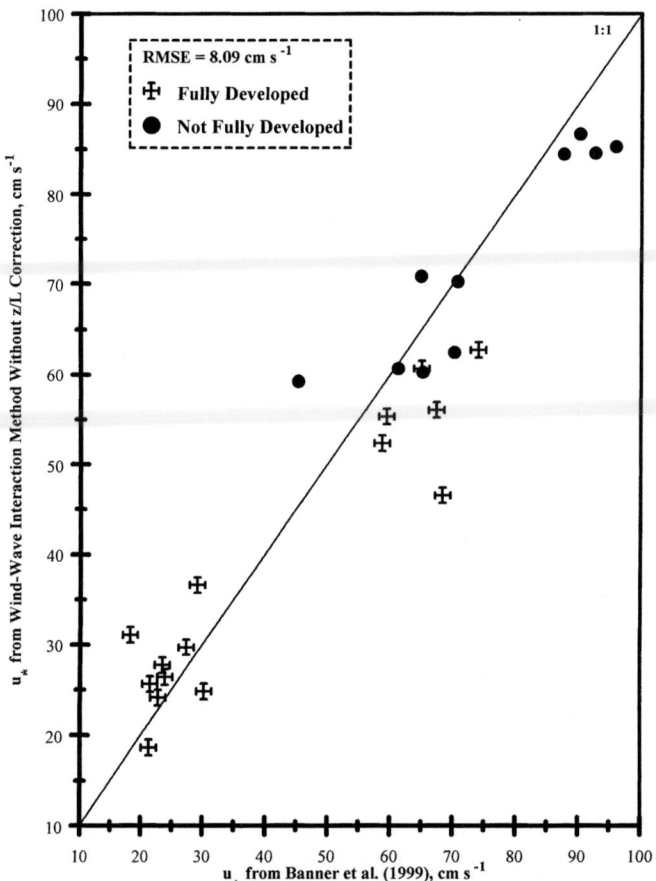

Figure 5

A comparison between Eqs. (26) through (28) and the measurements made by BANNER *et al.* (1999) to determine u_*.

aircraft, rawinsondings, and radar wind profilers and Radio Acoustic Sounding Systems (RASS), it has been shown by GARRATT (1992) that the mixing height h = LCL, the lifting condensation level under cumulus cloud conditions (where LCL = cloud base). The height of the LCL may be estimated by (see HSU, 1998)

$$H_{\text{LCL}} = 125(T_{\text{air}} - T_{\text{dew}}) \; , \tag{36}$$

where H_{LCL} is in meters and the dewpoint depression at the sea surface in degrees Celsius.

If T_{dew} is not available, it may be estimated by (HSU, 1988, p. 21)

$$T_{\text{dew}} = \frac{237.3 \log_{10}\left(\frac{e_{\text{air}}}{6.1078}\right)}{7.5 - \log_{10}\left(\frac{e_{\text{air}}}{6.1078}\right)} \; . \tag{37}$$

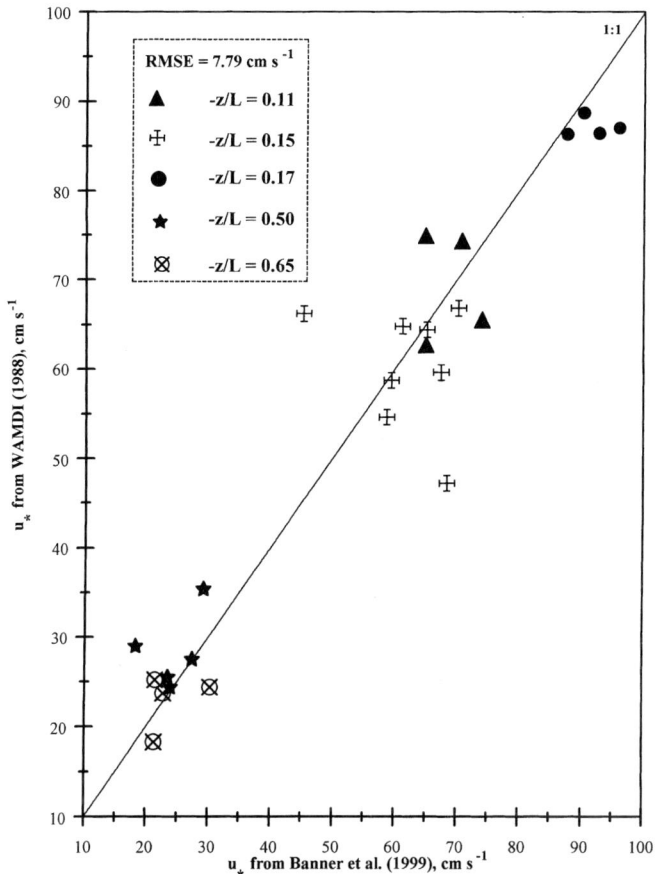

Figure 6

A comparison between Eqs. (34) and (6) and measurements made by BANNER *et al.* (1999) to determine u_*.

From HSU (1998)

$$e_{\mathrm{air}} = \frac{1}{0.62} P q_{\mathrm{air}}$$

and for operational applications (HSU, 1998, Fig. 3)

$$(q_{\mathrm{sea}} - q_{\mathrm{air}}) = 5.68 + 0.37(T_{\mathrm{sea}} - T_{\mathrm{air}}) \tag{38}$$

and

$$q_{\mathrm{sea}} = 0.62 \frac{e_{\mathrm{sea}}}{P} \tag{39}$$

where

$$e_{\mathrm{sea}} = 6.1078 \times 10^{[7.5 T_{\mathrm{sea}}/(237.3 + T_{\mathrm{sea}})]} \quad . \tag{40}$$

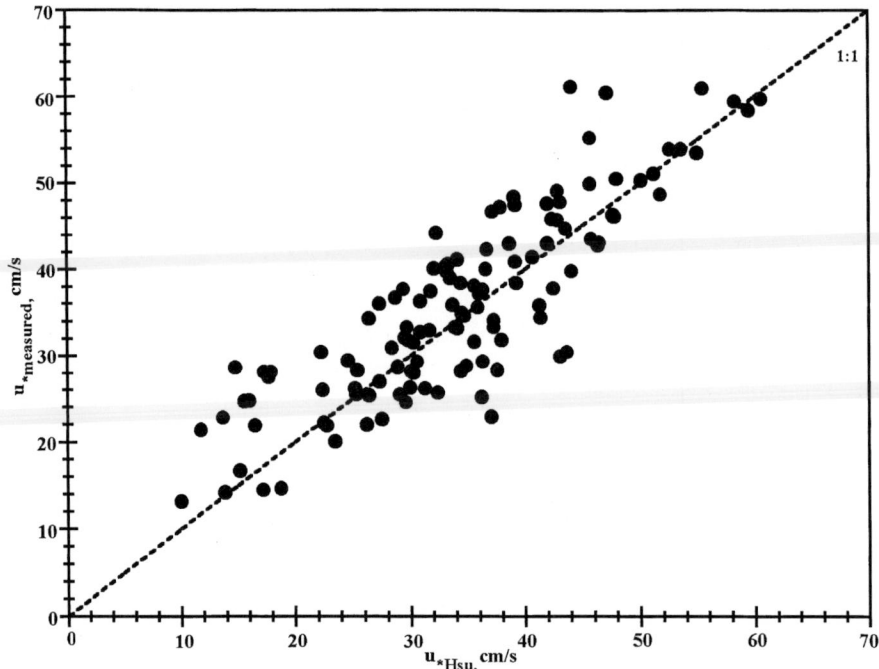

Figure 7

A comparison of u_* estimation based on Eqs. (25) and (34) with the incorporation of wave characteristics provided in Eqs. (27) and (28) against measurements made by DONELAN *et al.* (1997) under both unstable and stable conditions ($-7.69 \leq z/L \leq 0.28$). Root-mean-square error for 120 samples is 5.98 cm/s.

Since both T_{sea} and T_{air} can be obtained routinely by buoys, ships, and satellites, Eq. (36) can be used to estimate the mixing height if fair weather cumulus clouds are present.

On the eastern and Gulf coasts of the U.S., as well as over the East China Sea, cold air outbreaks are common in the winter season. Under these conditions, according to HSU (1997), the mixing height, Z_i, is convectively unstable that

$$Z_i = A + B\left(\overline{w'\theta'_v}\right)_0 \tag{41}$$

and from experiments conducted off the U.S. East Coast and over the East China Sea, Z_i is found to be

$$Z_i = 369 + 6004\left(\overline{w'\theta'_v}\right)_0, \tag{42}$$

where Z_i is in meters and $\left(\overline{w'\theta'_v}\right)_0$ is the buoyancy flux in meters per second Kelvin.

For operational applications, the buoyancy flux at the sea surface is found to be

$$\left(\overline{w'\theta'_v}\right)_0 = C_T U_{10}(T_{\text{sea}} - T_{\text{air}})\left(1 + \frac{0.07}{B}\right), \tag{43}$$

where $C_T = 1.10 \times 10^{-3}$ under unstable conditions (SMITH, 1980) and B is the Bowen ratio which has been provided in Eq. (5b).

Alternately, if we know the mixing height on land, H_{land} (in meters), and a barotropic boundary layer across the coastal zone exists, we may estimate the mixing height over the water, H_{sea} (in meters) by applying (HSU, 1988, p. 183)

$$H_{sea} = H_{land} \left(\frac{C_{dsea}}{C_{dland}} \right) \left(\frac{U_{sea}}{U_{land}} \right)^2 \tag{44}$$

where C_{dland} is the drag coefficient on land, C_{dsea} is based on Eq. (6), U_{sea} and U_{land} (both in m s^{-1}) are the wind speeds at sea and over land, respectively.

If the planetary boundary layer is baroclinic across the coastal zone, i.e., under the land and sea breeze effects, according to HSU (1988, p. 204)

$$H_{sea} = H_{land} - 123(T_{land} - T_{sea}) \ , \tag{45}$$

where T_{land} and T_{sea} (both in °C) are the air temperatures over land and sea, respectively.

In certain geographic regions and occasionally elsewhere, the sea-surface temperature is less than the air temperature. Therefore, the boundary layer is said to be stable. The mixing height under stable conditions, h_{stable}, is (see GARRATT, 1992)

$$h_{stable} = c^* \sqrt{\frac{u_* L}{f}} \ , \tag{46}$$

where f is the Coriolis parameter. From limited measurements under stable conditions, the coefficient c^* is found to be 0.11 as shown in Table 1. Certainly, more field experiments are needed to further substantiate Eq. (46) and the value of c^*.

6. Standard Deviations of Crosswind (σ_y) and Vertical (σ_z) Directions

Some dispersion models require σ_y and σ_z in which PASQUILL (1971) suggested that

$$\sigma_y = \sigma_v t S_y \left(\frac{t}{T_L} \right) \tag{47}$$

$$\sigma_z = \sigma_w t S_z \left(\frac{t}{T_L} \right) \tag{48}$$

or

$$\sigma_y = \left(\frac{\sigma_v}{U_{10}} \right) \times S_y \left(\frac{t}{T_L} \right) \tag{49}$$

Table 1

Measurements of the Height of the Stable Marine Atmospheric Boundary Layer over the Northeast Gulf of Mexico by a Research Aircraft

Profile #	u_* m/s	L m	h_{stable} m	c^*	z/L
4	0.29	233	200	0.208	0.043
5	0.24	420	60	0.051	0.024
6	0.18	120	60	0.110	0.083
7	0.34	403	120	0.088	0.025
10	0.25	390	75	0.065	0.026
11	0.23	171	100	0.136	0.058
Mean	0.26	290	103	0.110	0.043

Notes:
1) Data source: FAIRALL et al., 1980.
2) According to GARRATT (1992, p. 166)

$$\therefore h_{stable} = c^* \sqrt{\frac{u_* L}{f}}$$

$$\therefore c^* = \frac{h_{stable}}{\sqrt{\frac{u_* L}{f}}} \quad .$$

3) For comparison, c^* varies from 0.13 to 0.43 on land (GARRATT, 1992).

$$\sigma_z = \left(\frac{\sigma_w}{U_{10}}\right) \times S_z\left(\frac{t}{T_L}\right) \quad . \tag{50}$$

For overwater applications, the turbulence intensities are found to be (GEERNAERT et al., 1987)

$$\frac{\sigma_v}{U_{10}} = 0.0586 + 0.0012U_{10} \pm 0.015 \tag{51}$$

and

$$\frac{\sigma_w}{U_{10}} = 0.0369 + 0.0010U_{10} \pm 0.005 \quad . \tag{52}$$

All symbols are conventional in air-pollution meteorology (see, e.g., ZANNETTI, 1990).

7. The Vertical Eddy Diffusivity

In some K-diffusion models, the vertical eddy diffusivity, K_z, is required, e.g. (HANNA, 1984)

$$K_z = cu_*Z\left(1 - \frac{Z}{h}\right) \,,$$

(53)

where u_* is the friction velocity and h is the mixing height. For example, the Urban Airshed Model (UAM), the preferred model of the U.S. Environmental Protection Agency for ozone studies, has a meteorological preprocessing subroutine which employs the gradient transport (K) modeling (ZANNETTI, 1990, p. 234). Since in the K-theory, u_* is required for the computation of vertical diffusivity, K_z, and concentration, χ, it is recommended that proper formulas to reflect the overwater variations in aerodynamic roughness length, Z_0, and drag coefficient, C_d, be made in light of recent advances in the air–sea interaction field as shown in Sec. 4.

8. Summary

In air quality modeling, many parameters are required. Therefore, parameterization is needed. Because the sea surface is mobile due to various wind waves and ocean currents, this review attempts to synthesize those parameterizations specifically applied to the marine environment for the determination of stability length, variation of the wind speed with height, overwater friction velocity, the mixing height, and vertical eddy diffusivity. Many more field experiments are needed in order to further substantiate these results. It is felt that the review provided here offers a guide for practical applications since many *in situ* air–sea interaction parameters may not be available over vast regions of the ocean.

Acknowledgements

This study was supported in part by the U.S. Minerals Management Service under Contract 14-35-0001-30660 Task Order 19925. The first author also gratefully acknowledges the support of Dr. Eric L. Abraham in the form of an Endowed Professorship to LSU.

Appendix

B Bowen ratio
C_d Surface drag coefficient
C_{dn} Drag coefficient under neutral stability
C_p Wave speed at the spectral peak
C_{ph} Specific heat of air at constant pressure

C_T Heat transfer coefficient
c^* A coefficient for mixing height under stable conditions
e_{air} Water vapor pressure at the surface of air
e_{sea} Water vapor pressure at saturation of sea water
f Coriolis parameter
g Acceleration due to gravity
H Sensible heat flux
H_s Significant wave height
H_{land} Mixing height onshore
H_{sea} Mixing height offshore
H_{LCL} Height of lifting condensation level
h_{stable} Height of the surface layer under stable conditions
P Air pressure
p Exponent of power-law wind profile
q_{air} Specific humidity
q_{sea} Specific humidity of water
Ri Richardson number
S_y Lagrangian spectrum in y direction
S_z Lagrangian spectrum in z direction
T_{air} Air temperature
T_{sea} Sea-surface temperature
T_{dew} Dew-point temperature
T_L Lagrangian integral time scale
t Time
$\bar{U}$ Mean wind speed
U_z Wind speed at height z
U_{10} Wind speed at 10 m above sea surface
U_{10n} U_{10} under neutral conditions
u_* Friction velocity
w Instantaneous velocity component in z (vertical) direction
w' Fluctuating velocity component in z (vertical) direction
Z_i The mixing height
Z_0 Aerodynamic roughness length
z Coordinate position along z axis
ξ Monin-Obukhov stability parameter
$\bar{\theta}$ Mean potential temperature of air
θ_v Virtual potential temperature
θ'_v Fluctuating virtual potential temperature
ρ Mass density of air
σ RMS wave height
σ_v Standard deviation of velocity fluctuations in y direction
σ_w Standard deviation of velocity fluctuations in z direction

σ_y Standard deviation of wind speed in y direction
σ_z Standard deviation of wind speed in z direction
Φ_m Dimensionless wind shear
Ψ_m The Monin-Obukhov similarity function for normalized velocity

REFERENCES

ARYA, S. P., *Air Pollution Meteorology and Dispersion* (Oxford University Press, Oxford, 1999).

BANNER, M. L., CHEN, W., WALSH, E. J., JENSEN, J. B., LEE, S., and FANDRY, C. (1999), *The Southern Ocean Waves Experiment. Part I: Overview and Mean Results*, J. Phys. Oceanogr. *29*, 2130–2145.

BUSINGER, J. A., WYNGAARD, J. C., IZUMI, Y., and BRADLEY, E. F. (1971), *Flux-profile Relationships in the Atmospheric Boundary Layer*, J. Atmos. Sci. *28*, 181–189.

DONELAN, M. A., DOBSON, F., SMITH, S. D., and ANDERSON, R. A. (1993), *On the Dependence of Sea Surface Roughness on Wave Development*, J. Phys. Oceanogr. *23*, 2143–2149.

DONELAN, M. A., DRENNAN, W. M., and KATSAROS, K. B. (1997), *The Air–sea Momentum Flux in Conditions of Wind Sea and Swell*, J. Phys. Oceanogr. *27*, 2087–2099.

FAIRALL, C. W., MARKSON, R., SCHACHER, G. E., and DAVIDSON, K. (1980), *An Aircraft Study of Turbulence Dissipation Rate and Temperature Structure Function in the Unstable Marine Atmospheric Boundary Layer*, Boundary-Layer Meteorol. *19*, 453–469.

GARRATT, J. R., *The Atmospheric Boundary Layer* (Cambridge University Press, Cambridge, 1992).

GEERNAERT, G. L., LARSEN, S. E., and HANSEN, F. (1987), *Measurements of the Wind Stress, Heat Flux, and Turbulence Intensity During Storm Conditions Over the North Sea*, J. Geophys. Res. *92* (C12), 13,127–13,139.

HANNA, S. R., BRIGGS, G. A., and HOSKER, R. P., Jr., *Handbook on Atmospheric Diffusion* (Technical Information Center, U.S. Dept. of Energy, DOE/TIC-11223, 1982).

HANNA, S. R., *Applications in air pollution modeling*, In: *Atmospheric Turbulence and Air Pollution Modeling* (eds. F. T. M. Nieuwstadt and H. Van Dop) (D. Reidel Pub. Co. 1984) pp. 275–310.

HANNA, S. R., SCHULMAN, L. L., PAINE, R. J., PLEIM, J. E., and BAER, M. (1985), *Development and Evaluation of the Offshore and Coastal Dispersion Model*, J. Air Poll. Contr. Assoc. *35*, 1039–1047.

HSU, S. A., *Coastal Meteorology* (Academic Press, San Diego, CA, 1988).

HSU, S. A. (1992), *An Overwater Stability Criterion for the Offshore and Coastal Dispersion Model*, Boundary-Layer Meteorol. *60*, 397–402.

HSU, S. A., MEINDL, E. A., and GILHOUSEN, D. B. (1994), *Determining the Power-law Wind-profile Exponent under Near-neutral Stability Conditions at Sea*, J. Appl. Meteor. *33*, 757–765.

HSU, S. A. (1997), *Estimating Overwater Convective Boundary Layer Height from Routine Meteorological Measurements for Diffusion Applications at Sea*, J. Appl. Meteor. *36*, 1245–1248.

HSU, S. A. (1998), *A Relationship between the Bowen Ratio and Sea–air Temperature Difference under Unstable Conditions at Sea*, J. Phys. Oceanogr. *28*, 2222–2226.

HSU, S. A. (1999), *On the Estimation of Overwater Bowen Ratio from Sea–air Temperature Difference*, J. Phys. Oceanogr. *29*, 1372–1373.

HSU, S. A., BLANCHARD, B. W., and YAN, Z. (1999), *A Simplified Equation for Paulson's Ψ_m (Z/L) Formulation for Overwater Applications*, J. Appl. Meteor. *38*, 623–625.

KOMEN, G. J., CAVALERI, L., DONELAN, M., HASSELMANN, K., HASSELMANN, S., and JANSSEN, P. A. E. M, *Dynamics and Modelling of Ocean Waves* (Cambridge University Press, 1994).

LYONS, T. and SCOTT, B., *Principles of Air Pollution Meteorology* (CRC Press, Boca Raton, FL, 1990).

NATIONAL DATA BUOY CENTER (1990), *Climatic Summaries for NDBC Buoys and Stations, Update 1.* Available through NDBC, Stennis Space Center, MS 39529.

PASQUILL, F. (1971), *Atmospheric Dispersion of Pollution*, Quart. J. Roy. Meteor. Soc. *97*, 369–395.

PANOFSKY, H. A. and DUTTON, J. A., *Atmospheric Turbulence* (Wiley, New York, 1984).

SIMPSON, R. H. and RIEHL, H., The Hurricane and its Impact (Louisiana State University Press, Baton Rouge, LA, 1981).

SMITH, S. D. (1980), *Wind Stress and Heat Flux Over the Ocean in Gale Force Winds*, J. Phys. Oceanogr. *10*, 709–726.

SMITH, S. D. (1988), *Coefficients for Sea Surface Wind Stress, Heat Flux, and Wind Profiles as a Function of Wind Speed and Temperature*, J. Geophys. Res. *93*, 15,467–15,474.

SORBJAN, Z. (1986), *On Similarity in the Atmospheric Boundary Layer*, Boundary-Layer Meteorol. 34, 377–397.

THE WAMDI GROUP (1988), *The WAM Model – A Third Generation Ocean Wave Prediction Model*, J. Phys. Oceanogr. *18*, 1775–1810.

YELLAND, M. and TAYLOR, P. K. (1996), *Wind Stress Measurements From the Open Ocean*, J. Phys. Oceanogr. *26*, 541–558.

YELLAND, M. J., MOAT, B. I., TAYLOR, P. K., PASCAL, R. W., HUTCHINGS, J., and CORNELL, V. C. (1998), *Wind Stress Measurements from the Open Ocean Corrected for Airflow Distortion by the Ship*, J. Phys. Oceanogr. *28*, 1511–1526.

ZANNETTI, P., *Air Pollution Modeling* (Van Nostrand Reinhold, New York, 1990).

(Received March 1, 2000, accepted Sept. 17, 2000)

To access this journal online:
http://www.birkhauser.ch

Pure appl. geophys. 160 (2003) 317–324
0033–4553/03/020317–08

▌**Pure and Applied Geophysics**

Temperature Variation in the Urban Canopy with Anthropogenic Energy Use

HIROAKI KONDO[1] and YUKIHIRO KIKEGAWA[2]

Key words: Urban canopy, urban warming, cooling energy, anthropogenic heat.

Introduction

One of the detrimental effects caused by the urban warming is the increase of energy consumption due to the air conditioning of buildings in summer. In the cities of United States, the urban warming is surmised to increase the peak electric energy demand by 3 to 6% with 1.0 °C temperature rise (BRETZ et al., 1998). This increased rate of demand is estimated up to 3%/°C in recent years in Tokyo, and about 1.6 GW of new demand is required as the daily maximum temperature increases by 1.0 °C in the greater Tokyo area (SAKAI and NAKAMURA, 1999). Most of this huge demand of summer electricity is caused by the air-conditioning systems, and is considered to be one of the common characteristics in big cities of Asian countries. From the viewpoint of the reduction of CO_2 emission to mitigate the global warming, this huge demand should be reduced through the control of the urban warming. There were model results that analysed the relation between anthropogenic heat and temperature increase in Tokyo (URANO et al. (1999), ICHINOSE et al. (1999) etc.). However, they used mesoscale model only with static data of anthropogenic heat. Their results are too coarse, because the temperature in the city block highly depends on its structure (MURAKAMI et al., 2000) and the anthropogenic heat release dynamically depends on the ambient temperature. In the present study, a multi-scale numerical simulation system is developed to evaluate dynamically the increase of energy demands caused by the urban warming, and a case study is carried out for an urban canopy over a densely urbanized area in Tokyo.

---

[1] National Institute of Advanced Industrial Sciences and Technology, Tsukuba, Ibaraki 3058569, Japan. E-mail: kondo-hrk@aist.go.jp

[2] Fuji Research Institute Corporation, 2–3, Kanda-Nishikicho, Chiyoda-ku, Tokyo 1018443, Japan. E-mail: kike@cyg.fuji-ric.co.jp

Models

To carry out the investigation, a combination of multi-scale model system was developed as shown in Figure 1. This system consists of three numerical models: MM, CM, and BEM. MM is a three-dimensional mesoscale meteorological model developed at the National Institute for Resources and Environment (KONDO, 1995). MM was used mainly to generate the initial and upper boundary conditions for CM. For the urban canopy layer (UCL), we also developed a new one-dimensional urban canopy model (CM).

The basic equations of CM are one-dimensional diffusion equations,

$$\frac{\partial u}{\partial t} = \frac{1}{m}\frac{\partial}{\partial z}\left(K_m \cdot m \cdot \frac{\partial u}{\partial z}\right) - cau\left(\sqrt{u^2 + v^2}\right) + F_u \tag{1}$$

$$\frac{\partial v}{\partial t} = \frac{1}{m}\frac{\partial}{\partial z}\left(K_m \cdot m \cdot \frac{\partial v}{\partial z}\right) - cav\left(\sqrt{u^2 + v^2}\right) + F_v \tag{2}$$

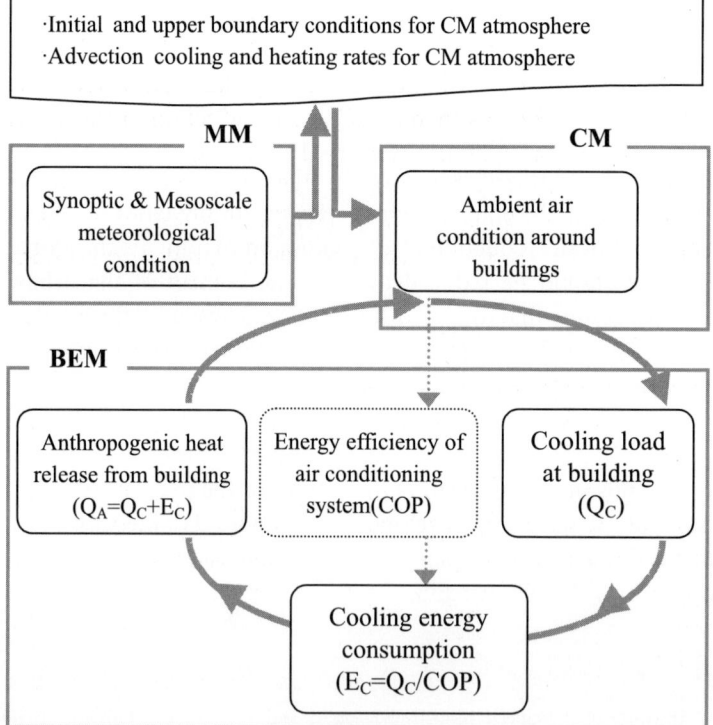

Figure 1
Methodology of modeling.

$$\frac{\partial\theta}{\partial t} = \frac{1}{m}\frac{\partial}{\partial z}\left(K_h \cdot m \cdot \frac{\partial\theta}{\partial z}\right) + \frac{Q_A}{c_p\rho} + F_\theta \ , \tag{3}$$

where u and v are the wind velocity component of x direction (positive towards the east) and y direction (positive towards the north). We consider an urban block of 0.5 to 1 km square in which the building bottom has the same area of the square (Fig. 2). The length of the building side is assumed as b and the distance between the buildings is assumed as w. We assume that four side walls of buildings are exactly directed north, south, east and west. The height of the building is not uniform, but the distribution of the height can be considered. We define the floor density ($P_w(z)$) of the considering area at level z ($0 \le P_w(z) \le 1$). Then,

$$m = 1 - \frac{b^2}{(w+b)^2} \cdot P_w(z) \tag{4}$$

$$a = \frac{b \cdot P_w(z)}{(b+w)^2 - b^2 \cdot P_w(z)} \ , \tag{5}$$

and c is a constant ($=0.1$). K_m and K_h are vertical turbulent diffusion coefficients based on GAMBO (1978), in which the scale length was modified with a similar idea in the plant canopy (WATANABE and KONDO, 1990). F_u, F_v and F_θ are effects of mesoscale advection which were obtained from MM calculation. The effect of complicated radiation processes among the buildings was considered to some extent (KONDO and LIU, 1998), such as shading effect and reflection of short wave, and re-emission of long wave from building and ground surfaces.

For the consideration of the dynamical variations of the anthropogenic heat released from air-conditioning systems in the buildings, BEM was developed. BEM is a simple sub-model for building energy analysis coupled with CM, and the

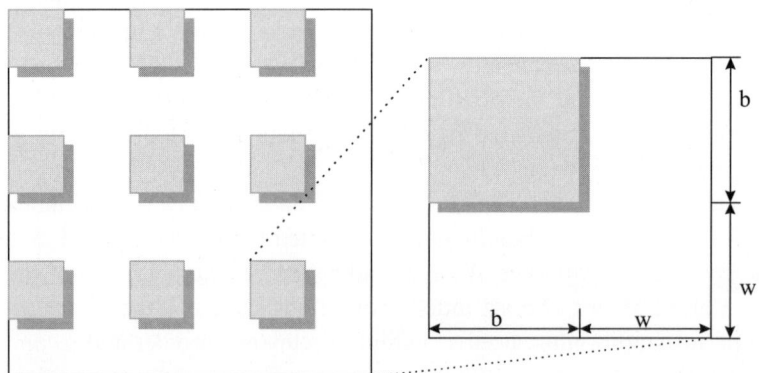

Figure 2
Assumed configuration of buildings in CM.

anthropogenic heat from buildings was calculated from the air temperature and humidity in the time integration loop of CM. BEM is a box-type heat budget model where a building in the urban block is treated as a box, and the thermal load in the buildings is calculated for the sensible and the latent heat components, separately.

To calculate the sensible heat component, we consider the heat exchange through the walls between indoor and outdoor, transmission of solar insolation through the windows, sensible heat exchange through ventilation, and the internal heat generation from machines and occupants. The outdoor condition is calculated from CM at each time step. For the latent heat component, we consider the water vapor intrusion through ventilation and evaporation from occupants. We also consider the overall heat capacity of the air in the building including interior equipment such as furniture and the overall volume of the air in the building. We simply assume that sensible and latent heat, which should be extracted by the air-conditioning systems, is a product of the thermal load and γ ($\gamma < 1.0$). Here, γ is the ratio of floor area under air-conditioning to the total floor area in a building, and depends upon the operational schedule of the air-conditioning systems.

Subsequently, energy demand required by the air-conditioning systems in a building is calculated using COP, which stands for the coefficient of performance of the heat pump system which represents the overall energy efficiency of air-conditioning. In BEM, the dependency of COP upon the ambient air temperature around the outdoor heat exchanger can be considered for several typical heat pump systems. Finally, the exhaust heat from air-conditioning systems of buildings, Q_A, is dynamically computed, which is equivalent to the sum of the demanded energy by the systems and the extracted thermal load itself.

Computational Results

Simulations were conducted for the Ootemachi area, the central business district in Tokyo (Fig. 3). The CM requires the parameters related with structures of the urban blocks; the averaged length of a building's base (b), the averaged distance between buildings (w) and the distribution of the floor density of buildings ($P_w(z)$). These parameters were calculated from the GIS data over an area of 500 m square including the building from which the observation of ambient temperature in Figure 4 was carried out on the roof. As for the structures of the buildings, the parameters of typical office buildings were adopted for the widths and materials of walls, the coverage of windows, thermal and radiative characteristics of the surface (Table 1). For BEM, we also set indoor conditions such as target temperature and humidity of air-conditioning, hourly building occupancy profiles and so on. As heat source machines in the air-conditioning systems, we employed two popular types of heat pumps. One is the air source heat pump which releases the exhaust heat from outdoor heat exchangers as sensible heat, and the other is the absorption type hot

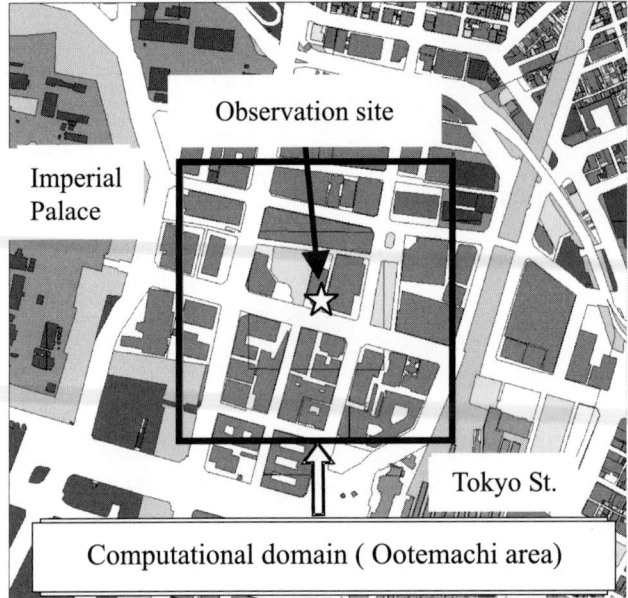

Figure 3
Computational domain.

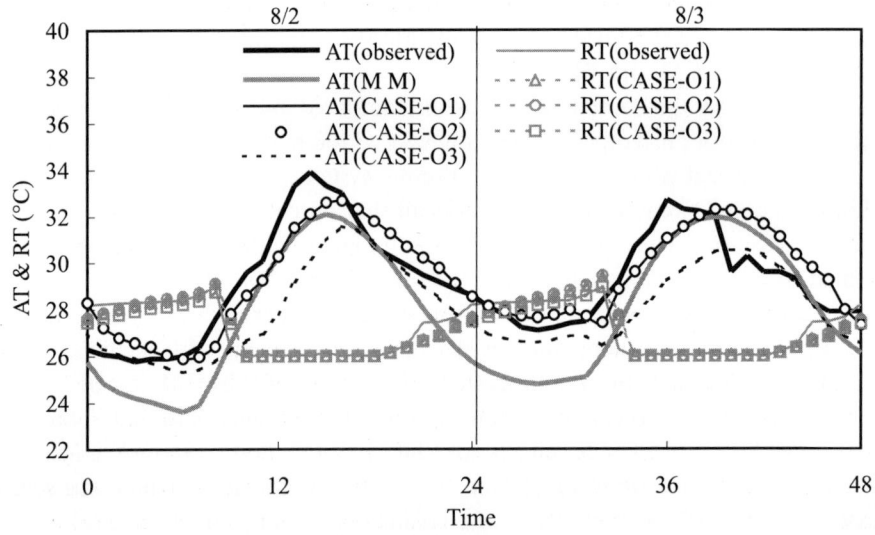

(AT : Air temperature , RT : Room temperature)

Figure 4
Computed and observed room temperatures, and those of ambient air temperature at 100 m above the
ground. Observation of AT was carried out at a roof of a building in which RT was observed.

Table 1

Parameters of the materials used in the calculation

	Surface albedo	Volumetric heat capacity (J m^{-3} K^{-1})	Thermal conductivity (J m^{-1} s^{-1} K^{-1})	Note
Ground (0–16 cm)	0.2 (0.15)	1.93×10^6	1.39	Concrete (with 10% of vegetation)
Ground (16–48 cm)	–	1.74×10^6	1.00	Loam
Roof materials (insulator)	–	0.06×10^6	0.04	Poli-ethylene foam
Roof materials (other part)	0.2	1.93×10^6	1.39	Concrete
Wall materials (insulator)	–	0.06×10^6	0.04	Poli-ethylene foam
Wall materials (other part)	0.2 (0.4)	1.93×10^6	1.39	Concrete (with 30% of window)

and chilled water generator which mainly releases the latent exhaust heat from cooling towers. The former is driven by electricity, and the latter by town gas. We assumed the ratio of them in a building to be 1:1, based on their market share for the Japanese office buildings in recent years. To obtain the initial and upper boundary conditions of the CM, simulations were performed using MM under the condition of typically two consecutive summer days, the 2nd and 3rd of August, 1998. The cooling and heating rates due to the mesoscale advection were also calculated in MM for CM. With the above conditions, the calculations were executed for two days with the coupled model of CM and BEM.

Figure 4 shows the comparison between computed and measured room temperature, and outdoor air temperature at a height of 100 m on the roof of the building. Anthropogenic heat sources of air-conditioning systems were put on the roofs of the buildings in case-O1(control run), whereas all the sources were set at 3 m above the ground in case-O2. The roof height of each building in the considered region was obtained from the GIS data. In case-O3, anthropogenic heat was assumed not to be released into the atmosphere but elsewhere such as sewage or ground water.

The calculated room temperatures show good agreements with the observations in all cases. As for outdoor air temperature at 100 m, results by CM and BEM (case-O1~O2) show better agreements with the observations compared to that obtained by using only MM, where the effects of the urban canopy are not considered. Particularly, CM and BEM reproduce the nocturnal course of temperature more realistically than MM does. Among the results by CM and BEM, case-O1 and O2 manifest better correspondence to the observations. This is reasonable, because many buildings in Ootemachi have heat exchangers of air-conditioning systems on their roofs as heat sources. When the release of anthropogenic heat is fully cut off in case-O3, the daily averaged temperature decreases by 1 °C compared to those in case-O1

and O2 at 100 m. At this height of upper UCL, there is scant difference in daily courses of the air temperature between case-O1 and O2 in spite of their differences in the levels of heat sources. However, their levels remarkably affect the air temperatures at the lower UCL. At 3 m above the ground, anthropogenic heat from this height increases the daily averaged temperature by 0.6 °C in case-O2 compared to that in case-O1. The temperature differences between case-O1 and O3 are also slightly increased by up to 1.3 °C in comparison with those at 100 m. This influence of anthropogenic heat on the air temperature in the lower UCL is caused by worse ventilation there.

Conclusions

A multi-scale simulation model system (mesoscale to building scale) was developed to estimate the increase of the cooling energy demands produced by urban warming in the summer. The system was applied to the Ootemachi area, a central business district in Tokyo. The computed cooling demand of electricity is also verified from the viewpoint of their sensitivity to daily maximum temperatures. If we take results of case-O1, O2 and O3, we can estimate the temperature sensitivity of the peak electricity demand to be 6.6%/°C. This sensitivity shows good agreement with actual regional averaged sensitivity over the Tokyo metropolitan area, which is reported as 6.51%/°C by the TEPCO (TEPCO, 1998). Preliminary verification of the models with observational data of outdoor and indoor conditions and of the cooling thermal load at a building in Ootemachi showed good results. In the near future of our study, we intend to apply the models to an evaluation of the countermeasures against the urban warming, from the viewpoint of urban energy savings, to stop the increase of anthropogenic CO_2 emission.

Acknowledgment

This study was supported by the Proposal-Based New Industry Creative Type Technology R&D Promotion Program from the New Energy and Industrial Technology Development Organization (NEDO) of Japan.

REFERENCES

BRETZ, S., AKBARI, H., and ROSENFELD, A. (1998), *Practical Issues for Using Solar-reflective Materials to Mitigate Urban Heat Islands*, Atmos. Environ. *32*(1), 95–101.
GAMBO, K. (1978), *Notes on the Turbulence Closure Model for Atmospheric Boundary Layers*, J. Meteor. Soc. of Japan *56*, 466–480.

ICHINOSE, T., SHIMODOZONO, K., and Hanaki, K. (1999), *Impact of Anthropogenic Heat on Urban Climate in Tokyo*, Atmos. Environ. *33*, 3897–3909.

KONDO, H. (1995), *The Thermally Induced Local Wind and Surface Inversion over the Kanto Plain on Calm Winter Nights*, J. Appl. Meteor. *34*(6), 1439–1448.

KONDO, H. and LIU, F. H. (1998), *A Study on the Urban Thermal Environment Obtained through One-dimensional Urban Canopy Model*, J. of Japan Soc. Atmos. Environ. *33*, 179–192 (in Japanese).

MURAKAMI, S., MOCHIDA, A., KIM, S., OOKA, R., YOSHIDA, S., KONDO, H., GENCHI Y., and SHIMADA, A. (2000), *Software Platform for the Total Analysis Wind Climate and Urban Heat Island. –Integration of CWE Simulation from Human Scale to Urban Scale-*. Proc. 3rd International Symposium on Computational Wind Engineering, pp. 23–26.

SAKAI, H. and NAKAMURA, K. (1999), *Utilization of Weather Information at an Electric Power Company*, SHASE *73*, 555–561 (in Japanese).

TOKYO ELECTRIC POWER COMPANY COMMUNICATIONS DEPT. (1998), TEPCO Illustrated.

URANO, A., ICHINOSE, T., and HANAKI, K., (1999), *Thermal Environment Simulation for Three-dimensional Replacement of Urban Activity*, J. Wind Eng. Ind. Aerodyn. *81*, 197–210.

WATANABE, T. and KONDO, J. (1990), *The Influence of Canopy Structure and Density upon the Mixing Length within and above Vegetation*. J. Meteor. Soc. Japan *68*, 227–235.

(Received March 1, 2000, accepted November 2, 2000)

To access this journal online:
http://www.birkhauser.ch

Pure appl. geophys. 160 (2003) 325–339
0033–4553/03/020325–15

© Birkhäuser Verlag, Basel, 2003

▌Pure and Applied Geophysics

Air Pollution Studies in Metromanila and Catalysis Technology Towards Clean Air Philippines

Susan M. Gallardo[1]

Abstract—Considerable air quality and emission data gathered in Metropolitan Manila (MM) led to the development of automobile exhaust treatment catalysts as well as their continued improvement. Findings of a 5-year (1993–1998) collaborative work on the development of base metal oxide catalysts for automobile exhaust are summarized here.

One study in 1991 reveals an average 16% increase in the number of motor vehicles in MM where 16% are new and the rest are old ones. Another study in 1992 shows the CO and hydrocarbon emission levels from different types of motor vehicles in MM as a function of the age of the vehicle, type of fuel, and the operating condition. Reports of the Department of Environment and Natural Resources (DENR) and other related studies also provided data showing the quality of air in MM.

Currently, there are several requirements to further improve the catalyst performance towards the reduction of NO_X and to develop catalyst-sorbent for simultaneous NO_X-SO_X removal. This is so because of the present condition of rain acidification that is found in certain places in MM. These air quality and emission data are needed not only to establish practical emission standards for motor vehicles and the stationary industries and power plants but also in the development of technologies for air pollution control and other clean technologies for cleaner air in the country.

Key words: Air quality, catalysis, emission data, acid rain, oxidation catalysts, air pollution.

Introduction

Air pollution is the presence of undesirable substances in the air; the quantities of which are large enough to produce harmful or deleterious effects on human health, vegetation, materials and visibility. There are several types of air pollutants, mainly inorganic gases, organic gases and particulate matter (PM). Among the inorganic gases, the oxides of nitrogen (NO_X), oxides of sulfur (SO_X), and the oxides of carbon (CO_X) are common. Hydrocarbons and oxygenated hydrocarbons are some of the organic gases. Particulate matter may be solid fumes, dusts, smoke, lead and ash or liquid type such as mist, oil and grease.

Five major air pollutants such as carbon monoxide, sulfur oxides, hydrocarbons, particulate matter, and nitrogen oxides are found in urban atmosphere. They may be

[1] Chemical Engineering Department, De la Salle University, 2401 Taft Avenue, Manila, Philippines. E-mail: coesmg@mail.dlsu.edu.ph, gallards@philonline.com.ph

derived from mobile sources and from stationary sources. In the United States there are about 123 million autos and about 40 million light trucks that are similar to autos in use (DE NEVERS, 1995). Half of the total CO emissions and about 35 percent of hydrocarbon and NO_X emissions emanate from motor vehicles. Motor vehicles also emit particles and SO_2 although their contribution is much less than other sources. Since the removal of tetraethyl lead from gasoline in 1970, the contribution of motor vehicles to atmospheric lead particles is considerably smaller now. Autos are found mostly in urban areas and thus air pollution concern in these areas is mostly with automobiles. This is the same trend that may be shown in other big cities such as Tokyo (TOKYO DATA BOOK ON ENVIRONMENT, 1989).

In developing countries the situation is somewhat different because of the differences in economic conditions, social, political and cultural attitudes. In this regard, surveys of the air pollution problem in Metropolitan Manila were conducted.

The 1991 Air Pollution Study in Metropolitan Manila

Metropolitan Manila or Metromanila (MM) is the busiest and thickly populated part of the Philippines. It has a total land area of 636 km² and a population of 7,928,867. It has 17 municipalities. Table 1 shows the 1990 population and land area of each municipality. Manila is shown to be the most densely populated part of MM

Table 1

Population (1990) and land area of Metromanila (Source: National Statistics Office)

	Land Area (km²)	Population
Philippines	300,000	60,546,009
Metromanila	636	7,928,867
Caloocan	55.8	761,001
Manila	38.3	1,598,918
Pasay City	13.9	366,623
Quezon City	166.2	1,666,766
Las Pinas	41.5	296,851
Makati	29.9	452,734
Malabon	23.4	278,380
Mandaluyong	26.0	244,538
Marikina	38.9	310,010
Navotas	2.6	186,799
Paranaque	38.3	307,717
Pasig	13.0	397,309
Pateros	10.4	51,401
San Juan	10.4	126,708
Taguig	33.7	266,080
Valenzuela	47.0	340,050

followed by Quezon City. Air pollution in these cities is derived mainly from mobile sources. Some are the effect of power generating plants and industrial plants. Table 2 presents an accounting of several pollutants from various sources done by the Department of Environment and Natural Resources. Particulate matter (PM), CO, NO_X and total organic gases (TOG) are derived mainly from mobile sources while SO_X originate from stationary sources. Thus, most of the harmful pollutants found in MM come from the transport industry or mobile sources. The Land Transportation Office (LTO) provides the data of the number of registered vehicles in MM. Table 3 indicates the number of motor vehicles that are old (renewal) and new. In the table, UV is for utility vehicles while MC/TC are motorcycles and tricycles. Tricycles are motorcycles that have sidecars for two passengers.

The number of motor vehicles registered in 1990 increased by 16.41% as compared to 1989. Of these, only 16% are new and the rest are old ones. 73.4% are gasoline fed and 25.2% are diesel fed.

DENR data shown in Table 4 gives an estimate of the emissions of the different air pollutants from motor vehicles classified by the fuel used. Here, gasoline-run

Table 2

Total estimate for 1989 of air pollution emissions in tons/year from various sources (Source: DENR-NCR)

Sources	PM	CO	SO_X	NO_X	TOG
Mobile	24,368	280,279	15,217	48,699	68,041
Industry	1,171	1,385	4,013	1,196	339
Power Plants	5,278	703	73,454	7,732	40

Table 3

Number of motor vehicles registered as new and renewal for 1990 in NCR (Source: LTO)

Type	New	Renewal
Car	35,880	271,079
UV	34,772	216,863
Trucks	9,980	34,912
Buses	1,297	3,950
MC/TC	25,035	7,089

Table 4

The 1988 estimate of air pollution emissions in tons/year from mobile sources (Source: DENR-NCR)

Fuel Type	PM	CO	SO_X	NO_X	TOG
Gasoline	1,369	246,578	1,369	13,693	41,537
Diesel	19,575	10,440	11,745	28,927	19,575

vehicles emit mostly CO and organic gases while diesel-fed ones give off PM, SO_X and NO_X.

The DENR also provided data of the air quality in different parts of MM from 1975 to 1983 in four monitoring stations for CO and SO_2. Although the levels were all below the standard, an increasing trend was shown (GALLARDO, 1993).

In this survey made, it was concluded that motor vehicles are mostly responsible for air pollution in urban areas like MM. However, the extent of emissions of the different pollutants from these vehicles was not available at that time. Thus, a separate study was conducted to determine the level of CO and hydrocarbon (HC) emissions from the different types of motor vehicles in MM in 1992.

A Study of Motor Vehicle Emissions in Metromanila: CO and HC Measurements

This study was conducted so that the extent of CO and HC emitted from different types of vehicles in MM could be measured. The motor vehicles were either gasoline run or diesel fed. Aside from this, the age of the vehicles was correlated with the pollutant's concentration since most cars are old and ill-maintained. Also, due to the heavy peak hour traffic, the emission levels were also determined during idling condition and during simulated operating condition.

An important factor that affects CO and HC emissions is the air-to-fuel ratio or A/F. For gasoline engines, an A/F ratio of 14.5 is typical stoichiometric A/F for many individual hydrocarbons or hydrocarbon mixtures. A well-tuned car operates at this ratio. The ratio A/F used may be larger or smaller than stoichiometric. When the ratio of the stoichiometric to actual is less than 1, more air is being supplied than is required for complete burning. Consequently excess air exists and the mixture is said to be lean. Conversely, when this ratio is greater than 1, the mixture is called a rich one.

Carbon monoxide results from the incomplete combustion of the fuel. The conditions which promote complete combustion are those which reduce the emission of carbon monoxide. Figure 1 shows the results of tests at a different A/F ratio. As the A/F ratio increases from 11 to 16, the CO in the exhaust gas decreases from 7.5% to 0.2%. The percentage of CO decreases more at even higher A/F ratios. This demonstrates that one way to decrease CO is to operate the engine with lean mixture.

The A/F ratio also affects the amount of unburned HC emitted by an engine. From the above figure, it is shown that the HC concentration first decreases as the A/F ratio increases from 11 to 16; then it increases to an A/F ratio of 22.5. This increase is attributed to what is known as misfire. The mixture is so lean that combustion does not always proceed from the ignition spark.

For the diesel engine, the load and speed of the engine control the amount of fuel injected. The A/F ratio of the diesel engine ranges from 15 to 100 (KATES and LUCK, 1974). This represents a very lean mixture.

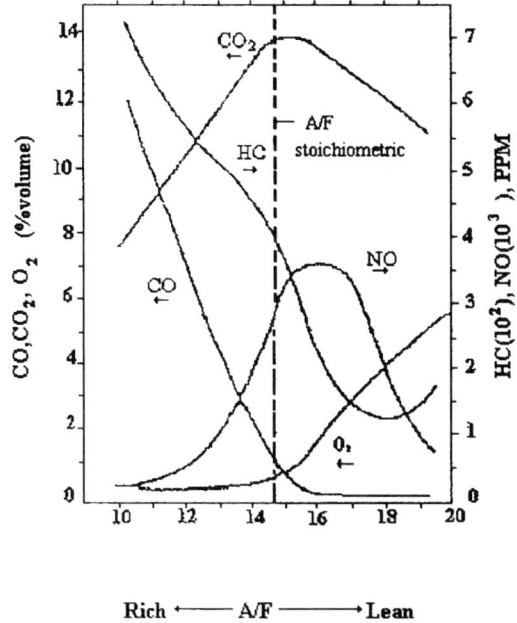

Figure 1
Concentration of Exhaust Gases as a function of A/F (McEvoy, 1975).

Sampling Procedures

Motor vehicles such as cars, tricycles, taxis, buses, and jeepneys were tested at random. Iyasaka CO-HC tester found in the LTO was used. During the test, the engine was first put in idle. The CO and HC levels were measured by a nondispersive infrared technique. Next, the accelerator pedal was used to simulate a running condition of 1200 to 1500 RPM.

Discussion of Results

Figure 2 depicts the extent of CO emission resulting from different types of motor vehicles. It must be noted that only cars and tricycles are run by gasoline, and the rest such as buses, taxis and jeepneys are fuelled by diesel. This figure shows that gasoline-run vehicles emit a higher concentration of CO compared to diesel type engines. An average of 7% CO and 5.2% CO is emitted by cars and tricycles, respectively.

Figure 3 shows the HC concentration of these different types of vehicles. It illustrates that gasoline-run vehicles emit higher HC concentration than diesel fed ones. Tricycles emit very high HC concentration with an average of 6600 ppm while cars emit an average of 540 ppm. Gasoline-run vehicles emit higher CO concentration because the A/F ratio used in gasoline engines is near stoichiometric while diesel

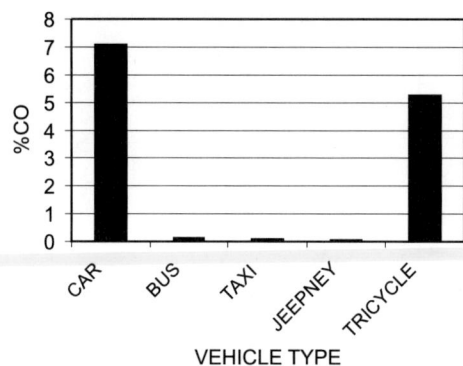

Figure 2
Type of vehicle vs. CO concentration.

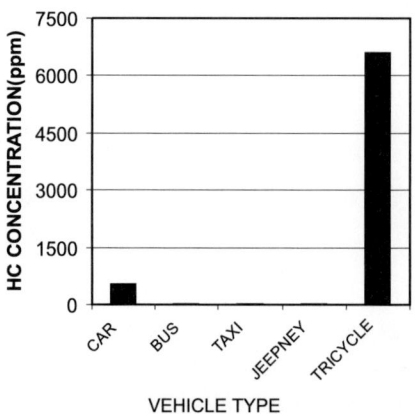

Figure 3
Type of vehicle vs. HC concentration.

engines are operated under the lean condition or use excess air. Diesel engine emits less HC than gasoline engines because diesel engines are operated at higher compression ratios. Aside from this, diesel engines are equipped with larger volume cylinders and more air intake can occur. Tricycles emit very high HC and CO. This is due to the small displacement of their engine, which prevents large air intake. Also, combustion is slower because of the limited number of cylinders. Tricycles tested have only two cylinders while cars have four cylinders. Jeepneys are manufactured locally which have four cylinders and displacement of 2400 cc. Cars have displacement from 1000 cc to 2000 cc while tricycles have only 125 cc.

In Figures 4 and 5, the solid line represents the data under idling condition while the broken line designated the running condition. Figure 4 expresses the age of cars as a function of CO and HC concentrations both during idling and running

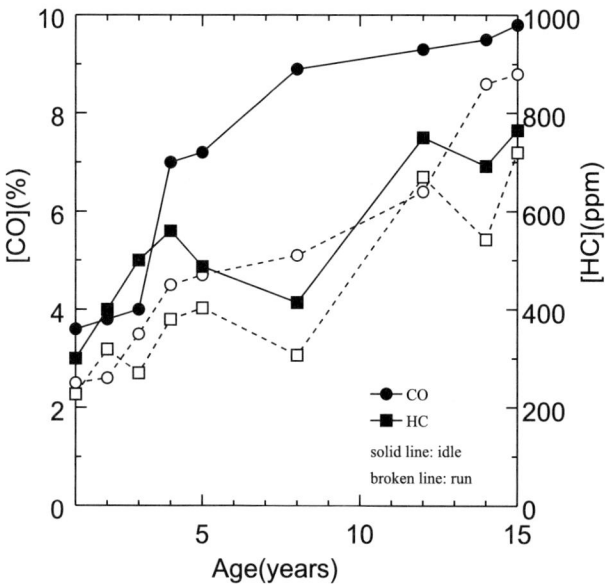

Figure 4
Age vs. CO and HC level of cars.

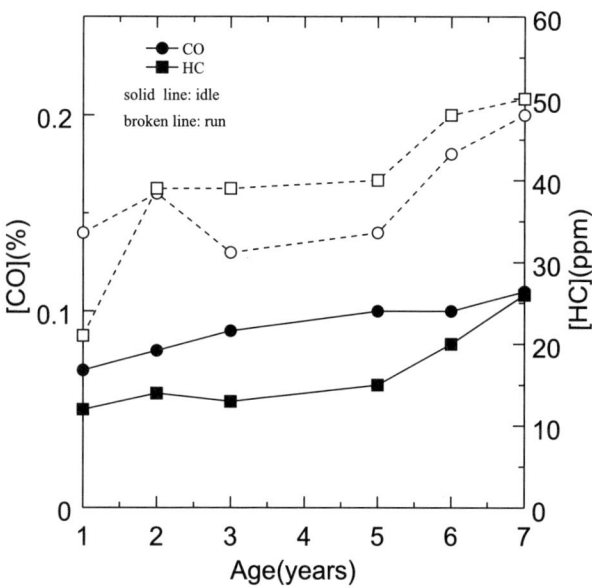

Figure 5
Age vs. CO and HC level of taxis.

condition. Figure 5 presents the age of taxis as a function of CO and HC concentrations also during idling and running condition. It is shown that the level of both CO and HC increases with the age of the motor vehicles. This is so because these vehicles are not kept well or are ill-maintained.

From the same figures, the CO and HC concentrations are higher during the idling than the running condition for gasoline fed cars. However, an opposite trend is true for diesel vehicles. This is expected since a higher temperature of the engine is attained during the running condition for gasoline run cars.

Number of Registered Vehicles in Metromanila

LTO data on the number of registered vehicles in MM was updated in this study to include 1990, 1991 and 1992 data. They are shown in Table 5. An annual increase in the number of motor vehicles is noted from the table.

Conclusions and Recommendations

The extent of CO emissions is quite high for gasoline run vehicles, especially for older ones. It exceeded the maximum permissible level set by LTO. The extent of CO emission is higher during idling condition for both cars and tricycles. This implies that the traffic conditions in Metromanila aggravate the problem of CO pollution. Hydrocarbon emission is also a problem to be addressed with regards to gasoline fed vehicles. Although the HC emission level of cars is tolerable, that of tricycles is exceedingly above the standard.

Although NO_X level was not checked due to the unavailability of a measuring instrument, it is expected to be low for gasoline-run vehicles and higher for diesel exhaust. Thus, for developing countries like the Philippines, the need to lower the level of CO as well as HC in gasoline exhaust must be considered.

Concerted efforts of the government and the private sectors are needed to propose a feasible short- and long-term solution to address the problem of vehicular emissions. DENR in collaboration with LTO should start reviewing their standards for CO and HC emission levels. At the same time, standards for NO_X emission must also be considered in the future. An NO_X emission tester should be made available to

Table 5

Number of registered vehicles in Metromanila (Source: LTO)

Year	Cars		UV		Trucks		Buses		MC/TC	
	Gas	Diesel	Gas	Diesel	Gas	Diesel	Gas	Diesel	Gas	Diesel
1990	292527	14432	137357	114278	5898	36994	349	4898	66577	0
1991	292619	16643	144818	132715	5221	40079	288	6030	73760	91
1992	311674	21539	157240	155948	5132	47030	513	9693	79958	529

monitor reliably, together with a chassis dynamometer to test vehicles on simulated trips.

Aside from improving the traffic conditions in MM, proper care and maintenance of vehicles must be taught to the public. Reducing the number of vehicles moving around the urban areas also must be planned.

Other Related Studies

The air quality in Metromanila was investigated in three traffic related sites like Ermita (Taft Avenue), Pasay City (Harrison) and Quezon City (Agham Road) with carbon monoxide as one of the air pollutants (ALEJANDRINO, 1989). The measurements at the roadside and at the rooftop of the monitoring site showed tolerable levels but the center island portion along Taft Avenue was found to be a risk spot with concentrations reaching 40 ppm. The national ambient air quality standard for 1 hour for CO is only 30 ppm. An "on-stream" mobile survey conducted, showed that carbon monoxide mean value along Taft Avenue to Quezon City was 36.7 ppm, with maximum concentration of 100 ppm. Also, from Quezon City to Manila via EDSA, a mean value of 26.7 ppm with a maximum value of 60–100 ppm was recorded.

Monitoring stations for CO were installed by First Signs Inc. in Elliptical Road (Quezon City) and White Plains (EDSA, Q.C.). The latest data provided for March 1997 show an average of 9 ppm with a maximum value of 16 ppm in White Plains while in Elliptical Road, the average value was 4 ppm with a maximum of 8 ppm.

Previous study in MM (ENGINEERING SCIENCE INC., 1992) reveals that the exposure to air pollutants has caused several respiratory symptoms and ailments. In an occupational health survey conducted, it revealed twelve leading signs and symptoms on workers exposed to air pollution, most of them working near heavy traffic conditions. They are headache, cough, nasal discharge, abdominal pains, fever, low back pains, sore throat, tachycardia, muscle cramps, nape pains, frequent sneezing, and decreased vision. In the same study conducted by SUBIDA et al. four chronic respiratory symptoms were found in drivers, commuters, policemen and traffic aides. They are wheezing, shortness of breath, chronic cough, and chronic phlegm production. It was also shown that jeepney drivers were more than twice affected by exposure to these emissions than the others sampled.

Recent Developments and Future Trends

To date, the number of registered vehicles in Metromanila has increased, with old vehicles predominating over the new ones. At the same time, the number of gasoline-fed vehicles is also increasing as shown in Figure 6. The solid line is for the number of

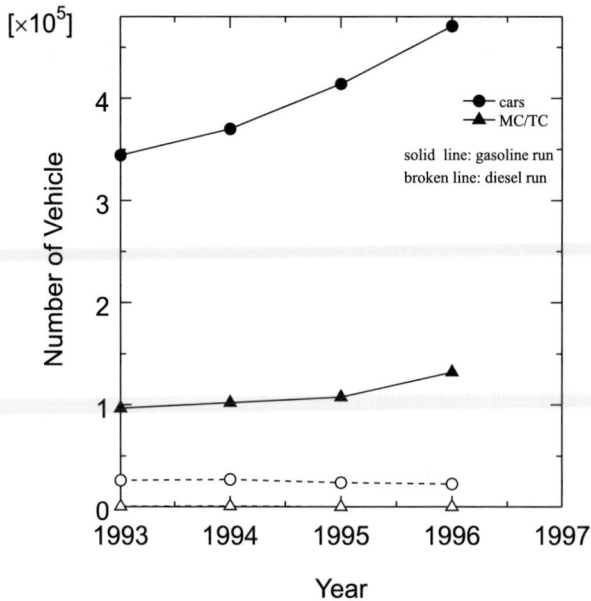

Figure 6
Trends in the number of motor vehicles in MM (1993–1996).

gasoline run vehicles and the broken line is for diesel run vehicles. This trend is expected to increase since the population projection in Metromanila (National Economic Development Authority) is also increasing with a growth rate of 3.3% as of September 1995.

DENR released a report covering the period (1990–1995) which shows the latest inventory for MM of motor vehicle airborne pollutants. They are shown in Table 6. PM_{10} are particulates that have diameters equal to or less than 10 microns. The increase of the emissions from the 1988 estimate to the 1990 estimate is shown in Table 7.

The Asian Development Bank (ADB) reported the year 2005 projections for vehicle population and emission trends with 1990 as the base year. From the report, the number of cars is projected to increase 106% from 1990 to 2005. The projected increase in CO emission is 129% while for the HC it is 101% if no control measures are taken.

Table 6

1990 estimate of total emissions from motor vehicles in MM in tons/year (DENR, 1995)

Fuel Type	PM_{10}	CO	SO_X	NO_X	TOG
Gasoline	2,869.05	528,605.61	237.66	27,855.08	88,124.30
Diesel	8,581.05	44,020.55	10,112.59	38,360.96	12,830.04

Table 7

Percent increase of emissions from 1988 estimate to 1990 estimate

Fuel Type	PM_{10}	CO	SO_X	NO_X	TOG
Gasoline	110	114	−117	103	112
Diesel	−56	322	−14	33	−34

In 1997 Congress passed a bill known as the 1997 Clean Air Act of the Philippines. In 1998 the senate approved it. Finally, on June 23, 1999, this act was signed by President Joseph Estrada and it has been named RA 8749 or The Philippine Clean Air Act of 1999. This is the first comprehensive environmental law related to air pollution control. This law consists of air quality improvement action plans which includes air pollution research and development programs. This research program is directed toward fuel reformulation, efficient techniques for control of by-products of combustion and emissions from evaporation of fuels and removal of potential pollutants from fuels. This also included the total ban of lead in the year 2000 and the reduction of the sulfur content of fuel. Emission standards are also included in the law as shown in Table 8.

Consideration of installing catalytic converters in all new cars in the Year 2000 is also planned, with studies to be made of adapting catalytic converters to old cars without engine modification. It is in this light where the development of suitable catalysts is necessary.

Catalyst Requirements of Developing Countries

The results of the air pollution studies done in Metromanila reveal that the needs of developing countries such as the Philippines are not identical to the third world countries. Even regulations on auto emissions are not yet fully implemented because implementing rules and regulations for RA 8749 are still being planned.

Today, leaded gasoline is being sold parallel to unleaded gasoline in Metromanila. In addition, the sulfur content of fuel used is high. Thus, a more chemically

Table 8

Emission standards for motor vehicles specified in RA 8749 in g/km

Type of Vehicle	Ref. Weight (KG)	CO	$HC + NO_X$	PM
Light Duty Vehicle	–	2.72	0.97	0.14*
Light Commercial	1250 < RW	2.72	0.97	0.14
	1250 < RW < 1700	5.17	1.4	0.19
	RW > 1700	6.9	1.7	0.25

stable catalyst is needed which is tolerant to lead and sulfur poisoning. Although lead will be totally banned by the year 2000 and sulfur content will be reduced, the use of TWC in the Philippines is still not practical because it is an expensive material. Pt, Rh and Pd belong to the Platinum group metals (PGM), which are precious metals. PGM consists of 6 elements under Group VIII of the periodic table. They are closely related in chemical and physical properties however their geological abundance, applications and supply and demand are significantly different (STEEL, 1991). A study (NIIYAMA and NAKAMURA, 1991) shows that the supply of these metals is dwindling although the demand for them, especially of rhodium is increasing. In fact, the primary usage of these precious metals derives from the auto-industries as auto-catalyst. The shortage of these metals is a serious problem and will surely make them increasingly more expensive.

Finally, the worsening traffic conditions in major Asian cities and the increasing number of gasoline-run vehicles such as cars and tricycles (or motorcycles) which are mostly long-used make CO and HC emissions increasingly high. Consequently, development of an oxidation catalyst is necessary in developing countries.

Development of Base Metal Oxide Catalysts

Realizing the need to develop an auto-emission control catalyst that is suitable to the conditions prevailing in developing countries, a 5-year collaborative project was initiated in FY 1993 by the Tokyo Institute of Technology (Japan) and De la Salle University (Philippines). This project aimed to improve the catalytic performance of alumina towards the oxidation of carbon monoxide and hydrocarbons. Alumina is known to have a De-NO_X activity (DIXON et al., 1960) and is less expensive since it is used as a support material only in TWC. One study (GALLARDO et al., 1998a) indicates that the oxidation activity of alumina should be improved towards low temperature oxidation and higher selectivity for CO_2 formation during hydrocarbon oxidation. Further, carbon formed is found to act as an in situ active site that promoted CO oxidation during low temperature (GALLARDO et al., 1998b). Also, the addition of a base metal oxide such as CeO_2 more highly activated alumina during CO oxidation, methanol oxidation, and propylene oxidation. A field performance study using CeO_2-Al_2O_3 catalysts shows that this performs fairly well as an automotive exhaust treatment catalyst that may require secondary air for older cars without engine modification.

Current Studies

Currently, there is a need to further improve the catalyst performance towards the reduction of NO_X. This is so because of the present condition of rain acidification found in certain places in MM. In an on-going study, the quality of rain is monitored using an automated acid rain monitor acquired through the Department of Science and Technology. Rain samplers were installed in five different places in Metromanila including a power plant in the south. The objective of the study is to verify acid deposition in MM and create an acid rain map for Metromanila. These data are useful in verifying acid rain models being developed for East Asia. In the model, meteorological data such as wind speed and wind direction are needed. These are taken into account in this present study and a separate report will be made regarding these. Table 9 shows partial results only. The monthly average pH, sulfates and nitrates of the rain samples collected at these sites from March to June, 1999 are shown.

Results obtained show that some areas in the metropolis suffer from acid rain. The pH values were mostly below 5.6. Similar values were obtained by a separate study conducted by the Department of Environment and Natural Resources (DENR Report, 1999). Two sampling areas were monitored: one representing an urban area and another, a rural area. Table 10 presents their results.

Table 9

Monthly average pH, sulfate concentration and nitrate concentration of rain samples collected in Metropolitan Manila in 1999

Location	Month	pH	$SO_4^=$ (µg/ml)	NO_3^- (µg/ml)
North MM	March	6.04	34.1	14.09
	April	5.2	20.98	11.67
	May	4.78	28.3	11.39
	June	5.04	14	2.72
East MM	March	6.09	27.62	11.14
	April	6.11	37.9	22.34
	May	6.14	22.89	16.6
	June	–	–	–
South MM	March	6.24	26.7	11.9
	April	5.95	22.9	13.02
	May	5.69	30.05	14.6
	June	4.72	12.1	5.31
Center MM	March	6.28	32.8	17.57
	April	5.96	16.74	10.42
	May	4.9	26.94	15.13
	June	–	–	–
West MM	March	5.39	14.96	5.81
	April	5.12	10.97	7.53
	May	4.18	23.94	14.01
	June	4.83	13.66	5.07
South Power Plant	March	4.3	173.97	6.79
	April	3.95	–	–
	May	3.97	750.2	19.39
	June	4.69	20.3	4.18

Table 10

Average monthly rain quality in 1999 (DENR)

Area	Month	pH	$SO_4^=$ (mg/l)	NO_3^- (mg/l)
Urban	April	6.29	11	0.442
	May	4.44	9	0.558
	June	4.64	Less than 5	0.379
Rural	April	6.24	7	0.118
	May	4.78	Less than 5	0.140
	June	4.59	Less than 5	8.95

Concluding Remarks

Air quality and emission data are needed not only in establishing practical emission standards for motor vehicles and the stationary industries and power plants but also in the development of technologies for air pollution control. They are also useful tools in dispersion modeling needed for forecasting air pollution problems and solutions.

For stationary power plants, a selective catalytic reduction process has been developed in Japan and is commercially available. Other catalytic materials are being improved for simultaneous SO_X and NO_X removal, diesel exhaust treatment, low temperature combustion, NO_X reduction, VOC destruction, and the conversion of greenhouse gases. Catalysis is one way to reduce emissions and minimize gaseous emission through cleaner fuels and clean technologies. Thus, catalysis plays a vital role is sustaining a cleaner air environment.

Acknowledgments

The writer wishes to thank the Tokyo Institute of Technology, the University Research and Coordinating Office of De la Salle University, the Science Foundation of De la Salle University, the Department of Environment and Natural Resources, and the Land Transportation Office of Quezon City for their support given in completing several studies discussed here. The author gratefully acknowledges Professor Hiroo Niiyama, Dr. Takashi Aida and her research assistants who assisted her in data gathering and analysis.

References

ALEJANDRINO, A. (1989), *Environmental Assessment of Traffic-related Air Pollutants in MM*, Ph.D. Dissertation, submitted to the University of the Philippines,. Diliman, QuezonCity, 1101 Philippines, 45–50.

DE NEVERS, N., *Air Pollution Engineering* (McGraw Hill Inc., 1995) 403 pp.

DENR (1995), *The Philippine Environmental Quality Report 1990–1995*, Environment Management Bureau Library, Department of Environment and Natural Resources, Visayas Avenue, Quezon City, Philippines.

DENR (1999), *Preliminary Report on Acid Deposition in the Philippines*, Environment Management Bureau, Department of Environment and Natural Resources, Visayas Avenue, Quezon City, Philippines.

Engineering Science Inc. and Basic Technology and Management Corporation (1992), Final Report for Vehicular Emission Control Planning in MM, submitted to Department of Environment and Natural Resources, Visayas Avenue, Quezon City, Philippines.

DIXON, J. K. *et al.*, *Catalysis*, vol. VII (Reinhold Publishing Corporation, 1960), pp. 303–321.

GALLARDO, S. (1993), *Air Pollution Study in Metromanila*, DLSU Engin. J. *IX*, 24–39.

GALLARDO, S., AIDA, T., and NIIYAMA, H. (1998a), *Development of Base Metal Oxide Catalyst For Automotive Emission Control*, Korean J. Chem. Eng. *15*(5) 480–485.

GALLARDO, S., AIDA, T., and NIIYAMA, H. (1998b), *In Situ Active Site Formation in CO Oxidation on Alumina*, Res. Chem. Intermed. *24*(4), 401–410.

KATES, E. and LUCK W. (Chicago, 1974), *Diesel and High Compression Gas Engines*, Am. Techn. Soc., 160–161.

MC EVOY, J. (Washington, D.C., 1975), *Catalysis for the Control of Automobile Pollutants*, Am. Chem. Soc.

NIIYAMA, H. and NAKAMURA, R. (1991), *Recent Progress in Automobile Catalyst*, Proc. Conf. Environ. Protect. of Big Asian Cities, Nov., Manila, 38 pp.

STEEL, M.C.F. (1991), *Supply and Demand of Precious Metals for Automotive and other uses*, Catalysis and Automotive Pollution Control II (Elsevier Science Publishers, U.S.A. 1991), 105 pp.

TOKYO METROPOLITAN RESEARCH INSTITUTE FOR ENVIRONMENTAL PROTECTION (1989), *Data Book of the Environment of Tokyo*, pp. 33–38

(Received March 1, 2000, accepted January 21, 2001)

Pure appl. geophys. 160 (2003) 341–348
0033–4553/03/020341–08

❙Pure and Applied Geophysics

Air Pollution Studies and Management Efforts in Nepal

P. K. Jha[1] and H. D. Lekhak[1]

Background

Air pollution is considered to be one of the serious and prominent types of environmental pollution that is prevalent in most industrial towns and cosmopolitan cities of the world. It had been a general impression in the past that air pollution is exclusively a problem of the industrially developed nations, however, recent studies have shown that air pollution is a growing problem in developing countries as well, and hence, attention should be paid to this evil before it is too late.

Nepal, a relatively small country with 147,181 sq km area inhabited by 22 million people, is known for exquisite environment. However, the real scenario is quite different because urban areas are environmentally degrading due to rapid unplanned urbanization and industrialization. Increasing numbers of human population, industries and automobiles, a high rate of deforestation, decreasing agricultural productivity, the frequent occurrence of floods in the lowlands, landslides in the midlands and forest fires are major environmental issues, and recent studies reveal that even the glorious mountain peaks of the high Himalayas have also undergone incipients pollution (Dokiya et al., 1992; Shrestha et al., 1997).

Considering the gravity of the issue of deteriorating air quality, Nepal must acutely prioritize the control of air pollution. The current article is written to assess the present state of air quality and review the efforts being made to mitigate air pollution in Nepal by different government and non-government organisations.

State of Air Pollution

Nepal's air is relatively unpolluted but automobiles, increasing at the rate of 14% per year in the last few years, urbanization growing at the rate of 4–5% per

[1] Central Department of Botany, Tribhuvan University, Kirtipur, Kathmandu, Nepal.
E-mail: pkjha@ecos.wlink.com.np

year and unplanned industrialization, have affected the air quality unfavorably in urban settlements (Table 1). In rural areas people, particularly women and children, suffer from indoor air pollution resulting from biomass burning for domestic purposes (PANDEY et al., 1987). SHARMA et al. (1999) have recorded a significant relationship between the air pollutants (NO_x, $PM_{7.07}$ and fungal elements) with the respiratory problems of children admitted to Kanti Children Hospital, Kathmandu.

The first air pollution study was conducted by BHATTARAI and SHRESTHA (1981). Several studies have been conducted in the last decade in the Kathmandu valley (Table 1) with the most widely studied air pollution parameter in Nepal being particulate matter. These studies have indicated high concentrations of dust particles in the air as a major air pollution problem in Nepal. In some areas in Kathmandu valley, particulate matter is ten times higher than the WHO standard. There are diverse reports and opinions regarding the chemical pollutants. KARMACHARYA and SHRESTHA (1993) reported SO_2, NO_x and Pb concentrations in the central part of Kathmandu. The WHO standards for respirable particles less than 10 μm in diameter are: 150 $\mu g/m^3$ (24 h average) (WHO/UNEP, 1992). SILWAL et al. (2001) reported gaseous concentrations of NO_x, SO_2 and O_3 reach in 60 ppb in general, but NO_x and O_3 reached 180 and 155 ppb at a few stations in Kathmandu valley in January and April, respectively.

Studies have also been conducted in the Himalayan mountains above 5000 m (DOKIYA et al., 1992, SHRESTHA et al., 1997). These studies indicate that there are transboundary air pollutants migrating to the high Himalayas polluting the glaciers. It is a nascent problem, and if it accelerates there may be serious consequences.

The Royal Nepal Academy of Science and Technology, and Environment Protection Council in 1993 initiated an extensive discussion (SHAH and JHA, 1995) and positive steps have been taken by the government to prevent pollution in Kathmandu Valley (JHA, 1995).

Air Pollution Control Efforts

Nepal in the early 90s realized the severity of the air pollution problem, particularly in Kathmandu. Nepal initiated air pollution control/management after 1992, through the establishment of the Environment Protection Council under the chairmanship of the Prime Minister, and the formation of the Ministry of Population and the Environment in 1994. His Majesty's Government (HMG) of Nepal enacted an environment protection act in 1996 and environment protection regulation in 1997. The environment protection act asks for institutionalization of the environment protection council, pollution control, and environment impact assessment of developmental and economic activities in forest, industry, mining, road, water resource and energy, tourism, drinking water, solid waste,

Table 1

Studies and major observations addressing air pollution in Nepal

S.N.	Study	Major observations	Reference
1.	Lead (Pb) concentration high in Kathmandu valley	In light traffic zone: 10–51 µg In heavy traffic zone: 323–574	BHATTARAI and SHRESTHA, 1981
2.	Indoor and outdoor air pollutants in the Himalayas	Aerosol particles reported in sediments, and river flow.	DAVIDSON et al., 1986
3.	Domestic smoke pollution and acute respiratory problem	High respiratory, chronic bronchitis and prevalence rate (5.6%) in Jumla because of domestic air pollution	PANDEY et al., 1987
4.	High level of particulates in Kathmandu, Biratnagar and Pokhara	Particulates high (> 1000 µg/m^3) in Kathmandu	CEDA, 1989
5.	Air pollutant and microbial studies	PM_{10} 197–775 µg/m^3 average 309 µg/m^3	RONAST, 1992
6.	Energy utilization and air pollution in Kathmandu	CO: 32 kg/km/h HC: 4.5 kg/km/h NO_x: 3.5 kg/km/h SO_x : 370 kg/km/h Particulates: 0.86 kg/km/h	DEVKOTA, 1992
7.	Chemical species in the deposition of some peaks of the Himalayas	Glaciers polluted NO_3: 0.02–0.52 µg/l SO_4: 0.14–3.20	DOKIYA et al., 1992
8.	Ambient air quality of Kathmandu valley	**i. (24h av)** – TSP: 194–535 µg/m^3 – PM_{10}: 59–127 µg/m^3 – SO_2: < 13 µg/m^3 – NO_x: 12–36 µg/m^3 – CO: < 12.0 µg/m^3 – Pb: 0.18–0.53 µg/m^3 **ii. (9h av)** – TSP: 789–2258 µg/m^3 – PM_{10}:102–498 µg/m^3 – SO_2: < 13–22 µg/m^3 – NO_x: 17–69 µg/m^3 – CO: < 11.0 µg/m^3 – Pb: 0.2–0.12 µg/m^3	KARMACHARYA and SHRESTHA, 1993
9.	Motor Vehicle Pollution	TSP: 319–876 µg/m^3 in Traffic sites. 273–350 µg/m^3 in residential sites SO_2: 100–225 µg/m^3 NO_x: 14–126 µg/m^3	KVVECP, 1993
10.	Chemical composition of aerosol of snow in the Himalayas (at 5050 m)	Snow rich in NH_4^+ and NO_3^- than aerosol. Aerosol samples show the presence of acidic gases, Ca^{++}, Na^+, K^+, Mg^{++}, Cl^-, NO_3^- and SO_4^-	SHRESTHA et al., 1997
11.	Deterioration in visibility in Kathmandu	Visibility days decreased from 115/days in early 1970 to 20 days in recent years	DMH, 1994
12.	Atmospheric composition of Nguzompa (at 5430 m)	Water soluble aerosol	WAKE et al., 1994

Table 1

Continued

S.N.	Study	Major observations	Reference
13.	Extent and dimension of lead pollution in Kathmandu	637.5 $\mu g/m^3$ in air	NESS, 1995
14.	Air pollution climatological study	More visibility when less fuel consumption, CO concentrations were 6–7 ppm at heavy traffic load	UPADHYAY and GHIMIRE, 1996
15.	Biological air borne particles were observed at different locations in Kathmandu	37 Fungal species belonging to 26 genera were isolated	SHARMA, 1997
16.	Urban air quality management strategy in Asia (Kathmandu valley report)	TSP concentration 800 $\mu g/m^3$. Respiratory problems and impact due to pollution estimated	URBAIR, 1997
17.	NO_x, SO_x and O_3 every month at 11 locations in Kathmandu Pollutant	NO_x: 20 – 180 ppb NO_2: 18 – 140 ppb NO: 10 – 80 ppb SO_2: 5–30 ppb	SILWAL et al., 1998
18.	Ambient air quality observed	PM_{10}: 100–190 $\mu g/m^3$ TSP: 308–401 $\mu g/m^3$ (460 to 3640 $\mu g/Nm^3$ near stacks) SO_2: 7.8–10.4 $\mu g/m^3$ NO_2: 16.0–25.3 $\mu g/m^3$	ENPHO, 1999
19.	NO_x concentration at six different sites of Kathmandu	Highest (0.035 ppm) at bus park area and lowest (0.012 ppm) at International airport area	LEADERS NEPAL, 1998
20.	$PM_{7.07}$ and NO_x concentration and fungal elements at different sites in Kathmandu	Highest (400.9 mg/m^3) at bus park area and lowest (42.02 mg/m^3) at airport Relationship of pollutants with respiratory problems was significantly high	SHARMA et al., 1999
21.	Spatial and temporal trends of aerosols in the Himalayas were studied	SO_4^{-2}, NH_4^+ and Ca^{2+} show clear spatial trend while Na^+, Mg^{2+} and Cl^- show no trend aerosol concentration peaks between April and June (SO_4^- and NO_3^- 16 times and NH_4^+ 20 times greater than the overall average)	SHRESTHA, 1999
22.	Air pollution in Kathmandu, Nepalgunj and Biratnagar	SPM and Pb were analyzed. Pb concentration ranged from 0.31 to 0.69 $\mu g/m^3$	LEADERS NEPAL, 1999

agriculture and health sectors. In January 1998, HMG, Nepal specified standards for petrol and diesel engine vehicles as follows: for petrol engines vehicles 4.5 percent carbon monoxide by volume and for diesel engines 65 HSU (Hartridge

Smoke Unit) for vehicles manufactured before 1994 and 75 HSU for vehicles manufactured after 1994. Recently, HMG Nepal has banned the movement of diesel-powered three wheelers in the Kathmandu Valley and restricted the import of two-stroke motorcycles. As a result, 600 diesel-powered three wheelers are transformed into electrified vehicles. HMG, Nepal has revised the vehicle pollution regulation (1999), meeting the Euro 1 standard to be imported into Nepal. These steps have improved the air pollution scenario in Kathmandu Valley. Some of the major air pollution control efforts being undertaken in Nepal are listed in Tables 2 and 3. These efforts are not completely successful, but definite progress has been achieved in improving air quality.

Conclusion

Most of the pollution studies conducted in Nepal were focussed on traffic emissions, whereas industrial, domestic and other sources of air pollutants are not seriously undertaken in Nepal. There are several Non Governmental Organizations (NGOs), government offices and institutes working in the field of air pollution. However, their activities are scattered and limited. The HMG Nepal should analyze, coordinate and encourage these organizations to work towards a pollution-free environment. Environmental research in Nepal needs financial and technical support from developed countries. HMG, Nepal should also emphasize environmental

Table 2

List of symbols/acronyms used in the text and tables

Acronyms	Full Name
CEDA	Centre for Economic Development and Administration (Tribhuvan University, Kathmandu)
DMH	Department of Meteorology and Hydrology, Kathmandu
ENPHO	Environment and Public Health Organization, Kathmandu
EPA	Environment Protection Act of Nepal
EPC	Environment Protection Council, HMG, Nepal
HMG	His Majesty's Government of Nepal
HSU	Hartridge Smoke Unit
KVVECP	Kathmandu Valley Vehicular Emission Control Project
MEIP	Metropolitan Environmental Improvement Programme of the World Bank and UNDP launched in six Asian Metropolitan Areas in 1992
NESS	Nepal Environmental and Scientific Services (P) Ltd., Kathmandu
NILU	Norwegian Institute of Air Research
$PM_{7.07}$	Respirable Particles less than 7.07 μm in aerodynamic diameter
RONAST	Royal Nepal Academy of Science and Technology, Kathmandu
URBAIR	Urban Air Quality Management Strategy
WHO	World Health Organization

Table 3

Air pollution control/management efforts in Nepal

Serial No.	Efforts/Activities
1.	Industrial Pollution Control Management Project (1992) by Ministry of Industry, HMG, Nepal.
2.	Kathmandu Valley Vehicle Emission Control Project (1993): Testing of vehicles and reported air pollution status. (UNDP Funded Project)
3.	URBAIR Programme (1992). The World Bank through Metropolitan Environment Improvement Program initiated the urban air quality management strategy in 1992–1995. A detailed draft of urban air quality management strategy was produced for Kathmandu in collaboration with the Norwegian Institute for Air Research (NILU).
4.	Nepal Environmental Policy and Action Plan (NEPAP) was prepared by His Majesty's Government of Nepal in 1993, suggesting short-, medium- and long-term plans for air pollution control (EPC, 1993).
5.	Installation of Air Pollution Control Device in Himal Cement Factory in 1995. This factory was one of the major dust polluters in Kathmandu Valley.
6.	Introduction of electrical charge technology to operate Vikram Tempo with an electric battery as a substitute for diesel fuel (introduced in 1994). HMG/Nepal has banned the use of Vikram Tempo (diesel-powered three wheeler) in Kathmandu Valley since September 14, 1999. This has resulted in the improvement of air quality.
7.	Lead-free gasoline was introduced in Kathmandu Valley during July 1997. Nepal Oil Corporation is concerned with gasoline quality.
8.	Environment Protection Act was passed by the parliament in 1997. This act asks for EIA (Environment Impact Assessment). Pollution control, Protection of National Heritage, Protected Areas, establishment of laboratories to support environmental conservation and pollution control, and institutionalization of Environment Protection Council as an advisory body to HMG of Nepal.
9.	Tribhuvan University has introduced B.Sc. Environmental Science in 1997, and hopefully M.Sc. from 2001 AD.
10.	For the last few years, every year 2–3 important national/international symposia are held in Kathmandu on pollution control.
11.	Nepal vehicular pollution emission regulation has approved the EURO 1 standard for Nepal in December 1999.
12.	Institute for Environmental Management has been established in 2000 AD with the financial support of 2.5 million U.S. $ from the Danish government.

education, awareness and decentralization policy. Pollution in urban areas is increasing, and if not controlled in time, will endanger people, nature and the overall economy of the country.

Acknowledgements

Sincere thanks to AQM Conference organizers and University Brunei Darussalam for travel support; and University Grants Commission, Nepal for Conference registration fee support to P.K. Jha. Thanks are due to Tribhuvan University, Nepal for encouragement.

REFERENCES

BHATTARAI, D. R. and SHRESTHA, P. R. (1981), *Lead Contents in the Dust of Kathmandu City Road*, Nep. Chem. Soc. Proc. *1*, 47–50.

CEDA (1989), *A study on the environmental problems due to urbanization in some selected Nagar Panchayats of Nepal*. Report submitted by Centre for Economic Development and Administration, Tribhuvan University, to UNDP, Kathmandu, 248 pp.

DAVIDSON, C. L., LIN, S., OSBORN, J. P., PANDEY, M. R., RASMUSSEN, R. A., and KHALIL, M. A. K. (1986), *Indoor and Outdoor Air Pollution in the Himalaya*, Environ. Science Techn. *20*, 561–567.

DEVKOTA, S. (1992), *Energy Utilization and Air Pollution in Kathmandu Valley, Nepal*, M.S. Thesis (EV-92-09) submitted to Asian Institute of Technology, Bangkok, 112 pp.

DMH (1994), *Climatological Record of Nepal*, Department of Meteorology and Hydrology (DMH), HMG, Nepal.

DOKIYA, Y., MARUTA, E., YOSHIKAWA, T., ISHIMORI, H., and TSURUMI, M. (1992), *Chemical Species in the Deposition at Some Peaks of the Himalaya*, Environ. Sci *5*(2), 109–114.

ENPHO (1999), *Reduction of Pollution and Waste in the Himal Cement Company Limited, Kathmandu, Nepal*. Project Report by Environment and Public Health Organization (ENPHO), Kathmandu, 52 pp.

EPC (1993), *Nepal Environmental Policy and Action Plan: Integrating Environment and Development*, Environment Protection Council (EPC), HMG, Kathmandu, 80 pp.

JHA, P. K. (1995), *Pollution Preventing Efforts and Strategies for the Kathmandu Valley*, Air, Water and Soil Pollution *85*, 2643–2648.

KARMACHARYA, A. P. and SHRESTHA, R. K. (1993), *Air Quality Assessment in Kathmandu Valley*, Environment and Public Health Organization, Kathmandu, 28 pp.

KVVECP (1993), *Ambient Air Quality Monitoring in Kathmandu Valley*, Kathmandu Valley Vehicular Emission Control Project Report, UNDP/92-034, Nepal.

LEADERS NEPAL (1998), *A Citizen Report on Air Pollution in Kathmandu: Children's Health at Risk*, Research Report, Leaders Nepal and Japan Environmental Corporation, Kathmandu, 106 pp.

LEADERS NEPAL (1999), *Air Pollution in the Face of Urbanization* (Citizens Report 99). Leaders Nepal, Kathmandu, 116 pp.

NESS (1995), *Assessment of the Applicability of Indian Cleaner Process Technology for Small Scale Brick Kiln Industries of Kathmandu Valley*, Nepal Environmental and Scientific Services (NESS) (P) Ltd., Nepal, 30 pp.

PANDEY, M. R., NEUPANE, R. P., GAUTAM, A., and SHRESTHA, I. P. (1987), *Domestic Smoke Pollution and Acute Respiratory Infection in Rural Community of the Hill Region*, Environ. Pollution *15*, 337–340.

RONAST (1992), *A Study on Traffic Volume in Busy Streets of Kathmandu Valley*, Report submitted by Royal Nepal Academy of Science and Technology (RONAST) to MEIP/URBAIR, 13 pp.

SHAH, R. and JHA, P. K. (eds) (1995), *Proc. Discussion Forum on Pollution Preventing Strategies for Kathmandu Valley*. Royal Nepal Academy of Science and Technology, Kathmandu, 50 pp.

SHARMA, B. K. (1997), *Study on Air Microflora of Kathmandu Valley and its Seasonal and Location Variation*, M.Sc. Dissertation, Central Department of Microbiology, Tribhuvan University, Nepal, 97 pp.

SHARMA, B. K., SHARMA, A. P., and POKHAREL, A. (1999), *Air Pollution and its Effect on Respiratory Diseases in Kathmandu*, Nepal J. Sci. and Tech. *1*, 115–122.

SHRESTHA, A. B. (1999), *Aerosol Research in Nepal Himalayas: Present Status and Future Prospects*, Scientific World *1*(1), 75–84.

SHRESTHA, A. B., WAKE, C. P., and DIBB, J. E. (1997), *Chemical Composition of Aerosol and Snow in the High Himalaya during the Summer Monsoon Season*, Atmos. Environ. *31*, 2815–2826.

SILWAL, S., HORIE, K., AOKI, M., TOTSUKA, T., and JHA, P. K., *Measurements of NOx, SO_2 and O_3 concentrations in Kathmandu and vicinities using sampling devices*, In *Environment and Agriculture: Biodiversity, Agriculture and Pollution in South Asia* (eds., Jha, P.K., Baral, S.B., Karmacharya, S.B., Lekhak, H. D., Lacoul, P. and Baniya, C.B) (Publ. Ecological Society (ECOS), Kathmandu, Nepal 2001) 459–464.

UPADHYAY, B. and GHIMIRE, B. R. (1996), *A study of air pollution climatology of Kathmandu Valley*. In *Environment and Biodiversity: In the Context of South Asia* (eds., Jha, P.K., Ghimire, G.P.S., Karmacharya, S.B., Baral, S.R. and Lacoul, P.) (Ecological Society (ECOS), Nepal) pp. 82–90.

URBAIR (1997), *Urban Air Quality Management Strategy in Asia: Kathmandu Valley Report*. The World Bank Technical Paper No. 378, Washington D.C., USA, 155 pp.

WAKE, C. P., DIBB, J. E., MAYEWSKI, P. A., ZHONGQIN, L., and ZICHU, X. (1994), *The Chemical Composition of Aerosols over the Eastern Himalayas and Tibetan Plateau during Low Dust Period*, Atmos. Environ. *28*(4), 695–704.

WHO/UNEP, *Urban Air Pollution in Megacities of the World*. World Health Organization and United Nations Environment Programme, (Blackwell Publishers, Oxford, UK 1992).

(Received July 3, 2000, accepted February 1, 2001)

To access this journal online:
http://www.birkhauser.ch

Pure appl. geophys. 160 (2003) 349–355
0033–4553/03/020349–07

© Birkhäuser Verlag, Basel, 2003

Pure and Applied Geophysics

Monthly Dispersion Characteristics over the South China Sea for Air Quality Modeling

S. A. Hsu[1]

Abstract—Monthly dispersion characteristics for air quality modeling over the South China Sea offshore the west coast of Borneo are studied using long-term ship measurements. It is found from monthly averages that the stability condition is nearly neutral throughout the year with the exception of April, May, and November which are slightly unstable. The lifting condensation level ranged from 338 to 450 m. The lowest value of the ventilation factor occurred in April and the highest in January. The friction velocity for each month is also provided to determine the vertical eddy diffusivity and horizontal and vertical dispersion coefficients.

Key words: Dispersion characteristics, South China Sea, mixing height, ventilation factor, friction velocity.

1. Introduction

In order to study the air quality in the offshore region, atmospheric dispersion models may be used. However, nearly all models require the following basic input parameters: Atmospheric stability classification; the dispersion coefficients in the horizontal (crosswind) and vertical directions; the vertical eddy diffusivity (which requires friction velocity formulation) in some diffusion models; and the atmospheric mixing height.

Over the South China Sea off the west coast of Borneo there are long-term records of monthly climatological data including air temperature, sea temperature, relative humidity, and wind speeds (see Table 1). The data source is based on the publication by U.S. NAVAL WEATHER SERVICE COMMAND (1975). The purpose of this paper is to employ these data to determine those basic input parameters for air quality modeling over the South China Sea. Some discussion of dispersion characteristics onshore along the west coast of Borneo are provided in HSU (2003).

---

[1] Coastal Studies Institute, 308 Geoscience Building, Louisiana State University, Baton Rouge, Louisiana 70803, U.S.A. E-mail: sahsu@antares.esl.lsu.edu

Table 1

Measured and Computed Parameters for Monthly Overwater Dispersion Estimates off the West Coast of Borneo in the South China Sea

Month	T_{air} (°C)	T_{sea} (°C)	RH%	U_Z (m s^{-1})	B	Z/L	Stability Class	LCL (m)	VF (m^2 s^{-1})	u_* (m s^{-1})
Jan	26.8	27.2	84	5.3	0.093	−0.078	D	363	2571	0.18
Feb	26.7	26.9	84	5.0	0.066	−0.051	D	363	2425	0.18
Mar	27.2	27.6	85	3.9	0.093	−0.143	D	338	1749	0.16
Apr	28.2	28.7	82	2.5	0.104	−0.414	C	413	1398	0.14
May	28.4	29.3	81	2.4	0.139	−0.727	C	450	1474	0.15
Jun	28.2	28.9	82	3.0	0.123	−0.378	D	413	1677	0.15
Jul	28.2	28.7	81	3.1	0.104	−0.270	D	450	1905	0.15
Aug	28.0	28.8	82	3.4	0.131	−0.329	D	425	1962	0.16
Sep	27.9	28.7	81	3.1	0.131	−0.396	D	450	1905	0.15
Oct	27.7	28.7	83	3.7	0.146	−0.335	D	400	1997	0.16
Nov	27.3	28.4	84	3.3	0.153	−0.457	C	363	1601	0.16
Dec	26.9	27.8	85	4.8	0.139	−0.183	D	350	2223	0.18

T_{air} = Air temperature; Z/L = Monin-Obukhov stability parameter; T_{sea} = Sea temperature; LCL = Lifting condensation level; RH = Relative humidity; VF = Ventilation factor; U_Z = Wind speed measured by ships; u_* = Friction velocity; B = Bowen ratio

2. Determining the Stability Classification

The atmospheric stability for dispersion estimates can be classified by the parameter Z/L, where Z is the height above the ground and L is the Monin-Obukhov length, which is defined as (PANOFSKY and DUTTON, 1984, p. 132)

$$L = -\frac{u_*^3 \rho C_p T_{air}}{g\kappa H\left(1 + \frac{0.07}{B}\right)} \tag{1}$$

where u_* is the friction velocity, g is the gravitational acceleration, κ is the von Karman constant, ρ is the air density, C_p is the specific heat of air at constant pressure, T_{air} is the air temperature, H is the surface layer sensible heat flux, and B is the Bowen ratio (the ratio of sensible to latent heat flux). Note that T_{air} should be T_v, the virtual temperature. However, since $T_v = T_{air}(1 + 0.68\ q)$ and q is at most 5% (see, e.g., KOMEN et al., 1994, p. 59), we use $T_v \cong T_{air}$.

From Eq. (1) and SMITH (1980, Eqs. 3 and 4)

$$\frac{H}{\rho C_p} = C_T U_{10}(T_{sea} - T_{air}) \tag{2}$$

and

$$C_d = \left(\frac{u_*}{U_{10}}\right)^2 \tag{3}$$

or $u_*^3 = U_{10}^3 \, C_d^{3/2}$. Since, from monthly averages $T_{sea} > T_{air}$, i.e., unstable conditions prevail,

$$L = -\frac{T_{air} \, C_d^{3/2} \, U_{10}^2}{g\kappa \, C_T (T_{sea} - T_{air})\left(1 + \frac{0.07}{B}\right)} \tag{4}$$

where C_T is the heat flux coefficient ($= 1.1 \times 10^{-3}$ for unstable conditions (SMITH, 1980 and 1988)), C_d is the drag coefficient, U_{10} is the wind speed at 10 m above the sea surface, and T_{sea} is the sea-surface temperature.

On the basis of thermodynamic conditions, a relationship between B and $(T_{sea} - T_{air})$ under unstable conditions (i.e., $T_{sea} > T_{air}$) has been proposed by HSU (1998) that

$$B = a \, (T_{sea} - T_{air})^b \tag{5a}$$

where a and b are to be determined by field experiments. Based on the availability of additional data sets from tropical oceans and coastal seas, HSU (1999) found

$$B = 0.146 \, (T_{sea} - T_{air})^{0.49} \tag{5b}$$

with a high correlation coefficient of 0.94 between B and $(T_{sea} - T_{air})$. We can now estimate Z/L from Eq. (4) by employing this B parameterization along with a proven C_d formulation used successfully in the third generation wave model (see WAMDI, 1988, p. 1784) that

$$C_d = \begin{bmatrix} 1.2875 \times 10^{-3}, & U_{10} < 7.5 \text{ m s}^{-1} \\ (0.8 + 0.065 \, U_{10}) \times 10^{-3}, & U_{10} \geq 7.5 \text{ m s}^{-1} \end{bmatrix} . \tag{6}$$

Since the location of anemometers on each ship may vary from approximately 10 m to 20 m depending on the vessel size, we can estimate the difference in wind speed between these two heights. According to HSU (1988) and HSU et al. (1994), the power-law wind profile may be applied for offshore applications. That is

$$\frac{U_{20 \text{ m}}}{U_{10 \text{ m}}} = \left(\frac{20 \text{ m}}{10 \text{ m}}\right)^{0.1} = 1.07 . \tag{7}$$

Therefore the difference is within 10%. On the other hand, according to the NATIONAL DATA BUOY CENTER (1990), the total system accuracy of wind speed measurement on buoys can be 10%. Since the wind speed difference between 10 and 20 m height is within 10%, we set $Z = 10$ m here as routinely done for air-sea interaction studies.

Now, values of Z/L can be computed from Eq. (4). Since $T_{sea} > T_{air}$ throughout the year, Z/L is negative. The results are listed in Table 1. According to the stability criterion for "L" from the "Offshore and Coastal Dispersion (OCD) Model" set by HANNA et al. (1985), and converted to "Z/L" categories by HSU (1992), the letter designation of stability class is also provided in the table. It can be seen that with the exception of April, May, and November which are in the slightly unstable (C) category, the rest of the year exhibits near-neutral (D) stability conditions. Note that for Class D, $|Z/L| < 0.4$ and for Class C, $-1.0 \leq Z/L \leq -0.4$ (for more detail, see Hsu (1992)).

3. Determining the Mixing Height

Due to actual evaporation, the air over the water is usually moister than that over land, and the top of the marine boundary layer is often times capped by clouds. On the basis of analysis of vertical soundings by research aircraft, rawinsondings, and radar wind profilers and Radio Acoustic Sounding Systems (RASS), it has been shown by Garratt (1992) that the mixing height can be approximated by the lifting condensation level (LCL) under cumulus cloud conditions (where LCL = cloud base). Note that the LCL is defined as the level in the atmosphere at which an unsaturated air parcel lifted dry adiabatically (i.e., at the rate $\approx -1\,°C/100$ m) would become saturated, i.e., for formation of clouds. According to Hsu (1998) under near-neutral conditions, the height of the LCL may be estimated by

$$H_{sea} = 125(T_{air} - T_{dew}) \tag{8}$$

where H_{sea} (in meters) is the mixing height over the water surface, and T_{air} and T_{dew} (in °C) are the air and dewpoint temperatures, respectively. However, if the clouds are stratiform (as is often the case over the U.S. West Coast), the height of the cloud top rather than the base is the mixing height.

From the monthly averages of air temperature and relative humidity as provided in Table 1, T_{dew} can be computed using the Smithsonian Meteorological Tables (List, 1984). Using these monthly T_{air} and T_{dew} values, the mixing height is obtained via Eq. (8) and listed in Table 1. It can be seen that the LCL ranges from 338 m to 450 m with lowest values occurring in March and highest in May, July, and September, respectively.

4. Determining the Ventilation Factor

The ventilation factor, VF, is defined as the product of the wind speed and mixing height; that is

$$\text{Ventilation Factor} = \text{LCL} \times U_{\text{mixed layer}} \tag{9}$$

where $U_{\text{mixed layer}}$ is the mean wind speed in the mixed layer, i.e., in the mid-level of LCL, or 0.5 LCL,

$$\frac{U_{\text{mixed layer}}}{U_{10\,m}} = \left[\frac{(1/2\text{LCL})}{10\text{ m}}\right]^{0.1} = \left(\frac{\text{LCL}}{20}\right)^{0.1}. \tag{10}$$

From data provided in Table 1 for $U_Z = U_{10\,m}$ and LCL, values of VF are determined from Eq. (9) via Eq. (10) and compiled in Table 1. From the viewpoint of monthly averages, Table 1 shows that the lowest VF occurred in April and the highest in January.

5. Determining the Friction Velocity

In some K-diffusion models (see, e.g., ZANNETTI, 1990), the vertical eddy diffusivity, K_z, is required,

$$K_z = \kappa\, u_* \, Z \left(1.1 - \frac{Z}{H_{sea}} \right) ,$$ (11)

where κ is the von Karman constant, u_* is the friction velocity and H_{sea} is the mixing height. For example, the Urban Airshed Model (UAM), one of the preferred models of the U.S. EPA for ozone studies, has a meteorological preprocessing subroutine which employs the gradient transport (K) modeling. In order to estimate the standard deviations of lateral and vertical winds, i.e. σ_v and σ_w, u_* is also needed along with Z/L and the mixing height (see, e.g., PANOFSKY and DUTTON, 1984). They are, when $Z/L < 0$,

$$\frac{\sigma_v}{u_*} = \left(12 - 0.5 \frac{Z_i}{L} \right)^{1/3}$$ (12)

and

$$\frac{\sigma_w}{u_*} = 1.25 \left(1 - 3\frac{Z}{L} \right)^{1/3}$$ (13)

where Z_i is the mixing height.

In the atmospheric surface boundary layer, u_* can be estimated by (see, e.g., GARRATT, 1992)

$$u_* = \kappa\, U_Z \left[\frac{\kappa}{\sqrt{C_{DN}}} - \psi_m \left(\frac{Z}{L} \right) \right]^{-1} .$$ (14)

According to HSU et al. (1999), for overwater applications under unstable conditions when $Z/L < 0$,

$$\psi_m \left(\frac{Z}{L} \right) = 1.05 \left(-\frac{Z}{L} \right)^{0.46}$$ (15)

and from Table 1 and YELLAND and TAYLOR (1996), when $U_Z \leq 6$ m s^{-1},

$$10^3\, C_{DN} = 0.29 + \frac{3.1}{U_{10}} + \frac{7.7}{U_{10}^2} .$$ (16)

Monthly averaged values of u_* as computed from above formulas are provided in Table 1. They range from 0.14 to 0.18 m s^{-1}. These u_* values along with the mixing height and Z/L estimates as provided in Table 1 can be used to compute σ_v and σ_w via Eqs. (12) and (13), respectively. Furthermore, these σ_v and σ_w can be input into dispersion equations to calculate σ_y and σ_z, the horizontal (lateral) and vertical dispersion coefficients used in Gaussian distribution modeling.

6. Conclusions

In order to improve air quality modeling over the South China Sea offshore from the west coast of Borneo, long-term meteorological measurements (on the time scale of monthly averages) from ships are employed. Those monthly averages used here include air temperature, sea temperature, wind speed, and relative humidity. For climatological estimation, monthly data are analyzed. Determination of the stability parameter shows that, in our study area, Z/L ranges from -0.05 to -0.7. Application of the U.S. EPA approved "Offshore and Coastal Dispersion (OCD) Model" stability classification indicates that "D" class for near-neutral conditions generally prevails except for April, May, and November at which time "C" class for slightly unstable is observed.

Using these near-neutral stability conditions, the mixing height can then be estimated based on the lifting condensation level (LCL), determined from the surface dewpoint depression. The range of LCL heights are found to be between 338 and 450 m. For pollution dispersion estimation, the ventilation factor, VF, for each month is also determined. The VF is the product of the wind speed in the middle of the mixing height (calculated from the power-law wind profile) and the LCL height. It is found that the pollution dispersal ability in our study area is lowest in April and highest in January.

In order to provide the modeller with recent advances in air-sea interaction for the drag coefficient formulation, the monthly averaged friction velocity, u_*, is also estimated. It can be used along with the LCL and Z/L for each month to estimate the vertical diffusion eddy diffusivity, K_z, and the standard deviation of crosswind and vertical dispersion coefficients, σ_y and σ_z.

Acknowledgment

This study is supported in part by an endowed professorship from Dr. Eric Abraham to the Center for Coastal, Energy and Environmental Resources of Louisiana State University. Comments from the reviewers are appreciated.

REFERENCES

GARRATT, J. R., *The Atmospheric Boundary Layer* (Cambridge University Press, Cambridge, 1992), 316 pp.

HANNA, S. R., SCHULMAN, L. L., PAINE, R. J., PLEIM, J. E., and BAER, M. (1985), *Development and Evaluation of the Offshore and Coastal Dispersion Model*, J. Air Poll. Contr. Assoc. 35, 1039–1047.

HSU, S. A., *Coastal Meteorology* (Academic Press, San Diego, CA, 1988), 260 pp.

HSU, S. A. (1992), *An Overwater Stability Criterion for the Offshore and Coastal Dispersion Model*, Boundary-Layer Meteorol. 60, 397–402.

HSU, S. A., MEINDL, E. A., and GILHOUSEN, D. B. (1994), *Determining the Power-law Wind-profile Exponent under Near-neutral Stability Conditions at Sea*, J. Appl. Meteor. 33, 757–765.

HSU, S. A. (1998), *A Relationship Between the Bowen Ratio and Sea-air Temperature Difference under Unstable Conditions at Sea*, J. Phys. Oceanogr. 28, 2222–2226.

HSU, S. A. (1999), *On the Estimation of Overwater Bowen Ratio from Sea-air Temperature Difference*, J. Phys. Oceanogr. 29, 1372–1373.

HSU, S. A., BLANCHARD, B. W., and YAN, Z. (1999), *A Simplified Equation for Paulson's Ψ_m (Z/L) Formulation for Overwater Applications*, J. Appl. Meteor. 38, 623–625.

HSU, S. A. (2003), *Thermodynamic Characteristics of the Sub-Cloud Layer Affecting Haze Dispersion along the West Coast of Borneo*, Pure Appl. Geophys. (this issue).

KOMEN, G. J., CAVALERI, L., DONELAN, M., HASSELMANN, K., HASSELMANN, S., and JANSSEN, P. A. E. M., *Dynamics and Modelling of Ocean Waves* (Cambridge University Press, 1994), 532 pp.

LIST, R. J., *Smithsonian Meteorological Tables* (Smithsonian Institution Press, Washington, D. C., 1951/1984), 527 pp.

NATIONAL DATA BUOY CENTER, *Climatic Summaries for NDBC Buoys and Stations – Update 1* (National Climatic Data Center, NSTL, MS, 1990).

PANOFSKY, H. A. and DUTTON, J. A., *Atmospheric Turbulence* (Wiley, New York, 1984), 397 pp.

SMITH, S. D. (1980), *Wind Stress and Heat Flux over the Ocean in Gale Force Winds*, J. Phys. Oceanogr. 10, 709–726.

SMITH, S. D. (1988), *Coefficients for Sea Surface Wind Stress, Heat Flux, and Wind Profiles as a Function of Wind Speed and Temperature*, J. Geophys. Res. 93, 15,467–15,474.

THE WAMDI GROUP (1988), *The WAM Model – A Third Generation Ocean Wave Prediction Model*, J. Phys. Oceanogr. 18, 1775–1810.

U.S. NAVAL WEATHER SERVICE COMMAND, *Summary of Synoptic Meteorological Observations (SSMO)*, Indonesian Coastal Marine Areas, vol. 1, Area 7 – Sarawak (National Climatic Center, Asheville, N.C., 1975).

YELLAND, M. and TAYLOR, P. K. (1996), *Wind Stress Measurements from the Open Ocean*, J. Phys. Oceanogr. 26, 541–558.

ZANNETTI, P., *Air Pollution Modeling* (Van Nostrand Reinhold, New York, 1990), 444 pp.

(Received January 15, 2000, accepted January 12, 2001)

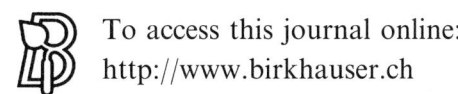

To access this journal online:
http://www.birkhauser.ch

Pure appl. geophys. 160 (2003) 357–394
0033–4553/03/020357–38

❘ **Pure and Applied Geophysics**

Mathematical Modeling of Diffusion and Transport of Pollutants in the Atmospheric Boundary Layer

MAITHILI SHARAN[1] and S. G. GOPALAKRISHNAN[2]

Abstract — The process of dispersion of air pollutants, in a broad sense, can be considered as the net outcome of various mechanisms involved in the transport of air pollutants from the source to the receptor. The major mechanisms are: (1) advection of pollutants by mean air motion, (2) mixing of pollutants by atmospheric turbulence and (3) mass diffusion due to concentration gradients. In addition, the physical and chemical nature of the effluent, the location of the stack and the nature of the terrain downwind from the stack, effect the dispersion of the pollutants. Various physical and mathematical aspects related to the transport and diffusion of air pollutants in the atmospheric boundary layer are discussed here. Further, some aspects of dispersion in a weak wind stable boundary layer are described. Finally, the current issues in the modeling of weak wind boundary layer are illustrated.

Key words: Stable boundary layer, weak wind conditions, dispersion, radiative cooling.

1. Introduction

The problem of air pollution is not of recent origin. It started as early as the 14th century, when coal began to replace wood as the primary source of energy. However, the alarming rise in the air pollution episodes associated with the accidental toxic gas and radioactive releases towards the turn of the 19th century, deterioration of environmental air quality due to industrialization and urbanization, and the growing risk of climate change associated with long-term adverse effects of unclean air have raised serious concern, in the recent past, for the need of cleaner environment. Lessons learned from Three Mile Island accident in the United States, Mihama SG tube rupture in Japan, Chernobyl disaster in Russia, Bhopal Methyl isocyanate gas leak in India and forest fires in Indonesia have resulted in a long-term debate for the operational safety and emergency response management. Mathematical models are being increasingly used for the regulatory and impact assessment studies.

---

[1] Centre for Atmospheric Sciences, Indian Institute of Technology, Hauz Khas, New Delhi 110016, India. E-mail: mathilis@cas.iitd.ernet.in

[2] Centre for Atmospheric Physics, Science Applications International Corporation, MS 2-3-1, 1710 SAIC Drive Mclean, VA 2210, U.S.A.

To facilitate the decision-making process for the sitting of industrial and residential complexes, and for the purpose of emergency response management and impact assessment studies, atmospheric dispersion models are necessary and indispensable tools. This is especially true in developing countries, where it is not feasible to operate the conventional and expensive monitoring/observational network to measure the concentration of air pollutants at various vulnerable locations on a regular basis.

The process of dispersion of air pollutants, in a broad sense, can be considered as the net outcome of various mechanisms involved in the transport of air pollutants from the source to the receptor. The major mechanisms are: (1) advection of pollutants by mean air motion, (2) mixing of pollutants by atmospheric turbulence and (3) mass diffusion due to concentration gradients. In addition, the chemical and physical nature (i.e., transformation and removal) of the effluent, the location of the stack and the nature of the terrain downwind from the stack, effect the dispersion of the pollutants.

The dispersion of the pollutants in the planetary boundary layer (PBL) is greatly influenced by the stability of the atmosphere near the surface. The structure of the boundary layer and the associated dispersion of air pollutants become complex with the degree of stability and weakening of the winds. Here, we describe briefly the various aspects related to the dispersion of pollutants in the boundary layer.

1.1 Various Scales of Dispersion

There are several phenomena in the atmosphere like the diffusing puff of smoke, the dust devil, acid rain or the climate change associated with CO_2 accumulation. Each one of these encompasses a characteristic length of space and interval of time called the space and time scales of motion of the phenomena under consideration. The air pollutants emitted from a source can become dispersed over a wide range of scales depending on the duration of release, quantity of release, height of release, the state of the atmosphere during the transport, etc. A preliminary distinction between the different transport scales of air pollution phenomena can be made as follows (ZANNETTI, 1990):

(a) Near-field phenomena (<1 km from the source); e.g., down-wash effects of plume caused by building aerodynamics.

(b) Short-range transport (<10 km from the source); e.g., the area in which the maximum ground-level impact of primary pollutants from an elevated source is generally found.

(c) Intermediate transport or mesoscale transport (between 10 km and 100 km); e.g., the area in which the chemical and the physical properties of the pollutants may become important.

(d) Long-range (or regional or interstate transport > 100 km); e.g., the area in which meteorological effects and deposition and transformation rates play key roles.

(e) Global effects; i.e., phenomena affecting the entire earth atmosphere; e.g., CO_2 accumulation.

1.2 The Hierarchy of Atmospheric Motions

The spectrum of scales of motions in the atmosphere is very large and complex. However, a preliminary classification is generally made on the basis of ORLANSKI's (1975) nomenclature. According to this nomenclature, the spectrum consists of the scale of motions, spawning from macro-α-scale, which are on the order of the circumference of the earth, at one end, to the microscale motions on the order of 1 mm or less which eventually dissipate the kinetic energy of a moving air parcel, converting it into heat energy at the other end. Between these scales lies the meso-β-scale that encompasses motions of length scales of about 20–200 km and a time scale of approximately 100 min and the meso-Γ-scale that has a length scale of about 2 to 20 km and a time scale of about 10 min. Dispersion over various scales of space and time (section 1.1) is largely influenced by the corresponding scales of atmospheric motions. While microscale motions at the lower end of the atmospheric spectra influence the diffusing puff of smoke, the large scale and regional (i.e., macro-β- and meso-α-) scale motions influence the long-range dispersion of air pollutants. The dispersion of air pollutants in the intermediate range of the transport scale (given in section 1.1), the focus of this work, is influenced by mesoscale (specifically the meso-β and Γ-scales) motions of the atmospheric spectra and is termed as mesoscale dispersion. In this, the terrain and the urban forcing influence the 'local flows' (PEARSON, 1980), local meteorological conditions and hence the transport and dilution of air pollutants.

1.3 Atmospheric Boundary Layer

The lowest region of the atmosphere which experiences surface effects through vertical exchanges of heat, momentum and moisture is called the planetary boundary layer or the PBL (PANOFSKY and DUTTON, 1984). Since most of the industrial stacks are located within this layer, it plays a pivotal role in the dispersion of air pollutants.

Depending on the delicate balance between shear production caused by vertical variations in mean wind and buoyancy production/suppression due to surface warming/cooling, the structure of the PBL varies from day to night-time. In a convective boundary layer (CBL), which often develops during the day time over land under fair weather conditions, convective turbulence dominates the mechanical turbulence and both these processes aid mixing. Due to the asymmetries associated with the convective updrafts and downdrafts, dispersion of passive materials in the CBL is strongly influenced by the vertical structure of turbulence, and consequently, the height of release. Although there exists a few unsolved problems particularly near the surface (NIEUWSTADT, 1980; VENKATRAM and DU, 1997), past studies, including the laboratory convection tank experiments by DEARDORFF and WILLIS

(1974) and WILLIS and DEARDORFF (1981); CONDORS field experiment (BRIGGS, 1993); and pioneering Large Eddy Simulations (LES; a modeling technique in which the model grid resolution of the order of 100 m or less is used to study turbulence) of DEARDORFF et al. (1969) and DEARDORFF (1970, 1972, 1974) and LAMB (1978, 1982) have all led to an excellent understanding of the dispersion in the CBL that develops over fairly flat and homogeneous domains. More recently, GOPALAKRISHNAN and AVISSAR (2000) and GOPALAKRISHNAN et al. (2000) made a systematic analysis of the impacts of heat patches and topographical features on the dispersion of passive materials in a shear-free CBL. LES and a Lagrangian particle model were used for that purpose. They found that the dispersion patterns produced in earlier numerical studies of LAMB (1982) and DEARDORFF (1972) were only limiting case over flat and homogeneous domains and that horizontal pressure gradients created by surface heterogeneities with a characteristic length scale larger than about 5 km generate atmospheric circulations, which impede vertical mixing and, as a result, remarkably influence particle dispersion in the CBL (Fig. 1). In a neutral boundary layer, which is often formed during evening transition, mechanical turbulence is the only mixing mechanism. The prevailing structure of the neutral Ekman PBL has been relatively well established (DEARDORFF, 1972; BROWN, 1974) under steady state, windy and cloudy conditions.

The structure of the stable boundary layer (SBL), which often evolves on a clear night over land, has not been clearly understood. The balance between mechanical generation of turbulence and damping by stability varies from one situation to another, creating SBL that ranges from being well mixed to non-turbulent (STULL, 1988). The influence of long-wave radiation (GARRATT and BROST, 1981; ESTOURNEL and GUEDALIA, 1985; GOPALAKRISHNAN et al., 1998), gravity waves (FINNIGAN and EINAUDI, 1981) and intermittence (KONDO et al., 1978) makes the study of SBL even over flat and homogeneous domains very complex. The complexity grows with the degree of stability and weakening of winds. Nonetheless, as an outcome of numerical modeling (GARRATT and BROST, 1981; TJEMKES and DUYNKERKE, 1989; MCNIDER et al., 1995; SHARAN and GOPALAKRISHNAN, 1997a,b; GOPALAKRISHNAN et al., 1998) and a few essential observational studies (ANDRÈ and MAHRT, 1982; GARRATT, 1982; NAPPO, 1991) a better understanding of the SBL has been attained within recent years.

2.1 Basic Governing Equations for Diffusion and Transport in the Atmosphere

The deterministic models for the dispersion of air pollutants are based on a set of conservation relations, namely: (i) conservation of mass; (ii) conservation of heat; (iii) conservation of momentum; (iv) conservation of water species; and (v) conservation of gaseous and aerosol materials. Equations based on these relations (PIELKE, 1981; PIELKE, 1984; BUSINGER, 1982; STULL, 1988; BACON et al., 2000) describe the instantaneous values of the properties of the air parcel that are dispersed in the atmosphere. Assuming that the concentration of the pollutants is very small

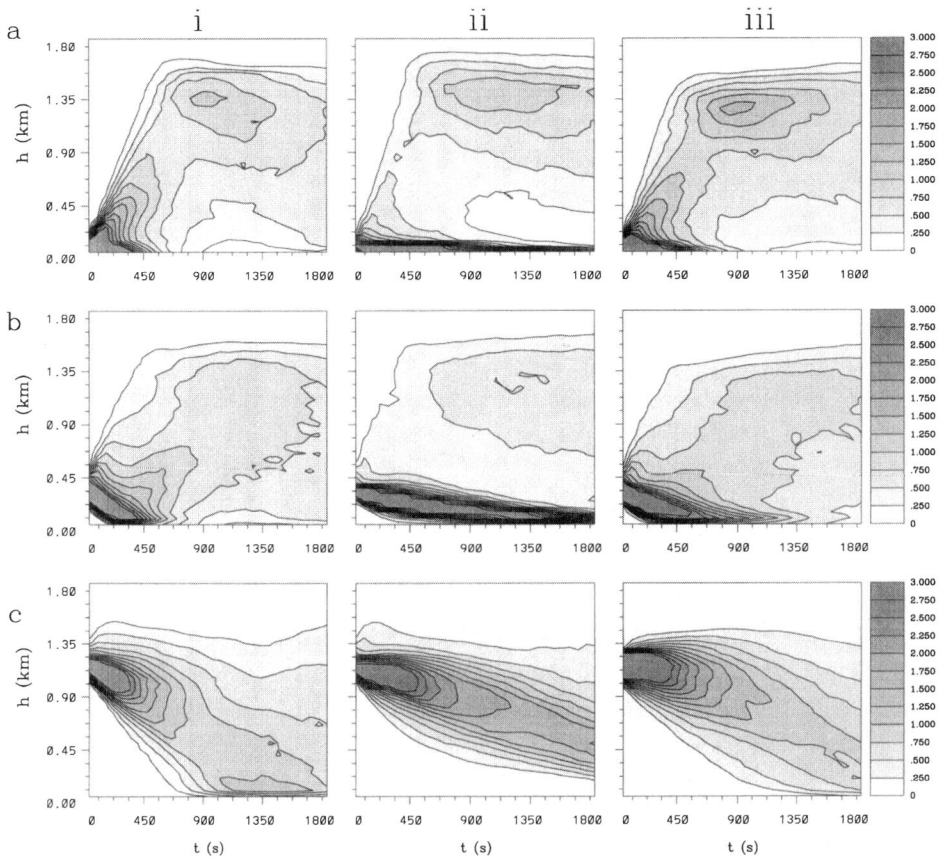

Figure 1

Dimensionless crosswind-integrated concentration contours obtained from instantaneous release from an area source located at (a) 40 m, (b) 300 m, and (c) 1200 m above the ground surface in the CBL that develops over (i) a flat homogeneous domain; (ii) a heterogeneous domain forced by a heat flux wave with a wavelength of 4.8 km and an amplitude of 0.2 k/m/s; and (iii) a domain with a hill 400-m high and 4.8-km long. The mean heat flux over the domain was set to 0.2 k/m/s. Here h is the height above the ground surface (adopted from GOPALAKRISHNAN and AVISSAR, 2000, J. Atmos. Sci. 57, 359).

and does not influence the flow and further neglect the effects of molecular diffusion, the governing equations in the Eulerian frame of reference are given by:

$$\frac{\partial \rho}{\partial t} + \frac{\partial}{\partial X_j}(\rho U_j) = 0 \tag{1}$$

$$\frac{\partial \theta}{\partial t} + U_j \frac{\partial \theta}{\partial X_j} - S_\theta = 0 \tag{2}$$

$$\frac{\partial U_i}{\partial t} + U_j \frac{\partial U_i}{\partial X_j} + \theta \frac{\partial \pi}{\partial X_i} + g\delta_{i3} + 2\varepsilon_{ijk}\Omega U_k = 0 \tag{3}$$

$$\frac{\partial q_m}{\partial t} + U_j \frac{\partial q_m}{\partial \pi_j} - S_{q_m} = 0 \tag{4}$$

$$\frac{\partial C_n}{\partial t} + U_i \frac{\partial C}{\partial X_j} - S_{c_n} = 0 \ , \tag{5}$$

where i and $m = 1, 2, 3$ and $n = 1, 2, 3, \ldots, M$.

In the above system of equations, ρ is the density of the air parcel; U_i represent the velocity components along the east-west, north-south and vertical directions respectively; θ is the potential temperature; S_θ is the source or sink of heat (for example: contributions due to short-wave and long-wave radiation); g is the acceleration due to gravity; Ω is the angular velocity of the earth's rotation; π is the scaled pressure; q_m ($q_1 = q_{solid\ water}$, $q_2 = q_{liquid\ water}$ and $q_3 = q_{water\ vapour}$) are the amounts of water substances; S_{q_m} is the source or sink of water substances (for example: contributions due to freezing, condensation and evaporation); C_n represents the concentration of the nth species and S_{c_n} is the source or sink of each of the pollutants (for example: dry fallout, scavenging and chemical transformation). The variables q_m and C_n have units of mass of substance to mass of air in the same volume. The independent variables are time (t) and the three space coordinates ($X_1 = X$, $X_2 = Y$ and $X_3 = Z$). In order to close the system of equations (1) to (5), four additional relationships are required.

The scale pressure that is also called the Exner's function is given by

$$\pi = C_p \left[\frac{P\,(\text{millibars})}{1000.0\ \text{millibars}} \right]^{R/C_p} , \tag{6}$$

where P is the unscaled hydrodynamic pressure, R is the gas constant for dry air and C_p is the specific heat capacity at constant pressure.

The unscaled pressure is determined from the equation of state

$$P = \rho R T_v \ , \tag{7}$$

where T_v is the virtual temperature and is given by

$$T_v = T(1 + 0.61 q_s) \tag{8}$$

in which q_s is the specific humidity.

Finally, the potential temperature corresponding to the temperature T_v is given by

$$\theta = T_v \left[\frac{1000.0\ \text{millibars}}{P\,(\text{in millibars})} \right]^{R/C_p} . \tag{9}$$

Equations (1) to (9) form a closed system of nonlinear partial differential equations that have to be solved to predict the fate of the air pollutants. However, it is not feasible to find an analytical solution of such a system. Further, due to the existence of a range of scales of motion from 300 m down to 1 mm (in the case of the typical turbulent flows), all being coupled nonlinearly; it is not possible to solve the system of equations numerically. An estimate by WYNGAARD (1982) shows that to resolve such a range of scales over a numerical domain of 10 km × 10 km, the area must be divided in roughly 10^{20} grid points and the equations (1)–(9) must be solved over each grid point. However, with suitable assumptions and relevant boundary and initial conditions, an approximate solution can be obtained for the system of equations which may be valid for a particular scenario.

2.2 Governing Equations for Dispersion in the Boundary Layer

Equations (1) to (8) can be simplified for specific boundary layer studies. In this section, the commonly made assumptions are listed below:

• *Reynolds averaging*: Although the instantaneous values of the properties of air parcel cannot be studied, the mean values of properties can be readily examined and easily measured. Consequently the equations are approximated for the mean values of variables. An instantaneous value of a variable is defined as the sum of its mean value ($\bar{\phi}$) over a period of time and a fluctuation term (φ') i.e.,

$$\phi = \bar{\phi} + \phi' \ , \tag{10}$$

where φ is any one of the nine dependent variables in the equations (1) to (9).

• *Anelastic and incompressibility approximations*: The system of equations becomes anelastic or soundproof (OGURA and PHILLIPS, 1962) if the local variation in fluid density with time is neglected. In this case, the equation of conservation of mass reduces to the deep continuity equation. Further, in some cases, the atmosphere is treated as an incompressible fluid in which the equation of conservation of mass reduces to a shallow continuity equation.

• *Hydrostatic approximation* (PIELKE, 1981): Atmosphere can be considered as a more or less stratified fluid wherein an approximate balance prevails between the pressure gradient forces and the force of acceleration due to gravity.

• *Boussineq approximation* (JENSEN and BUSCH, 1982; BUSINGER, 1982): The variation in fluid density is negligible with regard to inertia, and important only as it enters through the gravitational body force.

• *Geostrophic approximation*: The ambient synoptic fields are assumed to be in geostrophic balance and the ageostrophic effects are produced by the mesoscale perturbations (MCNIDER and PIELKE, 1981).

The simplified equations (1) to (9) for the mean variables in the component form are given by:

$$\frac{\partial \bar{U}}{\partial X} + \frac{\partial \bar{V}}{\partial Y} + \frac{\partial \bar{W}}{\partial Z} = 0 \tag{11}$$

$$\frac{d\bar{\theta}}{dt} = -\frac{\partial}{\partial X}\left(\overline{U'\theta'}\right) - \frac{\partial}{\partial Y}\left(\overline{V'\theta'}\right) - \frac{\partial}{\partial Z}\left(\overline{W'\theta'}\right) + F_\theta \tag{12}$$

$$\frac{d\bar{U}}{dt} = f\bar{V} - \bar{\theta}\frac{\partial\bar{\pi}}{\partial X} - \frac{\partial}{\partial X}\left(\overline{U'U'}\right) - \frac{\partial}{\partial Y}\left(\overline{U'V'}\right) - \frac{\partial}{\partial Z}\left(\overline{W'U'}\right) \tag{13}$$

$$\frac{d\bar{V}}{dt} = -f\bar{U} - \bar{\theta}\frac{\partial\bar{\pi}}{\partial Y} - \frac{\partial}{\partial X}\left(\overline{U'V'}\right) - \frac{\partial}{\partial Y}\left(\overline{V'V'}\right) - \frac{\partial}{\partial Z}\left(\overline{W'V'}\right) \tag{14}$$

$$\frac{\partial\bar{\pi}}{\partial Z} = -\frac{g}{\bar{\theta}} \tag{15}$$

$$\frac{d\bar{q}_m}{dt} = -\frac{\partial}{\partial X}\left(\overline{U'q'_m}\right) - \frac{\partial}{\partial Y}\left(\overline{U'q'_m}\right) - \frac{\partial}{\partial Z}\left(\overline{W'q'_m}\right) + F_{qm} \tag{16}$$

$$\frac{d\bar{C}_n}{dt} = -\frac{\partial}{\partial X}\left(\overline{U'C'_n}\right) - \frac{\partial}{\partial Y}\left(\overline{V'C'_n}\right) - \frac{\partial}{\partial Z}\left(\overline{W'C'_n}\right) + F_{C_n} \tag{17}$$

where

$$\bar{\pi} = C_p\left[\frac{\bar{P}}{1000.0}\right]$$

and

$$\frac{d}{dt} = \frac{\partial}{\partial t} + \bar{U}\frac{\partial}{\partial X} + \bar{V}\frac{\partial}{\partial Y} + \bar{W}\frac{\partial}{\partial Z}\ . \tag{18}$$

The source terms F_θ, F_{qm} and F_{cn} represent the integrated contribution of instantaneous source or sink terms appearing in the equations (2), (4) and (5), respectively.

In the above system of equations, overbars represent the mean field variables and f represents the Coriolis parameter. The cross-correlation terms of the form $U'W'$, $W'\theta'$, etc., appearing in equations (12)–(17) are known as the Reynolds stress terms (in kinematics form). These new unknowns have been introduced due to the Reynolds averaging of the instantaneous field variables in equations (2) to (5). It is to be noted that the above system of equations cannot be solved without further assumptions since the number of unknowns exceed the number of equations.

2.3.1 Parameterization of Reynolds stresses

The problem of finding relationships/equations for the unknown Reynolds stresses is called the 'closure problem.' Boussinesq proposed the analogy of molecular transport to model the Reynolds stress terms. According to this theory:

$$-(\overline{U'W'}, \overline{U'W'}) = \left[K_M\left(\frac{\partial\bar{U}}{\partial Z}\right), K_M\left(\frac{\partial\bar{V}}{\partial Z}\right)\right]$$

$$-(\overline{W'\theta'}) = K_H\left(\frac{\partial\bar{\theta}}{\partial Z}\right) \tag{19a}$$

$$-(\overline{W'q'}) = K_q\left(\frac{\partial\bar{q}}{\partial Z}\right)$$

$$-(\overline{W'C_n'}) = K_Z\left(\frac{\partial\overline{C_n}}{\partial Z}\right) \;,$$

where K_M, K_H, K_q and K_Z represent the vertical exchange coefficients of momentum, heat, moisture and pollutants, respectively. In analogy with equation (19a) it is possible to construct such relationships for the horizontal components also. For example,

$$-(\overline{U'C_n'}) = K_X\left(\frac{\partial\overline{C_n}}{\partial X}\right)$$
$$-(\overline{V'C_n'}) = K_y\left(\frac{\partial\overline{C_n}}{\partial Y}\right) \;.$$

(19b)

The central problem of closure modeling is to find suitable forms of eddy exchange coefficients which must account for the complexities of turbulence. A number of studies available in the literature (for example, O'BRIEN, 1970; BLACKADAR, 1979; ESTOURNEL and GUEDALIA, 1987; ENGLAND and MCNIDER, 1995) suggest possible forms of vertical eddy exchange coefficients in the atmosphere. Yet there seems to be no unique scheme of specifying the turbulent exchange coefficients that work well for all atmospheric conditions.

Based on the relative importance of surface heat flux and near-surface wind stress, the vertical eddy exchange coefficients in the surface layer may be computed using either Monin-Obukhov similarity or free convection theory. While flux-profile relationships required to evaluate these exchange coefficients developed by BUSINGER et al. (1971) and DYER (1974) are widely used to model moderately unstable surface layer, mixed layer relationships developed by WYNGAARD et al. (1971) may be applied in a purely buoyancy driven surface layer. Recently, SHARAN et al. (2002) found that the similarity functions proposed by BELJAARS and HOLTSLAG (1991) perform well in stable conditions in comparison with those of BUSINGER et al. (1971).

In a well-mixed layer, vertical eddy exchange coefficients may be parameterized using a simple O'Briens profile (O'BRIEN, 1970; YU, 1977). Such a profile requires only values of fluxes in the surface layer. More complex, nonlocal closure formulations recognizing the premise of down-gradient transport have also been proposed (STULL, 1988, 1993). However, after sunset, as the stability grows, the size of the turbulent eddies becomes small, and with increasing stability these eddies may no longer feel the effect of the surface. Under these conditions, NIEUWSTADT (1984) pointed out that it is considerably more appropriate to use a local scheme (BLACKADAR, 1979; MCNIDER and PIELKE, 1981; ESTOURNEL and GUEDALIA, 1987) for the parameterization of eddy exchange coefficients. Further, local parameterization schemes may be able to capture the turbulent characteristics aloft which have been found to influence the long-range transport of pollutants (MCNIDER et al., 1988). In view of the above facts, local closures are being used in various mesoscale/

boundary layer models (PIELKE, 1984) to parameterize the turbulence exchange processes in the SBL. In a simple, first order local-K closure, the eddy exchange coefficients are expressed in terms of the local shear, mixing length and the local stability function (BLACKADAR, 1979) depending upon the gradient Richardson number.

Some of the fundamental problems with the first-order or K-closure may be partly resolved with higher order closures. Several investigators (WYNGAARD, 1975; ANDRÈ et al., 1978; BLACKADAR, 1979) have studied the turbulence features in the SBL by using second or higher order closure schemes which involve explicit solution of the prediction equations for the boundary layer fluxes. ANDRÈ et al. (1978), for instance, studied the mean and turbulence structure of the PBL that developed over Wangara (days 33–34) using a third-order closure model. Observed mean profiles of winds matched quite well with the simulation. Unfortunately, turbulence observations were not available at that time. Later, MCNIDER and PIELKE (1981) using a simple first-order closure model demonstrated that their model was capable of simulating the mean features of the boundary layer that developed over Wangara on days 33–34. LEWELLEN and TESKE (1975) used a second-order closure model (DONALDSON, 1973) to simulate free-convective diffusion of materials from continuous ground-level line source. Their model predicted the surface-plume "lift-off" phenomenon, although it was not as pronounced as that observed by Deardorff and Willis in the Laboratory (LAMB, 1982). On the contrary, LAMB (1982) demonstrated that LES coupled with a Lagrangian particle model is capable of simulating the dispersion of passive materials released in the CBL far better than higher order closure models. Recently, SHARAN et al. (1999) used a second-order closure model to derive an expression for the standard deviation of vertical velocity fluctuations (σ_w) as a function of gradient Richardson number. SYKES (2001) developed a second-order closure model for concentration fluctuations and demonstrated the dramatic effects of wind shear on the horizontal dispersion rates. Although higher order closure models are expected to produce better results than first-order closure models, there is very few boundary layer turbulence data to evaluate these models. Also, the requirement of enormous computational resources and a poor understanding of complex equations for turbulence have limited the application of these schemes.

As a compromise between higher order closure schemes, which have limited practical applications and first-order schemes, which grossly accounts for turbulence, models based on TKE closure schemes are being used in the boundary layer studies (HOLT and RAMAN, 1988). YU (1977) tested the performance of 14 parameterization schemes using the O'Neill fifth period and Wangara day 32 experimental data. HOLT and RAMAN (1988) made an exhaustive review and comparative evaluation of multilevel boundary layer parameterizations for first-order and TKE closure schemes. MONEX 79 data (see for instance, MEYER and RAO, 1985) were used to test the performance of the schemes and they had pointed out that a TKE closure is preferable to first-order closure in predicting the overall turbulence structure of the boundary layer. GENON (1995) made a comparative evaluation of three closure schemes which included a simple first-order closure formulation proposed by LOUIS (1979) and two TKE formulations. The performance of the schemes was tested for three experiments: the Wangara experiment for clear sky conditions, the Cabauw and the Joint Air–Sea Interaction experiments concerning the interactions between turbulent and radiative processes in a cloud layer. More recently, SHARAN and GOPALAKRISHNAN (1997a) tested the performance of five local K formulations and a TKE mixing length closure formulation in strong and weak wind SBL. The Cabauw (Netherlands) and EPRI-Kincaid site (United States) observations were used for this purpose. Results indicate that for the strong wind case study, although the magnitudes of mixing are different for various closure schemes, the profiles of turbulent diffusivities in terms of shape, height above the surface where diffusion reaches maximum, and height to which significant diffusion takes place, are more or less comparable between most of the closure schemes (Fig. 2(i)). The differences in the magnitudes of diffusion produced by various closure schemes cause minor but noticeable changes in the thermodynamic structure and wind fields that increase with the evolution of SBL. However, although the profiles of turbulent diffusivities become weak, variable and poorly defined under weak wind conditions (Fig. 2(ii)), diffusion is so decreased that the mean profiles become insensitive to the differences in the diffusion that arise due to various parameterization schemes.

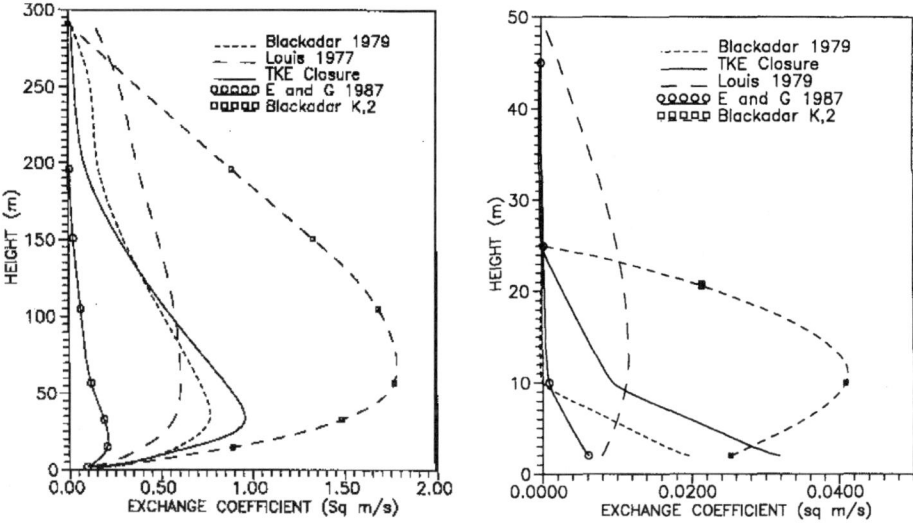

Figure 2
Typical variation of eddy diffusivities with height produced by various closure schemes in (i) a strong wind and (ii) a weak wind SBL (adopted from SHARAN and GOPALAKRISHNAN, 1997, J. Appl. Meteor. *550* and *551*).

3. Simple Dispersion Models

3.1 Analytical Dispersion Models

Although the system of equations (11) to (18) with an appropriate closure scheme (HOLT and RAMAN, 1988), initial and boundary conditions (PHYSIK, 1976; PIELKE, 1984) can only be solved numerically, simplified/idealized forms with suitable approximations/assumptions can be made amenable to analytical treatment. Their applications in predicting boundary layer flows are limited. An analytical model was proposed (SINGH et al., 1993) for studying variations in the diurnal wind structure in the PBL and the evolution of the low-level nocturnal jet. A time-dependent eddy-diffusivity coefficient corresponding to solar input was used. This analytical framework was used to examine the role of boundary-layer shear in the dispersion process (MCNIDER et al., 1993). BANNON and SALEM (1995) have used a two-layer model of the PBL to elucidate the basic features of the baroclinic boundary layer. Local and advective accelerations were neglected in the linear model dynamics. The model assumed constant eddy diffusivity and a linear variation of the geostrophic wind with height. SINGH et al. (1999) have obtained an analytical solution by considering a time-dependent geostrophic wind in their earlier model (SINGH et al., 1993).

Over the years, simplified equations have successfully been applied in the study of the dispersion of air pollutants. This is especially true when the flow is known. In such a situation, the equation for conservation of a pollutant can be treated independently. Equation (17) with the assumption of incompressibility and K-closure for the transport of a pollutant can be rewritten as:

$$\frac{\partial C}{\partial t} + U\left(\frac{\partial C}{\partial X}\right) + V\left(\frac{\partial C}{\partial Y}\right) + W\left(\frac{\partial C}{\partial Z}\right) = \frac{\partial}{\partial X}\left(K_x \frac{\partial C}{\partial X}\right) + \frac{\partial}{\partial Y}\left(K_y \frac{\partial C}{\partial Y}\right)$$

$$+ \frac{\partial}{\partial Z}\left(K_z \frac{\partial C}{\partial Z}\right) + Q_S + R_S , \qquad (20)$$

where Q_S and R_S represent the integrated forms of source and removal terms, respectively and the overbars appearing in equation (17) have been dropped for the sake of convenience.

The simplest analytical solution (CSANADY, 1972) is obtained from equation (20) for an instantaneous point source by assuming constant eddy diffusivities, orienting X-axis along the mean wind and neglecting the removal terms (SEINFELD, 1986). By expressing the eddy diffusivities in terms of the dispersion parameters in the analytical solution, it becomes the Gaussian puff solution. In addition to the assumptions made for obtaining the puff solution, if the downwind diffusion is assumed to be negligible in comparison to advection and the average conditions are considered stationary, this yields the more commonly used Gaussian plume solution (TURNER, 1970; PASQUILL, 1974). The solutions for variable eddy diffusivities are usually not Gaussian (SHARAN et al., 1996a). Even for constant eddy diffusivities, the steady-state solution of equation (20) with downwind diffusion term is not Gaussian (SHARAN et al., 1996b).

The Gaussian solutions (SMITH, 1984; WEIL et al., 1992) are presently used around the world in air pollution control and decision-making. Chemical transformations (SEINFELD, 1986), dry/wet depositions (VOLDNER et al., 1986) and removal processes can easily be incorporated in these models. Plume rise (BRIGGS, 1972), plume trapping (TURNER, 1970) and shoreline fumigation effects (VENKATRAM, 1977; VAN DOP et al., 1979) have been successfully accounted within the framework of Gaussian models. Also, these models have been extended to account for asymmetry and skewness in the CBL that results in abnormalities in ground level concentrations of pollutants dispersed in the atmosphere during the day-time (VENKATRAM, 1980). SINGH and GHOSH (1987) applied a model in the Gaussian framework for computing the concentration distribution of methyl isocyanate (MIC) released from the Union Carbide insecticide plant located at Bhopal.

Analytical models have been developed (YEH and HUANG, 1975; CHRYSIKOPOULOS et al., 1992; LIN and HILDEMANN, 1996, 1997) by taking U and K_z as power-law functions of height:

$$U(Z) = aZ^m; \quad K_z(Z) = bZ^n , \tag{21}$$

where the parameters a, b, m and n depend on the atmospheric conditions and on the ground roughness. Additionally, LIN and HILDMANN (1996) parameterized K_y as $f(X)Z^\gamma$ where f is a function of X and γ is a parameter, to obtain an analytical solution of the resulting advection diffusion equation. Recently, SHARAN et al. (1996a) and SHARAN and YADAV (1998) considered K_x, K_y and K_z as linear functions of downwind distance from the source to obtain a closed form solution of the advection-diffusion equation. Attempts have been made in the literature to obtain analytical solutions for the following forms of K_z (PASQUILL and SMITH, 1983):

(i) $K_z = $ constant
(ii) $K_z \propto Z^\gamma, \quad 0 \leq Z \leq h$
(iii) $K_z \propto (h - Z)^\gamma, \quad 0 \leq Z \leq h$
(iv) $K_z \propto Z(h - Z), \quad 0 \leq Z \leq h$
(v) $K_z \propto \begin{cases} Z, & 0 \leq Z \leq h/2 \\ (h - Z), & h/2 \leq Z \leq h \end{cases}$

where $\propto$ is the sign of the proportionality, γ is the exponent and h is the depth of the boundary layer.

Although the Gaussian type of models are sometimes applied to study the dispersion of air pollutants over complex terrain (STRIMAITIS et al., 1986), these models often fail over the terrain especially when the variations in wind with time and space become substantial. Most of the modifications to incorporate the changes in wind speed and direction in space are based on *ad hoc* assumptions and experimental conclusions which at best may be valid at a given time for a specific location (SHARAN et al., 1995a,b). Also, the dispersion parameters required for computing the concentration distribution in Gaussian models have their own limitations, particularly in weak wind and stable conditions. Hence, when the local flow determines the dispersion of air pollutants, some alternative approaches such as numerical modeling need to be adopted for reliable air quality analysis.

3.2 Simple Numerical Models for Dispersion

When the basic meteorological parameters such as the flow fields, the vertical thermal stratification and mixing depth can be obtained from regular meteorological observations (PACK et al., 1978; ARTZ et al., 1985), the advective diffusion equation (20) can be solved numerically using finite difference (HALTINER and WILLIAM, 1980), finite element (HUEBNER, 1975; ZIENKIEWICZ, 1979), boundary element methods (WROBEL and BREBBIA, 1981) and scattered grid approximation schemes (SHARAN et al., 1997). However, as pointed out by SEGAL et al. (1982), the meteorological observations required to numerically solve the equation (20) are usually available only from a synoptic network, and, unfortunately, in general lack sufficient temporal and spatial resolution to adequately resolve small-scale atmospheric features. This loss of resolution becomes a serious problem especially when mesoscale processes such as; the land and sea/lake breeze, mountain valley winds and urban circulation considerably influence the local meteorological conditions.

Several comprehensive observational studies (for example: ANGELL et al., 1972; LYONS and OLSSON, 1973; KEEN et al., 1979) have been performed to overcome the problem of inadequate resolution. However, such studies are site-specific and the conclusions of these studies may not be applicable at some other location. Further, due to the considerable expenses involved in such observational studies, they cannot be carried out for an indefinitely long period of time, or at every location where local air quality assessment is required.

3.2.1 Eulerian models for dispersion over complex terrain

One approach to overcome the limitation of the existing meteorological observations is to use mesoscale/boundary layer numerical models where the system of equations (11)–(16) is solved numerically to obtain the flow variables. The information so obtained is used as the input to solve the concentration equation (20). These models are known as Eulerian dispersion models.

Eulerian models have been successfully employed to study the dispersion over complex terrain. BORNSTEIN et al. (1975, 1987a,b), for instance, used such models to study the urban boundary layer structure of New York City. More recently, WESTPHAL et al. (1999) used a simplified version of the Eulerian aerosol and chemistry model of TOON et al. (1988) driven by dynamic and thermodynamic fields from the 15-km resolution Coupled Ocean–Atmospheric Mesoscale Prediction

System (COAMPS) and studied the release of chemical agents from destruction of an Iraqi ammunition depot at Khasmisiyah during the Gulf War. However, Eulerian dispersion models are more appropriate for the study of dispersion from area sources.In these models, the problem arises in the resolution of the point sources.

The problem of grid resolution becomes more serious in dealing with accidental toxic gas releases from a point source where the behavior of the plume close to the stack becomes very essential. Consider for example, the Bhopal Methyl Isocyanate gas accident: the gas leaked from a point source and affected an area of 50 sq km (SHARAN et al., 1995a). However, the worst affected area was about 300 to 400 m downwind of the stack (SINGH and GHOSH, 1985; SHARAN et al., 1995a). In order to study such dispersion, the horizontal grid size of Eulerian models should be a few tens of meters and such a resolution may not be possible within the framework of hydrostatic models. Nonhydrostatic models (MENGALKEMP, 1991) may be used in such studies; however, they are computationally expensive. On the other hand, Lagrangian particle dispersion (LPD) models, in which the behavior of pollutant species is examined relative to a moving coordinate system,are becoming efficient tools in simulating the characteristics of localized emissions.

3.2.2 Lagrangian particle dispersion models

In a Lagrangian particle model, a fictitious plume-particle is a small parcel of air and pollutant mixture and this parcel is assumed very small such that its expansion rate is insignificant relative to the scale of turbulence. The instantaneous position $(X_i(t + \Delta t))$ of the plume-particle at time $(t + \Delta t)$ can be described in terms of its position $(X_i(t))$ at a time t and is given by

$$X_i(t + \Delta t) = X_i(t) + (U_i + U_i')\Delta t; \quad i = 1, 2, 3 \qquad (22)$$

where U_i is the i-th component of the mean wind (averaged over a given period (HANNA, 1982)) and U_i' is the corresponding fluctuating part.

If a network of wind monitoring stations is used, the observed mean wind can be altered so that they are mass consistent (SHERMAN, 1978). However, in the absence of such a network, the mean fields are obtained from the meteorological model equations (1 to 16 and 19).

Two types of particle models have been evolved in the past which essentially differ in the determination of U_i'. They are:

- Deterministic particle in-cell models
- Statistical such as Monte Carlo-type models

In particle in-cell models, the domain of computation is subdivided into cells and the concentration of pollutants within each cell is computed. The K-theory of diffusion is used to compute the fluctuating components U_i' (LANGE, 1978).

The statistical models, which are more widely used on account of their simplicity and flexibility (ZANNETTI, 1990), utilize a semi/pseudo-random approach for the computation of U_i'. The turbulent velocity fluctuations U_i' are related to the fluctuations at previous time step (SMITH, 1968) and they are calculated as:

$$U_i'(t + \Delta t) = R_{U_i}(\Delta t)U_i'(t) + \sigma_{U_i}\Gamma_i[1 - R_{U_i}^2(\Delta t)]^{1/2} \, , \tag{23}$$

where $R_U(\Delta t)$ is the autocorrelation function (HANNA, 1978), σ_{U_i} is the standard deviation of fluctuations in i-th velocity component and Γ is the random normal variate with mean zero and standard deviation one. Details for computing σ_{U_i} are given in MCNIDER (1981). In the equation corresponding to vertical velocity component, MCNIDER et al. (1988) added a drift correction for simulating dispersion in the CBL.

A number of simple, statistical, random walk type of models have been developed and applied to study dispersion under neutral (HALL, 1975 and LEY, 1982) as well as convective conditions (WILSON et al., 1981; LEGG and RAUPACH, 1982; THOMSON, 1984). More complex statistical models based on Markov chain principle of correlation between particle velocities at successive time steps have also been developed (HANNA, 1982; REID, 1979; MCNIDER, 1981; BACON et al., 2000) and are being applied to study the dispersion of air pollutants in coastal regions (PIELKE et al., 1983; SEGAL et al., 1982; ROBINSON et al., 1992), over complex terrain (MCNIDER, 1981; ARRITT, 1985) and in urban environments (BOYBEYI et al., 1995; SHARAN et al., 1995a, 1996c). ANFOSSI et al. (1990) and BRUSASCA et al. (1992) demonstrated the feasibility of operating Markov chain type of particle models in near calm, stable and meandering atmospheric conditions. It should be noted that the simple Gaussian models fail to operate under such atmospheric conditions and one of the successes of particle models lies in simulating dispersion under such conditions.

Most of the earlier studies on dispersion were more or less restricted to tracer particles and non-buoyant smoke (LEY, 1982). However, in the late 80s and early 90s, the particle models have also been extended for the study of (a) heavy particles (WALKLATE, 1986, 1987; SAWFORD and GUEST, 1991), (b) Buoyant dispersion (COGAN, 1985; VAN DOP, 1992; ANFOSSI et al., 1993) and (c) dispersion of marginally heavy gas vapors (Fig. 3; GOPALAKRISHNAN and SHARAN, 1997).

Despite the fact that Markov chain type of particle models have become an efficient and versatile tool to study the dispersion of air pollutants, the computation of the number of variables is still based on indirect Eulerian-Lagrangian relationships (HANNA, 1978) which do not have adequate experimental support/validation. Extensive observations on Lagrangian turbulence may improve the quality of predictions obtained by such models.

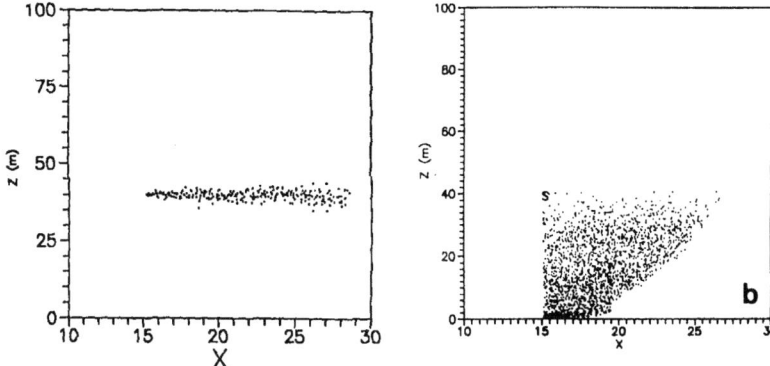

Figure 3
Modeled instantaneous positions of (i) passive and (ii) marginally heavy MIC Lagrangian, plume particles in the X–Z plane after 45 minutes of release (at around 0045 LST) during the Bhopal gas accident. The figure shows the impact of marginal heaviness on the dispersion scenario. In the above figures along the X axis, 1 unit = 750 m (adopted from GOPALAKRISHNAN and SHARAN, 1997, Atmos. Environ. *31*, 3369).

4. Weak Wind Dispersion

Weak and variable wind conditions have been observed to occur for a considerable period of time during the day as well as during the night in most parts of the world. However, they assume greater significance in tropical environments for a variety of reasons, greater frequency of occurrence being one of them. These conditions are sensitive as they have great potential for the occurrence of pollution episodes. The turbulent structure of the atmospheric boundary layer is poorly described for weak wind conditions. One of the major reasons for this being limited observational data. The complexity of the boundary layer grows with the weakening of the winds and the degree of atmospheric stability. Thus, it becomes important to study the dispersion in weak wind conditions.

4.1 Characteristic Features

The diffusion of a pollutant released from various emission sources is irregular and indefinite in weak and variable wind conditions. As a consequence, no single plume centerline is obvious and the observed concentration distribution is multipeaked and non-Gaussian (SAGENDORF and DICKSON, 1974). Also, it has been observed that the plume is subject to a great deal of horizontal undulations (SHARAN *et al.*, 1996d). Such an undulation is called "plume meandering." Quite often, highly localized concentration occurs near the ground in a stable atmosphere, especially when the winds are low.

Light wind stable conditions were excluded by Pasquill (PASQUILL, 1961) from the original stability classification because the diffusing plume is unlikely to have any definable travel. SAGENDORF (1975) suggested on the basis of diffusion experiments at Idaho (U.S.A.) that under the G conditions the plume is subject to considerable horizontal meandering. A review of several sets of diffusion data (VAN DER HOVEN, 1976) and the dispersion schemes (SHARAN et al., 1995b; YADAV and SHARAN, 1996) indicate that the effective sigma value can correspond to anything between A and F stability.

The traditional techniques are not applicable to the limiting conditions of weak wind and stable conditions. For example, the Gaussian plume model produces unreasonable results when applied to diffusion in low wind cases (BASS et al., 1979; ZANNETTI, 1986) because (i) down-wind diffusion is neglected in comparison with the advection in equation (20), (ii) the concentration is inversely proportional to U and therefore, the concentration approaches infinity as the wind speed tends to zero, (iii) the average conditions are stationary (ANFOSSI et al., 1990) and (iv) the non-availability of dispersion parameters in low winds. From the application point of view, the modeling approaches to low-wind speed diffusion problems are based on Gaussian plume concepts with some methods involving modification to dispersion parameters to explain the meandering effects (SAGENDORF and DICKSON, 1974; CIRILLO and POLI, 1992; YADAV, 1995; SHARAN et al., 1995b; YADAV et al., 1996b).

4.2 Plume Meandering

In a stable atmosphere, passive pollutants emitted from a continuous source located above the surface layer, when viewed along a vertical plane, exhibit a 'fanning' kind of behavior. In this case, the evolution of the plume is governed by small-scale eddies. Larger eddies with length and time scale larger than the plume width have also been observed to control the evolution of the plume (HANNA, 1986; SKUPNIEWICZ, 1987). However, these eddies have been found to influence only the horizontal motion of the plume, causing an undulation (meandering) in its evolution. Plume meandering may occur under different atmospheric conditions but is most effective for stable stratification and low wind speeds (HANNA, 1981; LEAHEY et al., 1988). Motions cause plume meander with length scales typically in the range of 100 m–10 km and meander periods range from a few minutes to a few hours (ETLING, 1990). ETLING (1990) has throughly discussed some of the physical reasons responsible for the observed plume meandering under low-wind stable conditions.

Extensive field observations of wind direction fluctuations under stable conditions over flat terrain (SAGENDORF and DICKSON, 1974); over slightly irregular terrain (HANNA, 1981; LEAHEY et al., 1988) and at a coastal site (RAYNER and HAYES, 1984) suggest that the standard deviation of wind direction fluctuations (σ_Θ) could be quite large under low-wind speed conditions in contrast to σ_Θ related to P-G stability categories. This implies that the ability of the atmosphere to diffuse the gaseous tracer does not decrease with the increase in atmospheric stability under low-wind conditions.

Some of the test runs in the diffusion experiments conducted at Idaho National Engineering Laboratories (INEL) in 1974 (SAGENDORF and DICKSON, 1974) in light wind stable conditions were characterized by large plume spread and showed peculiar concentration patterns. It appears that the tracer was transported in a particular direction, only briefly and then moved in an almost opposite direction (turning slightly less than 180°) for the remaining duration of the test period, thus resulting in non-zero concentrations in that particular direction only on a 100 m arc and not on 200 and 400 m arcs. This is due to deflection of the plume during the course of its travel. The inhomogeneity present in the actual flow is also an important factor in dispersing the plume.

GIFFORD (1959, 1960) was the first to recognize the important effect of plume meandering on concentration fluctuations. He developed a fluctuating plume model which accounts for the effects of large-scale meandering of the plume segments together with the small-scale turbulence inside the segments. HANNA (1986), while discussing the problem of concentration fluctuations, emphasized that although a continuous spectrum of motion exists, the assumption of a two-scale system for turbulent motion based on GIFFORD's (1959) fluctuating plume theory is reasonable. However, this model assumes the lateral and vertical concentration distribution across a plume to be Gaussian, which is likely to yield poor results under weak wind conditions.

4.3 Weak Wind Dispersion Models

Weaknesses in the predictive capability of the classical approaches in relation to low-wind conditions exist more or less at two levels: the application level (for example, non-availability of dispersion parameters) and the formulation level (which involves approximations and assumptions) Keeping both of the aspects in mind, observational and modeling studies in low-wind conditions were initiated at the Indian Institute of Technology (IIT), Delhi, India.

The first set of dispersion experiments were conducted at IIT Delhi sports ground during Feb. 1991 in weak wind conditions (SINGH et al., 1991; AGARWAL et al., 1995). The experimental site which lies in the city is characterized by flat terrain but surrounded by short buildings and other concrete structures by and large on all sides, and hence was chosen to study dispersion in the urban atmosphere. Sulphur Hexafluoride (SF_6) was released at a height of 1 m above the ground and the samplers were also placed approximately at the same height. Thus, for computational purposes the release can be considered from a ground-level source. The sampling grid involved receptors on 50, 100, 150 m (in some cases on 200 m) circular arcs with 45 degree angular spacing between them. Micrometeorological data (wind speed and wind direction etc.) were obtained from a 30 m multilevel micrometeorological tower installed close to the experimental site. Some of the experimental limitations were: (a) less number of samplers, (b) angular spacing between the samplers too large,

(c) 30 m tower not on-site, (d) poor response of the wind measuring instruments during light winds and (e) lack of turbulent measurements. Details regarding instrumentation, data collection, etc. are documented in a report by SINGH et al. (1991).

A year later, a similar field program was carried out by making the provision for the measurements of turbulent data using a sonic anemometer. However, the concentration data were not obtained due to contamination and the problems in the gas chromatograph. Nevertheless, turbulent data were analyzed to study the surface layer turbulence (AGARWAL et al., 1995) and spectral characteristics (YADAV et al., 1996a) in weak wind conditions.

A steady-state mathematical model was proposed (SHARAN et al., 1996b) for low-wind conditions by taking into account the diffusion in all three coordinate directions and the advection along the mean wind. Assuming constant eddy diffusivities, an analytical solution was obtained for the resulting system of equations for a ground/elevated source (SHARAN et al., 1995b, 1996b). The analytical solution thus obtained proves to be non-Gaussian. The traditional solution that leads to the Gaussian plume formula is only a limiting case of the model when the concentration close to the plume centerline is desired. This is equivalent to the condition when the downwind diffusion is neglected ($K_x \rightarrow 0$). The tracer data collected at IIT Delhi were used to validate the model.

In the above model and, in general, in any simple Gaussian type models (TURNER, 1970), atmospheric diffusion equation is solved by assuming K's to be constant. However after obtaining a solution, since it is well known that K's may vary in space and time (BERKOWICZ and PRAHM, 1979; TIRABASSI et al., 1987; BARTZIS, 1989; ARYA, 1995), they are further expressed as a function of downwind distance through dispersion parameters. This approach is widely accepted from an application point of view, although mathematically it is inconsistent (LLEWELYN, 1983; SHARAN et al., 1996a). With this limitation in mind, SHARAN et al. (1996a) and SHARAN and YADAV (1998) rederived the solution to the advection-diffusion equation in which eddy diffusivities were assumed to be linear functions of downwind distance. However, runs from this variable K-model (SHARAN et al., 1996a) were found to underestimate concentration distribution relative to the observations in IIT-SF$_6$ convective diffusion tests. Nevertheless, the results approach those based on Arya's model (ARYA, 1995) as well as the Gaussian approach using parameterization of turbulent intensities in terms of convective velocity. The reasons attributed to the underpredicting trend are the uncertainties involved in (i) the empirical constants appearing in the parameterization of turbulence intensity in terms of w_*, (ii) the estimation of w_*.

Another extensive data set in low-wind conditions is available from the diffusion experiments conducted by INEL in 1974 at Idaho, U.S.A. (SAGENDORF and DICKSON, 1974). The receptors were located on concentric circles of radii 100, 200 and 400 m at an angular distance of 6°. In some of the test runs, the non-zero concentrations were observed around 360°. The existing models overpredict the peak and underpredict the plume spread and fail to predict the observed concentration distribution around the circle. SHARAN et al. (1995b) and YADAV and SHARAN (1996) have analyzed the various schemes for the dispersion parameters for the treatment of dispersion in the low-wind stable conditions. The scheme based on short-term averaging was proposed to compute the concentration distribution around the circular arcs.

SHARAN and YADAV (1998) used their variable K-model to simulate these diffusion experiments in low-wind stable conditions by parameterizing the turbulence intensities in terms of u_*. The computed results near the observations in terms of peak values and plume spread. Also, the computed results are comparable with the model of CIRILLO and POLY (1992). However, this model fails when U becomes zero. Also, the model does not provide for the upstream diffusion. ARYA (1995) has shown the importance of upstream diffusion in weak wind near-source dispersion. Although there has been progress in the modeling of dispersion in low-wind conditions, a more realistic dispersion model valid for all wind speeds remains to be developed.

In order to treat the plume meandering, a time-dependent model based on coupled plume segment and Gaussian puff approaches was formulated (SHARAN et al., 1996d). The model is used to simulate non-homogeneous dispersion conditions and it can also be used to describe both non-homogeneous, non-stationary dispersion conditions.

In all these analytical models for the dispersion from the near-surface releases, the dependence of eddy diffusivities and mean wind on height has been ignored. However, it is important to consider the height-dependence of eddy diffusivities and mean wind in the dispersion models.

5. Complex Numerical Models for Dispersion

Over the past thirty years, dispersion models have undergone a decade-by-decade advance. While the 1970s saw the initial success of simple Gaussian models, the introduction of the university-born Regional Atmospheric Modeling System, RAMS (PIELKE et al., 1992), and the Penn State/NCAR Mesoscale Model now in its fifth version, MM5 (GRELL et al., 1994) in the 1980s for dispersion applications improved the quality of predictions over complex and heterogeneous domains. In the 1990s these complex models for flow and dispersion have driven towards increasingly finer resolution, mostly through the use of nested grids and better physics. For instance, NICHOLLS et al. (1995) used the RAMS model to study airflow around buildings using a fine grid increment of 1 m. It should be emphasized that in order to understand dispersion in an urban canopy, very high resolutions on the order of a few tens of meters may be required.

At the same time dispersion modeling was benefiting from the research and technology boom, computational fluid dynamics (CFD) researchers were creating new innovative numerical techniques designed to model fluid flows around complex boundaries. In the 1970s and early 1980s, the models developed for aerospace engineering and plasma physics were surprisingly similar to their counterparts in atmospheric sciences. The grids were composed of regular, rectangular cells extending from no-slip or free-slip surfaces. As more computational power became available and atmospheric modelers were adding more physics into their models, CFD researchers resorted to unstructured grids. The Operational Multi-scale Environmental model with Grid Adaptivity (OMEGA; BACON et al., 2000) is a novel atmospheric modeling system developed at Science Application International Corporation (SAIC) with support from the Defense Threat Reduction Agency (DTRA). OMEGA was developed for real-time weather and airborne hazard prediction. Conceived to link the latest computational fluid dynamics and high resolution gridding technologies with numerical weather prediction, OMEGA permits unstructured horizontal grids of continuously varying spatial resolutions ranging from about 100 km down to about 1 km to better resolve the gradients in concentration of air pollutants (Fig. 4), local terrain effects and/or important physical features of atmospheric circulation and cloud dynamics. This unique capability provides not only a higher resolution in the region of evolving weather systems but also allows a natural interaction with the larger scale flow without the need for multiple computational nests or grid motion.

Apart from atmospheric models for flow and transport, air quality models have also undergone a sea of changes. Newly released models-3 or its prototype Community Multiscale Air Quality (CMAQ) modeling system by the United States Environmental Protection Agency is aimed to improve (1) the environmental management community's ability to evaluate the impact of air quality management practices for multiple pollutants at multiple scales and (2) the general ability to better probe, understand, and simulate chemical and physical interactions in the atmosphere.

6. Dispersion Modeling in the SBL

The study of the SBL over land under fair weather conditions is of major concern in air pollution meteorology. Most of the industrial stacks are located within this layer and hence the dispersion of air pollutants is adversely affected by the state of the SBL. Complex numerical models (section 5) along with certain observations have offered insights on the structure and dispersion of pollutants in the SBL. In this section, we describe the structure of the SBL over a homogeneous terrain, the flow modifications over a non-homogeneous terrain, the associated forcing and an overview of the existing observational and modeling status of the SBL.

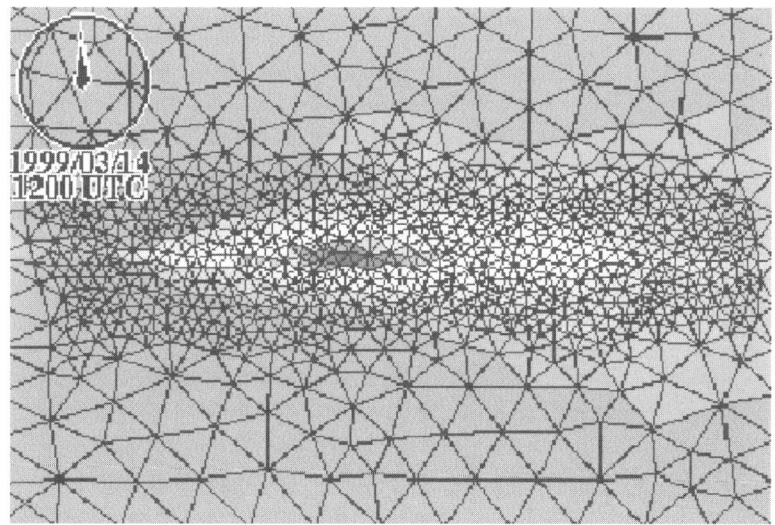

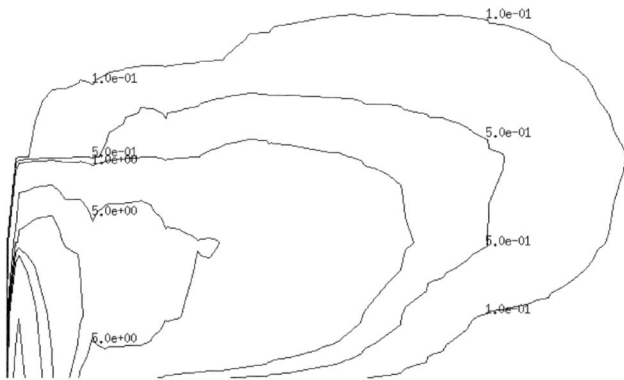

Figure 4
Unstructured, dynamically adaptive grid technology used to study plume concentration distribution. The top figure shows a horizontal slice of a plume resolved using adaptive grid at an altitude of about 100 m (AGL). The bottom figure shows a cross section along the plume centerline. The vertical dimension has been exaggerated and represents a total height of 250 m. The contours represent values of 0.1 (outermost contour), 0.5, 1, 5, 10, 50, 100, 500, and 1000 (arbitrary units; adopted from SARMA *et al.*, 1999, Air Pollution VII, WIT Press, Southampton, 59–68).

6.1 Structure of the SBL over Homogeneous Terrain

On a clear night, the surface inversion layer, the depth of which signifies the height of the stable layer, is formed close to the surface as a result of surface cooling. Most of the pollutants released from the industrial stack are trapped within this layer. The inversion layer is observed to grow throughout the night and its growth is influenced by three major processes namely (i) turbulence, (ii) radiation and (iii) advection. Turbulence generated by the wind shear cools the inversion layer by the process of downward transfer of heat from the warmer layer above it. Heat is eventually transferred to the cooling surface. The radiative cooling due to long-wave radiative flux divergence leads to the progressive thickening of the surface inversion layer. The advective cooling is generated due to horizontal inhomogeneities e.g., katabatic flows. However, in the absence of synoptic gradients and over a horizontally homogeneous surface, the third factor that cools the surface inversion layer becomes insignificant.

ANDRÈ and MAHRT (1982) examined the SBL data from the Wangara and Voves experiments and inferred that the lower part of the nocturnal inversion layer normally appears to be turbulent but strongly stratified. The thicker part of the inversion layer is characterized by weaker stratification which appears to be mostly generated by clear-air radiative cooling. The radiatively cooled layer thickens significantly as night proceeds. However, the thickness of the turbulent layer normally varies slowly during the night, but differs significantly from night to night. GARRATT and BROST (1981) and TJEMKES and DUYNKERKE (1989) studied the effect of radiative cooling in the SBL numerically and pointed out that the SBL grows into a three-layer structure; for layers close to the surface and the uppermost part of the turbulent layer, radiative cooling mechanism dominates, whereas in the central part occupying most of the turbulent layer, turbulent cooling dominates.

Apart from the thermodynamic structure, the wind profiles in the SBL may have a complex behavior. While the wind speed is governed by buoyancy, friction and entrainment in the lowest 2 to 10 m, the wind direction in this layer is determined by the local topography (STULL, 1988). Over a flat terrain the wind can become calm, resulting in adverse air pollution problems. MAHRT (1999) has extensively reviewed stratified boundary layers.

Higher in the SBL, as thermal stratification progresses, a low-level wind maxima (BLACKADAR, 1957) or a low-level nocturnal jet (MAHRT et al., 1979; KURZEJA et al., 1991) generally appears. The wind-maxima occurs at the top of the layer of significant turbulence and is usually referred to as the top of the momentum layer (MAHRT et al., 1979). The wind maxima influences the long-range transport of air pollutants (MCNIDER et al., 1988). Most of the numerical studies (ANDRE et al., 1978; ZEMAN, 1979; MCNIDER and PIELKE, 1981; STULL and DRIEDONKS, 1987) except that of DELAGE (1974) have been more or less successful in modeling these phenomena under windy conditions.

Yet, a weak wind condition prevails globally for a considerable period of time and assumes special importance in a stable atmosphere due to its high air pollution potential (see, for instance, SHARAN et al., 1995a, 1996c and GOPALAKRISHNAN and SHARAN, 1997 for the impacts of weak wind on the dispersion of Methyl Isocyanate during the infamous Bhopal gas accident). Recent theoretical studies of GOPALAKRISHNAN et al. (1998) and ESTOURNEL and GUEDALIA (1985) indicate that, over a fairly flat and homogeneous domain, the radiative and turbulent processes that control the evolution of the SBL differ considerably, between strong and weak wind conditions. While ESTOURNEL and GUEDALIA (1985) illustrated that the stable inversion layer evolves in a weak wind SBL, whereas, the depth of the layer varies little under strong wind conditions, GOPALAKRISHNAN et al. (1998) numerically showed that when shear driven turbulence is weak, clear air radiative cooling dominantly influences the integrated cooling budget within the SBL (Fig. 5). These findings raise an interesting and fundamental question: Is the mean structure of a weak wind SBL very different from those under windy conditions? More recently, RAMAKRISHNA et al. (2002) studied the effects on the SBL of geostrophic wind. The major objective of this study was to ascertain if the mean structure and evolution of the weak wind SBL is very different from those under windy conditions. Some meteorological data collected during the plume validation experiment conducted by Electric Power Research Institute (EPRI) over a nearly flat-homogeneous terrain at Kincaid, USA (39°35′ N and 89°25′ W) and an improved version of one-dimensional meteorological boundary layer model originally developed by PIELKE (1974) and further modified by SHARAN and GOPALAKRISHNAN (1997a) with TKE mixing length closure and a layer-by-layer emissivity based radiation scheme (MAHRER and PIELKE, 1977) was used for that purpose. The study revealed that while larger shear resulting from stronger wind produced SBL with an average depth exceeding 300 m in which the shear production term was larger than the buoyancy consumption term in the TKE budget within the lower 100 m (layer cooled by turbulence), a weak geostrophic wind produced shallow SBL with an average depth of approximately 100 m in which shear production term was weak and comparable to buoyancy consumption even within the turbulence layer. Also, the wind maxima which was minimally above 200 m altitude under windy conditions was located at an altitude of less than 100 m for the weak wind case, perhaps due to weaker diffusion in the boundary layer during transition. Finally, in contrast to a strong wind SBL where cooling within the surface inversion layer is dominated by turbulence, radiative cooling becomes larger than turbulent cooling under weak wind conditions, which is consistent with our earlier findings for an idealized SBL (GOPALAKRISHNAN et al., 1998).

6.2 Flow Modification over Non-homogeneous Terrains

In the stable nocturnal boundary layer where the effect of turbulence is generally low, the advective forcing due to horizontal inhomogeneities causes a significant

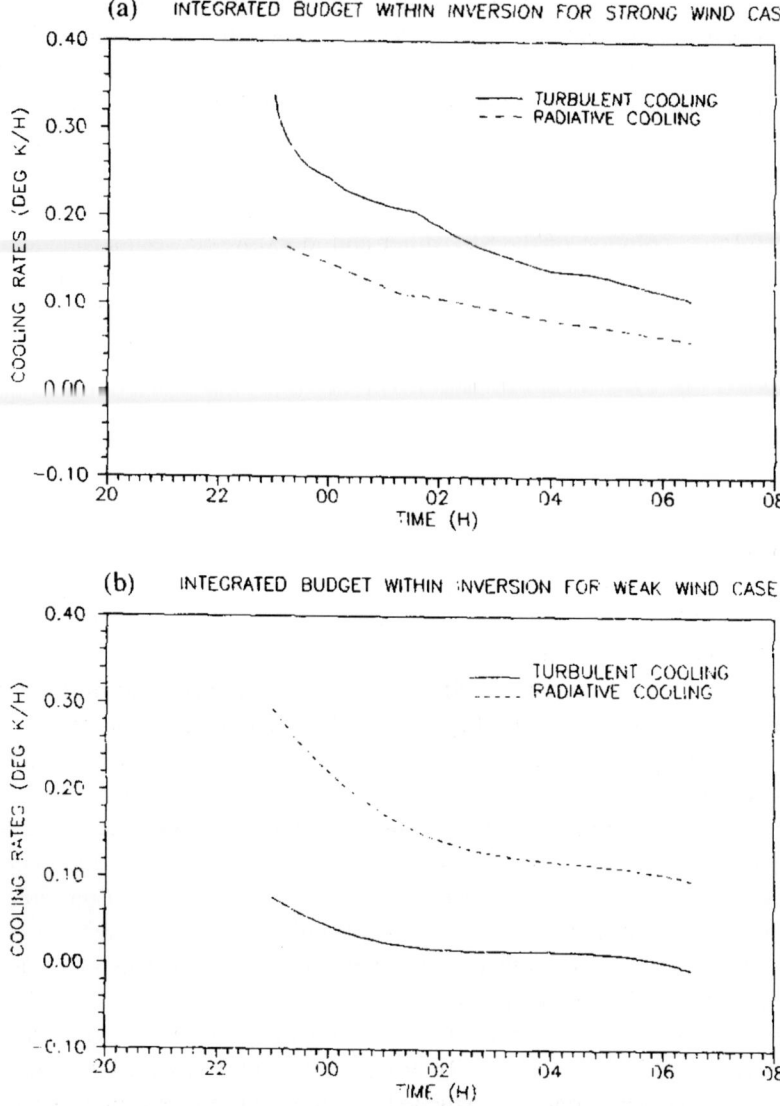

Figure 5
Integrated cooling budgets within the inversion layer for (a) strong and (b) weak wind SBL (adopted from
GOPALAKRISHNAN *et al.*, 1998, J. Atmos. Sci. *55*, 959).

modification in the structure of the SBL. The inhomogeneities may arise due to the
differential heating of a sloping terrain as in the case of slope flows and mountain
valley flows, or be the effect of differences in the radiative, thermodynamic and
mechanical properties of the underlying terrain as in the case of the urban
circulation, or result from the differential heat capacity of the underlying terrain as

ANDRÉ, J. C. and MAHRT, L. (1982), *The Nocturnal Surface Inversion and Influence of Clear-air Radiational Cooling*, J. Atmos. Sci. *39*, 864–878.

ANDREN, A. (1995), *Stably Stratified Atmospheric Boundary Layers*, Q. J. R. Meteorol. Soc. *121*, 961–985.

ANFOSSI, D., BRUSASCA, G., and TINARELLI, G. (1990), *Simulation of Atmospheric Diffusion in Low Windspeed Meandering Conditions by a Monte-Carlo Dispersion Model*, II Nuovo Cimento *13c*, 995–1006.

ANFOSSI, D., FERRERO, E., BRUSASCA, G., MARZORATI, A., and TINARELLI, G. (1993), *A Simple Way of Computing Buoyant Plume Rise in Lagrangian Stochastic Dispersion Models*, Atmos. Environ. *27A*, 1443–1451.

ANGELL, J. K., PACK, D. H., MACHTA, L., DICKSON, C. R., and HOECKER, W. H. (1972), *Three-dimensional Air Trajectories Determined from Tetroon Flights in the Planetary Boundary Layer of Los Angeles Basin*, J. Appl. Meteorol. *11*, 451–471.

ARRITT, R. W., *Numerical Studies of Thermally and Mechanically Forced Circulations over Complex Terrain*, Ph.D. dissertation, Cooperative Institute for Research in the Atmosphere, Colorado State University, Fort Collins, 1985.

ARTZ, R., PIELKE, R. A., and GALLOWAT, J. (1985), *Comparison of the ARL/ATAD Constant Level and the NCAR Isentropic Trajectory Analyses for Selected Case Studies*, Atmos. Environ. *19*, 47–63.

ARYA, S. P. S. (1995), *Modeling and Parameterization of Near-source Diffusion in Weak Winds*, J. Appl. Meteorol. *34*, 1112–1122.

BACON, D. P., AHMAD, N. N., BOYBEYI, Z., DUNN, T. J., HALL, M. S., LEE, P. C. S., SARMA, R. A., TURNER, M. D., WAIGHT, K. T., YOUNG, S. H., and ZACK, J. W. (2000), *A Dynamically Adapting Weather and Dispersion Model: The Operational Multi-scale Environment Model with Grid Adaptivity (OMEGA)*, Mon. Wea. Rev. *128*, 2044–2076.

BANNON, P. R. and SALEM, S. L. (1995), *Aspects of the Baroclinic Boundary Layer*. J. Atmos. Sci. *52*, 574–596.

BARTZIS, J. G. (1989), *Turbulent Diffusion Modeling for Wind Flow and Dispersion Analysis*, Atmos. Environ. *23*, 1963–1969.

BASS, A., BENKLEY, C. W., SCIRE, J. S., and MORRIS, C. S., *Development of mesoscale air quality simulation models*. vol. 1, *Comparative Sensitivity Studies of Puff, Plume and Grid Models for Long-distance Dispersion* (U.S. Environmental Protection Agency, EPA 600/7-80-056, Research Triangle Park, NC. 1979).

BELJAARS, A. C. M. and HOLTSLAG, A. A. M. (1991), *Flux Parameterization over Land Surfaces for Atmospheric Models*, J. Appl. Meteorol. *30*, 327–341.

BERKOWICZ, R. and PRAHM, L. P. (1979), *Generalization of K Theory for Turbulent Diffusion. Part I: Spectral Turbulent Diffusivity Concept*, J. Appl. Meteorol. *13*, 266–272.

BLACKADAR, A. K. (1957), *Boundary Layer Wind Maxima and their Significance for the Growth of Nocturnal Inversion*, Bull. Am. Meteorol. Soc. *38*, 283–290.

BLACKADAR, A. K., *High resolution models of the planetary boundary layer*. In *Advances in Environmental and Scientific Engineering*, vol. I (Gordon and Breach, New York 1979).

BORNSTEIN, R. D. (1975), *The Two-dimensional URBMET Urban Boundary Layer Model*, J. Appl. Meteorol. *14*, 1459–1477.

BORNSTEIN, R. D., PECHINGER, U., SALVADOR, R., SHIEH, L. J. and LUDWIG F., *Modeling the Polluted Coastal Urban Environment, vol. 2, Dispersion Model (EPRI Report*, EA- 5091 for contract no. 1630–13, 1987a), 153 pp.

BORNSTEIN, R. D., PECHINGER, U., MILLER, R., KLOTZ, S. and STREET, R. L. (1987b): *Modeling the Polluted Coastal Urban Environment*, vol. 1: PBL Model, EPRI report EA-5091 for contract no. 1630–13, 172 pp.

BOYBEYI, Z., RAMAN, S., and ZANNETTI, P. (1995), *Numerical Investigation of Possible Role of Local Meteorology in Bhopal Gas Accident*, Atmos. Environ. *29-4*, 479–496.

BRIGGS, G. A. (1972), *Discussion of Chimney Plumes in Neutral and Stable Surroundings*, Atmos. Environ. *6*, 507–510.

BRIGGS, G. A. (1993), *Plume Dispersion in the Convective Boundary Layer. Part II: Analyses of CONDORS Field Data*, J. Appl. Meteorol. *32*, 1388–1425.

BROST, R. A. and WYNGAARD, J. C. (1978), *A Model Study of the Stably Stratified Planetary Boundary Layer*, J. Atmos. Sci. *35*, 1427–1440.

BROWN, M. J., *Urban parameterizations for mesoscale meteorological models*. In *Mesoscale Atmospheric Dispersion* (Z. Boybeyi, ed.) Advances in Air Pollution (WIT Press, UK 2001).

BROWN, R. A., *Analytical Methods in the Planetary Boundary-layer Modelling* (Adam Hilger, London 1974).

BRUSASCA, G., TINARELLI, G., and ANFOSSI, D. (1992), *Particle Model Simulation of Diffusion in Low-wind Speed Stable Conditions*, Atmos. Environ. *26A*, 707–723.

BUSINGER, J. A., WYNGAARD, J. C., IZUMI, Y., and BRADLEY, E. F. (1971), *Flux-profile Relationships in the Atmospheric Surface Layer*, J. Atmos. Sci. *28*, 181–189.

BUSINGER, J. A., *Equations and Concepts*. In *Atmospheric Turbulence and Air Pollution Modelling* (Nieuwstadt, F.T.M. and Van Dop, H. eds) (D. Reidel Publ., Dordrecht, Holland. 1982) pp 1–36.

CHRYSIKOPOULOS, C. V., HILDEMANN, L. M., and ROBERTS, P. V. (1992), *A Three-dimensional Steady-state Atmospheric Dispersion Model for Emissions from a Ground-level Area Source*, Atmos. Environ. *26A*, 747–757.

CIRILLO, M. C. and POLI, A. A. (1992), *An Intercomparison of Semi- empirical Diffusion Models Under Low-windspeed, Stable Conditions*, Atmos. Environ. *26A*, 765–774.

COGAN, J. L. (1985), *Monte Carlo Simulation of Buoyant Dispersion*, Atmos. Environ. *21*, 867–878.

CSANADY, G. T., *Turbulent Diffusion in the Environment* (D. Reidel Publ., Dordrecht, Holland. 1973).

DEARDORFF, J. W., WILLIS, G. E., and LILLY, D. K. (1969), *Laboratory Investigation of Nonsteady Penetrative Convection*, J. Fluid Mech. *35*, 7–31.

DEARDORFF, J. W., (1970), *A Three-dimensional Numerical Investigation of the Idealized Planetary Boundary Layer*, Geophys. Fluid Dyn. *1*, 377–410.

DEARDORFF, J. W. (1972), *Numerical Investigation of Neutral and Unstable Planetary Boundary Layers*, J. Atmos. Sci. *29*, 91–115.

DEARDORFF, J. W. (1974), *Three-dimensional Study of the Height and Mean Structure of a Heated Planetary Boundary Layer*, Boundary-Layer Meteorol. *7*, 81–106.

DEARDORFF, J. W. and WILLIS, G. E., *Computer and laboratory modeling the mixed layer*. In *Advances of Geophysics*, vol. 18 (Academic Press, New York 1974).

DELAGE, Y. (1974), *A Numerical Study of the Nocturnal Atmospheric Boundary Layer*, Quart. J. Roy. Meteorol. Soc. *100*, 351–364.

DONALDSON, C. duP. (1973), *Construction of a dynamic model of the production of atmospheric turbulence and the dispersal of atmospheric pollutants*. In *Workshop of Micrometeorology*, (D.A. Haugen, ed) (American Meteorological Society Publ.), pp 392.

DYER, A. J. (1974), *A Review of Flux-profile Relationships*, Boundary-Layer Meteorol. *7*, 363–372.

ENGLAND, D. E. and McNIDER, R. T. (1995), *Stability Functions Based upon Shear Function*, Boundary-Layer Meteorol. *74*, 113– 130.

ESTOURNEL, C. and GUEDALIA, D. (1985), *Influence of Geostrophic Wind on Atmospheric Nocturnal Cooling*, J. Atmos. Sci. *42*, 2695–2698.

ESTOURNEL, C. and GUEDALIA, D. (1987), *A New Parameterization of Eddy Diffusivities for Nocturnal Boundary-layer Modeling*. Boundary-Layer Meteorol. *39*, 191– 203.

ETLING, D. (1990), *On Plume Meandering under Stable Stratification*, Atmos. Environ. *24A*, 1979–1985.

FAST, J. D., ZHONG, S., and WHITEMAN, D. (1996), *Boundary Layer Evolution Within a Canyonland Basin. Part II: Numerical Simulation of Nocturnal Flows and Heat Budgets*, J. Appl. Meteorol. *35*, 2162–2178.

FINNIGAN, J. J. and EINAUDI, F. (1981), *The Interaction between an Internal Gravity Wave and the Planetary Boundary Layer, Part II: The Effect of the Wave on the Turbulence Structure*. Quart. J. Roy. Meteorol. Soc. *107*, 807–832.

GARRATT, J. R. and BROST, R. A. (1981), *Radiative Cooling Effects within and above the Nocturnal Boundary Layer*, J. Atmos. Sci. *38*, 2730–2746.

GARRATT, J. R. (1982), *Observations in the Nocturnal Boundary Layer*, Boundary-Layer Meteorol. *22*, 21–48.

GENON, M. L. (1995), *Comparison of Different Simple Turbulence Closures with a One-dimensional Boundary Layer Model*, Mon. Wea. Rev. *123*, 163–180.

GIFFORD, F. A. (1959), *Statistical Properties of a Fluctuation Plume Dispersion Model*, Adv. Geophys. *6*, 117–138.

GIFFORD, F. A. (1960), *Peak to Average Concentration Ratios According to Fluctuation Plume Dispersion Model*, Int. J. Air Water Pollut. *3*, 253–260.

GOPALAKRISHNAN, S. G., *Mesoscale Dispersion Modelling in a Weak Wind Stable Boundary Layer with a Special Reference to the Bhopal Gas Episode* (Ph.D. Thesis, Indian Institute of Technology, Delhi, India 1996) 177 pp.

GOPALAKRISHNAN, S. G. and SHARAN, M. (1997), *A Lagrangian Particle Model for Marginally Heavy Gas Dispersion*, Atmos. Environ. *31*, 3369–3382.

GOPALAKRISHNAN, S. G., SHARAN, M., McNIDER, R. T., and SINGH, M. P. (1998), *Study of Radiative and Turbulent Processes in the Stable Boundary Layer under Weak Wind Conditions*, J. Atmos. Sci. *55*, 954–960.

GOPALAKRISHNAN, S. G. and AVISSAR, R. (2000), *A LES Study of the Impacts of Land Surface Heterogeneity on Dispersion in the Convective Boundary Layer*, J. Atmos. Sci. *57*, 352–371.

GOPALAKRISHNAN, S. G., ROY, S. B., and AVISSAR, R. (2000), *An Evaluation of the Scale at which Topographical Features Affect the Convective Boundary Layer Using Large Eddy Simulations*, J. Atmos. Sci. *57*, 334–351.

GRELL, G. A., DUDHIA, J., STAUFFER, D. R., *A Description of the Fifth-generation Penn State/NCAR Mesoscale Model (MM5)*, (NCAR/TN-398 + IA, National Center for Atmospheric Research, Boulder, CO, 1994) 107 pp.

HALL, C. D. (1975), *The Simulation of Particle Motion in the Atmosphere by a Numerical Random-walk Model*, Quart. J. Roy. Meteorol. Soc. *101*, 235–244.

HALTINER, G. J. and WILLIAMS, R. T., *Numerical Prediction and Dynamic Meteorology*, 2nd ed. (John Wiley and Sons, New York, 1980) 477 pp.

HANNA, S. R. (1978), *Some Statistics of Lagrangian and Eulerian Wind Fluctuations*, J. Appl. Meteorol. *18*, 518–525.

HANNA, S. R. (1981), *Diurnal Variation of Horizontal Wind Direction Fluctuations in Complex Terrain at Geysers*, Bound.-Layer Meteorol. *58*, 207–213.

HANNA, S. R., *Applications in air pollution modeling*, In *Atmospheric Turbulence and Air Pollution Modelling* (Nieuwstadt, F.T.M. and Van Dop, H., eds. 1982), pp. 275–310.

HANNA, S. R. (1986), *Spectra of Concentration Fluctuations: The Time Scales of Meandering Plumes*, Atmos. Environ. *20*, 1131–1137.

HILDERBRAND, P. H. and ACKERMAN, B. (1984), *Urban Effects on the Convective Boundary Layer*, J. Atmos. Sci. *41*, 76–91.

HOLT, T. and RAMAN, S. (1988), *A Review and Comparative Evaluation of Multilevel Boundary Layer Parameterizations for First-order and Turbulent Kinetic Energy Closure Schemes*, Revi. Geophys. *26*, (4), 761–780.

HUEBNER, K. H., *The Finite Element Method for Engineers* (John Wiley and Sons, New York 1975).

JENSEN, N. O. and BUSCH, N. E., *Atmospheric Turbulence, Engineering Meteorology* (Plate, E.J., ed.) (Elsevier, Amsterdam 1982) pp. 179–231.

JOHNSON, G. T., OKE, T. R., LYONS, T. J., STEYN, D. G., WATSON, I. D., and VOOGT, J. A. (1991), *Simulation of Surface Urban Heat Islands under 'Ideal' Conditions at Night, Part 1: Theory and Tests Against Field Data*, Boundary-Layer Meteorol. *56*, 275–294.

KEEN, C. S., LYONS, W. A., and SCHUH, J. A. (1979), *Air Pollution Transport Studies in Coastal Zone Using Kinematic Diagnosis Analysis*, J. Appl. Meteorol. *18*, 606–615.

KONDO, J., KANECHIKA, O., and YOSUDA, N. (1978), *Heat and Momentum Transfers under Strong Stability in the Atmospheric Surface Layer*, J. Atmos. Sci. *35*, 1012–1021.

KURZEJA, R. J., BERMAN, S., WEBER, A. H. (1991), *A Climatological Study of the Nocturnal Planetary Boundary Layer*, Boundary-Layer Meteorol. *54*, 105–128.

LAMB, R. G. (1978), *A Numerical Simulation of Dispersion from an Elevated Point Source in the Convective Planetary Boundary Layer*, Atmospheric Environment. *12*, 1297–1304.

LAMB, R. G., *Diffusion in the convective boundary layer*. In *Atmospheric Turbulence and Air Pollution Modelling* (Nieuwstadt, F.T.M. and Van Dop, H. eds.) (D. Reidel Publ., Dordrecht, Holland 1982) pp. 69–106.

LANGE, R. (1978), *ADPIC—A Three-dimensional Particle-in-cell Model for the Dispersal of Atmospheric Pollutants and its Comparison to Regional Tracer Studies*, J. Appl. Meteorol. *17*, 320.

LEAHEY, D. F., HANSEN, M. C., and SCHROEDER, M. B. (1988), *An Analysis of Wind Fluctuation Statistics Collected under Stable Atmospheric Conditions at Three Sites in Alberta, Canada*, J. Appl. Meteorol. *27*, 774–777.

LEGG, B. J. and RAUPACH, M. R. (1982), *Markov Chain Simulations of Particle Dispersion in Inhomogeneous Flows: The Mean Drift Velocity Induced by a Gradient in Eulerian Velocity Variance*, Boundary-Layer Meteorol. *24*, 3–13.

LEWELLEN, W. S. and TESKE, M. (1975), *Turbulence Modeling and its Application to Atmospheric Diffusion. Part I: Recent Program Development, Verification and Application; Part II: Critical Review of the Use of Invariant Modeling*, EPA-600/4-75-16a,b. Part I, 79 pages; part II, 50 pages.

LEY, A. J. (1982), *A Random Walk Simulation of Two-dimensional Turbulent Diffusion in the Neutral Surface Layer*, Atmos. Environ. *16*, 2799–2808.

LIN, J. S. and HILDEMANN, L. M. (1996), *Analytical Solutions of the Atmospheric Diffusion Equation with Multiple Sources and Height-dependent Wind Speed and Eddy Diffusivities*, Atmos. Environ. *30*, 239–254.

LIN, J. S. and HILDEMANN, L. M. (1997), *A Generalized Mathematical Scheme to Analytically Solve the Atmospheric Diffusion Equation with Dry Deposition*, Atmos. Environ. *31*, 59–71.

LLEWELYN, R. P. (1983), *An Analytical Model for the Transport, Dispersion and Elimination of Air Pollutants Emitted from a Point Source*, Atmos. Environ. *17*, 249–256.

LOUIS, J. F. (1979), *A Parametric Model of Vertical Eddy Fluxes in the Atmosphere*, Boundary-Layer Meteorol. *17*, 187–202.

LYONS, W. A. and OLSSON, L. E. (1973), *Detailed Mesometeorological studies of Air Pollution Dispersion in the Chicago Lake Breeze*, Mon. Weath. Rev. *101*, 387–403.

MAHRER, Y. and PIELKE, R. A. (1977), *A Numerical Study of Air Flow over Irregular Terrain*, Beitrage zur Physik der Atmosphäre, *50*, 98–113.

MAHRT, L., HERALD, R. C., LENSCHOW, D. H., and STANKOV, B. B. (1979), *An Observational Study of the Structure of the Nocturnal Boundary Layer*, Boundary-Layer Meteorol. *17*, 247–264.

MAHRT, L. (1999), *Stratified Atmospheric Boundary Layers*, Boundary-Layer Meteorol. *90*, 375–396.

MALKUS, J. S. and STERN, M. E. (1953), *The Flow of a Stable Atmosphere over a Heated Island* Part I, J. Meteorol. *10*, 30–41.

MASON, P. J. and DERBYSHIRE, S. H. (1990), *Large Eddy Simulation of the Stably–stratified Atmospheric Boundary Layer*, Boundary-Layer Meteorol. *53*, 117–162.

MCNIDER, R. T., *Investigation of the Impact of Topographic Circulations on the Transport and Dispersion of Air Pollutants* (Ph.D. Dissertation, University of Virginia, Charlottesville, 1981).

MCNIDER, R. T. and PIELKE, R. A. (1981), *Diurnal Boundary-layer Development over Sloping Terrain*, J. Atmos. Sci. *38*, 2198–2212.

MCNIDER, R. T., MORAN, M. D., and PIELKE, R. A. (1988), *Influence of Diurnal and Inertial Boundary Layer Oscillations on Long-range Dispersion*, Atmos. Environ. *22*, 2445–2462.

MCNIDER, R. T., SINGH, M. P., and LIN, J. T. (1993), *Diurnal Wind-structure Variations and Dispersion of Pollutants in the Boundary Layer*, Atmos. Environ. *27A*, 2199–2214.

MCNIDER, R. T., ENGLAND, D. E., FRIEDMAN, M. J., and XINGZHONG SHI (1995), *Predictability of the Stable Atmospheric Boundary Layer*, J. Atmos. Sci. *52*, 1602–1614.

MENGELKAMP, H. T. (1991), *Boundary Layer Structure over an Inhomogeneous Surface: Simulation with a Non-hydrostatic Mesoscale Model*, Boundary-Layer Meteorol. 323–341.

MEYER, W. D. and RAO, G. V. (1985), *Structure of the Monsoon Low-level Flow and Monsoon Boundary Layer over the East Central Arabian Sea*, J. Atmos. Sci. *42*, 1929–1943.

NAPPO, C. J. (1991), *Sporadic Breakdowns of Stability in the PBL over Simple and Complex Terrain*, Boundary-Layer Meteorol. *54*, 69–87.

NICHOLLS, M. E., PIELKE, R. A., EASTMAN, J. L., FINLEY, C. A., LYONS, W. A., TREMBACK, C. J., WALKO, R. L., and COTTON, W. R. (1995), *Applications of the RAMS Numerical Model to Dispersion over Urban Areas*. In: Wind Climate in Cities (Cermarh, J. E. *et al* eds.) (Kluwer Academic Publ. The Netherlands) 703–732.

NIEUWSTADT, F. T. M. (1980), *Application of Mixed-layer Similarity to the Observed Dispersion from Ground-level Source*, J. Appl. Meteorol. *19*, 157–161.

NIEUWSTADT, F. T. M. (1984), *The Turbulent Structure of the Stable Nocturnal Boundary Layer*, J. Atmos. Sci. *41*, 2202–2216.

O'BRIEN, J. (1970), *A Note on the Vertical Structure of the Eddy Exchange Coefficient in the Planetary Boundary Layer*, J. Atmos. Sci. *27*, 1213–1215.

OGURA, Y. and PHILLIPS, N. A. (1962), *Scale Analysis of Deep and Shallow Convection in the Atmosphere*, J. Atmos. Sci. *19*, 173–179.

OKE, T. R. *Boundary Layer Climates* (2nd edition), (Methuen: London and New York (1987)) 435 pp.

ORLANSKI, I. (1975), *A Rational Subdivision of Scales for Atmospheric Processes*, Bull. Am. Meteorol. Soc. *56*, 527–530.

PACK, D. H., FERBER, G. J., HEFFTER, J. L., TELEGADAS, K., ANGELL, J. K., HOECKER, W. H., and MACHTA, L. (1978), *Meteorology of Long-range Transport*, Atmos. Environ. *12*, 425–444.

PANOFSKY, H. A. and DUTTON, J. A., *Atmospheric Turbulence Wiley* (Inter Science, New York 1984).

PASQUILL, F. (1961), *The Estimation of the Dispersion of Windborne Material*, Meteorol. Mag. *90*, 33–49.

PASQUILL, F., *Atmospheric Diffusion* (John Wiley and Sons, New York 1974).

PASQUILL, F. and SMITH, F. B., *Atmospheric Diffusion* (Ellis Horwood Ltd. Halstead Press, Chichester, England 1983).

PEARSON, R. A., *Local Flows, Developments in Atmospheric Science 11, Atmospheric Planetary Boundary Layer Physics*, (Longhetto, A. ed.) (Elsevier Scientific Publ. New York, 1980) pp. 95–157.

PHYSIK, W. (1976), *A Numerical Model of the Sea-breeze Phenomenon over a Lake or Gulf*, J. Atmos. Sci. *33*, 2107– 2135.

PIELKE, R. A. (1974), *A Three-dimensional Numerical Model of Sea Breeze over South Florida*, Mon. Weather. Rev. *102*, 115–139.

PIELKE, R. A., *Mesoscale numerical modeling*. In (*Advances in Geophysics*) vol. 23 (Academic Press, New York. 1981).

PIELKE, R. A., MCNIDER, R. T., SEGAL, M., and MAHRER, Y. (1983), *The Use of a Mesoscale Numerical Model for Evaluations of Pollutant Transport and Diffusion in Coastal Regions and over Irregular Terrain*, Bull. Am. Meteorol. Soc. *64*, 243–249.

PIELKE, R. A., *Mesoscale Meteorological Modeling* (Academic Press, New York, N.Y. (1984)).

PIELKE, R. A., COTTON, W. R., WALKO, R. L., TREMBACK, C. J., LYONS, W. A., GRASSO, L. D., NICHOLAS, M. E., MORON, M. D., WESLEY, D. A., LEE, T. J., and COPERLAND, J. H. (1992), *A Comprehensive Meteorological Modeling System-RAMS*, Meteorol. and Atmos. Physics *49*, 69–91.

RAMAKRISHNA T. V. B. P. S., SHARAN, M., GOPALAKRISHNAN, S. G., and ADITI (2002), *Mean Structure of the Nocturnal Boundary Layer in Strong and Weak Wind Conditions: EPRI Case Study*, submitted for publication.

RAYNOR, G. S. and HAYES, J. V. (1984), *Wind Direction Meander at a Coastal Site during Onshore Flows*, J. Clim. and Appl. Meteorol. *23*, 967–978.

REID, J. D. (1979), *Morkov Chain Simulations of Vertical Dispersion in the Neutral Surface Layer for Surface and Elevated Releases*, Boundary-layer Meteorol. *16*, 3–22.

ROBINSON, J., MAHRER, Y., and WAKSHAL, E. (1992), *The Effects of Mesoscale Circulation on the Dispersion of Pollutants (SO_2) in the Eastern Mediterranean, Southern Coastal Plain of Israel*, Atmos. Environ. *26B*, 271–277.

SAGENDORF, J., *Diffusion under Low Windspeed and Inversion Conditions* (NOAA, Environ. Res. Labs, Air Res. Lab., Technical Memorandum *52*. 1975)

SAGENDORF, J. and DICKSON, C. R., *Diffusion under Low Windspeed Inversion Conditions*, (NOAA Tchnical Memo-ERL-ARL-52, Air Resources Labs, Silver Spring. 1974)

SAIKI, E. M., MOENG, C.-H., and SULLIVAN, P. P., *Large eddy simulation of the stably stratified planetary boundary layer* (13th Symp. on *Boundary Layer and Turbulence*, pp. 211–214, 10–15 January, Dallas, Texas, U.S.A. 1999).

SARMA, R. A., AHMAD, N. N., BACON, D. P., BOYBEYI, Z., DUNN, T. J., HALL, M. S., and LEE, P. C. S. (1999), *Application of Adaptive Grid Refinement to Plume Modeling*, Air Pollution VII, WIT Press, Southampton, 59–68.

SAWFORD, B. L. and GUEST, F. M. (1991), *Lagrangian Statistical Simulation of the Turbulent Motion of Heavy Particles*, Boundary-Layer Meteorol. *54*, 147–166.

SCHULTZ, P. and WARNER, T. T. (1982), *Characteristics of Summer-time Circulations and Pollutant Ventilation in Los Angeles Basin*, J. Appl. Meteorol. *21*, 672–682.

SEGAL, M., MCNIDER, R. T., PIELKE, R. A., and MCDOUGAL, D. S. (1982), *A Numerical Model Simulation of the Regional Air Pollution Meteorology in the Greater Chesapeake Bay Area — Summer Day Case Study*, Atmos Environ. *16*, 1381–1397.

SEINFELD, J. H., *Atmospheric Chemistry and Physics of Air Pollution* (John Wiley and Sons, New York 1986).

SHARAN, M., MCNIDER, R. T., GOPALAKRISHNAN, S. G., and SINGH, M. P. (1995a), *Bhopal Gas Leak: A Numerical Simulation of Episodic Dispersion*, Atmos. Environ. *29*, 2061–74.

SHARAN, M., YADAV, A. K., and SINGH, M. P. (1995b), *Comparison of Various Sigma Schemes for Estimating Dispersion of Air Pollutants in Low Winds*, Atmos. Environ. *29*, 2051–59.

SHARAN, M., SINGH, M. P., and YADAV, A. K. (1996a), *A Mathematical Model for the Atmospheric Dispersion in Low Winds with Eddy Diffusivities as Linear Functions of Downwind Distance*, Atmos. Environ. *30*, 1137–45.

SHARAN, M., SINGH, M. P., YADAV, A. K., AGGARWAL, P., and NIGAM, S. (1996b), *A Mathematical Model for the Dispersion of Pollutants in Low Wind Conditions*, Atmos. Environ. *30*, 1209–20.

SHARAN, M., GOPALAKRISHNAN, S. G., MCNIDER, R. T., and SINGH, M. P. (1996c), *Bhopal Gas Leak: A Numerical Investigation of the Prevailing Meteorological Conditions*, J. Applied Meteorol. *35*, 1637–1657.

SHARAN, M., YADAV, A. K., and SINGH, M. P. (1996d), *A Time-dependent Mathematical Model Using Coupled Puff and Segmented Approaches*, J. Appl. Meteorol. *35*, 1625–1631.

SHARAN, M. and GOPALAKRISHNAN, S. G. (1997a), *Comparative Evaluation of Turbulent Exchange Coefficients for Strong and Weak Wind Stable Boundary Layer Modelling*, J. Appl. Meteorol. *36*, 545.

SHARAN, M., and GOPALAKRISHNAN, S. G. (1997b), *Bhopal Gas Accident: A Numerical Simulation of the Gas Dispersion Event*, Environ. Software *12*, 135–141.

SHARAN, M., KANSA, E. J., and GUPTA, S. (1997), *Application of Multiquadric Method for Numerical Solution of Elliptic Partial Differential Equations*, Appl. Mathematics and Computation *84*, 275–302.

SHARAN, M. and RAMAKRISHNAN, T. V. B. P. S. (1998), *Thermodynamic and Turbulent structure in a weak wind nocturnal boundary layer*. Proc. 25th National and 1st International Conf. on Fluid Mechanics and Fluid Power (eds. Veeravalli *et al.*) vol. 1 pp. 493–502, New Delhi.

SHARAN, M. and YADAV, A. K. (1998), *Simulation of Diffusion Experiments under Light Wind Stable Conditions by a Variable K-Theory Model*, Atmos. Environ. *32*, 3481–92.

SHARAN, M., GOPALAKRISHNAN, S. G., and MCNIDER, R. T. (1999), *A Local Parameterization Scheme for σ_w under Stable Conditions*, J. Appl. Meteorol. *38*, 617–622.

SHARAN, M., GOPALAKRISHNAN, S. G., MCNIDER, R. T., and SINGH, M. P. (2000), *Bhopal Gas Leak: A Numerical Investigation on the Possible Influence of Urban Effects on the Prevailing Meteorological Conditions*, Atmos. Environ. *34*, 539–552.

SHARAN, M., RAMAKRISHNA, T. V. B. P. S., and ADITI (2002), *On the Bulk Richardson Number and Monin-Obukhov Similarity Theory in an Atmospheric Surface Layer under Weak Wind Stable Conditions*, submitted for publication.

SHEA, D. M. and AUER, A. H. (1978), *Thermodynamic Properties and Aerosol Patterns in the Plume Downwind of St. Louis*, J. Appl. Meteorol. *17*, 689–698.

SHERMAN, A. C. (1978), *A Mass — Consistent Model for Wind Fields over Complex Terrain*, J. Appl. Meteorol. *17*, 312–319.

SINGH, M. P. and GHOSH, S., *Perspectives in Air Pollution Modelling with Special Reference to the Bhopal Gas Tragedy* (CAS, Tech. Report IIT, New Delhi 1985).

SINGH, M. P. and GHOSH, S. (1987), *Bhopal Gas Tragedy: Model Simulation of Dispersion Scenario*, J. Hazardous Materials *17*, 1–22.

SINGH, M. P., AGARWAL, P., NIGAM, S., and GULATI, A., *Tracer Experiments — A Report*, (Tech. Report, CAS, IIT Delhi 1991).

SINGH, M. P., MCNIDER, R. T., and LIN, J. T. (1993), *An Analytical Study of Diurnal Wind-structure Variations in the Boundary Layer and the Low Level Nocturnal Jet*, Boundary-Layer Meteorol. *59*, 441–460.

SINGH, M. P., MCNIDER, R. T., MEYERS, R., and GUPTA, S. (1997), *Nocturnal Wind Structure Overland and Dispersion of Pollutants: An Analytical Study*, Atmos. Environ. *31*, 105–115.

SINGH, M. P., YADAV, A. K., MCNIDER, R. T., MEYERS, R., SHARAN, M., and LATIF, A. (1999), *Nocturnal dispersion of plume-effect of baroclinicity and geostrophic wind varying with time*. Presented in International Conference cum Workshop on Air Quality Management, University of Brunei, Darussalam

SKUPNIEWICZ, C. E. (1987), *Measurements of Overwater Diffusion: The Separation of Relative Diffusion and Meander*, J. Clim. and Appl. Meteorol. *26*, 949–958.

SMITH, F. B. (1968), *Conditioned Particle Motion in a Homogeneous Turbulent Field*, Atmos. Environ. *2*, 491–508.

SMITH, M. E. (1984), *Review of the Attributes and Performances of 10 Rural Diffusion Models*, Bull. Am. Meterol. Soc. *65*, 554–558.

STERN, M. E. and MALKUS, J. S. (1953), *The Flow of a Stable Atmosphere over a Heated Island*, Part II, J. Meteorol. *10*, 105–120.

STRIMAITIS, D. G., DICRISTOFARO, D. C., and LAVERY, T. F. (1986), *The complex terrain dispersion model.* EPA Document (EPA-600- D-85/220, *Atmospheric Sciences Research Laboratory*) (Research triangle Park, North Carolina, 1986).

STULL, R. *An Introduction to Boundary Layer Meteorology* (Kluwer Academic Publ., Netherlands, 1988) 666 pp.

STULL, R. B. and DRIEDONKS, A. G. M. (1987), *Applications of the Transilient Turbulence Parameterization to Atmospheric Boundary-Layer Simulations*, Boundary-layer Meteorol. *40*, 209–239.

STULL, R. (1993), *Review of Transilient Turbulence Theory and Non-local Mixing.* Boundary-layer Meteorol. *62*, 21–96.

SULLIVAN, P. P., MCWILLIAMS, J. C., and MOENG, C.-H. (1994), *A Subgrid-scale Model for Large-Eddy Simulation of Planetary Boundary Layer Flows*, Boundary-layer Meteorol. *71*, 247–276.

SYKES, R. I., LAGRANGIAN, *Puff Dispersion Modeling and Uncertainty Estimation Using Second-order Closure. Mesoscale Atmospheric Dispersion* (Z.Boybeyi ed.) *Advances in Air Pollution* (WIT Press, UK 2001).

THOMSON, D. J. (1984), *Random Walk Modelling of Diffusion in Inhomogeneous Turbulence*, Quart. J. Roy. Meteorol. Soc. *110*, 1107–1120.

TIRABASSI, T., TAGLIAZUCCA, M., and GALLIANI, G. (1987), *Easy to Use Air Pollution Model for Turbulent Shear Flow*, Environ. Software *2*, 37–44.

TJEMKES, S. A. and DUYNKERKE, P. G. (1989), *The Nocturnal Boundary Layer: Model Calculations Compared with Observations*, J. Appl. Meteorol. *28*, 161–175.

TOON, O. B., TURCO, R. P., WESTPHAL, D. L., MALONE, R., and LIU, M. S. (1988), *A Multidimensional Model for Aerosols: Description of Computational Analogs*, J. Atmos. Sci. *45*, 2123–2143.

TURNER, D. B., *Workbook of Atmospheric Dispersion Estimates* (U.S. Environmental Protection Agency, North Carolina., 1970).

VAN DOP, H., STEENKIST, R., and NIEUWSTADT, F. T. M. (1979), *Revised Estimates for Continuous Shoreline Fumigation*, J. Appl. Meteorol. *18*, 133–137.

VAN DOP, H. (1992), *Buoyant Plume Rise in a Lagrangian Framework*, Atmos. Environ. *26A*, 1335–1346.

VAN DER HOVEN, I. (1976), *A Survey of Field Measurements of Atmospheric Diffusion Under Low-Wind-speed Inversion Conditions*, Nuclear Safety. *17*, 223–230.

VENKATRAM, A. (1977), *A Model of Internal Boundary-layer Development*, Boundary-layer Meteorol. *11*, 419–437.

VENKATRAM, A. (1980), *Dispersion from an Elevated Source in a Convective Boundary Layer*, Atmos. Environ. *14*, 1–10.

VENKATRAM, A. and DU, S. M. (1997), *An Analysis of the Asymptotic Behavior of Cross-wind-integrated Ground-level Concentrations Using Lagrangian Stochastic Simulation*, Atmos. Environ. *31*, 1467–1476.

VISKANTA, R., BERGSTROM, R. W., and JOHNSON, R. O. (1977), *Radiative Transfer in a Polluted Urban Planetary Boundary Layer*, J. Atmos. Sci. *34*, 1091–1103.

VOLDNER, E. C., BARRIE, L. A., and SIROIS, A. (1986), *A Literature Review of Dry Deposition of Oxides of Sulphur and Nitrogen with Emphasis on Long-range Transport Modeling in North America*, Atmos. Environ. *20*, 2101–2123.

WALKLATE, P. J. (1986), *A Markov-Chain Particle Dispersion Model Based on Air Flow Data: Extension to Large Water Droplets*, Boundary-Layer Meteorol. *37*, 313–318.

WALKLATE, P. J. (1987), *A Random Walk Model for Dispersion of Heavy Particles in Turbulent Air Flow*, Boundary-Layer Meteorol. *39*, 175–190.

WEIL, J. C., SYKES, R. I. and VENKATRAM, A. (1992), *Evaluating Air-quality Models: Review and Outlook*, J. Appl. Meteorol. *31*, 1121–1145.

WESTCOTT, N. (1989), *Influence of Mesoscale Winds on the Turbulent Structure of the Urban Boundary Layer over St. Louis*, Boundary-Layer Meteorol. 283–292.

WESTPHAL, D. L., HOLT, T. R., CHANG, S. W., BAKER, N. L., HOGAN, T. F., BRODY, L. R., GODFREY, R. A., GOERSS, J. S., CUMMINGS, J. A., LAWS, D. J., and HINES, C. W. (1999), *Meteorological Reanalyses for the Study of Gulf War Illness: Khamisiyah Case Study*, Weather and Forecasting. *14*, 215–241.

WILLS, G. E. and DEARDORFF, J. W. (1981), *A Laboratory Study of Dispersion from a Source in the Middle of the Convective Mixed Layer*, Atmos. Environ. *15*, 109–117.

WILSON, J. D., THURTELL, G. W., and KIDD, G. E. (1981), *Numerical Simulation of Particle Trajectories in Inhomogeneous Turbulence I: Systems with Constant Turbulent Velocity Scales*, Boundary-Layer Meteorol. *21*, 295–313.

WONG, R. A. and DIRKS, R. A. (1978), *Mesoscale Perturbation on Airflow in the Urban Mixing Layer*, J. Appl. Meteorol. *17*, 677–678.

WROBEL, L. and BREBBIA, C. A., *Time-Dependent Potential Problems, Progress in Boundary Element Methods* (Pentech Press, London. 1980).

WYNGAARD, J. C., COTE, O. R., and IZUMI, Y. (1971), *Local Free Convection, Similarity, and the Budgets of Shear Stress and Heat Flux*, J. Atmos. Sci. *28*, 1171–1182.

WYNGAARD, J. C. (1975), *Modeling the Planetary Boundary Layer: Extension to the Stable Case*, Boundary-Layer Meteorol. *9*, 441–460.

WYNGAARD, J. C., *Boundary layer modeling*. In *Atmospheric Turbulence and Air Pollution Modelling* (Nieuwstadt, F.T.M. and Van Dop, H. eds.) (D. Reidel Publ., Dordrecht, Holland. 1982) pp. 69–106.

YADAV, A. K., *Mathematical Modelling of Dispersion of Air Pollutants in Low Wind Conditions* (Ph.D. Thesis, Indian Institute of Technology, Delhi, India 1995) 151 pp.

YADAV, A. K. and SHARAN, M. (1996), *Statistical Evaluation of Sigma Schemes for Estimating Dispersion in Low Wind Conditions*, Atmos. Environ. *30*, 2595–2606.

YADAV, A. K., RAMAN, S., and SHARAN, M. (1996a), *Surface Layer Turbulence Spectra and Eddy Dissipation during Low Winds in Tropics*. Boundary-Layer Meteorol. *79*, 205–224.

YADAV, A. K., SHARAN, M., and SINGH, M. P. *Atmospheric Dispersion in Low Winds*. Proc. of *First World Congress of Nonlinear Analysts* (eds.) V. Lakshmikantham, Walter de Gruyter, (Berlin, 1996b) pp. 3567–3593.

YAMADA,T., BUNKER, S., and MOSS, M. (1992), *Numerical Simulations of Atmospheric Transport and Diffusion over Coastal Complex Terrain*, J. Appl. Meteorol. *31*, 565–578.

YEH, G. T. and HUANG, C. H. (1975), *Three-dimensional Air Pollution Modeling in the Lower Atmosphere*, Boundary-Layer Meteorol. *9*, 381–390.

YU, T. W. (1977), *A Comparitive Study on Parameterization of Vertical Exchange Processes*, Mon. Wea. Rev. *105*, 57–66.

ZANNETTI, P. (1986), *A New Mixed Segmented-puff Approcach for Dispersion Simulation*, J. Appl. Meteorol. *20*, 1023–1211.

ZANNETTI, P., *Air Pollution Modeling* (Van Nostrand Reinhold, New York. 1990), 444 pp.

ZEMAN, O. (1979), *Parameterization of the Dynamics of Stable Boundary Layer and Nocturnal Jets*, J. Atmos. Sci. *36*, 792–804.

ZIENKIEWICZ, O. C., *The Finite Element Method* (Tata McGraw-Hill Publ. 1979).

(Received March 1, 2000, accepted January 31, 2001)

To access this journal online:
http://www.birkhauser.ch

Pure appl. geophys. 160 (2003) 395–404
0033–4553/03/020395–10

© Birkhäuser Verlag, Basel, 2003

Pure and Applied Geophysics

A Note on the Estimation of Eddy Diffusivity and Dissipation Length in Low Winds over a Tropical Urban Terrain

Anil Kumar Yadav[1], Sethu Raman[2],
and Dev Dutta S. Niyogi[2]

Abstract — Urban terrain poses a challenge for modeling air pollutant diffusion. In tropics, because of the dominant low wind speed environment, the importance of understanding the turbulence diffusion is even more critical, and uncertain. The objective of this study is to estimate the vertical eddy diffusivity of an urban, tropical atmosphere in low–wind speeds. Turbulence measurements at 1 Hz were made at 4-m level over an urban terrain with a roughness length of 0.78 m during winter months. Eddy diffusivity is estimated from spectral quantities of the turbulence data involving turbulent kinetic energy (E) and its dissipation rate (ε). The spectral information of the vertical velocity fluctuations is used to estimate the vertical length scale which provides information on the eddy diffusivity. In addition, the product of friction velocity and the vertical length scale has been used to non-dimensionalize the eddy diffusivity, which is shown to increase with increasing instability. Using the eddy diffusivity (K) estimates from the $E - \varepsilon$ approach, a relation is suggested for the mixing length based eddy diffusivity models of the form: $K = c_w \cdot [2.5 - 0.5(z/L)]$, where z is the measurement height, L is the Obukhov length, and c_w has an average value close to 1 for unstable and near 0.5 for stable conditions for the urban terrains.

Key words: Air quality, atmospheric boundary layer, dissipation length, eddy diffusivity, tropics, urban terrain.

Introduction

Eddy diffusivity is an important variable for planetary boundary layer (PBL) parameterizations. It is used as a dimensional parameter in turbulence schemes. Following the Fickian diffusion, turbulent energy flux is assumed to flow down the gradient (STULL, 1988). In various environmental applications, there is a need to predict eddy diffusivity from knowledge of turbulence, for which the idea of quantifying vertical mixing from the spectral quantities is a viable approach (LEE, 1996).

[1] Department of Applied Mathematics, Guru Jambheshwar University, Hisar-125001 Haryana, India.
[2] Department of Marine, Earth, and Atmospheric Sciences, North Carolina State University, Raleigh, NC 27695-7236, U.S.A.

Corresponding Author: Professor Sethu Raman, Department of Marine, Earth, and Atmospheric Sciences, and State Climate Office of North Carolina, North Carolina State University, Raleigh, NC 27695 - 7236, U.S.A. Email: sethu_raman@ncsu.edu

Recent research focussed towards understanding the turbulence structure of the atmospheric surface layer in low wind speeds (AGARWAL *et al.*, 1995; YADAV *et al.*, 1996). This research has significance, particularly in the tropics, because of the paucity of data and absence of such studies. The turbulence structure in the tropics can be different from that in the mid-latitudes, particularly under low wind conditions. In addition to developing an understanding of turbulence structure in the tropics, knowledge of eddy diffusivity is important for estimating pollutant dispersion in the lower atmosphere.

The observations used in this study are part of comprehensive turbulence and diffusion studies undertaken in the tropics as described in AGARWAL *et al.* (1995). Analyses of the turbulence data set, such as the variation of the turbulence intensities with atmospheric stability and wind speed, have revealed the presence of two characteristic regimes: one below, and the other above, 1 ms^{-1} wind speed. YADAV *et al.* (1996) studied the spectral characteristics of a turbulence data set with wind speeds more than 1 ms^{-1}. For this regime results showed that the inertial subrange features of the normalized power spectrum are generally consistent with Monin-Obukhov scaling. However, the well-established spectral relations (based mainly on the Kansas data) regarding the dependence of dimensionless dissipation rate on the stability parameter did not match with this tropical, urban observations.

There are various methods of parameterizing eddy diffusivity (cf., STULL, 1988). In a recent study, LEE (1996) examined the velocity and air temperature spectra and eddy diffusivity over forests. The study extended HANNA'S (1968) postulation that the vertical eddy diffusivity can be obtained from spectral information of the vertical velocity fluctuations. In addition to eddy diffusivity, dissipation length is another variable that is important in PBL and diffusion studies. Dissipation length scale is used to parameterize boundary layer turbulence in the models (cf. MELLOR and YAMADA, 1974; ANDRÈ *et al.*, 1978). The objective of this study is to examine, with the use of a data set covering low-wind urban stable and unstable conditions in the tropics, the stability dependence of dissipation length scale and eddy diffusivity using two different methods. Spectral and turbulence information of velocity fluctuations is used for the computation of eddy diffusivity values.

Data Description

Experimental details of the data used in this study can be found in AGARWAL *et al.* (1995), as well as YADAV *et al.* (1996). The experimental site is an urban terrain in New Delhi, India (28.43 N, 77.18 E) with an aerodynamic roughness length of 0.78 m (RAMAN *et al.*, 1990). The turbulent (1 Hz frequency) wind and temperature fluctuations were measured using a sonic anemometer and a fine wire thermocouple

(Campbell SWS-211/2 EK), respectively. The instruments were housed on a 30-m tower in a small field inside the Indian Institute of Technology, Delhi campus in the city. The site can be described as relatively flat, open, urban terrain with small bushes and trees at a distance greater than 500 m (YADAV *et al.*, 1996). Observations were made in stable, unstable and near-neutral atmospheric conditions. The original data set consists of 38 hourly test runs. These data were divided into two groups: one for wind speeds below 1 m s^{-1} and the other for wind speeds greater than 1 m s^{-1}, for turbulence analysis by AGARWAL *et al.* (1995). Based on the values of stability parameter and restricting the lower limit of mean wind speed to 1 m s^{-1}, 16 runs have been chosen for the present analysis. The lower limit of 1 m s^{-1} has been imposed principally due to the constraints in the instrumentation detection, and possible biases for wind speeds below 1 m s^{-1}. These 16 runs pertain to a continuous day-night period in a winter month. Figure 1 shows the mean wind speed (U) and stability parameter (z/L) variation for this period. The mean wind speed ranges between 1 and 3.5 m s^{-1}, at 4 m height. Typically unstable conditions correspond to higher winds ($\sim$ 2 to 3.5 m s^{-1}), while most stable or near–neutral conditions correspond to average wind speed between 1 and 2 m s^{-1}. Mean values of some of the relevant variables for the runs considered in this study are shown in Table 1. As seen from the data, the 16 cases present fairly diverse conditions that are representative of a tropical urban terrain.

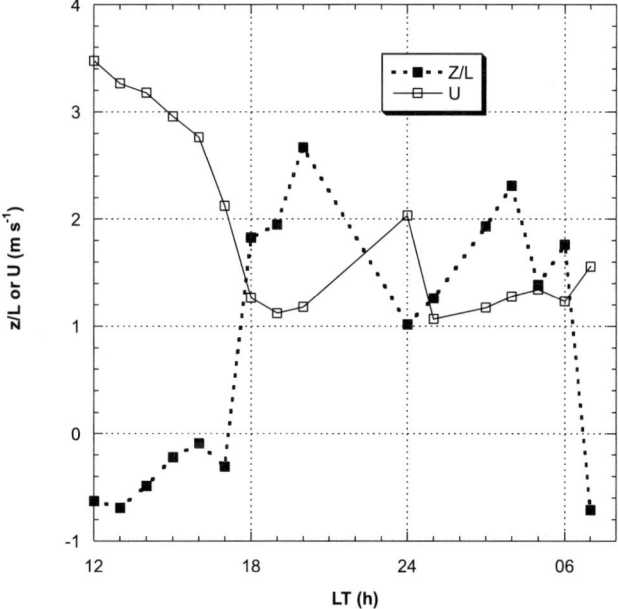

Figure 1
Variation of average wind speed (U m s^{-1}) and stability parameter (z/L) with local time.

Table 1

Some mean variables of the data set for each test run on February 14–15, 1992

Run number	Time (LST)	U (m s^{-1})	u_* (m s^{-1})	z/L	H_0 (Wm^{-2})	K (m^2 s^{-1})
1	1200	3.47	0.615	−0.63	138.42	3.337
2	1300	3.26	0.596	−0.69	147.08	3.589
3	1400	3.18	0.623	−0.49	127.54	4.559
4	1500	2.96	0.590	−0.22	47.79	3.485
5	1600	2.77	0.518	−0.09	13.87	2.682
6	1700	2.12	0.447	−0.31	29.31	1.728
7	1800	1.27	0.317	1.83	−56.42	1.194
8	1900	1.12	0.262	1.95	−31.71	0.702
9	2000	1.18	0.244	2.67	−34.01	0.794
10	0000	2.03	0.386	1.02	−44.83	1.56
11	0100	1.07	0.286	1.26	−20.25	0.79
12	0300	1.17	0.278	1.93	−24.16	0.612
13	0400	1.28	0.267	2.31	−21.86	0.669
14	0500	1.35	0.325	1.39	−23.52	0.898
15	0600	1.22	0.322	1.76	−27.58	0.668
16	0700	1.54	0.347	−0.71	10.14	1.16

Results and Discussion

Results of the spectral analysis of the 16 runs are presented in YADAV *et al.* (1996). The observed spectra were consistent with Monin-Obukhov scaling. Using derived parameters and spectral information, particularly the dissipation rate, the eddy diffusivity can be estimated.

Eddy Diffusivity

Eddy diffusivity for momentum was determined from turbulence measurements at $z = 4$ m by using two different formulations. The first is based on $E - \varepsilon$ closure and the other on HANNA'S (1968) formulation. In $E - \varepsilon$ closure, the eddy diffusivity K is parameterized as (cf. LEE, 1996),

$$K = c_\varepsilon \frac{E^2}{\varepsilon} \, , \tag{1}$$

where c_ε is a constant (generally considered to be 0.03 to 0.04, (STULL, 1988), E is the turbulent kinetic energy (TKE), and ε is the TKE dissipation rate. As seen in Figure 1, z/L is smaller for higher wind speed. This indicates better potential for mechanical turbulence over the rough, urban terrain. The dependence of the eddy diffusivity on surface friction velocity (u_*) is shown in Figure 2. As shown in the figure, the eddy diffusivity values follow a linear relation ($r = 0.975$) which can be approximated as,

$$K = -1.85 + 9u_* \, . \tag{2}$$

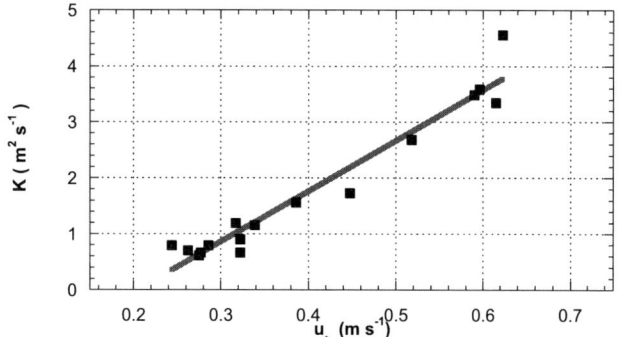

Figure 2
Variation of $E - \varepsilon$ closure based eddy diffusivity (K, m^2 s^{-1}) with friction velocity (u_*, m s^{-1}). The line
shown is the best fit given by $K = -1.85 + 9\ u_*$ (r = 0.975).

The stability dependence of eddy diffusivity from $E - \varepsilon$ closure is examined
further in Figure 3. The data indicate larger scatter for unstable conditions as
compared to the stable regimes. Overall as expected, eddy diffusivity decreases
with increasing stability. Eddy diffusivity increases rapidly for values of z/L
smaller than 0 (unstable conditions). Thus in general, vertical eddy diffusivity
estimation based on the $E - \varepsilon$ closure is sensitive to correct specification of the
stability parameter z/L, particularly for the unstable conditions. For the unstable
conditions, K ranges from about 1 m^2 s^{-1} to 5 m^2 s^{-1} while for stable conditions
its value changes slightly between 0.6 to 1 m^2 s^{-1}. A quadratic best fit is also
possible, of the form,

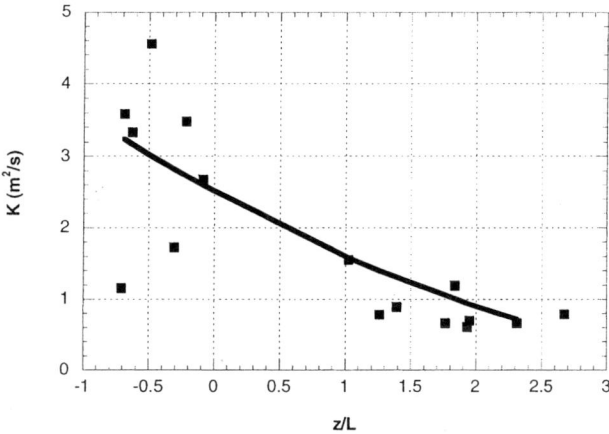

Figure 3
Variation of $E - \varepsilon$ closure based eddy diffusivity (K, m^2 s^{-1}) with the stability parameter z/L. The solid line
corresponds to the quadratic best–fit of the form $A + B \cdot (z/L) + C \cdot (z/L)^2$ with A, B, and C equaling 2.5,
-1.0, and 0.1, respectively.

$$K = A + B\left(\frac{z}{L}\right) + C\left(\frac{z}{L}\right)^2 \tag{3}$$

with the values of A, B, C equal to 2.5, -1.0, and 0.1, for our data set.

The other formulation for calculating eddy diffusivity is based on HANNA'S (1968) postulation. It is dependent upon the gross characteristics of the vertical velocity component of turbulent eddies. Accordingly, diffusivity K can be obtained from spectral features of the vertical velocity component, and can be expressed as,

$$K = c \cdot \frac{\sigma_w^4}{\varepsilon} \,, \tag{4}$$

where c is a proportionality constant. HANNA (1968) proposed a value of $c = 0.3$ while LEE (1996) suggested a broader range depending on the level of turbulence, averaged to a value close to 0.41. Equation (4) can also be written as,

$$K = c_w \cdot l_w \cdot \sigma_w \tag{5}$$

where,

$$l_w = \frac{\sigma_w^3}{\varepsilon} \,. \tag{6}$$

l_w is called the vertical integral scale and is a measure of the vertical size of the eddies responsible for most of the mixing (TENNEKES and LUMLEY, 1972). The variation of the constant c_w with stability has been studied by LEE (1996), using surface layer similarity functions and local similarity relations suggested by NIEUWSTADT (1984). In our study, the value of c_w for neutral conditions obtained by comparing HANNA'S (1968) formulation with $E - \varepsilon$ formulation was considerably higher than that which many investigators consider typical (0.3). We attribute this elevated value to the increased roughness and higher turbulence levels within the urban study site. LEE (1996) also found similar higher values over vegetation canopies with high surface roughness. Our c_w values range from 0.6 to 1.45 (with an average value around 1) for the unstable conditions, and from 0.37 to 0.73 (with a mean around 0.5) for the stable conditions.

Excluding the proportionality constant c_w, the product $l_w \cdot \sigma_w$ has been plotted for the present data against the stability parameter z/L in Figure 4. On the unstable side the scatter is relatively less, as compared to the $E - \varepsilon$ based diffusivity values shown in Figure 3. The variation for the stable conditions is nearly similar for both cases (Figs. 3 and 4). For the two parameters ($l_w \cdot \sigma_w$ and z/L) a linear variation ($r = 0.85$) can be obtained of the form,

$$l_w \sigma_w = 2.5 - 0.5\left(\frac{z}{L}\right) \,. \tag{7}$$

Combining equations (5) and (7), one can then obtain eddy diffusivity as,

$$K = c_w \cdot \left[2.5 - 0.5\left(\frac{z}{L}\right)\right] \tag{8}$$

with c_w around 1 for unstable and around 0.5 for stable conditions. Note that the value of the constant c_w is larger than that obtained by LEE (1996), since our

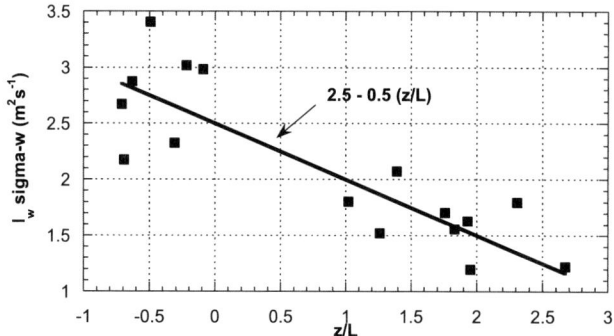

Figure 4

Variation of eddy diffusivity scale $l_w \cdot \sigma_w$ (m^2 s^{-1}) based on HANNA'S (1968) postulation with the stability parameter z/L. A best–fit of the form, $l_w \cdot \sigma_w = 2.5 - 0.5 \, (z/L)$ is obtained.

observations span an urban terrain which has larger roughness and turbulent exchanges as compared to the forest canopy in Lee's study. However, the values are also within the ranges obtained in LEE (1996) for a different atmospheric stability.

Further, since σ_w is not routinely available, the product of friction velocity (u_*) and the vertical (l_w) may form a surrogate parameter for estimating eddy diffusivity. Figure 5 shows a scatter plot of $[K/(u_* l_w)]$ versus z/L. The trends of dimensionless K are similar to those shown in Figure 3 (for $E - \varepsilon$ case) for both stable as well as unstable conditions. Overall, the scatter is less, and is more apparent for the unstable conditions in the case of the dimensionless K. Though not shown in the figure, a linear relation can be obtained from the data, of the form:

$$\frac{K}{(u_* \cdot l_w)} = \left[1 - 0.2\left(\frac{z}{L}\right)\right] . \tag{9}$$

The validity and robustness of the nondimensionalization $K/u_* l_w$ needs to be tested further with more data sets, particularly in the tropical urban domains.

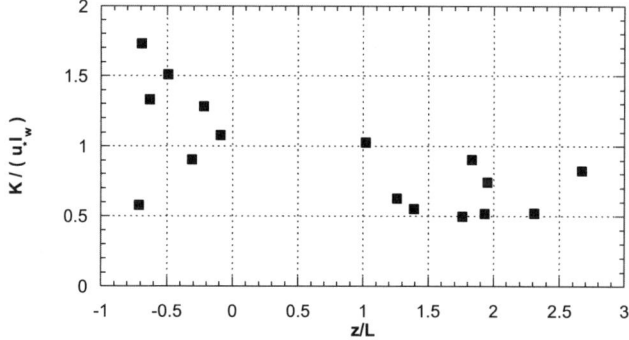

Figure 5

Variation of normalized $E - \varepsilon$ closure based eddy diffusivity $K/(u_* l_w)$ with the stability parameter z/L.

Dissipation Length

The dissipation length is an important parameter determining the level of turbulence. It provides a measure regarding the size (magnitude) of large energy-containing eddies. It varies as a function of stability and height above ground (LOUIS *et al.*, 1983), and is calculated as (TENNEKES and LUMLEY, 1972)

$$l = \frac{E^{1.5}}{\varepsilon} . \tag{10}$$

Figure 6 gives the time evolution of the dissipation length computed from equation (10) in terms of E and ε. When compared with the time evolution of turbulent kinetic energy and ε (not shown), a resemblance between the behaviors of these three variables is seen. At night, there appears to be some activity such as breaking of gravity waves leading to patchy turbulence. The general behavior of dissipation length with time exhibits variations principally due to the change in the stability. The computed mean values and standard deviations of the dissipation length for the urban terrain are 27.6 ± 5.6 m and 16.2 ± 3.1 m for the unstable and the stable cases, respectively. They correspond to the measurements at 4 m above the ground surface. Thus the mean value of the normalized ratio l/z is about 7 in unstable, and 4 in the stable atmospheric conditions. Over a flat and homogeneous site, LOUIS *et al.* (1983) report values of l/z around 5 for near-neutral, 10 for convective and 1 for stable conditions. Our value falls in between the near-neutral and convective values

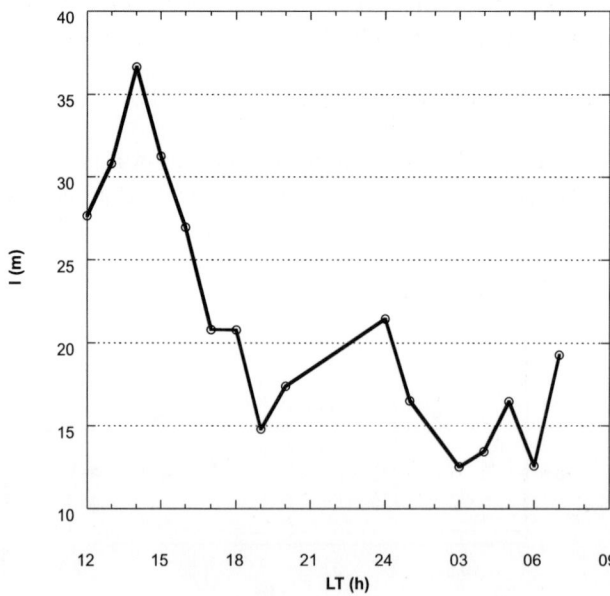

Figure 6
Diurnal variation of dissipation length (l, m) for February 14–15, 1992 for the tropical, urban terrain.

obtained by LOUIS *et al.* (1983) over flat and homogeneous terrain. On the other hand, the stable case value of l/z for the present data is higher in comparison to the corresponding value obtained by LOUIS *et al.* (1983). This could be due to roughness and other terrain characteristics, as well as wind speed conditions over the tropical, urban site in our study.

The normalized dissipation length (l/z) is plotted as a function of stability parameter z/L in Figure 7. The values show similar variations as seen for the eddy diffusivity in Figure 3.

Conclusions

Eddy diffusivity for momentum has been estimated from a data set representing low-wind, tropical, urban conditions. Two methods used for this purpose are $E - \varepsilon$ closure formulation and the HANNA'S (1968) mixing–length based postulation. The examination of stability dependence of eddy diffusivity for momentum reveals that diffusivity increases with the increase in the magnitude of the stability parameter z/L. The scatter, particularly on the unstable side, reduces when diffusivity is normalized by the product of friction velocity and vertical integral scale. A relation is suggested for eddy diffusivity estimation over urban terrain of the form: $K = c_w [2.5 - (z/L)]$, where c_w equals 1 for unstable and about 0.5 for stable conditions.

The dissipation length has also been estimated using the TKE and the eddy dissipation rate. Its variation with the stability indicates its sensitivity to the stability parameter z/L, particularly for the unstable conditions. The mean values of l/z in the

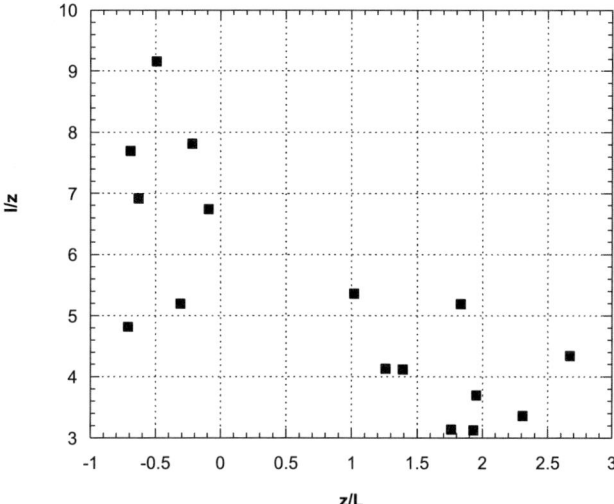

Figure 7
Variation of the normalized dissipation length (l/z) with the stability parameter z/L.

present study have been compared with those over flat and homogeneous terrain and are found to be more for the stable case and less for the unstable conditions.

Although the data are limited over a small stability range, they provide insight of the behavior of eddy diffusivity and the vertical length scale. This information can be adopted in modeling dispersion characteristics over tropical, urban regions.

Acknowledgements

This work was supported by the Division of International Programs and the Division of Atmospheric Sciences, National Science Foundation under grant ATM-0080088. We gratefully acknowledge detailed comments and kind suggestions by Professor S. A. Hsu, Louisiana State University, Prof. G. V. Rao, St. Louis University, and another anonymous reviewer that significantly enhanced the presentation of this paper.

REFERENCES

AGARWAL, P., YADAV, A. K., GULATI, A., RAMAN, S., RAO, S., SINGH, M. P., NIGAM, S., and REDDY, N. (1995), *Surface layer turbulence processes in low windspeeds over land*, Atmos. Environ. *29*, 2089–2098.

ANDRÈ, J.C., DE MOOR, G., LACARRER, P., THETRY, G., and DU VACHAT, R. (1978), *Modelling the 24-hour Evolution of the Mean and Turbulent Structures of the Planetary Boundary Layer*, J. Atmos. Sci. *35*, 1861–1883.

HANNA, S. R. (1968), *A Method of Estimating Vertical Eddy Transport in the Planetary Boundary Layer Using Characteristics of the Vertical Velocity Spectrum*, J. Atmos. Sci. *25*, 1026–1032.

LEE, X. (1996), *Turbulence Spectra and Eddy Diffusivity over Forests*, J. Appl. Meteor. 35, 1307–1318.

LOUIS, J. F., WEIL, A., and VIDAL-MADJAR, D. (1983), *Dissipation Length in Stable Layer*, Boundary-Layer Meteor. *26*, 229–243.

MELLOR, G. L. and YAMADA, T. (1974), *A Hierarchy of Turbulence Closure Models for Planetary Boundary Layers*, J. Atmos. Sci. *31*, 1791–1806.

NIEUWSTADT, F. T. M. (1984), *The Turbulent Structure of the Stable Nocturnal Boundary Layer*, J. Atmos. Sci. *41*, 2202–2216.

RAMAN, S., TEMPLEMAN, S., TEMPLEMANS B., HOLT, T., MURTHY, A. B., SINGH, M. P., AGARWAL, P., NIGAM, S., PRABHU, A., and AMEENULLAH, S. (1990), *Structure of the Indian Southwesterly Pre-monsoon and Monsoon Boundary Layers: Observations and Numerical Simulation*, Atmos. Environ. *24A* (4), 723–734.

STULL, R. B. *An Introduction to Boundary Layer Meteorology* (Kluwer Academic Press, 1988), 666 pp.

TENNEKES, H. and LUMLEY, J. L., *A First Course in Turbulence* (MIT Press, 1972), 300 pp.

YADAV, A. K., RAMAN, S., and SHARAN, M. (1996), *Surface Layer Turbulence Spectra and Eddy Dissipation during Low Wind in Tropics*, Boundary Layer Meteor. *79*, 205–217.

(Received August 7, 2000, accepted March 16, 2001)

To access this journal online:
http://www.birkhauser.ch

Pure appl. geophys. 160 (2003) 405–418
0033–4553/03/020405–14

© Birkhäuser Verlag, Basel, 2003

█ **Pure and Applied Geophysics**

Dispersion of Flue Gases from Power Plants in Brunei Darussalam

P. N. Tandon,[1] P. Ramalingam,[2] and A. Q. Malik[2]

Abstract — A series of mathematical and computer models describing alternative methods for the disposal of flue gases emitted from coal and gas fired power plants is discussed. Brunei Darussalam has three gas-fired power plants using approximately 3.106 m^3/day gas and emitting substantial flue gases into the atmosphere. After desulphurisation with sea water, carbon-dioxide, the gas primarily associated with global warming and the main constituent of flue gas, can be dissolved under pressure in seawater and injected into the sea at suitable depth. The injected solution constitutes a negatively buoyant plume in the sea, carried by currents to deeper regions. It has been noted that the solution mixes and reacts with other oceanic components and converts to carbonates and sulphates that can remain near the bottom for several hundreds of years. Until a better alternative is developed, this may be an immediate solution to the problem of dealing with flue gases. The feasibility and economics of this alternative have been discussed in the literature. For optimal design criteria for such disposal, numerous parameters (such as location, pipe diameter, type of diffuser, angles of discharge, etc.) are involved in the mathematical analysis. Many alternative sets of these parameters must be used as input parameters to arrive at final design parameters for optimal results.

Key words: Mathematical and computer models, flue gas disposal, effluent, carbon dioxide, solubility, concentration, dilution.

Introduction

Two major problems currently, acid rain and global warming, of special concern in Asia, are mainly caused by the high emissions of gaseous pollutants from a variety of industrial sources, the most important being fuel in combustion plants. A substantial amount of carbon dioxide is emitted into the atmosphere by fossil fuel combustion in various industrial and power plants and this is thought to result in a considerable increase in atmospheric temperature. There is a growing demand for the development of practical and economic methods to control air pollution from a wide range of industrial plants. The major environmental impact derives from

[1] Department of Mathematics University of Brunei Darussalam, Tungku Link, Gadong BE 1410, Negara Brunei Darussalam. E-mail: tandonpn@fos.ubd.edu.bn
[2] Department of Physics, University of Brunei Darussalam, Tungku Link, Gadong BE 1410, Negara Brunei Darussalam.

sulphur dioxide (SO_2), hydrochloric acid gas (HCl), and carbon dioxide CO_2 emitted as flue gases. These emissions contribute to the problems of acid rain and global warming (MARCHETTI, 1979). The average warming in the atmosphere due to doubling of the emission of carbon dioxide is predicted to be roughly from 1.5 °C to 4.5 °C (BAES *et al.*, 1980). The climate changes produced by excess emission of CO_2 could appear, grow and become unacceptable before effective counter measure are taken.

Absorption of these gases into seawater is one of the possible methods of controlling their emission into the atmosphere. A solution of the CO_2 gas in seawater may be injected into the sea, under pressure, at a suitable depth (TANDON and RAMALINGHAM, 1998). Once absorbed in seawater, flue gases are converted into sulphate, bicarbonate and chloride (FOFL *et al.*, 1995). In principle, these ions may be discharged into the oceans in considerable quantities since seawater has considerably more absorbing capacity.

MARCHETTI (1977) proposed that CO_2 might be separated from the flue gases and then injected into the oceans. Desulphurisation of flue gases by means of seawater scrubbing offers several advantages over the current popular lime/limestone methods (PAULEY, 1984); RADOJEVIC and TRESSIEDER, 1992). It helps to relieve strain on valuable fresh water resources as well as minimize adverse effects on seawater chemistry and the ecosystem (ABRAMS *et al.*, 1988). To reduce atmospheric pollution by SO_2 deriving from the combustion of fuels, several flue desulphurisation processes have been developed, based on scrubbing the gas with seawater. SO_2 is selectively absorbed and oxidized to sulphates. Sulphates are natural seawater components that, in themselves, have negligible negative effects on the marine ecosystem (HAUGEN and DRANGE, 1992). After removing sulphur dioxide from flue gases through scrubbing with seawater, carbon dioxide is dissolved in seawater which is then injected into the sea (HAGEN and KOLDERUP, 1979). The increase in density of this seawater, resulting from CO_2 dissolution, should be sufficient to transport the dissolved gas to lower depths. For shallow injections near the shore, gravity currents will carry the dense CO_2 rich water along the bottom slope towards deeper waters. Energy and capital costs for such shallow injection could be less than for deep ocean injection (HAROGH and ZEMBA, 1991).

The main purpose of the present study is to investigate alternative methods of disposal of flue gases in the deep ocean. The negatively buoyant CO_2-enriched seawater solution would transport the gas from emission sites to great depths. Sinking currents would transport and spread the injected CO_2 over a large areas (ISHITANI and MATUHASI, 1996). In real applications of this methodology, the extent of CO_2 desorption will depend on the dilution factor, the depth of discharge, and other parameters such as the length of the diffuser, number of ports, etc. Once the effluent is diluted to the background level, desorption of CO_2 from the plume would cease to be important (MUSTACHI *et al.*, 1979). In this way, the injected CO_2 is carried to deep water sediments already rich in $CaCO_3$, where it is sequestered for long periods (ARNOLD *et al.*, 1979; HAGEN and KOLDERUP, 1979; RADOJEVIC, 1989; RADOJEVIC and TRESSIEDER, 1992).

Mathematical and Computer Models for Plume Dynamics

Numerous mathematical and computer models have been developed for the disposal of various kinds of effluents into the sea (FAN and BROOKS, 1969; MUELLENHOFF *et al.*, 1985; WOOD, 1993; LEE and CHEUNG, 1990; BAUMGARTNER *et al.*, 1994; LORIN, 1999). These models follow either Euler's approach or the Lagrangian concept. In the Eulerian framework, numerical solutions of the governing diffusion equations are typically accompanied by spurious oscillations which depict nonphysical representations of transport phenomena. In the Eulerian approach, pollutant concentration is determined at certain fixed points (ROBERT *et al.*, 1989). Numerical solutions by Finite Difference and or Finite Element Methods introduce numerical diffusion and sometimes result in negative concentrations. Consequently, it does not depict the true conservation of the pollutants. On the other hand, the Lagrangian approach follows the plume element along the course of its trajectory for example JETLAG, developed by LEE and CHEUNG (1990). JETLAG considers a buoyant jet with a circular cross section directed into uniform cross flow. The jet discharges the effluent at angle (θ) with respect to the horizontal plane for positively and negatively buoyant plumes as shown in Figure 1.

The geometry of the injection is illustrated in Figures 2a and 2b, with an inclination angle θ of the jet with the horizontal plane and with the current direction σ in the horizontal plane. The trajectories in the turbulent flow field are determined by considering both advection and dispersion of the substance in water at desired

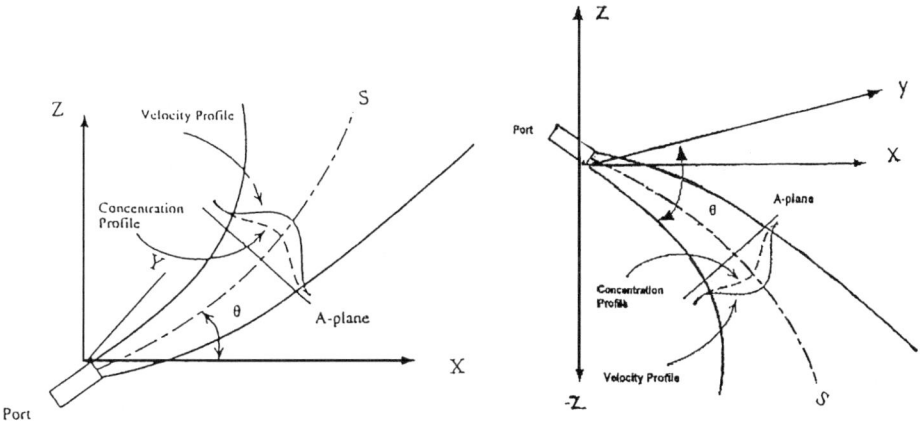

Figure 1
Upward and downward plumes emerging from a source.

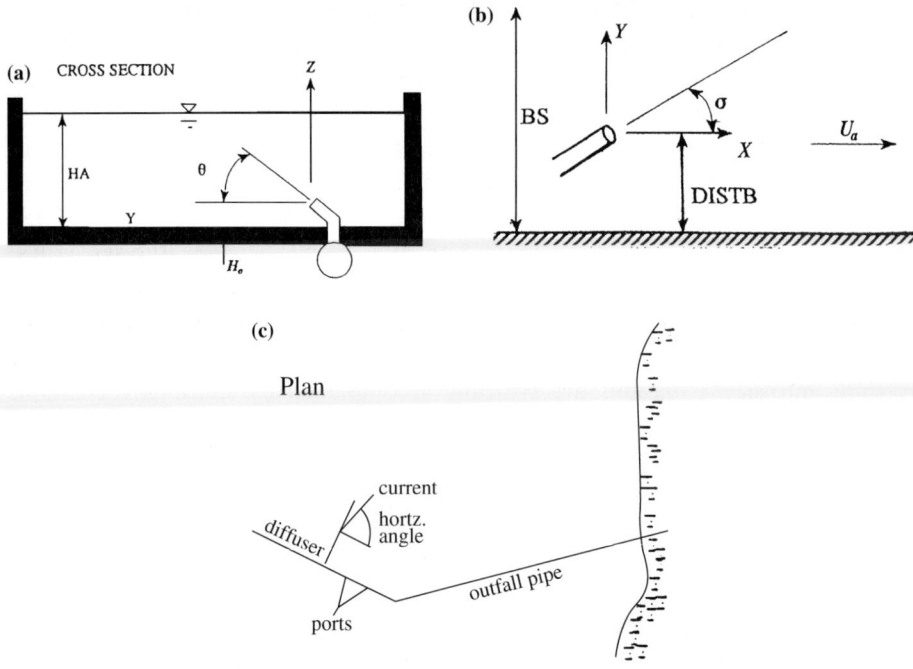

Figure 2
(a) Elevation sketch of the discharge. (b) Plane view sketch of the discharge. (c) Definition sketch of a multi-port diffuser attached to an outfall pipe.

time intervals by using local balance equations. This approach describes exact mass conservation with complete absence of numerical diffusion. A more recently developed model HYDROTRACK also uses the Lagrangian approach, with the pollutant represented by a large number of tracer particles (MAIER-REIMER, 1982; GARCIA and RODRIGUEZ, 1997). This transport and dispersion model calculates the particle trajectories in a turbulent flow field considering buoyancy, advection and dispersion. Several methods based on length scale concepts are included in Cornel Mixing Zone Expert Systems or simply CORMIX 1, 2 and 3.

CORMIX 1 (DONEKER and JIRKA, 1990) is intended for submerged single port discharges whereas CORMIX 2 (AKAR and JIRKA, 1991) calculates dilution and plume formation and its other characteristics for any of the diffuser types considered: staged, unidirectional and alternating. CORMIX 3 (JONES, 1990) is meant for surface discharges of effluents from channels, open channels or open pipes (TSANIS *et al.*, 1994; TSANIS and VALEO, 1994). Figure 2c shows a schematic of the geometrical features of a diffuser attached at the end of a long out-fall pipe through which the effluent is released for different types of discharges.

Results for the alternative methods through single port and multi-port diffusers have been obtained by using CORMIX 1 and 2, respectively. TANDON and

RAMALINGAM (1998) and TANDON (1999, 2000) have studied, in detail, the properties of the plume formation and its development for different types of effluents released through various types of diffusers attached to the main pipe in the sea. The program is run for several alternatives starting from the single port diffuser to multi-port diffusers to yield guidance on optimal design parameters. The results as observed from the analysis of the various types of discharges justify the need for further detailed studies in alternative designs of outfalls and on setting optimal design criteria for disposal of CO_2 in shallow water discharges. Therefore, the same input parameters are again used for running the computer programs CORMIX 1 and CORMIX 2 for a single port and multi-port diffusers at various possible vertical angles (θ) and horizontal angles (σ). A new method of calculation of the dispersion of gases emitted into the atmosphere from the stack has been developed by OOMS (1972) and modified by KHAN and ABBASI (1997) by the application of the recently modified plume path theory of dispersion of heavy gases.

There are three gas fired power plants serving in Bandar Seri Begawan, the capital of Brunei Darussalam. Each plant consumes 10^6 m^3 gas per day. The composition of the gas is CH_4 88%, C_2H_6 5%, CO_2 4%, C_3H_2 2% and the remaining 1% in other gases. The volume of CO_2 output per day is estimated to be 0.8×10^9 m^3 and hence requires a large amount of seawater for its dissolution at higher pressures. As deepwater injection is not possible in Brunei Darussalam, the present study deals with shallow water injection. Negative buoyancy of CO_2-enriched seawater would then cause the transport of the dissolved gas from the emission site to the deep ocean. The analysis presented here satisfies the need to safeguard marine life downstream of the injection point (LEONG et al., 1985, 1987, 1995; JACQUES and GANOULIS, 1994) and suggests design criteria for such disposal.

Methodology

Various alternative forms for CO_2 disposal into the sea (in the form of clathrates, liquid gas and in dissolved form) have been discussed in the literature for the disposal in sea as shown in Figure 3. The concentrated solution of CO_2 in seawater is preferable to other forms. As mentioned in the introduction, the most important aspect of this study is to find a preferred alternative for the disposal of flue gases emitted from power generation plants. One such plant is already working in Norway in the vicinity of existing oil and gas fields; another is in Bombay, India. We have used CORMIX 1 and 2 for various possibilities of disposal through single and multi-port diffusers at different combinations of angle of release, and a comparative study has been made with the results of other available models.

The data on gas composition for the three power plants in Brunei Darussalam used in this study are only approximate values. Based on the composition estimates given above, a simple calculation gives the total amount of CO_2, SO_2 and other

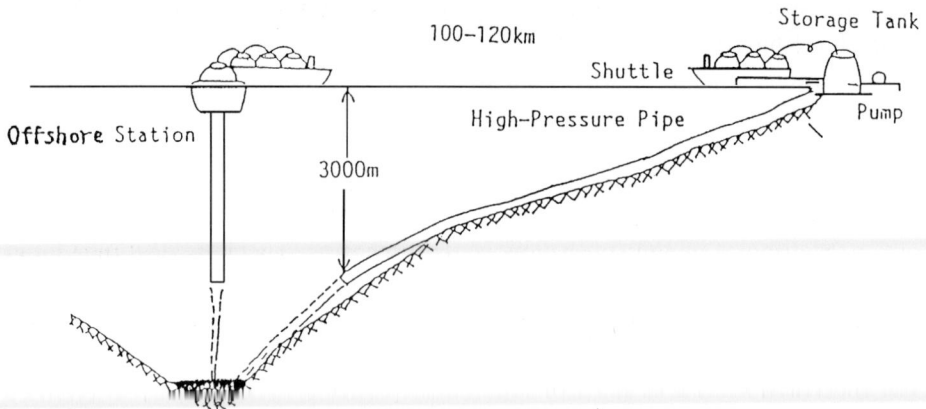

Figure 3
Basic concept of disposal systems to ocean.

components of the exhaust flue gas. After sequestering the SO_2 from the flue gas (HAUGEN and DRANGE, 1992; HAROGH *et al.*, 1991; ISHITANI *et al.*, 1996), one can calculate the amount of seawater required for CO_2 dissolution and finally arrive at the total amount of effluent that can be disposed into the sea. Figure 4 describes the variation of compressed gas density in seawater. This diagram shows the density of

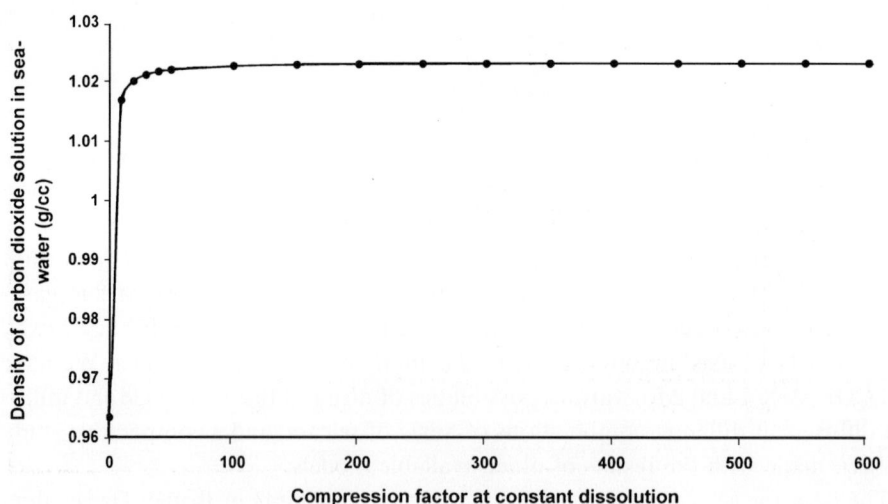

Figure 4
Density variation of CO_2 in seawater with compression.

the solution increasing and becoming more negatively buoyant as its density exceeds that of seawater.

Results and Discussion

The main gaseous pollutants emitted from the power plants are CO_2 and SO_2 which are readily converted into sulphate and carbonates in the sea. Seawater already contains enormous amounts of sulphates, and this process adds a small fraction as compared to the amount already present. It has been estimated that if all the sulphur present in the atmosphere is dumped into the sea, it will form an additional thin layer of sulphur of the order of the thickness of paper.

The emissions estimates were used as input parameters into the program CORMIX 1, for different angles of discharge along the direction of the main current of the sea (x axis). A comparative study of the results reveals that the depth within the near-field region decreases sharply with the increase in the angle of discharge (Figures 5a, b, c). The effect of increasing the ambient velocity is to decrease the axial distance required to arrive at the same concentration.

The corresponding Figures 6 (a, b, c) describe the results for dilution in the near- and far-field regions. Similar conclusions are drawn in support of concentration, i.e., for obtaining the same dilution, increasing ambient velocity requires shorter distances.

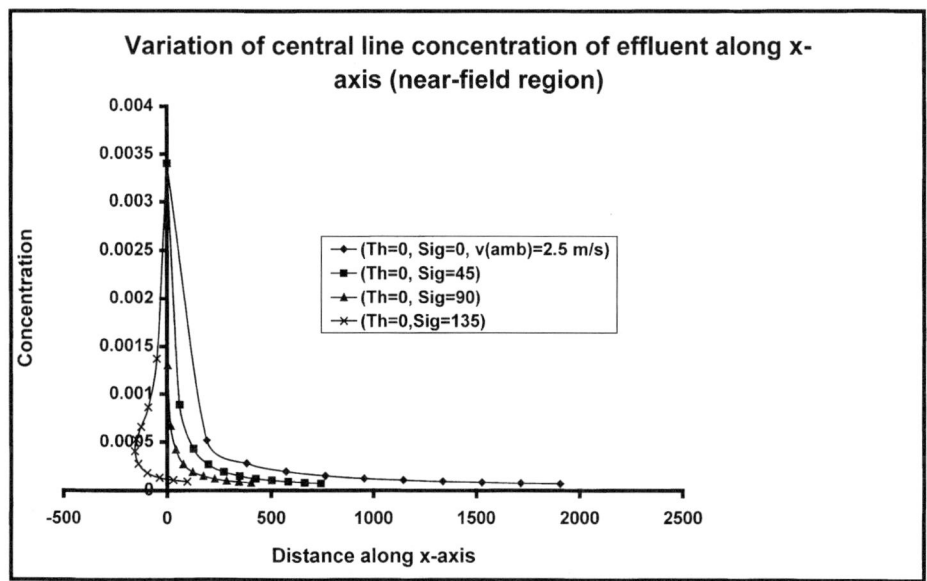

Figure 5

Variations of central line concentration with axial distance for different σ. (a): For velocity $v = 2.5$ m/s. (b): For velocity $v = 4$ m/s. (c): For velocity $v = 5$ m/s.

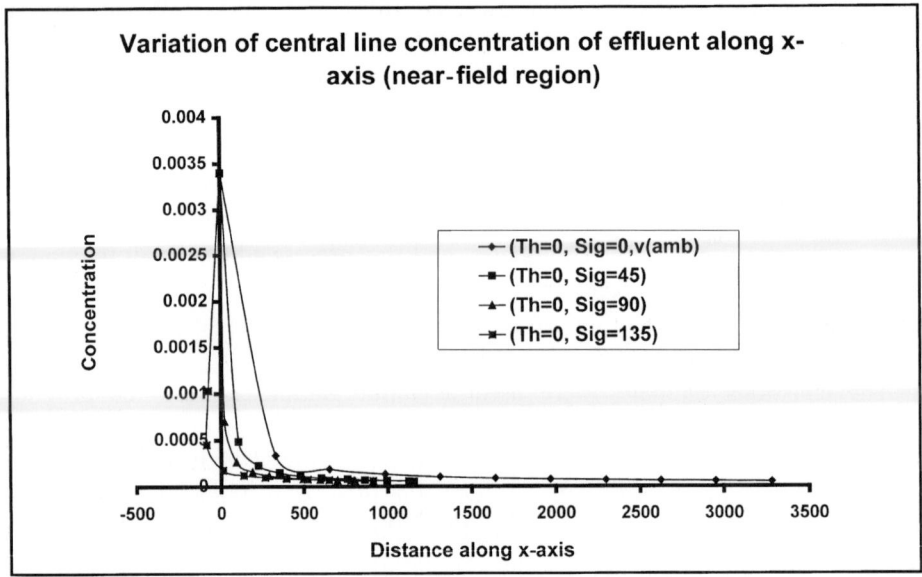

Figure 5b

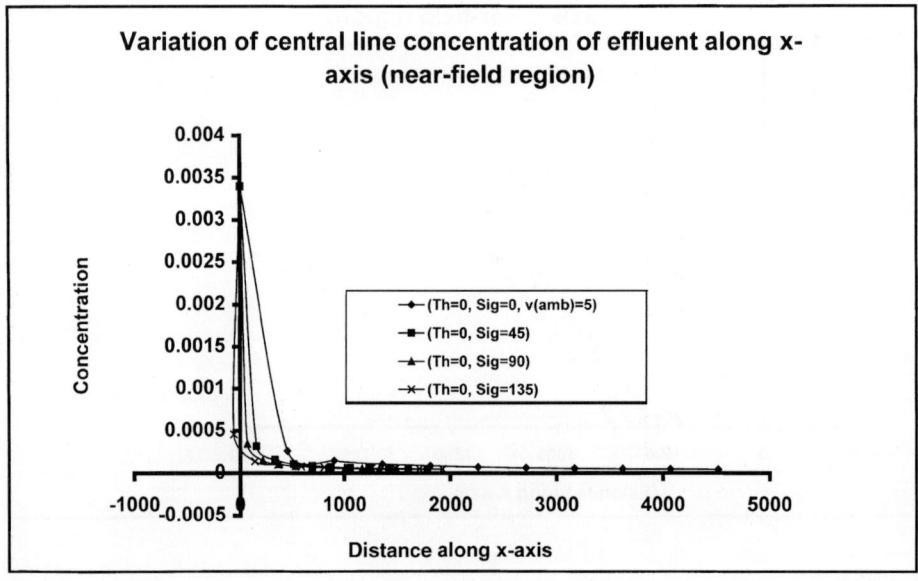

Figure 5c

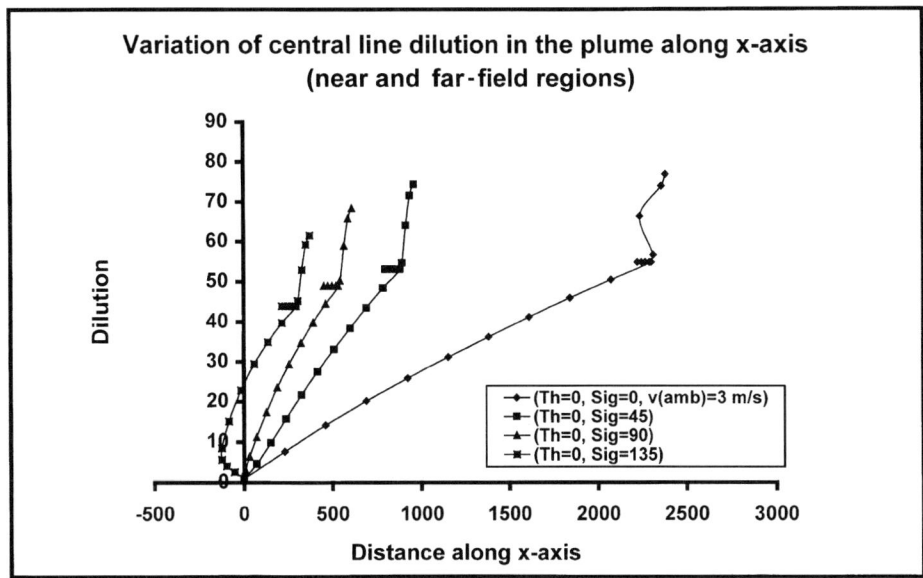

Figure 6
Variations for central line dilution with axial distance for different σ. (a): For velocity $v = 3$ m/s. (b): For velocity $v = 4$ m/s. (c): For velocity $v = 5$ m/s.

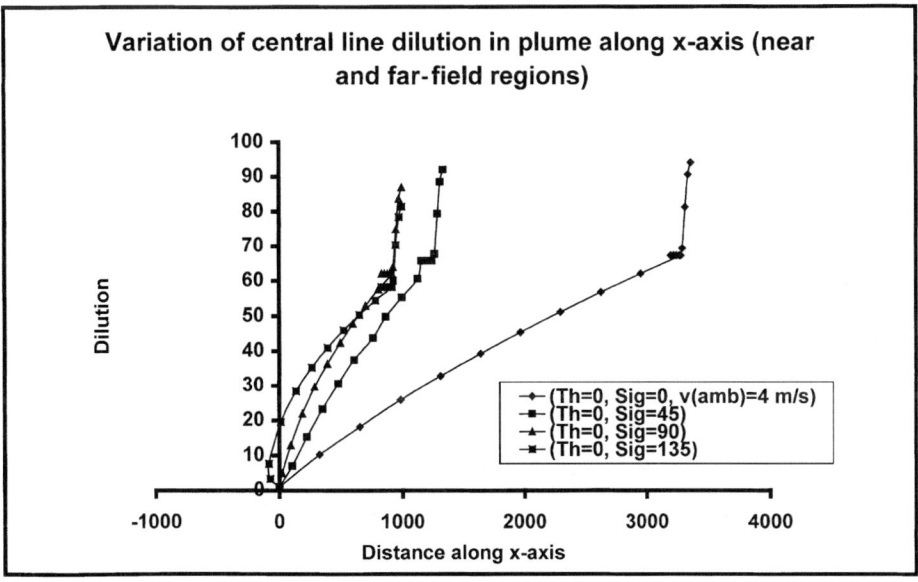

Figure 6b

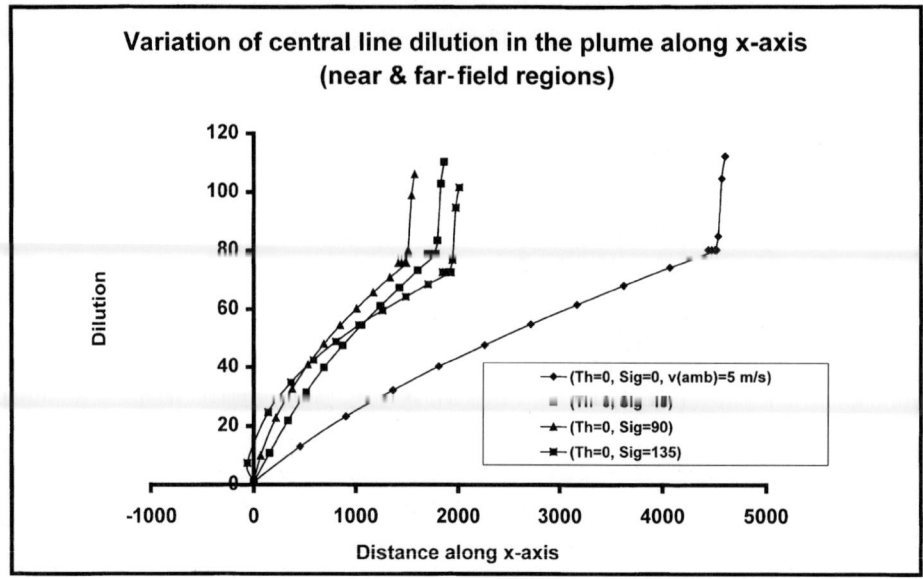

Figure 6c

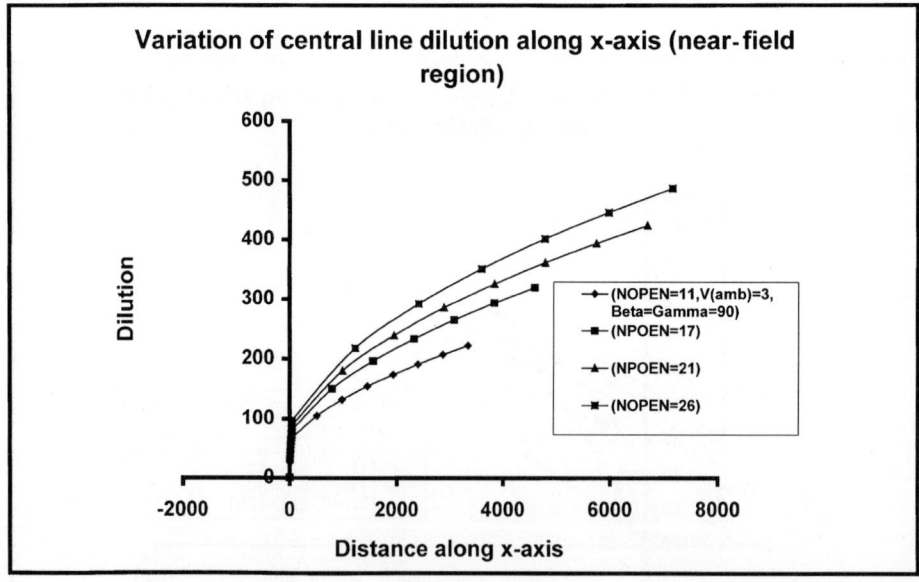

Figure 7

Variations of central line dilution with axial distance for different σ and number of ports. (a): For velocity $v = 3$ m/s. (b): For velocity $v = 4$ m/s. (c): For velocity $v = 5$ m/s.

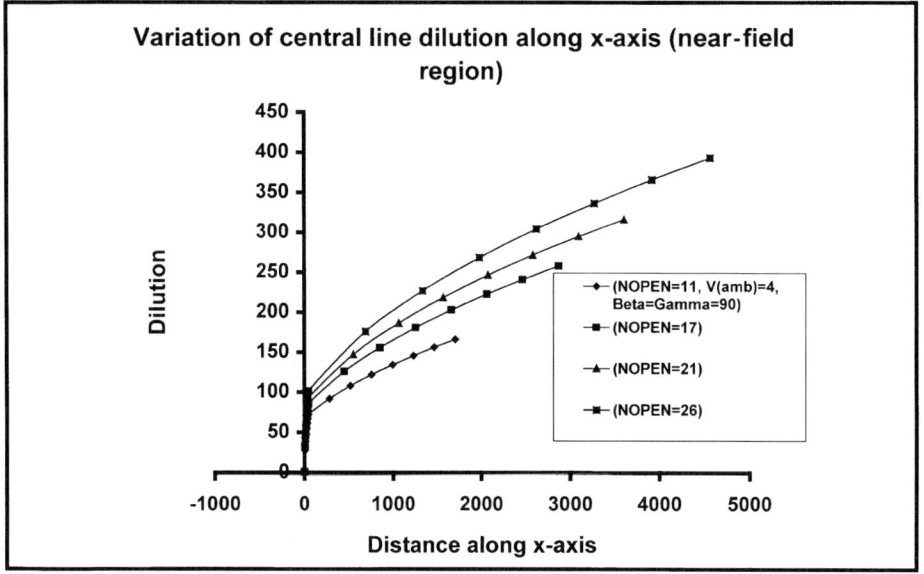

Figure 7b

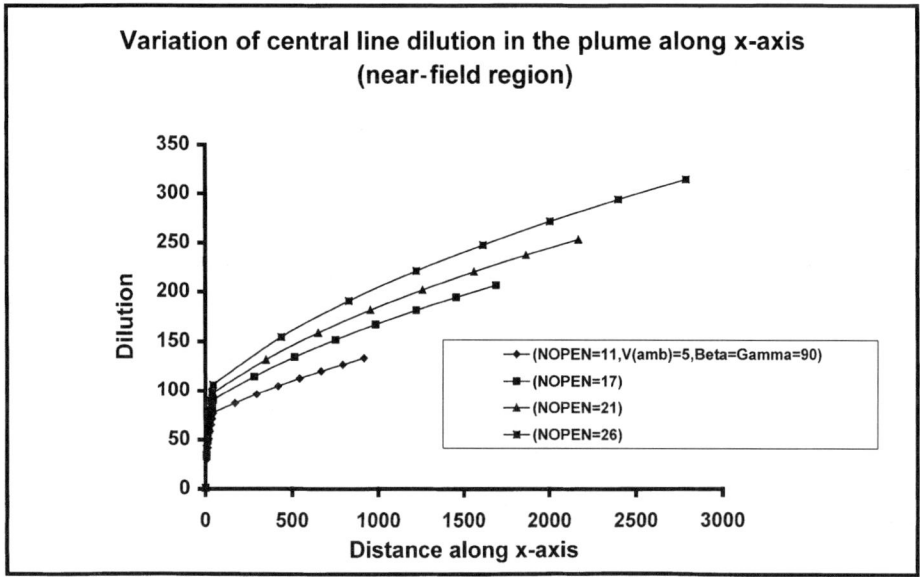

Figure 7c

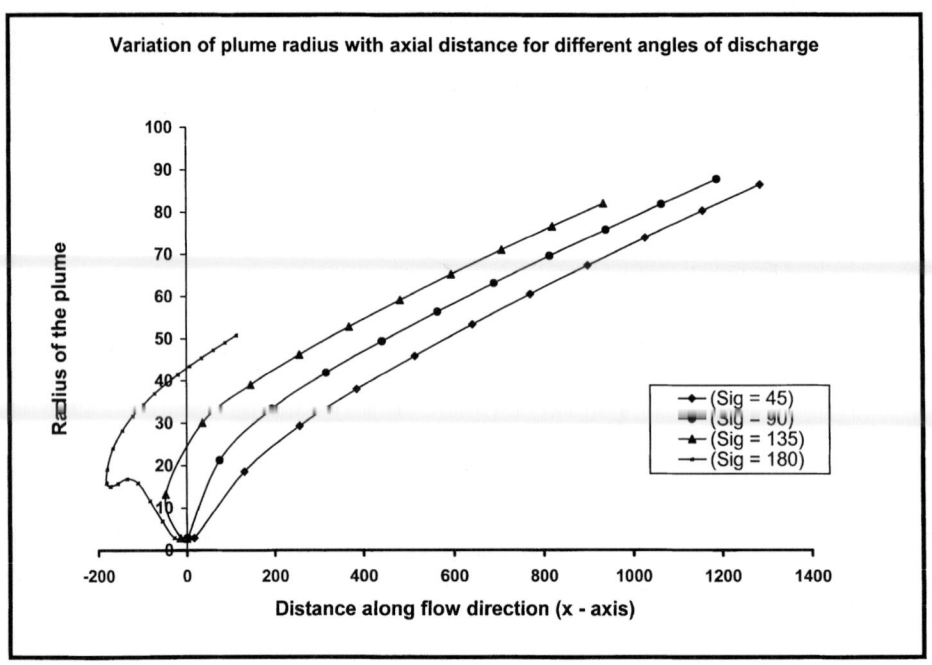

Figure 8
Variation of plume radius with axial distance for different σ.

Figures 7(a, b, c) depict the central line dilution along the axial direction with the number of ports for three different ambient velocities: (a) $v = 3$ m/s, (b) $v = 4$ m/s, and (c) $v = 5$ m/s.

It is concluded from these figures that increasing the number of ports from 11 to 26 reduces the axial distance to arrive at the same dilution. The effect of increasing ambient velocity remains the same as in the earlier two cases discussed above. Figure 8 describes the variation of plume radius with axial distances for different angles of discharge sigma. This figure concludes that at each point on the central line of the growing plume, it is diverging. This figure also depicts how well the CO_2 is spread, as carbonates, horizontally in the sea bottom.

A comparative study reveals that if all the ports on a diffuser are directed towards the main current, the desired dilution is reached in a relatively shorter distance from the disposal site.

Acknowledgement

Financial support for this work through the project 'Costal Water Dynamics' from University of Brunei Darussalam is greatly acknowledged.

REFERENCES

ABRAMS, J. Z., ZACZEK, S. J., BENZ, A. D., and AWERBUCH, L. (1988), *Use of Seawater in Sea-water Desulphurisation*, J. Air Pollution Control Assoc. *38*, 969–974.

AKAR, P. J. and JIRKA, G. H. (1991), *CORMIX 2: An Expert System for Hydrodynamic Mixing Zone Analysis for Conventional and Toxic Multiport Discharges*, EPA/Report No. EPA/600/3-90/073, ERL–U.S. Environmental Protection Agency, Athens, Georgia, 1–92.

ARNOLD, D. S., BARNET, D. A., and RSOM, R. H. (1982), CO_2 *can be Produced from Flue Gas*, Oil and Gas J. *80*, 130–136.

BAES, C. F. Jr., BEALL, S. E., and LEE, D. W. *The collection disposal and storage of carbon dioxide*. In *Interaction of Energy and Climate* (eds. Bach, W., Pankrath, J., and Williams, J.) (D. Reidel Publishing Co., Boston MA 1980) pp. 495–519.

BAUMGARTNER, D. J., FRICK, W. E., and ROBERTS, P. J. W. (1994), *Dilution Models for Effluent Discharges*, USEPA Report (Third Edition), 1–189.

DONEKER, R. L. and JIRKA, G. H. (1990), *Hydrodynamic Classification of Submerged Single Port Discharges*, J. Hydraulic Engin., ASCE, vol. 117, no. 9. 1095–1112.

FAN, L. N. and BROOKS, N. H. (1969), *Numerical Solutions of Turbulent Buoyant Jet Problems*, W. M. Keck Laboratory of Hydraulics and Water Resources, Report KH-R-18, California Institute of Technology, Pasadena, California, 1–94.

FOEL, W., GREEN, C., AMANN, M., BHATTACHARYA, S., CARMICHAEL, G., CHADWICK, M., HETTLINGH, J. P., HORDIJK, L., SHAH, J., SHRESTHA, R., STREETS, D., and ZHAO, D. (1995), *Energy Use, Emissions and Air Pollution Reduction Strategies in Asia, Water, Air and Soil Pollution*, Special Issue for Acid Rains *85*, 2283–2288.

GARCIA, R. and RODRIGUEZ, J. J., (1997), *HydroTrack: A visual tool to simulate hydrodynamics, Pollutant and Sediment transport in rivers and Coastal regions*. In *Hydroinformatics'98* (eds. Babovic, V. and Larsen, L. C.) pp. 583–589.

HAGEN, R. I. and KOLDERUP, H. (1979), *Flue Gas Desulphurisation Pilot Study Phase –1: Survey of Major Installations – Seawater Scrubbing and Gas Desulphurisation Process*, NATO-CCMS Study Phase –1, Appendix 25-D, U.S. Department of Commerce, National Technical Information Service, PB – 299005, D1–D15.

HAUGEN, P. M. and HELGE DRANGE (1992), *Sequestering CO_2 in Deep Sea by Shallow Injection*, Nature *357*, 318–320.

HAROGH, D. GOLOMB and ZEMBA, S. (1991), *Feasibility, Modeling and Economics of Sequestering Power Plant CO_2 Emission in Deep Sea*, Environ. Progress *10*(1), 64–72.

ISHITANI, H. and MATUHASI, R. *Possibility and Evaluation of CO_2 disposal in Deep Ocean*. In *Global Environmental Security*, (eds. Yujuki Sujiki, Kajuhiro Ueta Shunsuke Mori) (Springer–Verlag, Berlin, Heidelberg, Chapter 3, 79–129.

JACQUES, G. and GANOULIS (1994), *Engineering Risk Analysis of Water Pollution*, VCH Weinheim. N.Y., 1–330.

JONES, G. R. (1990), *CORMIX 3: An Expert System for the Analysis and Prediction of Buoyant Surface Discharges*, M.Sc. Thesis Cornel University, Ethaca, N.Y., 1–110.

KHAN, F. I. and ABBASI (1997), *Application of Recently Modified Plume Path Theory on Dispersion of Heavy Gases*, Environ. Res. Forum, vols. 7 and 8, 449–460.

LEE, J. H. and CHEUNG, V. (1990), *Discussion on Marine Outfall Design–Computer Models for Initial Dilution in a Current and Reply by Authors*, Proc. Instn. of Civ. Engineers, Part I, *88*, 481–486.

LEONG, T. S., LEONG, Y. K., HO, S.C., KHOO, K. H., KAM, S. P., SULAIMAN HANAPI, WONG, T. M., and LEGORE, RICHARD S. (1985), *An Environmental Baseline Study of the Macrobenthos in the Vicinity of Crude Oil Terminal, Sabah (Labuan) and Sarawak (Bintulu and Lutong) and Selected Off-shore Platforms*, Sabah Shell, Bhd./Sabah Shell Petroleum Co., 1–433.

LEONG, T. S., LEONG, Y. K., HO, S. C., KHOO, K. H., KAM, S. P., SULAIMAN HANAPI, WONG, T. M., and LEGORE, RICHARD S. (1987), *Effects of Crude Oil Terminal on Tropical Benthic Communities in Brunei*, Marine Pollution Bulletin *18*, 31–35.

LEONG, T. S., HAUT, K. K., CHAN, F. M., and ENG. L. P. (1995), *The Environmental Status of Sungai Bera, Seria, Brunei Darussalam*, Final Report (No. 005141/HSE/94) submitted to Brunei Shell Pertonium Company Seria, Brunei Darussalam, 1–103.

LORIN, R. D. (1999), *Fundamentals of Environmental Discharge Modeling*, CRC Press, Boca Raton, London New York Washington, Chapter 4, 53–80.

MAIER-REIMER, E. (1982), *On tracer methods in Computational Hydrodynamics*, Chap. 9. In *Engineering Applications of Computational Hydraulics*, vol. 1 (eds. Abbott, M. B. and Cunge, J. A. Pitman), 198–216.

MARCHETTI, C. *On Geoengineering and* CO_2 *Problem in Climatic Change*, I D (Reidel Publishing Company, Dordrecht, Holland, 1977), 1, pp. 59–68.

MARCHETTI, C. *Constructive Solutions to* CO_2 *Problem in Climate Change in "Man's Impact on Climate"* (eds. Bach, W., Pankrath, J., and Kellog, W.) (Elsevier Scientific Publications) pp. 85–107.

MUELLENHOFF, W. P., SOLDATE, Jr., A. M., BAUMGARTNER, D. J., SCHULDT, M. D., DAVIS, L. R., and FRICK, W. E. (1985), *Initial Mixing Characteristics of Municipal Discharges*, vol. I, EPA-60/3-85-073a, U. S. Environmental Protection Agency, Newport, Oregon, 1–42.

MUSTACCHI, C., ARMENATE, P., and CANA, V. (1979), *Carbon dioxide Removal from Power Plant Exhausts*, Environ. International *2*, 453–456.

OOMS, G. (1972), *A New Method for the Calculation of Plume Path of Gases Emitted by Stack*, Atmos. Environ. *6*, 899–905.

PAULEY, C. R. (1984), *Recovery of* CO_2 *from Flue Gas*, Chem. Engin. Progress *80*, 59–64.

RADOJEVIC, M. (1989), *The Use of Seawater for Desulphurisation*, Environ. Tech. Lett. *10*, 71–76.

RADOJEVIC, M. and TRESSIEDER, D. A. (1992), *Disposal of* SO_2 *in Seawater*, Nature *356*, 391–393.

ROBERTS, P. J. W., SNYDER, W. H., and BAUMGARTNER, D. H. (1989), *Ocean Outfalls I: Submerged Wastefield Formation; II: Spatial Evaluation of Submerged Wastefields; III: Effect of Diffuser Design on Submerged Wastefields*, J. of Hydraulic Engin. ASCE *115*(1), 1–70.

TANDON, P. N. and RAMALINGAM, P. *Alternative models for the disposal flue gases in deep sea*. In Proc. Second Internat. Conf. on *Environmental Coastal Regions* (ed. Brebbia, C. A.) (WIT Publications 1998) *1*, pp. 185–196

TANDON, P. N. (1999), *Mathematical and Computer Models for the Disposal of Effluents in Deep Sea with Possible Applications to Brunei Darussalam*, Science Bull., *1*, 37–44.

TANDON, P. N. *Mathematical and Computer models for the disposal of liquid effluents in Deep Sea with Applications to Brunei Darussalam*. An invited article published in *Mathematical Analysis and Applications* (ed. Dwivedi, A. P.) (Narosa Publishing Company 2000) pp. 136–150.

TSANIS, I. K., VALEO, C., and DIAO, Y. (1994), *Comparison of Near-field Mixing Zone Models for Multiport Diffusers in Great Lakes*, Canadian J. Civil Engin. *21*(1), 141–155.

TSANIS, I. K. and VALEO, C. (1994), *Mixing Zone Models for the Submerged Discharge*, Computat. Mechanics Publ., Southampton Boston, 1–168.

WOOD, I. R. (1993), *Asymptotic Solutions and Behaviour of Outfall Plumes*, J. Hydraulic Engin. *119*(5), 555–580.

(Received May 23, 2000, accepted January 21, 2001)

To access this journal online:
http://www.birkhauser.ch

Pure appl. geophys. 160 (2003) 419–427
0033–4553/03/020419–09

❙ Pure and Applied Geophysics

Thermodynamic Characteristics of the Subcloud Layer Affecting Haze Dispersion Along the West Coast of Borneo

Shih-Ang Hsu[1]

Abstract — Analyses of 5-year (1994–1998) monthly data from 4 upper-air and 7 surface meteorological stations along the west coast of Borneo showed that throughout the year the subcloud layer is convectively unstable, the lifting condensation level (LCL) is below 400 m, and the ventilation factor (VF) is within 2000 m^2 s^{-1}. Spatial distribution of LCL and VF indicated that the lowest values are found in the south and the highest in the north. However, the VF values were all in the "bad" category for pollution dispersion.

Key words: Haze, Borneo, mixing height, ventilation factor.

1. Introduction

Haze, according to GEER (1996), is defined as the suspension in the atmosphere of extremely small, dry aerosols which individually are invisible to the naked eye, but collectively give the sky an opalescent appearance; sometimes used in reference to the associated reduction in visibility. Haze usually indicates subsaturated air, whereas fog or mist indicates saturated conditions. The haze layer is usually bounded at the top by a temperature inversion (or cloud layer), and frequently extends downward to the ground.

Clearing and burning of land for agricultural use has contributed to the haze problem in Southeast Asia which has become a regional disaster. The purpose of this study is to provide thermodynamic information related to the characteristics of atmospheric mixing height and the ventilation factor along the west coast of Borneo.

2. Characteristics of the Equivalent Potential Temperature

In atmospheric thermodynamics, the parameter known as the equivalent potential temperature, θ_e, is a very useful parameter to combine both the temperature

---

[1] Coastal Studies Institute, 308 Geoscience Building, Louisiana State University, Baton Rouge, Louisiana 70803, U.S.A. E-mail: sahsu@antares.esl.lsu.edu

and moisture content, using the dewpoint for the measure of moisture, into one value so that (see, e.g., WALLACE and HOBBS, 1977, and for a graphical description of θ_e, see, e.g., JACOBSON, 1999)

$$\theta_e = \theta \exp\left(\frac{Lw}{C_p T}\right) \tag{1}$$

$$\theta = T\left(\frac{1000}{P}\right)^{0.286} \tag{2}$$

$$w = 0.622 \frac{e}{P - e} \tag{3}$$

and from HSU (1988),

$$e = 6.1078 \times 10^{[7.5 T_{\text{dew}}/(237.3 + T_{\text{dew}})]} \tag{4}$$

where θ is the potential temperature in degrees Celsius, L is the latent heat of condensation in J kg^{-1}, w is the mixing ratio in g kg^{-1}, C_p is the specific heat at constant pressure in J kg^{-1} K^{-1}, T is the temperature in degrees Celsius, P is the pressure in mb, e is the vapor pressure in mb, and T_{dew} is the dewpoint in degrees Celsius.

On the basis of "Monthly Climatic Data for the World" published by U.S. NOAA, θ_e values are computed from four upper-air stations along the west coast of Borneo. They are at Kuching (see Fig. 1 and abbreviated in Table 1 and Fig. 2 as Kuch), Bintulu (Bint), Brunei (Brun), and Kota Kinabalu (Kota). In order to understand the thermodynamics along the coast, θ_e at the surface and 850 mb level are analyzed. An example is provided in Table 1. It is clear that throughout the year over our study area $\partial \theta_e / \partial Z < 0$, indicating the atmospheric boundary layer is convectively unstable so that clouds are often formed. In order to see that our $\partial \theta_e / \partial Z$ values are reasonable, the mean tropical condition in September, as plotted in ANTHES (1982, p. 29) is compared to the values listed in Table 1. The mean values for both cases are approximately 9 K, indicating our values in Table 1 are consistent with other regions in the tropics. This is also supported by the long-term ship observations in the South China Sea off the west coast of Borneo that the average cloud cover for the year was 5.9 (out of 8.0, or 74%), with the minimum of 4.4 (55%) in May and maximum of 7.0 (88%) in November. These data are also provided in Table 1.

3. Characteristics of Mixing Height

Since the cloud base is usually the lifting condensation level, LCL (see, e.g., GARRATT, 1992), and the layer between the surface and LCL is the subcloud layer or the mixed layer from the viewpoint of air pollution meteorology, the LCL is computed as follows:

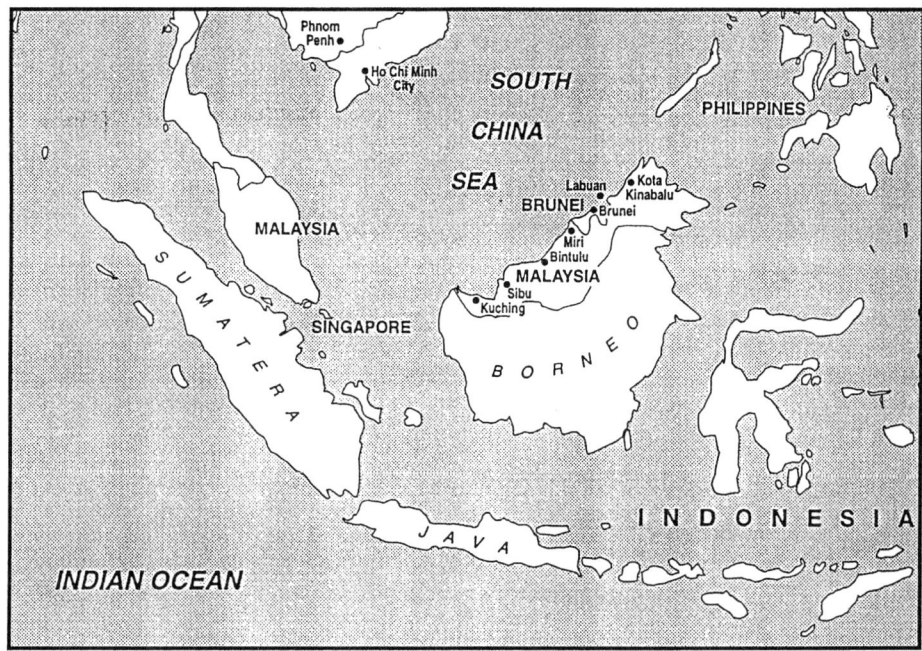

Figure 1
The study area and station locations.

Table 1

Characteristics of the convectively unstable layer ($\partial\theta_e/\partial z < 0$) between the surface and 850 mb (i.e., $\theta_{e\,SFC} - \theta_{e\,850}$, in °C) along the west coast of Borneo in 1997

Month	Kuch	Bint	Brun	Kota	Total Cloud Amount*		$U^{**}_{20\,m}$
Jan.	8	9	8	8	6.1	76	5.3
Feb.	8	8	7	6	6.9	86	5.0
Mar.	6	8	11	9	6.0	75	3.9
Apr.	5	7	10	7	5.9	74	2.5
May.	6	5	11	8	4.4	55	2.4
Jun.	7	8	9	8	5.1	64	3.0
Jul.	7	8	9	6	5.0	63	3.1
Aug.	6	11	14	12	5.9	74	3.4
Sep.	9	9	10	8	5.8	73	3.1
Oct.	5	7	8	6	6.5	81	3.7
Nov.	5	7	7	6	7.0	88	3.3
Dec.	4	5	9	7	6.3	79	4.8

* Based on long-term (1867–1973) data offshore from U.S. NAVAL WEATHER SERVICE COMMAND (1975) (in eighths in the first column and percent in the second).
** Same as * but for wind speed measurements (in m s^{-1}) at approximately 20 m above the mean sea level.

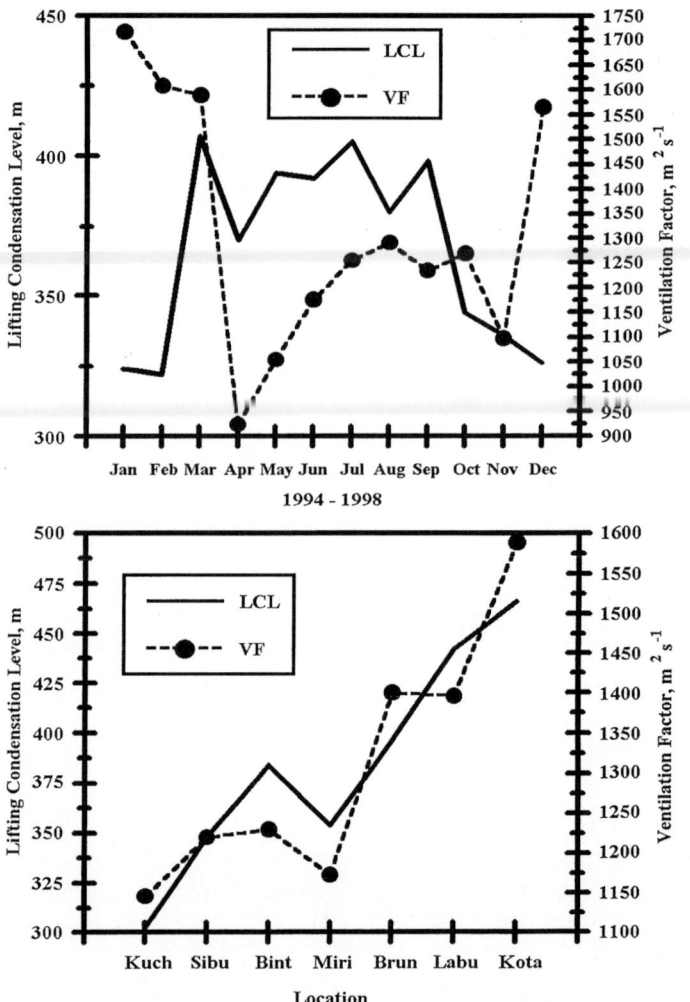

Figure 2
Temporal (monthly) and spatial (alongshore) variations of the lifting condensation level (LCL) and the
ventilation factor (VF) on the west coast of Borneo.

For the dry adiabatic lapse rate (see, e.g., HSU, 1988, pp. 23–24),

$$\frac{dT_{air}}{dz} = -\frac{g}{C_p} = -\frac{0.98\,\mathrm{K}}{100\,\mathrm{m}} \tag{5}$$

where z is the altitude. In a well-mixed atmospheric boundary layer from surface to
the lifting condensation level (LCL: HSU, 1988, pp. 26–27),

$$T_{air\,LCL} - T_{air\,SFC} = -\frac{0.98\,\mathrm{K}}{100\,\mathrm{m}} H_{LCL} \tag{6}$$

where H_{LCL} is the height of the LCL. For dewpoint lapse rate (see, e.g., MCILVEEN, 1986, p. 151),

$$\frac{dT_{dew}}{dz} = -\frac{gR_v}{L_T R} T_{dew} \; , \tag{7}$$

where R_v and R are the gas constants for water vapor and dry air, respectively. For typical low tropospheric T_{dew} between 283 K (or 10 °C) and 293 K (or 20 °C), Eq. (7) becomes approximately

$$\frac{dT_{dew}}{dz} = -\frac{0.18 \, K}{100 \, m} \tag{8}$$

or

$$T_{dew \; LCL} - T_{dew \; SFC} = -\frac{0.18 \, K}{100 \, m} H_{LCL} \; . \tag{9}$$

At the LCL, $T_{dew \; LCL} = T_{LCL}$, and Eqs. (9)–(6) become

$$T_{air \; SFC} - T_{dew \; SFC} = \frac{0.80 \, K}{100 \, m} H_{LCL} \tag{10}$$

$$\therefore H_{LCL} = 125(T_{air \; SFC} - T_{dew \; SFC}) \tag{11}$$

where H_{LCL} is in meters and the dewpoint depression at the surface is in degrees Celsius. Therefore, we have

$$\text{Mixing Height} = \text{LCL} = 125(T_{air} - T_{dew}) \; . \tag{12}$$

From the surface observation section in the same NOAA publication, monthly values of both T_{air} and e (and thus dewpoint from Eq. (4)) are given at three stations in addition to the four sites named previously. They are at Sibu, Miri, and Labuan (Labu) (see Fig. 1). Our results of LCL analysis are shown in Figure 2. In the top panel, the monthly variations of LCL is provided. It can be seen that between October and February, LCL along the west coast of Borneo is less than 350 m whereas in other months it is slightly higher, around 375 m. Spatial distribution of LCL is shown in the lower panel of Figure 2. The lowest LCL values are found in the southern region, the second minimum is found in the Miri area, and the highest values are in the north of our study area.

4. Characteristics of the Ventilation Factor

The ventilation factor is defined as the product of wind speed and mixing height (see, e.g., EAGLEMAN, 1996). For our purpose, the mixing height is the LCL and the wind speed is the mean speed in the mixing layer. In other words,

$$\text{Ventilation Factor} = \text{LCL} \times \text{Wind Speed in Mixed Layer} \; . \tag{13}$$

While the LCL can be estimated from Eq. (12), the wind speed in the mixed layer must be specified. This is done as follows:

From Figure 2 (lower panel), the LCL can be as low as 300 m. Therefore, we use 150 m as the minimal height for the mean wind speed in the mixed layer. According to DAVENPORT (1965) (see also Plate, 1971, P. 40), the wind speeds at 150 m, U_{150m}, for both onshore and offshore are related through that at the gradient wind level, $U_{gradient}$, as

$$U_{150\,m\text{ onshore}} = 0.76\, U_{gradient} \tag{14}$$

$$U_{150\,m\text{ offshore}} = 0.91\, U_{gradient}\ . \tag{15}$$

Dividing Eq. (14) by Eq. (15), we have

$$U_{150\,m\text{ onshore}} = 0.84\, U_{150\,m\text{ offshore}}\ . \tag{16}$$

Since the anemometers on ships are usually located at 20 m (see, e.g., ROLL, 1965 and KRAUS, 1972), we set $U_{20\,m}$ for the ship observations.

From HSU (1988) and HSU et al. (1994), for offshore applications the power-law wind profile at 150 m and 20 m heights may be employed so that

$$\frac{U_{150\,m\text{ offshore}}}{U_{20\,m\text{ offshore}}} = \left(\frac{150\,m}{20\,m}\right)^{0.1} = 1.2 \tag{17}$$

From Eqs. (16) and (17), we obtain

$$U_{150\,m\text{ onshore}} = 0.84 \times 1.2\, U_{20\,m\text{ offshore}}$$
$$= 1.0\, U_{20\,m\text{ offshore}} \tag{18}$$

or

$$U_{150\,m\text{ onshore}} = \text{Wind Speed Observed By Ships}\ . \tag{19}$$

Therefore, we use wind speed measurements from ships as provided in the U.S. NAVAL WEATHER SERVICE COMMAND (1975) without further height adjustment for the mean wind speed in the mixed layer. The wind speeds at 20 m, U_{20m}, are provided in Table 1. With both $U_{20\,m}$ and the LCL, the ventilation factor can be estimated from Eq. (13). Our results are shown in Figure 2. It is found that the highest VF is in January and the lowest in April. If we apply the VF category from the State of Colorado Department of Health in Denver, when VF ≤ 2000 m^2 s^{-1}, the pollution dispersion category is designated as "bad" (see Table 2). Since monthly VF values are well within this value, it is inferred that the air pollution dispersal index in this area is also "bad." This is to explain why widespread breathing problems are experienced in southeast Asia, since the likelihood for haze dispersion is poor.

Table 2

Pollution dispersion forecast categories related to atmospheric ventilation (product of wind speed and mixing depth) (after EAGLEMAN, 1996, based on the air pollution dispersal index used by the State of Colorado Dept. of Health in Denver)

Pollution dispersion	Ventilation (m^2/sec)
Bad	0–2000
Fair	2001–4000
Good	4001–6000
Excellent	6001 or more

Spatial distribution of VF is shown in the lower panel of Figure 2. It is seen that the lowest values of VF are found in the southern region, the second minimum is in the Miri area, and the highest values are in the north of the study area. However, the VF remains in the "bad" category from a pollution dispersal point of view.

5. Limitations of the Approach

Because this study contains both theoretical and empirical equations, there are two limitations to this approach. The first is the accuracies of the measurements and the second the diurnal variations in meteorological parameters. The measurement error is discussed briefly as follows:

According to MAZZARELLA (1985), for standard WMO stations (surface measurements), sensor resolutions for air dry-bulb and wet-bulb temperatures are $\pm 0.1\,°C$; for relative humidity, 1%; for wind speed 1 kt (or approximately 0.5 m s^{-1}); and for wind directions, $10°$. According to LALLY (1985), for standard WMO stations (upper-air measurements), for pressure measurements, the accuracy of the baroswitch element is ± 1 mb near the surface, ± 2 mb in the 500 mb region, and ± 1 mb at 10 mb. For temperature measurement, the acceptable accuracy is $\pm 0.5\,°C$. For humidity measurement, the error may be as much as 10%.

The diurnal variations are discussed as follows. According to ATKINSON (1971), above the layer affected by surface heating, the diurnal variation of temperature in the tropical atmosphere is very small being, as shown by radiosonde soundings, on the order of 1 °C or less with the daylight soundings registering slightly higher temperatures. The diurnal variation of moisture aloft, however, is somewhat greater, especially in the mid-troposphere over continental areas. For example, at Saigon, Vietnam, the mean 12 Z dewpoint for 950 mb to 250 mb ranged from 1.4 °C to 3.6 °C greater than those at 00 Z, the major differences occurring above 500 mb.

Because this study deals with the subcloud layer near the surface, and if one accepts the limitations of measurement errors and diurnal variations in temperature and dewpoint, then this study is useful from operational viewpoints, such as assessing the air quality potential as reported here.

6. Conclusions

Using monthly data from 4 upper-air and 7 surface meteorological stations from 1994 through 1998 along the west coast of Borneo, the thermodynamic characteristics of the subcloud layer affecting the haze dispersion are investigated. It is found that throughout the year the atmospheric boundary layer is convectively unstable so that clouds are usually formed. Characteristics of the mixing height as represented by the lifting condensation level show that between October and February, heights are less than 350 m whereas in other months they are slightly higher, around 375 m. Characteristics of the ventilation factor indicate the highest values in January and the lowest in April. However, since they are all below 2000 m^2 s^{-1}, the dispersal ability in our study area is categorized as "bad" based on the criteria set by the state of Colorado Department of Health in Denver. Characteristics of the spatial distribution show that lowest values of both mixing height and ventilation factor are found in the south and the highest in the north. Since this study is based on monthly averages, the day to day variations must be investigated so that forecasts of the air pollution index for haze can be improved for the protection of the public health in this region.

Acknowledgment

This study is supported in part by an endowed professorship from Dr. Eric Abraham to the Center for Coastal, Energy and Environmental Resources of Louisiana State University.

REFERENCES

ANTHES, R. A., *Tropical Cyclones, Their Evolution, Structure and Effects* (American Meteorological Society, Boston, MA, 1982) 208 pp.

ATKINSON, G. D., *Forecasters' Guide to Tropical Meteorology* (Air Weather Service, United States Air Force, Technical Report 240, 1971) pp. 5–10 to 5–14.

DAVENPORT, A. G., *The relationship of wind structure to wind loading*, In *Wind Effects on Buildings and Structures* (National Physical Laboratory, Symposium No. 16) (Her Majesty's Stationary Office, London, 1965) pp. 54–102.

EAGLEMAN, J. R., *Air Pollution Meteorology* (Trimedia Publishing Co., Lenexa, KS, 1996) 258 pp.

GARRATT, J. R., *The Atmospheric Boundary Layer* (Cambridge University Press, Cambridge, 1992) 316 pp.

GEER, I. W., (ed.), *Glossary of Weather and Climate* (American Meteorological Society, Boston, MA, 1996) 272 pp.

HSU, S. A., *Coastal Meteorology* (Academic Press, San Diego, CA, 1988) 260 pp.

HSU, S. A., MEINDL, E. A., and GILHOUSEN, D. B. (1994), *Determining the Power-law Wind-profile Exponent Under Near-Neutral Stability Conditions at Sea*, J. Appl. Meteor. *33*, 757–765.

JACOBSON, M. Z., *Fundamentals of Atmospheric Modeling* (Cambridge University Press, 1999) 656 pp.

KRAUS, E. B., *Atmosphere–Ocean Interaction* (Clarendon Press, Oxford, 1972) 275 pp.

LALLY, V. E., *Upper air in situ observing systems*, Chapter 8 in *Handbook of Applied Meteorology*, (D. D. Houghton, ed.) pp. 352–360 (John Wiley & Sons, New York, 1985).

MAZZARELLA, D. A., *Measurements Today*, Chapter 5, in *Handbook of Applied Meteorology* (D. D. Houghton, ed.), pp. 283–328 (John Wiley & Sons, New York, 1985).

MCILVEEN, J. F. R., *Basic Meteorology* (Van Nostrand Reinhold, (UK), Berkshire, 1986) 457 pp.

PLATE, E. J., *Aerodynamic Characteristics of Atmospheric Boundary Layer* (U.S. Atomic Energy Commission, Oak Ridge, TN, 1971) 190 pp.

ROLL, H. U., *Physics of the Marine Atmosphere* (Academic Press, New York, 1965) 426 pp.

U.S. NAVAL WEATHER SERVICE COMMAND, *Summary of Synoptic Meteorological Observations* (*SSMO*), Indonesian Coastal Marine Areas, vol. 1, Area 7 – Sarawak (National Climatic Center, Asheville, N.C., 1975) 553 pp.

WALLACE, J. M. and HOBBS, P. V., *Atmospheric Science, An Introductory Survey* (Academic Press, New York, 1977) 467 pp.

(Received February 15, 2000, accepted September 7, 2000)

To access this journal online:
http://www.birkhauser.ch

Pure appl. geophys. 160 (2003) 429–438
0033–4553/03/020429–10

© Birkhäuser Verlag, Basel, 2003

| Pure and Applied Geophysics

Three-dimensional Simulations of the Mean Air Transport During the 1997 Forest Fires in Kalimantan, Indonesia Using a Mesoscale Numerical Model

ORBITA ROSWINTIARTI[1] and SETHU RAMAN[1]

Abstract — This paper describes the meteorological processes responsible for the mean transport of air pollutants during the ENSO-related forest fires in Kalimantan, Indonesia from 00 UTC 21 September to 00 UTC 25 September, 1997. The Fifth Generation of the Pennsylvania State University-National Center for Atmospheric Research (PSU-NCAR) Mesoscale Model (MM5) is used to simulate three-dimensional winds at 6-hourly intervals. A nonhydrostatic version of the model is run using two nested grids with horizontal resolutions of 45 km and 15 km. From the simulated wind fields, the backward and forward trajectories of the air parcel are investigated using the Vis5D model.

The results indicate that the large-scale subsidence over Indonesia, the southwest monsoon low-level flows (2–8 m s^{-1}), and the shallow planetary boundary layer height (400–800 m) play a key role in the transport of air pollutants from Kalimantan to Malaysia, Singapore and Brunei.

Key words: Air quality, tropics, ENSO, long range transport, atmospheric boundary layer.

1. Introduction

El Niño/Southern Oscillation (ENSO) is one of the most prominent and important phenomenon in the tropical ocean-atmosphere system at the interannual time scale (with a range of 2 to 7 years). The oceanic component, El Niño, is associated with the anomalous warm sea-surface temperature (SST) in the eastern-central tropical Pacific Ocean. The atmospheric component, Southern Oscillation, is related to the strength of the zonal Walker circulation over the Pacific region, i.e., the difference in surface pressure between the southeast tropical Pacific and the Indonesian-Australian regions. During the ENSO events, the eastward shift of deep convection from the Indonesian-Australian regions, which manifests as low Southern Oscillation Index (SOI), often leads to abnormally dry conditions over Indonesia (ROPELEWSKI and HALPERT, 1987). Many studies have also documented that the ENSO event is linked to a weakening of the Asia-Australia monsoon (WEBSTER and YANG, 1992).

[1] Department of Marine, Earth and Atmospheric Sciences, North Carolina State University, Raleigh, North Carolina 27695-8208, U.S.A.

In the 1997/98 ENSO event, drought conditions in Indonesia that persisted from June 1997 to May 1998 resulted in large-scale forest fires, particularly in Kalimantan and Sumatera Islands. It was reported that 3 million hectares of the forests were burned in Kalimantan and 1.5 million hectares in Sumatera. The associated smoke and haze that blanketed the neighboring countries of Malaysia, Singapore, and Brunei had substantial impacts on human life and economic development in the region.

The objective of this study is to characterize the wind circulations responsible for the transport of air pollutants during forest fires in Kalimantan in September, 1997. The Defense Meteorological Satellite Program (DMSP) detected numerous fires over east and south Kalimantan on 21 September, 1997 (Fig. 1a). Figure 1b displays the associated smoke distributions observed from the Total Ozone Mapping Spectrometer (TOMS) aerosol data. It is shown that the smoke coverage with aerosol indices ≥3.5 spreads towards the neighboring countries of Malaysia, Singapore and Brunei. In this study, the Fifth Generation of the Pennsylvania State University-National Center for Atmospheric Research (PSU-NCAR) Mesoscale Model, known as MM5, is used to simulate the three-dimensional winds and the planetary boundary layer from 00 UTC, 21 September to 00 UTC 25 September, 1997. The backward and forward air mass trajectories determined from the simulated winds are examined to investigate the likely path of the mean air pollutants. The ability to anticipate the interactions between synoptic and regional meteorological processes during ENSO events in Indonesia, and improved predictions associated with the long-range transport of aerosols and pollutants would lead to better management decisions at the early stages of forest fire for future episodes.

This paper is arranged as follows: a brief description of the MM5 and the experimental design are described in section 2. Section 3 presents the model results and trajectory analyses. Conclusions of the study are given in section 4.

a)   b)

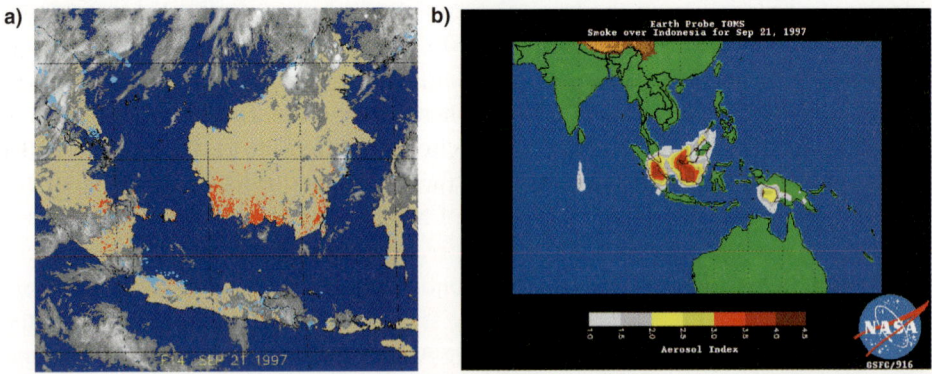

Figure 1
a) Forest fires (hotspots) in Kalimantan, Indonesia during the 1997/98 ENSO detected by the Defense Meteorological Satellite Program (DMSP) and b) associated smoke distributions observed by the Total Ozone Mapping Spectrometer (TOMS) aerosol data for 21 September, 1997.

2. Description of the Model and Experimental Design

The numerical model used in this study is a nonhydrostatic version of the MM5 (DUDHIA, 1993; GRELL et al., 1995). The model used here has 21 σ levels in the vertical (1.0, 0.999, 0.993, 0.971, 0.916, 0.888, 0.832, 0.777, 0.721, 0.666, 0.610, 0.499, 0.444, 0.388, 0.333, 0.277, 0.222, 0.166, 0.111, 0.055, 0.0) between surface and 100 hPa.

A simple-ice scheme (DUDHIA, 1989) is used for the nonconvective precipitation scheme. This scheme treats the microphysical processes of ice and snow identically to most other cloud-resolving models, except it does not include supercooled water and unmelted snow. The convective precipitation is parameterized using a Kuo-Anthes scheme (KUO, 1974; ANTHES, 1977). In this scheme, the amounts and the vertical distributions of the latent heat released and the sensible heat transported by the deep cumulus clouds are expressed solely in terms of the temperature difference between the environment and the convergence of moisture produced by the large-scale flow.

In the simulations the Blackadar high-resolution model (BLACKADAR, 1979; ZHANG and ANTHES, 1982) is used for planetary boundary layer physics. A simple radiative cooling model is used for the atmospheric radiation processes.

Figure 2 shows the model simulation domains in the Coarse-Grid Mesh (CGM) and Fine-Grid Mesh (FGM) which cover outer (15.5°N–15.3°S; 86.9°E–142.8°E) and inner regions (10.5°N–9.9°S; 105.0°E–124.7°E), respectively. The corresponding horizontal resolutions for the CGM and FGM are 45 km and 15 km, respectively. Thus there are (79 × 140) and (159 × 148) grid points for the CGM and FGM domains.

The European Centre for Medium-Range Weather Forecasts (ECMWF) analysis data with 2.5° × 2.5° resolution at 15 pressure levels on 00 UTC 21 September, 1997 are used to specify the initial conditions. Figures 3a–b show the analyzed streamlines and wind speeds at 850 hPa and mean sea-level pressure (MSLP) for 00 UTC 21 September, 1997. The weak southwest monsoonal flows can be clearly seen in Figure 3a where the low-level easterly winds over northern Australia (with magnitudes between 2 and 8 m s^{-1}) curve as the southerly/southwesterly winds over Indonesia. These features are caused by a weak mean sea level pressure (MSLP) gradient between northern Australia and Southern Asia (Fig. 3b).

3. Model Results

a. Streamlines and Wind Speeds

Figures 4a–b display the analyzed and day-1 simulated streamlines and wind speeds at 850 hPa over the CGM domain for 00 UTC 22 September, 1997. The model simulates the main features of the southwest monsoon flows reasonably well.

MODEL DOMAINS

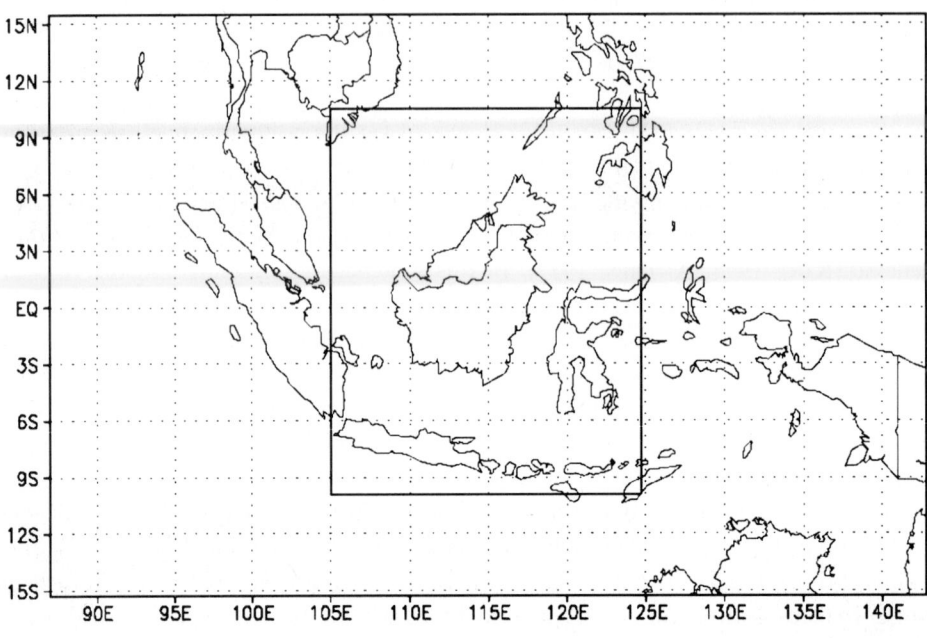

Figure 2
Model domains in the Coarse-Mesh Grid (CGM) and Fine-Mesh Grid (FGM) for the simulations.

These features include the easterly winds over the southeastern Indian Ocean and the cross-equatorial flows characterized by the deflected southeasterly winds into the southwesterly winds over the equator. Simulated wind speeds are also in the same range as the analyzed winds. However, high-pressure systems located over Philippines and the South Chine Sea to the west of Brunei are not simulated well.

Figures 4c–d display the analyzed and day-2 simulated streamlines and wind speeds at 850 hPa over the CGM domain for 00 UTC 23 September, 1997. The model simulates stronger easterly winds over the southeastern Indian Ocean and weaker southwesterly winds over the western Pacific Ocean. Distributions of winds are in close agreement with the analysis. However, the high-pressure system near the west coast of Brunei is missing in the simulation and wind speeds over the eastern part of Indonesia are not accurately simulated. Figures 5a–d show the analyzed and day-1 and day-2 simulated streamlines and wind speeds at 850 hPa over the FGM domain. A small improvement in the simulation of wind directions and wind speeds is gained as the resolution of the model increases. However, further analyses will focus on the coarse grid domain because of the transport over a larger area.

ANALYZED WINDS AT 850 hPA AND MSLP FOR 00 UTC 21 SEP 1997

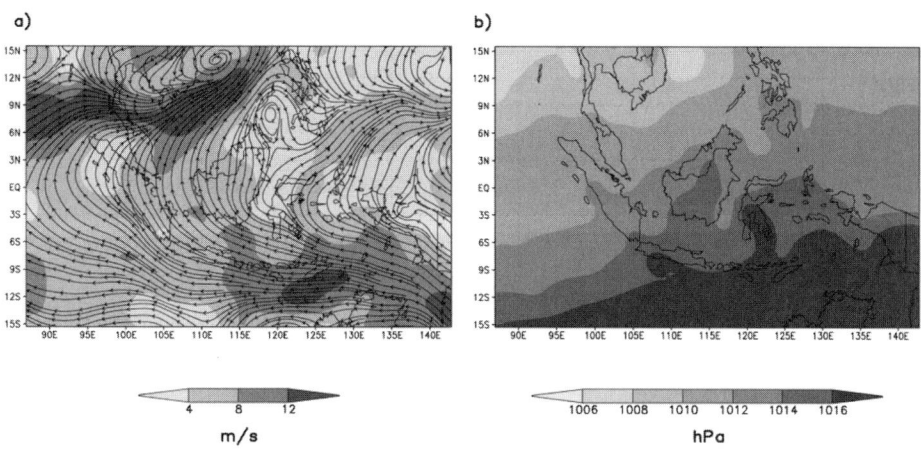

Figure 3
a) Analyzed streamlines and wind speeds at 850 hPa and b) mean surface pressure (MSLP) for 00 UTC 21 September, 1997 over the CGM domain as initial conditions.

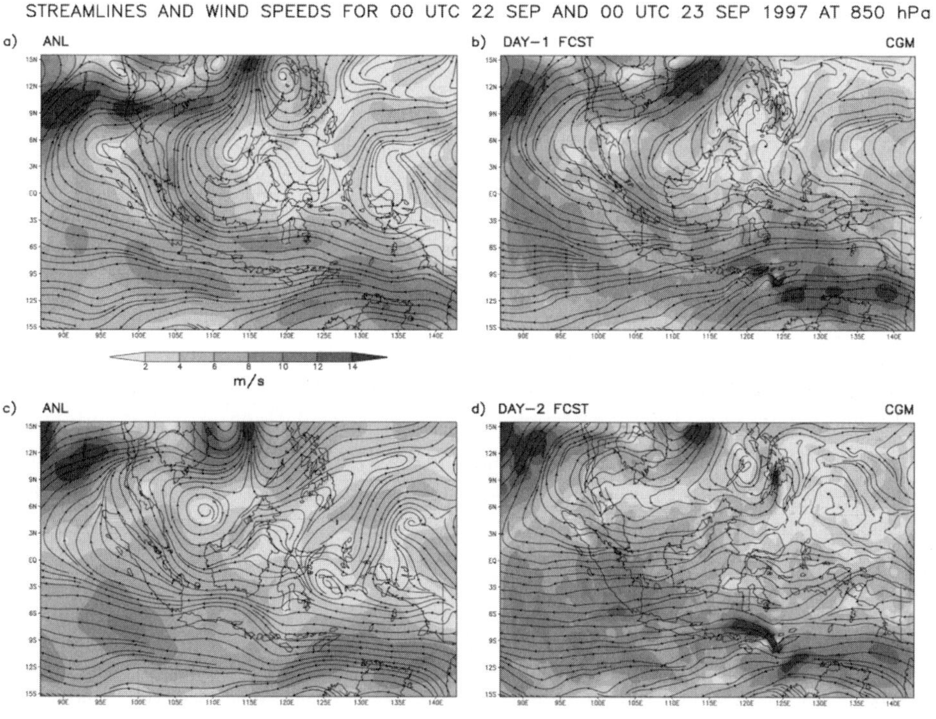

STREAMLINES AND WIND SPEEDS FOR 00 UTC 22 SEP AND 00 UTC 23 SEP 1997 AT 850 hPa

Figure 4
Analyzed and simulated streamlines and wind speeds at 850 hPa for 00 UTC 22 September, 1997 (a and b) and 00 UTC 23 September, 1997 (c and d) for the CGM domain.

STREAMLINES AND WIND SPEEDS FOR 00 UTC 22 SEP AND 23 SEP 1997 AT 850 hPa

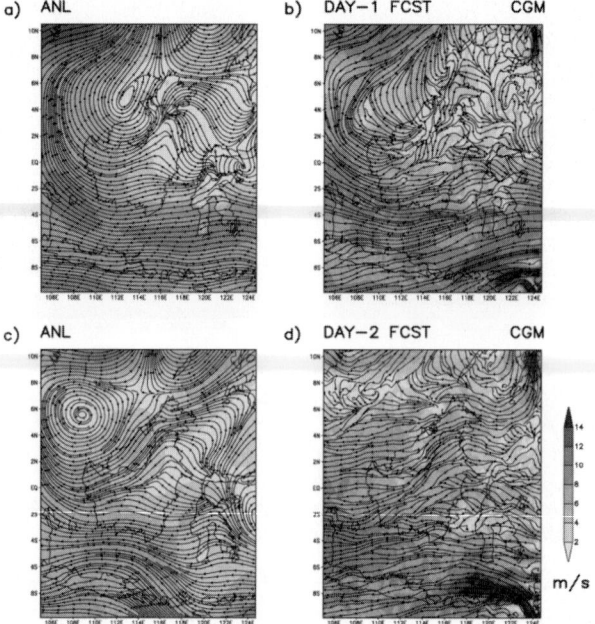

Figure 5
As in Figure 4, except for the FGM domain.

Figures 6a–b display the analyzed and day-3 simulated streamlines and wind speeds at 850 hPa over the CGM domain for 00 UTC 24 September, 1997. The low-level flow analysis indicates that easterly winds particularly over Kalimantan, Java, and Sumatera become the southeasterly winds and the simulated winds are in general agreement with the analyzed winds. Figures 6c–d display the analyzed and day-4 simulated streamlines and wind speeds at 850 hPa over the CGM domain for 00 UTC 25 September, 1997. Although the mean circulations can be simulated reasonably, the wind speeds over the southeastern Indian Ocean are slightly overestimated, while those over the western Pacific are underestimated. The forecast skills given by mean, bias, root-mean-square error, and correlation coefficient between the analyzed and simulated winds for the CGM are given in Table 1.

Figures 7a–d show the simulated planetary boundary layer heights for 06 UTC 21 September, 1997, 12 UTC 21 September, 1997, 18 UTC 21 September, 1997, and 00 UTC 22 September, 1997. It is clear that the simulated planetary boundary layer heights vary with time and space. However, under the influence of large-scale subsidence and low-level divergence the simulated boundary layer height is shallow with values less than 1200 m evn over land in locations such as Brunei. This condition leads to the entrapment of pollutants in the boundary layer.

STREAMLINES AND WIND SPEEDS FOR 00 UTC 24 SEP AND 00 UTC 25 SEP 1997 AT 850 hPa

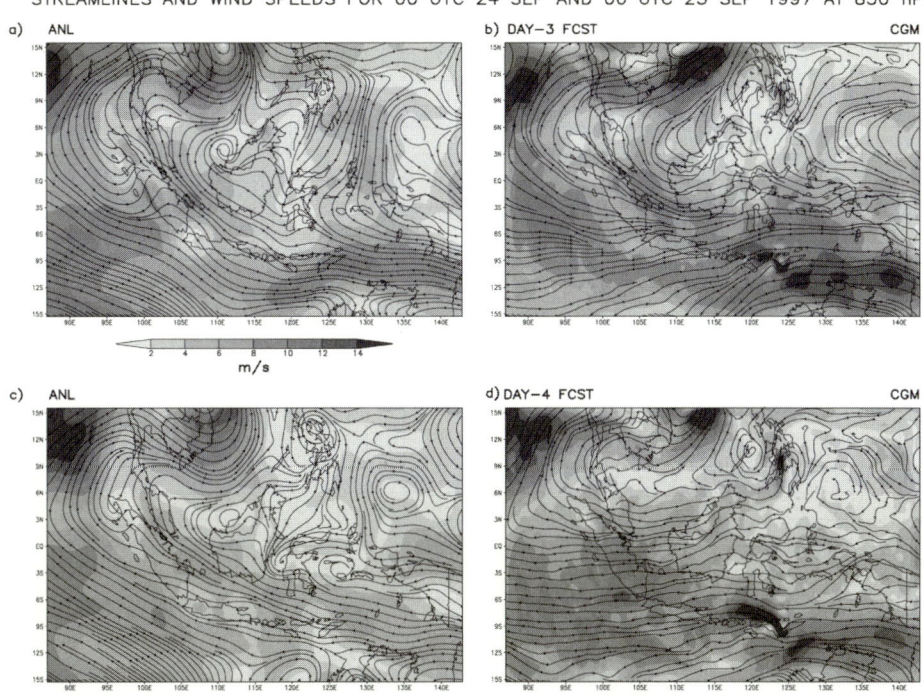

Figure 6

Analyzed and simulated streamlines and wind speeds at 850 hPa for 00 UTC 24 September, 1997 (a and b) and 00 UTC 25 September, 1997 (c and d) for the CGM domain.

Table 1

Forecast skill statistics for the CGM domain

Date	Var.	Mean (m s^{-1})		BIAS (m s^{-1})	RMSE (m s^{-1})	CORR
		ANL	MM5			
00 UTC	U	−3.1	−2.2	0.9	2.9	0.89
22 Sep. 97	V	1.3	1.3	0.0	2.1	0.77
00 UTC	U	−3.8	−2.2	1.4	3.2	0.85
23 Sep. 97	V	0.6	1.1	0.5	2.5	0.67
00 UTC	U	−4.2	−2.2	2.0	3.5	0.82
24 Sep. 97	V	1.3	1.9	0.6	2.9	0.54
00 UTC	U	−3.5	−1.4	1.9	3.4	0.83
25 Sep. 97	V	0.9	1.5	0.6	2.7	0.66

b. Trajectories Analysis

Since a pollutant spreads out both horizontally and vertically due to dispersion, a realistic representation of the mean transport can only be obtained from an ensemble

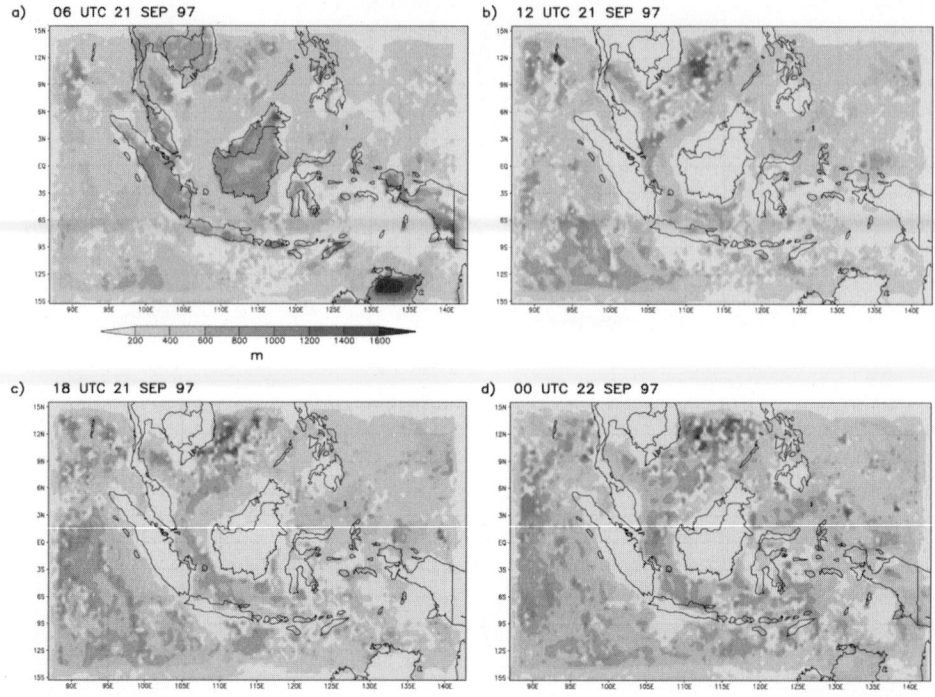

Figure 7
Simulated planetary boundary layer heights for a) 06 UTC 21 September, 1997, b) 12 UTC 21 September, 1997, c) 18 UTC 21 September, 1997, and d) 00 UTC 22 September, 1997 for the CGM domain.

of various trajectory pathways. Trajectories are examined between 900 hPa (~1 km) and 850 hPa (~1.5 km). These altitudes are chosen to represent average plume rises and to examine the coupling of flows in the lower boundary layer with flows at higher altitudes.

Figure 8 displays the backward trajectories ending in Malaysia and Singapore at 00 UTC 25 September, 1997 at 900 hPa. Backward trajectories are used to identify the source of air masses classified as polluted air. In general, air parcels arrive in Malaysia and Singapore at 900 hPa originating from east Kalimantan at 800 hPa. Because of persistently low-level easterly winds (with amplitude 6–7 m s^{-1}), transport from 116°E to 102°E takes place in only four days. Figure 9 illustrates the backward trajectories terminating in Malaysia and Singapore at 00 UTC 25 September, 1997 at 850 hPa. Air masses from Sarawak and the South China Sea at 800 hPa arrive in Malaysia at 850 hPa. Meanwhile, plumes from east Kalimantan make their way to Singapore.

These simulations illustrate that the long-range transports involve the entire depth of the troposphere. Moreover, the large-scale subsidence and dry atmosphere associated with the 1997/98 ENSO event cause aerosols and pollutants in the upper troposphere to move downward rapidly to the surface.

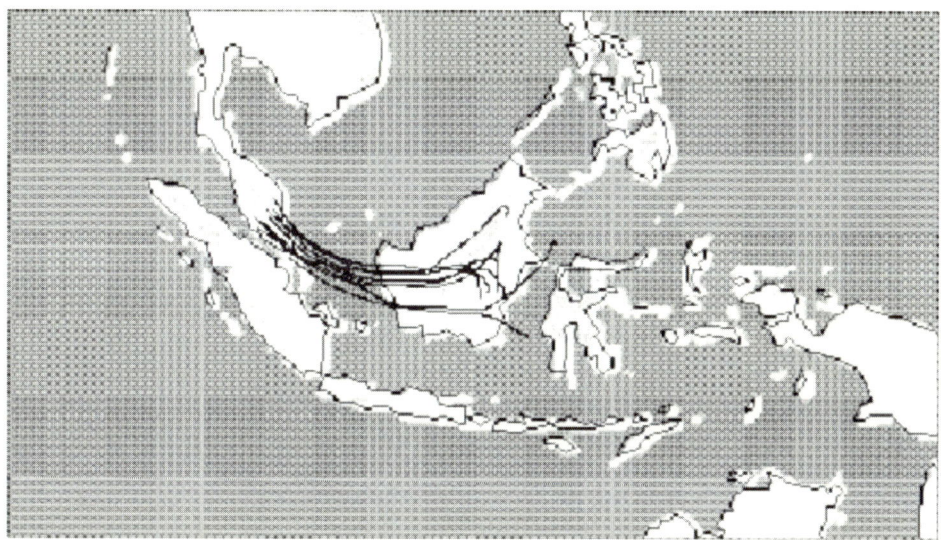

Figure 8
Backward trajectories terminating at 00 UTC 25 September, 1997 at 900 hPa in Malaysia and Singapore.

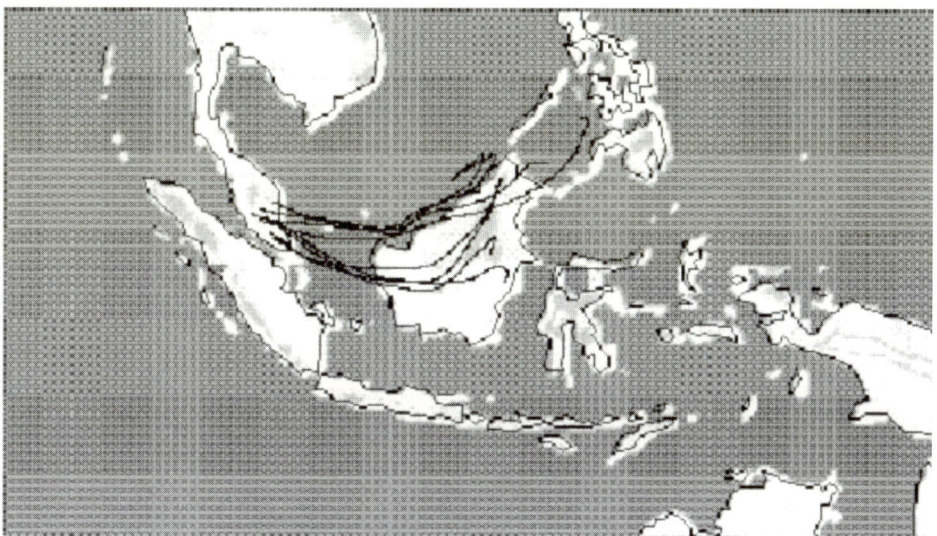

Figure 9
Backward trajectories terminating at 00 UTC 25 September, 1997 at 850 hPa in Malaysia and Singapore.

4. Conclusions

Advection by synoptic-scale winds and entrainment into the boundary layer are the primary mechanisms responsible for transporting air pollutants and aerosols during forest fires in Kalimantan between 00 UTC 21 September to 00 UTC 25 September, 1997. The MM5 model is able to simulate the general features of the transport. The winds in this region are generally easterly/southeasterly driven by the synoptic pressure gradient between the Indonesian-Australian regions and Southeast Asia. The transport of air masses is reasonably well coupled with mean flows through the lower and upper troposphere. Since there are large-scale subsidence and no precipitating clouds, there is no removal of aerosols through precipitation.

Acknowledgements

This work was supported by the Atmospheric Sciences Division, National Science Foundation under grant ATM–9632390 and ATM–0080088.

REFERENCES

ANTHES, R. A. (1977), *A Cumulus Parameterization Scheme Utilizing a One-dimensional Cloud Model*, Mon. Wea. Rev. *105*, 270–286.

BLACKADAR, A. K., *High resolution models of the planetary boundary layer*. In *Advances in Environmental Science and Engineering* (J. Pfafflin and E. Ziegler, eds.) vol. 1, (Gordon and Breach, 1979) pp. 50–85.

DUDHIA, J. (1989), *Numerical Study of Convection Observed During the Winter Monsoon Experiment Using a Mesoscale Two-dimensional Model*, J. Atmos. Sci. *46*, 3077–3107.

DUDHIA, Y. (1993), *A Nonhydrostatic Version of the Penn State-NCAR Mesoscale Model: Validation Tests and Simulations of an Atlantic Cyclone and Cold Front*, Mon. Wea. Rev. *121*, 1493–1513.

GRELL, G. A., DUDHIA, J., and STAUFFER, R. D. (1995), *A Description of the Fifth-generation Penn State/ NCAR Mesoscale Model (MM5)*, NCAR Tech. Note TN-398 + STR, 122 pp. [Available from UCAR Communications, P. O. Box 3000, Boulder, CO 80307.]

KUO, H. L. (1974), *Further Studies of the Parameterization of the Influence of Cumulus Convection of Large-scale Flow*, J. Atmos. Sci. *31*, 1232–1240.

ROPELEWSKI, C. F. and HALPERT, M. S. (1987), *Global and Regional Scale Precipitation Patterns Associated with the El Niño/Southern Oscillation*, Mon. Wea. Rev. *115*, 1606–1624.

WEBSTER, P. J. and YANG, S. (1992), *Monsoon and ENSO: Selective Interactive Systems*, Quart. J. Roy. Meteor. Soc. *118*, 877–926.

ZHANG, D. and ANTHES, R. A. (1982), *A High-resolution Model of the Planetary Boundary Layer-Sensitivity Tests and Comparisons with SESAME-79 Data*, J. Appl. Meteor. *21*, 1954–1609.

(Received July 1, 2000, accepted May 26, 2001)

 To access this journal online:
http://www.birkhauser.ch

Notes to Authors

PAGEOPH welcomes original contributions in English (and occasionally in French and German) on all aspects of geophysics. All manuscripts should be submitted to the Regular Issues Editor-in-Chief, in triplicate, formatted with double spacing and wide margins. For further details see the following paragraphs.

Format of Manuscripts

Length and Page Charges: A paper should not exceed 16 printed pages including tables and figures. For articles exceeding 16 printed pages the authors will be charged sFr. 80.00 for each additional page. No page charges, except those for color prints, are required for contributors to special issues.

Title Page: This should include the the complete title, full names and addresses of all authors. In addition, corresponding authors should provide their fax number and e-mail address if they are available.

Abbreviated Title: It is necessary to indicate an abbreviated title, which will be used as a running head (no more than 50 characters including spaces).

Abstract: The abstract should be in English, and in the language of the text, if different. It has to be of no more than 10 sentences and should be concise and self-contained.

Keywords: Up to 6 keywords should be listed, suitable for incorporation into information-retrieval systems.

Text: The text must include a citation for each item listed under References; the approximate position of each figure should be indicated in the text. The metric system should be used throughout the text, figures, and tables.

Tables: Tables are to be presented on separate pages, with a brief title for each.

Figures: Figure captions and legends are to be typed on a separate page or pages as the last element in the manuscript. Make sure that line thickness and lettering allow an adequate size reduction. Heliographic or photocopies are not suitable for reproduction. Highquality, glossy, photographic prints must be submitted. Color prints are permitted but authors will be charged for them.

References: They are to be listed in alphabetical order in the following style:
Journal article: Haurwitz, B., and Cowley, A.D. (1973), The Diurnal and Semidurnal Barometric Oscillations, Global Distribution and Annual Variation, Pure Appl. Geophys. 102, 193-222.

Whole book: Bath, M., Introduction to Seismology (Birkhäuser, Basel 1973).

Article in a book: Haurwitz, B., and Cowley, A.D., Barometric oscillations, In Introduction to Seismology (ed. Bath. M.) (Birkhäuser, Basel 1973) pp. 193-222.

Submission of Manuscripts

Manuscripts must be submitted in triplicate, formatted with double line spacing and wide margins. Copies of the figures should be attached at the end of the manuscript. High-quality, prints may be submitted later. All manuscript pages, including references, tables, and captions, should be numbered consecutively, starting with the title page as page one.

All manuscripts should be submitted to Regular Issues Editor-in-Chief
Brian Mitchell
Department of Earth & Atmospheric Sciences
Saint Louis University
3507 Laclede Avenue
St. Louis, MO 63103, USA
e-mail: mitchbj@eas.slu.edu

The final version of a manuscript accepted for publication in PAGEOPH should be sent as both a hard copy and on diskette to the Editor-in-Chief. Delivering manuscripts in electronic form may substantially facilitate the publication process provided certain points are taken into consideration:

- Texts should be delivered in Ms Word or Wordperfect
- Mathematics in Tex, LaTex and any Tex formats
- Figures in TIFF (high resolution) or EPS

The electronic and printed version must be absolutely identical. All pictorial and graphic illustrations should be delivered as hard copy originals. Do not fail to include a hard copy for ready vicwing. Back-up copies of the diskettes must be kept. Diskettes must be adequately protectedfor transport.

Galley Proofs
Unless indicated otherwise, galley proofs will be sent to the first-named author directly from Birkhäuser Verlag AG and should be returned with the least possible delay. Textual alterations made in the galley proof stage will be charged to the author. One copy of the corrected proof is to be returned immediately.
The editorial office assumes no responsibility for delayed proofs, errors in the original manuscript, or major alterations in proofs for any reason.

Reprints
The authors will receive 50 reprints of each article without charge. Additional reprints may be ordered in lots of 50 when the final corrected page proofs are returned. Orders submitted thereafter are subject to considerably higher rates.